LONDON MATHEMATICAL SOCIETY LECTURE NOTE SERIES

Managing Editor: Professor N.J. Hitchin, Mathematical Institute,
University of Oxford, 24–29 St Giles, Oxford OX1 3LB, United Kingdom

The titles below are available from booksellers, or from Cambridge University Press
at www.cambridge.org/mathematics

180 Lectures on ergodic theory and Pesin theory on compact manifolds, M. POLLICOTT
181 Geometric group theory I, G.A. NIBLO & M.A. ROLLER (eds)
182 Geometric group theory II, G.A. NIBLO & M.A. ROLLER (eds)
183 Shintani zeta functions, A. YUKIE
184 Arithmetical functions, W. SCHWARZ & J. SPILKER
185 Representations of solvable groups, O. MANZ & T.R. WOLF
186 Complexity: knots, colourings and counting, D.J.A. WELSH
187 Surveys in combinatorics, 1993, K. WALKER (ed)
188 Local analysis for the odd order theorem, H. BENDER & G. GLAUBERMAN
189 Locally presentable and accessible categories, J. ADAMEK & J. ROSICKY
190 Polynomial invariants of finite groups, D.J. BENSON
191 Finite geometry and combinatorics, F. DE CLERCK *et al*
192 Symplectic geometry, D. SALAMON (ed)
194 Independent random variables and rearrangement invariant spaces, M. BRAVERMAN
195 Arithmetic of blowup algebras, W. VASCONCELOS
196 Microlocal analysis for differential operators, A. GRIGIS & J. SJÖSTRAND
197 Two-dimensional homotopy and combinatorial group theory, C. HOG-ANGELONI *et al*
198 The algebraic characterization of geometric 4-manifolds, J.A. HILLMAN
199 Invariant potential theory in the unit ball of $\mathbf{C}^n$, M. STOLL
200 The Grothendieck theory of dessins d'enfant, L. SCHNEPS (ed)
201 Singularities, J.-P. BRASSELET (ed)
202 The technique of pseudodifferential operators, H.O. CORDES
203 Hochschild cohomology of von Neumann algebras, A. SINCLAIR & R. SMITH
204 Combinatorial and geometric group theory, A.J. DUNCAN, N.D. GILBERT & J. HOWIE (eds)
205 Ergodic theory and its connections with harmonic analysis, K. PETERSEN & I. SALAMA (eds)
207 Groups of Lie type and their geometries, W.M. KANTOR & L. DI MARTINO (eds)
208 Vector bundles in algebraic geometry, N.J. HITCHIN, P. NEWSTEAD & W.M. OXBURY (eds)
209 Arithmetic of diagonal hypersurfaces over infinite fields, F.Q. GOUVÉA & N. YUI
210 Hilbert C^*-modules, E.C. LANCE
211 Groups 93 Galway/St Andrews I, C.M. CAMPBELL *et al* (eds)
212 Groups 93 Galway/St Andrews II, C.M. CAMPBELL *et al* (eds)
214 Generalised Euler-Jacobi inversion formula and asymptotics beyond all orders, V. KOWALENKO *et al*
215 Number theory 1992–93, S. DAVID (ed)
216 Stochastic partial differential equations, A. ETHERIDGE (ed)
217 Quadratic forms with applications to algebraic geometry and topology, A. PFISTER
218 Surveys in combinatorics, 1995, P. ROWLINSON (ed)
220 Algebraic set theory, A. JOYAL & I. MOERDIJK
221 Harmonic approximation, S.J. GARDINER
222 Advances in linear logic, J.-Y. GIRARD, Y. LAFONT & L. REGNIER (eds)
223 Analytic semigroups and semilinear initial boundary value problems, KAZUAKI TAIRA
224 Computability, enumerability, unsolvability, S.B. COOPER, T.A. SLAMAN & S.S. WAINER (eds)
225 A mathematical introduction to string theory, S. ALBEVERIO, *et al.*
226 Novikov conjectures, index theorems and rigidity I, S. FERRY, A. RANICKI & J. ROSENBERG (eds)
227 Novikov conjectures, index theorems and rigidity II, S. FERRY, A. RANICKI & J. ROSENBERG (eds)
228 Ergodic theory of $\mathbf{Z}^d$ actions, M. POLLICOTT & K. SCHMIDT (eds)
229 Ergodicity for infinite dimensional systems, G. DA PRATO & J. ZABCZYK
230 Prolegomena to a middlebrow arithmetic of curves of genus 2, J.W.S. CASSELS & E.V. FLYNN
231 Semigroup theory and its applications, K.H. HOFMANN & M.W. MISLOVE (eds)
232 The descriptive set theory of Polish group actions, H. BECKER & A.S. KECHRIS
233 Finite fields and applications, S.COHEN & H. NIEDERREITER (eds)
234 Introduction to subfactors, V. JONES & V.S. SUNDER
235 Number theory 1993–94, S. DAVID (ed)
236 The James forest, H. FETTER & B.G. DE BUEN
237 Sieve methods, exponential sums, and their applications in number theory, G.R.H. GREAVES *et al*
238 Representation theory and algebraic geometry, A. MARTSINKOVSKY & G. TODOROV (eds)
240 Stable groups, F.O. WAGNER
241 Surveys in combinatorics, 1997, R.A. BAILEY (ed)
242 Geometric Galois actions I, L. SCHNEPS & P. LOCHAK (eds)
243 Geometric Galois actions II, L. SCHNEPS & P. LOCHAK (eds)
244 Model theory of groups and automorphism groups, D. EVANS (ed)
245 Geometry, combinatorial designs and related structures, J.W.P. HIRSCHFELD *et al*
246 p-Automorphisms of finite p-groups, E.I. KHUKHRO
247 Analytic number theory, Y. MOTOHASHI (ed)
248 Tame topology and o-minimal structures, L. VAN DEN DRIES
249 The atlas of finite groups: ten years on, R. CURTIS & R. WILSON (eds)
250 Characters and blocks of finite groups, G. NAVARRO
251 Gröbner bases and applications, B. BUCHBERGER & F. WINKLER (eds)
252 Geometry and cohomology in group theory, P. KROPHOLLER, G. NIBLO, R. STÖHR (eds)
253 The q-Schur algebra, S. DONKIN
254 Galois representations in arithmetic algebraic geometry, A.J. SCHOLL & R.L. TAYLOR (eds)
255 Symmetries and integrability of difference equations, P.A. CLARKSON & F.W. NIJHOFF (eds)
256 Aspects of Galois theory, H. VÖLKLEIN *et al*
257 An introduction to noncommutative differential geometry and its physical applications 2ed, J. MADORE
258 Sets and proofs, S.B. COOPER & J. TRUSS (eds)
259 Models and computability, S.B. COOPER & J. TRUSS (eds)
260 Groups St Andrews 1997 in Bath, I, C.M. CAMPBELL *et al*

261 Groups St Andrews 1997 in Bath, II, C.M. CAMPBELL *et al*
262 Analysis and logic, C.W. HENSON, J. IOVINO, A.S. KECHRIS & E. ODELL
263 Singularity theory, B. BRUCE & D. MOND (eds)
264 New trends in algebraic geometry, K. HULEK, F. CATANESE, C. PETERS & M. REID (eds)
265 Elliptic curves in cryptography, I. BLAKE, G. SEROUSSI & N. SMART
267 Surveys in combinatorics, 1999, J.D. LAMB & D.A. PREECE (eds)
268 Spectral asymptotics in the semi-classical limit, M. DIMASSI & J. SJÖSTRAND
269 Ergodic theory and topological dynamics, M.B. BEKKA & M. MAYER
270 Analysis on Lie Groups, N.T. VAROPOULOS & S. MUSTAPHA
271 Singular perturbations of differential operators, S. ALBEVERIO & P. KURASOV
272 Character theory for the odd order function, T. PETERFALVI
273 Spectral theory and geometry, E.B. DAVIES & Y. SAFAROV (eds)
274 Descriptive set theory and dynamical systems, M. FOREMAN *et al*
275 Computational and geometric aspects of modern algebra, M.D. ATKINSON *et al*
276 Singularities of plane curves, E. CASAS-ALVERO
277 The Mandelbrot set, theme and variations, TAN LEI (ed)
278 Global attractors in abstract parabolic problems, J.W. CHOLEWA & T. DLOTKO
279 Topics in symbolic dynamics and applications, F. BLANCHARD, A. MAASS & A. NOGUEIRA (eds)
280 Characters and automorphism groups of compact Riemann surfaces, T. BREUER
281 Explicit birational geometry of 3-folds, A. CORTI & M. REID (eds)
282 Auslander–Buchweitz approximations of equivariant modules, M. HASHIMOTO
283 Nonlinear elasticity, Y. FU & R.W. OGDEN (eds)
284 Foundations of computational mathematics, R. DEVORE, A. ISERLES & E. SÜLI (eds)
285 Rational points on curves over finite fields, H. NIEDERREITER & C. XING
286 Clifford algebras and spinors 2nd, P. LOUNESTO
287 Topics on Riemann surfaces and Fuchsian groups, E. BUJALANCE *et al.*
288 Surveys in combinatorics, 2001, J. HIRSCHFELD (ed)
289 Aspects of Sobolev-type inequalities, L. SALOFF-COSTE
290 Quantum groups and Lie theory, A. PRESSLEY (ed)
291 Tits buildings and the model theory of groups, K. TENT (ed)
292 A quantum groups primer, S. MAJID
293 Second order partial differential equations in Hilbert spaces, G. DA PRATO & J. ZABCZYK
294 Introduction to the theory of operator space, G. PISIER
295 Geometry and integrability, L. MASON & YAVUZ NUTKU (eds)
296 Lectures on invariant theory, I. DOLGACHEV
297 The homotopy category of simply connected 4-manifolds, H.-J. BAUES
298 Higher operads, higher categories, T. LEINSTER
299 Kleinian Groups and Hyperbolic 3-Manifolds, Y. KOMORI, V. MARKOVIC & C. SERIES (eds)
300 Introduction to Möbius Differential Geometry, U. HERTRICH-JEROMIN
301 Stable modules and the D(2)-problem, F.E.A. JOHNSON
302 Discrete and Continuous Nonlinear Schrödinger Systems, M.J. ABLOWITZ, B. PRINARI & A.D. TRUBATCH
303 Number Theory and Algebraic Geometry, M. REID & A. SKOROBOGATOV (eds)
304 Groups St Andrews 2001 in Oxford Vol. 1, C.M. CAMPBELL, E.F. ROBERTSON & G.C. SMITH (eds)
305 Groups St Andrews 2001 in Oxford Vol. 2, C.M. CAMPBELL, E.F. ROBERTSON & G.C. SMITH (eds)
306 Peyresq lectures on geometric mechanics and symmetry, J. MONTALDI & T. RATIU (eds)
307 Surveys in Combinatorics 2003, C.D. WENSLEY (ed.)
308 Topology, geometry and quantum field theory, U.L. TILLMANN (ed)
309 Corings and Comodules, T. BRZEZINSKI & R. WISBAUER
310 Topics in Dynamics and Ergodic Theory, S. BEZUGLYI & S. KOLYADA (eds)
311 Groups: topological, combinatorial and arithmetic aspects, T. W. MÜLLER (ed)
312 Foundations of Computational Mathematics, Minneapolis 2002, FELIPE CUCKER *et al* (eds)
313 Transcendental aspects of algebraic cycles, S. MÜLLER-STACH & C. PETERS (eds)
314 Spectral generalizations of line graphs, D. CVETKOVIC, P. ROWLINSON & S. SIMIC
315 Structured ring spectra, A. BAKER & B. RICHTER (eds)
316 Linear Logic in Computer Science, T. EHRHARD *et al* (eds)
317 Advances in elliptic curve cryptography, I. F. BLAKE, G. SEROUSSI & N. SMART
318 Perturbation of the boundary in boundary-value problems of Partial Differential Equations, D. HENRY
319 Double Affine Hecke Algebras, I. CHEREDNIK
321 Surveys in Modern Mathematics, V. PRASOLOV & Y. ILYASHENKO (eds)
322 Recent perspectives in random matrix theory and number theory, F. MEZZADRI & N.C. SNAITH (eds)
323 Poisson geometry, deformation quantisation and group representations, S. GUTT *et al* (eds)
324 Singularities and Computer Algebra, C. LOSSEN & G. PFISTER (eds)
325 Lectures on the Ricci Flow, P. TOPPING
326 Modular Representations of Finite Groups of Lie Type, J. E. HUMPHREYS
328 Fundamentals of Hyperbolic Manifolds, R. D. CANARY, A. MARDEN, & D. B. A. EPSTEIN (eds)
329 Spaces of Kleinian Groups, Y. MINSKY, M. SAKUMA & C. SERIES (eds)
330 Noncommutative Localization in Algebra and Topology, A. RANICKI (ed)
331 Foundations of Computational Mathematics, Santander 2005, L. PARDO, A. PINKUS, E. SULI & M. TODD (eds)
332 Handbooks of Tilting Theory, L. ANGELERI HÜGEL, D. HAPPEL & H. KRAUSE (eds)
333 Synthetic Differential Geometry 2ed, A. KOCK
334 The Navier–Stokes Equations, P. G. DRAZIN & N. RILEY
335 Lectures on the Combinatorics of Free Probability, A. NICA & R. SPEICHER
336 Integral Closure of Ideals, Rings, and Modules, I. SWANSON & C. HUNEKE
337 Methods in Banach Space Theory, J. M. F. CASTILLO & W. B. JOHNSON (eds)
338 Surveys in Geometry and Number Theory, N. YOUNG (ed)
339 Groups St Andrews 2005 Vol. 1, C.M. CAMPBELL, M.R. QUICK, E.F. ROBERTSON & G.C. SMITH (eds)
340 Groups St Andrews 2005 Vol. 2, C.M. CAMPBELL, M.R. QUICK, E.F. ROBERTSON & G.C. SMITH (eds)
341 Ranks of Elliptic Curves and Random Matrix Theory, J.B. CONREY, D.W. FARMER, F. MEZZADRI & N.C. SNAITH (eds)
342 Elliptic Cohomology, H.R. MILLER & D.C. RAVENEL

London Mathematical Society Lecture Note Series. 344

Algebraic Cycles and Motives

Volume 2

Edited by

JAN NAGEL
Université de Lille I

CHRIS PETERS
Université de Grenoble I

CAMBRIDGE UNIVERSITY PRESS
Cambridge, New York, Melbourne, Madrid, Cape Town, Singapore,
São Paulo, Delhi, Dubai, Tokyo, Mexico City

Cambridge University Press
The Edinburgh Building, Cambridge CB2 8RU, UK

Published in the United States of America by Cambridge University Press, New York

www.cambridge.org
Information on this title: www.cambridge.org/9780521701754

First published 2007

A catalogue record for this publication is available from the British Library

ISBN 978-0-521-70175-4 Paperback

Contents of Volume 2

Contents of Volume 1

Preface

These proceedings contain a selection of papers from the EAGER conference "Algebraic Cycles and Motives" that was held at the Lorentz Center in Leiden on the occasion of the 75th birthday of Professor J.P. Murre (Aug 30–Sept 3, 2004). The conference attracted many of the leading experts in the field as well as a number of young researchers. As the papers in this volume cover the main research topics and some interesting new developments, they should give a good indication of the present state of the subject. This volume contains sixteen research papers and six survey papers.

The theory of algebraic cycles deals with the study of subvarieties of a given projective algebraic variety X, starting with the free group $Z^p(X)$ on irreducible subvarieties of X of codimension p. In order to make this very large group manageable, one puts a suitable equivalence relation on it, usually *rational equivalence.* The resulting *Chow group* $CH^p(X)$ in general might still be very big. If X is a smooth variety, the intersection product makes the direct sum of all the Chow groups into a ring, the Chow ring $CH^*(X)$. Hitherto mysterious ring can be studied through its relation to cohomology, the first example of which is the cycle class map: every algebraic cycle defines a class in singular, de Rham, or ℓ-adic cohomology. Ultimately this cohomological approach leads to the theory of motives and motivic cohomology developed by A. Grothendieck, M. Levine, M. Nori, V. Suslin and A. Voevodsky, just to mention a few main actors.

There were about 60 participants at the conference, coming from Europe, the United States, India and Japan. During the conference there were 22 one hour lectures. On the last day there were three special lectures devoted to the scientific work of Murre, in honour of his 75th birthday. The lectures covered a wide range of topics, such as the study of algebraic cycles using Abel–Jacobi/regulator maps and normal functions, motives (Voevodsky's triangulated category of mixed motives, finite-dimensional motives), the conjectures of Bloch–Beilinson and Murre on filtrations on Chow groups and Bloch's conjecture, and results of a more arithmetic flavour for varieties defined over number fields or local fields.

Let us start by discussing the **survey papers**. The first, a paper by J. Ayoub is devoted to the construction of a motivic version of the vanishing cycle formalism. It is followed by a paper by L.Barbieri Viale who presents an overview of the main results of the theory of *mixed motives of level at most one*. In a series of recent papers, M. De Cataldo and L. Migliorini have made a detailed study of the *topological properties of algebraic maps using the theory of perverse sheaves*. Their survey provides an introduction to this work, illustrated by a number of low-dimensional examples. Déglise's paper contains a careful exposition of *Voevodsky's theory of sheaves with transfers over a regular base scheme*, with detailed proofs. The paper of M. Green and P. Griffiths contains an outline of an ambitious research program that centers around the *extension of normal functions over a higher-dimensional base*, and its applications to the Hodge conjecture. (The case where the base space is a curve is known by work of F. El Zein and S. Zucker.) A. Krishna and V. Srinivas discuss the theory of *zero-cycles on singular varieties* and its applications to algebra. The paper of D. Ramakrishnan is a brief survey of results concerning *algebraic cycles on Hilbert modular varieties.*

In discussing the **research papers** we have grouped according to the main research themes, although in the proceedings they are listed alphabetically according to the name of the authors.

One of the leading themes in the theory of algebraic cycles is the study of the conjectural *Bloch–Beilinson filtration* on Chow groups. In the course of his work on motives, J. Murre found an equivalent and more explicit version of this conjecture, which states that the motive of a smooth projective algebraic variety should admit a Chow–Künneth decomposition with a number of specific properties. The paper of B. Kahn, J. Murre and C. Pedrini contains a detailed exposition of these matters with emphasis on the study of the transcendental part of the motive of a surface. The paper of S. Bloch and H. Esnault is devoted to the construction of an algebraic cycle that induces the Künneth projector onto $H^1(U)$ for a quasi-projective variety U, and the paper of Miller *et al.* shows the existence of certain Chow–Künneth projectors for compactified families of abelian threefolds over a certain Picard modular surface studied by Holzapfel. Beauville studies the splitting of the Bloch–Beilinson filtration for certain symplectic projective manifolds. The notion of "finite–dimensionality" of motives, which recently attracted a lot of attention, is studied in the papers of S.-I. Kimura and U. Jannsen.

The latter paper uses this notion to verify Murre's conjectures in a number of examples.

Another important theme is the *study of algebraic cycles using Hodge theory.* The paper of C. Peters and J. Steenbrink deals with the motivic nearby fiber and its relation to the limit mixed Hodge structure of a family of projective varieties. Morihiko Saito constructs the total infinitesimal invariant of a higher Chow cycle, an object that lives in the direct sum of the cohompology of filtered logarithhmic complexes with coefficients. In the papers of M. Asakura, S. Saito and J. Nagel, infitesimal methods are used to study the regulator map on higher Chow groups. M. Asakura and S. Saito use these techniques to verify a conjecture of Beilinson ("Beilinson's Hodge conjecture with coefficients") in certain cases. J. Lewis defines a twisted version of Milnor K-theory and a corresponding twisted version of the regulator, which is shown to have a nontrivial image in certain examples.

The remaining papers deal with a variety of topics. The papers of C. Deninger and A. Werner and S. Brivio and A. Verra deal with *vector bundles.* C. Deninger and A. Werner study the category of degree zero vector bundles with "potentially strongly semistable reduction" on a p-adic curve. S. Brivio and A. Verra investigate the properties of the theta map defined on the moduli space of semistable vector bundles over a curve. T. Shioda studies the structure of the *Mordell–Weil lattice of certain elliptic K3 surfaces*, and the paper of J. Stienstra studies a potential link between the theory of *motives and string theory* using diffraction patterns.

The conference has been financed by the Lorentz Center, EAGER (European Algebraic Geometry Research Training Network), the KNAW (Royal Netherlands Academy of Arts and Sciences), and the Thomas Stieltjes Instituut. We heartily thank these institutions for their financial support.

It is a pleasure to dedicate this volume to Jacob Murre. The study of algebraic cycles and motives has been his life-long passion, and he has made a number of important contributions to the subject.

Chris Peters and Jan Nagel, May 2006.

Program

Day	Hour	Speaker	Title
Monday Aug 30	10:00–11:00	P. Griffiths	Algebraic cycles and singularities of normal functions
	11:15–12:15	A. Beauville	When does the Bloch–Beilinson filtration split?
	13:30–14:30	S. Müller-Stach	Higher Abel–Jacobi maps
	14:30–15:30	F. Déglise	Cycle modules and triangulated mixed motives
	16:00–17:00	O. Tommasi	Rational cohomology of the moduli space of genus 4 curves
Tuesday August 31	09:30–10:30	H. Esnault	Deligne's integrality theorem in unequal characteristic and rational points over finite fields
	11:15–12:15	U. Jannsen	Some remarks on finite dimensional motives
	13:30–14:30	L. Barbieri Viale	Motivic Albanese
	14:45–15:45	B. van Geemen	Some remarks on Brauer groups of elliptic fibrations on K3 surfaces
	16:15–17:15	L. Migliorini	Hodge theory of projective maps
Wednesday September 1	09:30–10:30	K. Künnemann	Extensions in Arakelov geometry
	11:15–12:15	J.-L Colliot-Thélène	Zero-cycles on linear algebraic groups over local fields
	13:30–14:30	C. Deninger	Vector bundles on p-adic curves and parallel transport
Thursday September 2	09:30–10:30	S. Saito	Finiteness results for motivic cohomology
	11:15–12:15	T. Shioda	Finding cycles on certain K3 surfaces
	13:30–14:30	D. Ramakrishnan	Cycles on Hilbert modular fourfolds
	14:45–15:45	A. Verra	Moduli of vector bundles on curves and correspondences: the genus two case
	16:15–17:15	J. Ayoub	Conservation of ϕ and the Bloch conjecture
Friday September 3	10:00–11:00	A. Conte	25 years of joint work with Jaap Murre
	11:30–12:30	V. Srinivas	Zero cycles on singular varieties
	13:30–14:30	F Oort	Geometric aspects of the scientific work of Jaap Murre, I
	14:45–15:45	S. Bloch	Geometric aspects of the scientific work of Jaap Murre, II

Participants

Last name and Initial	Institute
Amerik, E.	Univ. of Paris XI, Orsay, France
André, Y..	E.N.S. Paris, France
Ayoub. J.	Univ. of Paris VI, Ivry sur Seine, France
Barbieri Viale, L.	Univ. of Roma I, Italy
Beauville, A.	Univ. of Nice, France
Biglari, S.	Univ. of Leipzig, Germany
Bloch, S.	Univ.of Chicago, Ill., United States
Colliot-Thélène, J.-L.	Univ. of Paris XI, Orsay, France
Colombo, E.	Univ. of Milano, Italy
Conte, A.	Univ. of Turin, Italy
De Jeu, R.	Univ. of Durham, United Kingdom
Déglise, F.	Inst. Galilée, Villetaneuse, France
Del Angel, P.	CIMAT, Guanajuato, Mexico
Deninger, C.	Univ. Münster, Germany
Edixhoven, B.	Univ. of Leiden, Netherlands
Eriksson, D.	E.N.S. Paris, France
Esnault, H.	Univ. of Essen, Germany
Faber, C.	KTH, Stockholm, Sweden
Gordon, B.	Univ. Maryland, Washington, United States
Griffiths, P.	IAS, Princeton, United States
Grooten, M.	Univ. of Nijmegen, Netherlands
Guletskii, V.	Belar. St. Ar. Univ., Minsk, Belarus
Haran, S.	Technuion, Haifa, Israel
Höring, A.	Univ. of Bayreuth, Germany
Jannsen, U.	Univ. of Regensburg, Germany
Kahn, B.	Univ. of Paris VI, Paris, France
Kimura, S.	Hiroshima Univ., Japan
Kimura, K.	Tsukuba Univ., Japan
Kloosterman, R.	RUG, Groningen, Netherlands
Künnemann, K.	Univ.of Regensburg, Germany

Lemahieu, A.	Univ. of Leuven, Belgium
Lewis, J.	Univ. of Edmonton, Alberta, Canada
Looijenga, E.	Univ. of Utrecht, Netherlands
Lübke, M.	Univ. of Leiden, Netherlands
Marchisio, M.	Univ. of Turin, Italy
Migliorini, L.	Univ. of Bologna, Italy
Miller, A.	Univ. of Heidelberg, Germany
Müller-Stach, S.	Univ. of Mainz, Germany
Murre, J.	Univ. of Leiden, Netherlands
Nagel, J.	Univ. de Lille I, Villeneuve d'Asque, France
Nicaise, J.	Univ. of Leuven, Belgium
Oort, F.	Univ. of Utrecht, Netherlands
Pedrini, C.	Univ. of Genova, Italy
Peters, C.	Univ. of Grenoble I, Saint Martin d'Hères, France
Popovici, D.	Univ. of Paris VII, Orsay, France
Ramakrishnan, D.	Caltech, Pasadena, CA, United States
Ramon Mari, J.	Humbold Univ., Berlin, Germany
Reuvers, E.	Univ. of Nijmegen, Netherlands
Rydh, D.	Univ. of Gothenburg, Sweden
Saito, S.	Grad. school math., Tokyo, Japan
Schepers, J.	Univ. of Leuven, Belgium
Shioda, T.	Rikkyo Univ., Tokyo, Japan
Springer, T.	Univ. of Utrecht, Netherlands
Srinivas, V.	Tata Univ., Mumbai, India
Steenbrink, J.	Univ. of Nijmegen, Netherlands
Stienstr, J.	Univ. of Utrecht, Netherlands
Swierstra, R.	Univ. of Utrecht, Netherlands
Tommasi, O.	Univ. of Nijmegen, Netherlands
Van Geemen, B.	Univ. of Milano, Italy
Verra, A.	Univ. of Roma III, Italy
Veys,W.	Univ. of Leuven, Belgium

Volume 2

Research Articles

1

Beilinson's Hodge Conjecture with Coefficients for Open Complete Intersections

Masanori Asakura

Graduate School of Mathematics, Kyushu University, Fukuoka 812-8581, Japan

asakura@math.kyushu-u.ac.jp

Shuji Saito

Graduate School of Mathematical Sciences, Tokyo University, Tokyo, 153-8914, Japan

sshuji@ms.u-tokyo.ac.jp

Dedicated to Professor J.P. Murre on the occasion of his 75th birthday

1.1 Introduction

Let U be a smooth algebraic variety over $\mathbb{C}$ and let U^{an} be the analytic site on $U(\mathbb{C})$, the associated analytic space. An important object to study in algebraic geometry is the regulator map from the higher Chow group ([7]) to the singular cohomology of U (cf. [18])

$$\mathrm{reg}_U^{p,q} : CH^q(U, 2q-p) \otimes \mathbb{Q} \to (2\pi\sqrt{-1})^q W_{2q} H^p(U^{\mathrm{an}}, \mathbb{Q}) \cap F^q H^p(U^{\mathrm{an}}, \mathbb{C}),$$

where F^* and W_* denote the Hodge and the weight filtrations of the mixed Hodge structure on the singular cohomology defined by Deligne [8]. For the special case $p = q$, we get

$$\mathrm{reg}_U^q : CH^q(U, q) \otimes \mathbb{Q} \to H^q(U^{\mathrm{an}}, \mathbb{Q}(q)) \cap F^q H^q(U^{\mathrm{an}}, \mathbb{C}). \quad (\mathbb{Q}(q) = (2\pi\sqrt{-1})^q \mathbb{Q})$$

Beilinson's Hodge conjecture claims the surjectivity of reg_U^q (cf. [11, Conjecture 8.5]). In [4] we studied the problem in case U is an open complete intersection, namely U is the complement in a smooth complete intersection X of a simple normal crossing divisor $Z = \cup_{j=1}^s Z_j$ on X such that $Z_j \subset X$ is a smooth hypersurface section. One of the main results affirms that reg_U^q is surjective if the degree of the defining equations of X and Z_j are sufficiently large and if U is general in an appropriate sense. Indeed, under the assumption we have shown a stronger assertion that reg_U^q is surjective even restricted on the subgroup $CH^q(U, q)_{dec}$ of decomposable elements in $CH^q(U, q)$, which is not true in general. In order to explain this, let $K_q^M(\mathcal{O}(U))$ be the Milnor K-group of the ring $\mathcal{O}(U) = \Gamma(U, \mathcal{O}_{Zar})$ (see

§1.2.3 for its definition). We have the natural map

$$\sigma_U : K_q^M(\mathcal{O}(U)) \to CH^q(U,q)$$

induced by cup product and the natural isomorphism

$$K_1^M(\mathcal{O}(U)) = \Gamma(U, \mathcal{O}^*_{Zar}) \xrightarrow{\cong} CH^1(U,1)$$

and $CH^q(U,q)_{dec}$ is defined to be its image. Note that we have the following formula for the value of reg_U^q on decomposable elements;

$$\begin{aligned} \mathrm{reg}_U^q(\{g_1,\dots,g_q\}) &= [g_1]\cup\cdots\cup[g_q] \in H^q(U^{\mathrm{an}}, \mathbb{Q}(q)) \\ &= \operatorname{dlog} g_1 \wedge\cdots\wedge \operatorname{dlog} g_q \in H^0(X, \Omega^q_{X/\mathbb{C}}(\log Z)) \\ &= F^q H^q(U^{\mathrm{an}}, \mathbf{C}) \end{aligned}$$

where $g_j \in \mathcal{O}(U)^*$ for $1 \le j \le q$ and $[g_j] \in H^1(U^{\mathrm{an}}, \mathbb{Q}(1))$ is the image of g_j under the map $\mathcal{O}(U)^* \to H^1(U^{\mathrm{an}}, \mathbb{Z}(1))$ induced by the exponential sequence

$$0 \to \mathbb{Z}(1) \to \mathcal{O}_{U^{\mathrm{an}}} \xrightarrow{\exp} \mathcal{O}^*_{U^{\mathrm{an}}} \to 0.$$

In what follows we are mainly concerned with the map

$$\mathrm{reg}_U^q : K_q^M(\mathcal{O}(U)) \otimes \mathbb{Q} \to H^q(U^{\mathrm{an}}, \mathbb{Q}(q)) \cap F^q H^q(U^{\mathrm{an}}, \mathbb{C}) \tag{1.1}$$

which is the composition of the regulator map and σ_U.

Now we consider the following variant of the above problem. Assume that we are given a smooth algebraic variety S over $\mathbb{C}$ and a smooth surjective morphism $\pi : U \to S$ over $\mathbb{C}$. Let $\pi^{\mathrm{an}} : U^{\mathrm{an}} \to S^{\mathrm{an}}$ be the associated morphism of sites. Assume that the fibers of π are affine of dimension m. Then $R^b\pi_*^{\mathrm{an}}\mathbb{Q} = 0$ for $b > m$ and we have the natural map

$$\alpha : H^{m+q}(U^{\mathrm{an}}, \mathbb{Q}(m+q)) \to H^q(S^{\mathrm{an}}, R^m\pi_*^{\mathrm{an}}\mathbb{Q}(m+q))$$

which is an edge homomorphism of the Leray spectral sequence

$$E_2^{a,b} = H^a(S^{\mathrm{an}}, R^b\pi_*^{\mathrm{an}}\mathbb{Q}(m+q)) \Rightarrow H^{a+b}(U^{\mathrm{an}}, \mathbb{Q}(m+q)).$$

Note that $H^a(S^{\mathrm{an}}, R^b\pi_*^{\mathrm{an}}\mathbb{Q}(m+q))$ carries in a canonical way a mixed Hodge structure and α is a morphism of mixed Hodge structures ([17] and [1]). Let

$$\mathrm{reg}_{U/S}^{m+q} : K_{m+q}^M(\mathcal{O}(U)) \otimes \mathbb{Q} \to H^q(S^{\mathrm{an}}, R^m\pi_*^{\mathrm{an}}\mathbb{Q}(m+q)) \cap F^{m+q} \tag{1.2}$$

be the composition of reg_U^{m+q} and α where $F^t \subset H^q(S^{\mathrm{an}}, R^m\pi_*^{\mathrm{an}}\mathbb{C})$ denotes the Hodge filtration. In this paper we study $\mathrm{reg}_{U/S}^{m+q}$ in case U/S is a family of open complete intersections, namely in case that the fibers of π are open complete intersections. Roughly speaking, our main results affirm that

$\mathrm{reg}^{m+q}_{U/S}$ is surjective for $q = 0, 1$ if $\pi : U \to S$ is the pullback of the universal family of open complete intersection of sufficiently high degree via a dominant smooth morphism from S to the moduli space. Let $d_i, e_j \geq 0$ $(1 \leq i \leq r,\ 1 \leq j \leq s)$ be fixed integers. Let $\mathbf{M} = \mathbf{M}(d_1, \cdots, d_r; e_1, \cdots, e_s)$ be the moduli space of the sets $(X_{1,o}, \ldots, X_{r,o}; Y_{1,o} \ldots, Y_{s,o})$ of smooth hypersurfaces in $\mathbb{P}^n$ of degree $d_1, \cdots, d_r; e_1, \cdots, e_s$ respectively which intersect transversally with each other. Let $f : S \to \mathbf{M}$ be a morphism of finite type with $S = \operatorname{Spec} R$ nonsingular affine and let $X_i \to S$ and $Y_j \to S$ be the pullback of the universal families of hypersurfaces over $\mathbf{M}$. Put

$$X = X_1 \cap \cdots \cap X_r \quad \text{and} \quad U = X \setminus \bigcup_{1 \leq j \leq s} X \cap Y_j$$

with the natural morphisms $\pi : U \to S$. Put

$$\mathbf{d} = \sum_{i=1}^{r} d_i, \quad \delta_{\min} = \min_{\substack{1 \leq i \leq r \\ 1 \leq j \leq s}} \{d_i, e_j\}, \quad d_{\max} = \max_{1 \leq i \leq r} \{d_i\}.$$

Theorem 1.1.1. *(see §1.3) Assume f is dominant smooth.*

(1) Assuming $\delta_{min}(n - r - 1) + \mathbf{d} \geq n + 1$,

$$\mathrm{reg}^{m}_{U/S} : K^M_m(\mathcal{O}(U)) \otimes \mathbb{Q} \to H^0(S^{\mathrm{an}}, R^m \pi^{\mathrm{an}}_* \mathbb{Q}_U(m + 1))$$

is surjective.

(ii) Assuming

$$\delta_{min}(n - r - 1) + \mathbf{d} \geq n + 2,\ \delta_{min}(n - r) + \mathbf{d} \geq n + 1 + d_{\max},\ \delta_{min} \geq 2,$$

$$\mathrm{reg}^{m+1}_{U/S} : K^M_{m+1}(\mathcal{O}(U)) \otimes \mathbb{Q} \to H^1(S^{\mathrm{an}}, R^m \pi^{\mathrm{an}}_* \mathbb{Q}_U(m + 1)) \cap F^{m+1}$$

is surjective.

The method of the study is the infinitesimal method in Hodge theory and is a natural generalization of that in [3] and [4]. To explain this, we now work over an arbitrary algebraically field k of characteristic zero which will be fixed in the whole paper. Let $f : S \to \mathbf{M}$ and $\pi : U \to S$ be defined over k as above. Following Katz and Oda ([12]), we have the algebraic Gauss Manin connection on the de Rham cohomology (see §1.2.2)

$$\nabla : H^\bullet_{\mathrm{dR}}(U/S) \longrightarrow H^\bullet_{\mathrm{dR}}(U/S) \otimes_R \Omega^1_{R/k}. \tag{1.3}$$

The map ∇ is extended to $H^\bullet_{\mathrm{dR}}(U/S) \otimes_R \Omega^q_{R/k} \longrightarrow H^\bullet_{\mathrm{dR}}(U/S) \otimes_R \Omega^{q+1}_{R/k}$ by imposing the Leibniz rule

$$\nabla(e \otimes \omega) = \nabla(e) \wedge \omega + e \otimes d\omega \tag{1.4}$$

and it induces the complex

$$\begin{aligned} \mathrm{Gr}_F^{p+1} H^m_{\mathrm{dR}}(U/S) \otimes \Omega^{q-1}_{R/k} &\longrightarrow \mathrm{Gr}_F^{p} H^m_{\mathrm{dR}}(U/S) \otimes \Omega^{q}_{R/k} \\ &\longrightarrow \mathrm{Gr}_F^{p-1} H^m_{\mathrm{dR}}(U/S) \otimes \Omega^{q+1}_{R/k}, \end{aligned}$$

where $m = \dim(U/S)$ and $F^\bullet$ denotes the Hodge filtration:

$$F^p H^q_{\mathrm{dR}}(U/S) = H^q_{Zar}(X, \Omega^{\geq p}_{X/S}(\log Z)) \subset H^q_{\mathrm{dR}}(U/S).$$

The cohomology at the middle term of the complex has been studied in [3] when $1 \leq p \leq m-1$.

In the study of the variant of Beilinson's Hodge conjecture, a crucial role will be played by the kernel of the following map:

$$\overline{\nabla}_q : F^m H^m_{\mathrm{dR}}(U/S) \otimes \Omega^q_{R/k} \longrightarrow \mathrm{Gr}_F^{m-1} H_{\mathrm{dR}}(U/S) \otimes \Omega^{q+1}_{R/k} \quad (q \geq 0).$$

which arises as the special case $p = m$ in the above complex. The key result is, roughly speaking, that when $f : S \to \mathbf{M}$ factors as $S \xrightarrow{g} T \xrightarrow{i} \mathbf{M}$ where g is smooth and i is a regular immersion of small codimension, then the kernel of $\overline{\nabla}_q$ is generated by the image of

$$\mathrm{dlog} : K^M_{m+q}(\mathcal{O}(U)) \longrightarrow F^m H^m_{\mathrm{dR}}(U/S) \otimes \Omega^q_{R/k}$$

(see §1.2.3 for its definition). In case $k = \mathbb{C}$ it implies the surjectivity of $\mathrm{reg}^{m+q}_{U/S}$ (1.2) for $q = 0$ and 1 by using the known surjectivity of the map (1.1) for $U = S$.

The main tool for the proof of the above key result is the theory of generalized Jacobian rings developed by the authors in [3]. It describes the Hodge cohomology groups of U and the Gauss-Manin connection $\overline{\nabla}_q$ in terms of multiplication of the rings, so that the various problems can be translated into algebraic computations in Jacobian rings. We show several computational results on Jacobian rings in §1.4 and §1.5. The basic techniques for this were developed by M.Green, C.Voisin and Nori. We note that a key to the computational results is Proposition 1.5.5, which is proved in [3] as a generalization of Nori's connectivity theorem ([14]) to open complete intersections.

Notation and Conventions

For an abelian group M, we write $M_{\mathbb{Q}} = M \otimes_{\mathbb{Z}} \mathbb{Q}$.

1.2 The Main Theorem

Throughout the paper, we work over an algebraically closed field k of characteristic zero.

1.2.1 Setup

We fix integers $n \geq 2$, $r, s \geq 1$, $n \geq r$ and $d_1, \cdots, d_r$, $e_1, \cdots, e_s \geq 1$. We put

$$\mathbf{d} = \sum_{i=1}^{r} d_i, \quad \mathbf{e} = \sum_{j=1}^{s} e_j, \quad \delta_{\min} = \min_{\substack{1 \leq i \leq r \\ 1 \leq j \leq s}} \{d_i, e_j\},$$
$$d_{\max} = \max_{1 \leq i \leq r} \{d_i\}, \quad e_{\max} = \max_{1 \leq j \leq s} \{e_j\}.$$

Let $P = k[X_0, \cdots, X_n]$ be the polynomial ring over k and P^d denote the subspace of the homogeneous polynomials of degree d. Then the space $P^d - \{0\}$ parametrizes hypersurfaces in $\mathbb{P}^n$ of degree d with a chosen defining equation. Let

$$\widetilde{\mathbf{M}} = \widetilde{\mathbf{M}}(d_1, \cdots, d_r; e_1, \cdots, e_s) \subset \prod_{i=1}^{r} (P^{d_i} - \{0\}) \times \prod_{j=1}^{s} (P^{e_j} - \{0\})$$

be the Zariski open subset such that the associated divisor $X_{1,o} + \cdots + X_{r,o} + Y_{1,o} + \cdots + Y_{s,o}$ to any point $o \in \widetilde{\mathbf{M}}$ is a simple normal crossing divisor on $\mathbb{P}^n$, namely all $X_{i,o}$ and $Y_{j,o}$ are nonsingular and they intersect transversally with each other. Put $X_o = X_{1,o} \cap \cdots \cap X_{r,o}$ and $Z_{j,o} = X_o \cap Y_{j,o}$. Then X_o is a nonsingular complete intersection of dimension $n - r$, and $\sum_{j=1}^{s} Z_{j,o}$ is a simple normal crossing divisor on X_o.

Let $f : S \to \widetilde{\mathbf{M}}$ be a morphism of finite type with $S = \operatorname{Spec} R$ nonsingular affine. We write $P_R = P \otimes_k R$ and $P_R^\ell = P^\ell \otimes_k R$. Let

$$F_i \in P_R^{d_i} \ (1 \leq i \leq r) \quad \text{and} \quad G_j \in P_R^{e_j} \ (1 \leq j \leq s) \tag{1.5}$$

be the pullback of the universal polynomials over the moduli space. We denote by X, X_i, Y_j and Z_j the associated families of the complete intersections X_o, $X_{i,o}$, $Y_{j,o}$ and divisors $Z_{j,o}$ respectively. Thus we get the smooth morphisms:

$$\pi_X : X \longrightarrow S, \quad \pi_{X_i} : X_i \longrightarrow S, \quad \pi_{Y_j} : Y_j \longrightarrow S, \quad \pi_{Z_j} : Z_j \longrightarrow S. \tag{1.6}$$

We write

$$X_* = \sum_{i=1}^{r} X_i, \quad Y_* = \sum_{j=1}^{s} Y_j, \quad Z_* = \sum_{j=1}^{s} Z_j.$$

Put $U = X - Z_*$ and we get $\pi : U \to S$, a family of open complete intersections.

1.2.2 Gauss-Manin connection

For an integer $q \geq 0$ we have the *Gauss-Manin connection*

$$\nabla : H^{\bullet}_{\mathrm{dR}}(U/S) \longrightarrow H^{\bullet}_{\mathrm{dR}}(U/S) \otimes \Omega^1_{R/k}. \tag{1.7}$$

Here $H^{\bullet}_{\mathrm{dR}}(U/S)$ is the de Rham cohomology defined as

$$H^k_{\mathrm{dR}}(U/S) = H^k_{\mathrm{Zar}}(X, \Omega^{\bullet}_{X/S}(\log Z_*)) = \Gamma(S, R^k\pi_{X*}\Omega^{\bullet}_{X/S}(\log Z_*)),$$

where the second equality follows from the assumption that S is affine. It is an integrable connection and satisfies the Griffiths transversality:

$$\nabla(F^p H^{\bullet}_{\mathrm{dR}}(U/S)) \subset F^{p-1} H^{\bullet}_{\mathrm{dR}}(U/S) \otimes \Omega^1_{R/k} \tag{1.8}$$

with respect to the Hodge filtration

$$F^p H^{\bullet}_{\mathrm{dR}}(U/S) := H^{\bullet}_{\mathrm{Zar}}(X, \Omega^{\geq p}_{X/S}(\log Z_*)). \tag{1.9}$$

We are interested in $H^{n-r}_{\mathrm{dR}}(U/S)$ (Since X is a complete intersection, the cohomology in other degrees is not interesting). We denote by

$$\nabla_q : F^{n-r} H^{n-r}_{\mathrm{dR}}(U/S) \otimes \Omega^q_{R/k} \longrightarrow F^{n-r-1} H^{n-r}_{\mathrm{dR}}(U/S) \otimes \Omega^{q+1}_{R/k} \tag{1.10}$$

the map given by (1.4). Noting

$$F^p H^{n-r}_{\mathrm{dR}}(U/S)/F^{p+1} H^{n-r}_{\mathrm{dR}}(U/S) \simeq H^{n-r-p}(X, \Omega^p_{X/S}(\log Z_*)),$$

(1.8) implies that ∇_q induces

$$\overline{\nabla}_q : H^0(X, \Omega^{n-r}_{X/S}(\log Z_*)) \otimes \Omega^q_{R/k} \longrightarrow H^1(X, \Omega^{n-r-1}_{X/S}(\log Z_*)) \otimes \Omega^{q+1}_{R/k}. \tag{1.11}$$

Our main theorem gives an explicit description of $\mathrm{Ker}(\overline{\nabla}_q)$ under suitable conditions. For its statement we need more notations.

1.2.3 Milnor K-theory

We denote by $K^M_\ell(\mathcal{A})$ the Milnor K-group of a commutative ring $\mathcal{A}$ ([13, 19]). By definition, it is the quotient of $\mathcal{A}^{*\otimes \ell}$ by the subgroup generated by

$$a_1 \otimes \cdots \otimes a_\ell, \quad (a_i + a_j = 0 \text{ or } 1 \text{ for some } i \neq j).$$

The element represented by $a_1 \otimes \cdots \otimes a_\ell$ is called the Steinberg symbol, and denoted by $\{a_1, \cdots, a_\ell\}$. We have

$$\{a_1, \cdots, a_i, \cdots, a_j, \cdots, a_\ell\} = -\{a_1, \cdots, a_j, \cdots, a_i, \cdots, a_\ell\} \quad \text{for } i \neq j$$

following from the expansion $\{ab, -ab\} = \{a, b\} + \{b, a\} + \{a, -a\} + \{b, -b\}$.

Let $\mathcal{O}(U) = \Gamma(U_{Zar}, \mathcal{O}_U)$ be the ring of regular functions on U. We have the dlog map

$$\mathrm{dlog}: K_\ell^M(\mathcal{O}(U)) \longrightarrow H^0(\Omega^\ell_{X/k}(\log Z_*)), \quad \{h_1, \cdots, h_\ell\} \longmapsto \frac{dh_1}{h_1} \wedge \cdots \wedge \frac{dh_\ell}{h_\ell}. \tag{1.12}$$

Assuming $\ell \geq n - r = \dim(X/S)$, there is the unique map

$$\upsilon_X : \Omega^\ell_{X/k}(\log Z_*) \longrightarrow \Omega^{n-r}_{X/S}(\log Z_*) \otimes \Omega^{\ell-n+r}_{S/k}. \tag{1.13}$$

such that its composition with $\Omega^{n-r}_{X/k}(\log Z_*) \otimes \Omega^{\ell-n+r}_{S/k} \to \Omega^\ell_{X/k}(\log Z_*)$ is the identity map $\Omega^{n-r}_{X/k}(\log Z_*) \otimes \Omega^{\ell-n+r}_{S/k} \to \Omega^{n-r}_{X/S}(\log Z_*) \otimes \Omega^{\ell-n+r}_{S/k}$. Let

$$\psi_\ell^M : K_\ell^M(\mathcal{O}(U)) \longrightarrow H^0(\Omega^{n-r}_{X/S}(\log Z_*)) \otimes \Omega^{\ell-n+r}_{R/k} = F^{n-r} H^{n-r}_{\mathrm{dR}}(U/S) \otimes \Omega^{\ell-n+r}_{R/k}.$$

be the composition of υ_X and dlog. Its image is contained in $\mathrm{Ker}(\nabla_{\ell-n+r})$ since it lies in the image of $H^0(\Omega^\ell_{X/k}(\log Z_*))$. Thus we get the map

$$\psi_\ell^M : K_\ell^M(\mathcal{O}(U)) \longrightarrow \mathrm{Ker}(\nabla_{\ell-n+r}). \tag{1.14}$$

We will also consider the induced maps

$$K^M_{\ell+n-r}(\mathcal{O}(U)) \otimes_{\mathbb{Z}} \Omega^{q-\ell,d=0}_{R/k} \longrightarrow \mathrm{Ker}(\nabla_q); \quad \xi \otimes \omega \mapsto \psi^M_{\ell+n-r}(\xi) \wedge \omega,$$

$$K^M_{\ell+n-r}(\mathcal{O}(U)) \otimes_{\mathbb{Z}} \Omega^{q-\ell}_{R/k} \longrightarrow \mathrm{Ker}(\overline{\nabla}_q); \quad \xi \otimes \omega \mapsto \psi^M_{\ell+n-r}(\xi) \wedge \omega,$$

where $\Omega^{\bullet,d=0}_{R/k} = \mathrm{Ker}(d : \Omega^\bullet_{R/k} \to \Omega^\bullet_{R/k})$ is the module of closed forms.

Now we construct some special elements in $K_\ell^M(\mathcal{O}(U))$. Let $\ell \geq 1$ be an integer. We define $\bigwedge^\ell(G_j)$ as the $\mathbb{Q}$-vector space spanned by symbols v_J indexed by multi-indices $J = (j_0, \cdots, j_\ell)$ $(1 \leq j_k \leq s)$ with relations

$$v_{j_0 \cdots j_p \cdots j_q \cdots j_\ell} = -v_{j_0 \cdots j_q \cdots j_p \cdots j_\ell} \quad \text{for } 0 \leq p \neq q \leq \ell \tag{1.15}$$

and

$$\sum_{k=0}^{\ell+1} (-1)^k e_{j_k} v_{j_0 \cdots \widehat{j_k} \cdots j_{\ell+1}} = 0. \tag{1.16}$$

We formally put $\overset{0}{\bigwedge}(G_j) = \mathbb{Q}$. By convention, $\overset{\ell}{\bigwedge}(G_j) = 0$ if $s = 0$ or 1. We easily see

$$\dim_{\mathbb{Q}} \overset{\ell}{\bigwedge}(G_j) = \binom{s-1}{\ell} \text{ with basis } \{v_{1j_1\cdots j_\ell}\ ;\ 2 \le j_1 < \cdots < j_\ell \le s\},$$

and $\overset{\ell}{\bigwedge}(G_j) = 0$ if $\ell \ge s$. Let $G_j^{e_i}/G_i^{e_j}|_X$ be the restriction on X of a rational function $G_j^{e_i}/G_i^{e_j}$ on $\mathbb{P}^n_R = \mathrm{Proj}(R[X_0, \ldots, X_n])$. Then we have a natural homomorphism

$$sym_\ell : \overset{\ell}{\bigwedge}(G_j) \longrightarrow K_\ell^M(\mathcal{O}(U))_{\mathbb{Q}}. \tag{1.17}$$

$$v_J \mapsto g_J := e_{j_0}^{-\ell+1}\left\{G_{j_1}^{e_{j_0}}/G_{j_0}^{e_{j_1}}|_X, \cdots, G_{j_\ell}^{e_{j_0}}/G_{j_0}^{e_{j_\ell}}|_X\right\} \quad (J = (j_0, \ldots, j_\ell))$$

Putting $g_j = G_j/X_0^{e_j}|_X$, a calculation shows

$$\mathrm{dlog}(g_J) = \sum_{\nu=0}^{\ell}(-1)^\nu e_{j_\nu}\frac{dg_{j_0}}{g_{j_0}} \wedge \cdots \wedge \widehat{\frac{dg_{j_\nu}}{g_{j_\nu}}} \wedge \cdots \wedge \frac{dg_{j_\ell}}{g_{j_\ell}} \quad \text{on } \{X_0 \ne 0\}. \tag{1.18}$$

The maps $\psi_\bullet^M$ and sym_ℓ induce a homomorphism

$$\Psi^q_{U/S} : \bigoplus_{\ell=0}^{q} \overset{\ell+n-r}{\bigwedge}(G_j) \otimes_{\mathbb{Q}} \Omega^{q-\ell}_{R/k} \longrightarrow \mathrm{Ker}(\overline{\nabla}_q); \quad g_I \otimes \eta \mapsto \psi_\ell^M(g_I) \bigwedge \eta. \tag{1.19}$$

The main theorem affirms that this map is an isomorphism under suitable conditions. In order to give the precise statement we need to introduce some notations.

1.2.4 Statement of the Main Theorem

Let $T_{R/k}$ be the derivation module of R over k which is the dual of $\Omega^1_{R/k}$. A derivation $\theta \in T_{R/k}$ acts on $P_R = P \otimes_k R = R[X_0, \ldots, X_n]$ by $\mathrm{id}_P \otimes \theta$. Introducing indeterminants $\mu_1, \ldots, \mu_r, \lambda_1, \ldots, \lambda_s$, we define an R-linear homomorphism

$$\Theta = \Theta_{(F_i, G_j)} : T_{R/k} \longrightarrow A_1(0), \quad \theta \mapsto \sum_{i=1}^{r}\theta(F_i)\mu_i + \sum_{j=1}^{s}\theta(G_j)\lambda_j. \tag{1.20}$$

where

$$A_1(0) = \bigoplus_{i=1}^{r} P_R^{d_i}\mu_i \bigoplus \bigoplus_{j=1}^{s} P_R^{e_j}\lambda_j \quad (P_R^\ell = P^\ell \otimes_k R). \tag{1.21}$$

We note that Θ is surjective (resp. an isomorphism) if $f : S = \mathrm{Spec}(R) \to \widetilde{M}$ is étale (resp. smooth). Put

$$W = \mathrm{Im}(\Theta) \subset A_1(0).$$

It is a finitely generated R-module.

For an ideal $I \subset P_R$ we denote by $A_1(0)/I$ the quotient of $A_1(0)$ by the submodule

$$\bigoplus_{i=1}^{r}(I \cap P_R^{d_i})\mu_i \bigoplus \bigoplus_{j=1}^{s}(I \cap P_R^{e_j})\lambda_j.$$

For a variety V over k we denote by $|V|$ the set of the closed points of V. Let $\alpha \in |\mathbb{P}_R^n|$ and $x \in S = \mathrm{Spec}(R)$ be its image with $\kappa(x)$, its residue field. Let $\mathfrak{m}_{\alpha,x} \subset P_x := P \otimes_R \kappa(x)$ be the homogeneous ideal defining α in $\mathrm{Proj}(P_x)$ and let $\mathfrak{m}_\alpha \subset P_R$ be the inverse image of $\mathfrak{m}_{\alpha,x}$. The evaluation at α induces an isomorphism (note $\kappa(x) = k$)

$$v_\alpha : A_1(0)/\mathfrak{m}_\alpha \simeq \bigoplus_{i=1}^{r} k \cdot \mu_i \bigoplus \bigoplus_{j=1}^{s} k \cdot \lambda_j. \tag{1.22}$$

We now introduce the conditions for $\Psi_{U/S}^q$ to be an isomorphism. We fix an integer $q \geq 0$. Consider the following four conditions.

(I) Both W and $A_1(0)/W$ are locally free R-modules. We put

$$c = \mathrm{rank}_R(A_1(0)/W).$$

(II) W has no base points: $W \to A_1(0)/\mathfrak{m}_\alpha$ is surjective for $\forall \alpha \in |\mathbb{P}_R^n|$.

$\mathbf{(III)}_q$ One of the following conditions holds:

(a) $q = 0$ and $\delta_{\min}(n - r - 1) + \mathbf{d} - n - 1 \geq c$,

(b) $q = 1$, $\delta_{\min}(n - r - 1) + \mathbf{d} - n - 1 \geq c + 1$ and $\delta_{\min}(n - r) + \mathbf{d} - n - 1 - d_{\max} \geq c$,

(c) $\delta_{\min}(n - 1) - n - 1 \geq c + q$.

$\mathbf{(IV)}_q$ For any $x \in |S|$ and any $1 \leq j_1 < \cdots < j_{n-r} \leq s$, there exist $q + 1$ points $\alpha_0, \cdots, \alpha_q \in |X \cap Y_{j_1} \cap \cdots \cap Y_{j_{n-r}}|$ lying over x such that the map

$$W \to A_1(0)/(J' + \mathfrak{m}_{\alpha_0}) \bigoplus \cdots \bigoplus A_1(0)/(J' + \mathfrak{m}_{\alpha_q})$$

is surjective. Here $J' \subset A_1(0)$ denotes the R-submodule generated by the elements

$$L \cdot (\sum_{i=1}^{r} \frac{\partial F_i}{\partial X_k} \mu_i + \sum_{j=1}^{s} \frac{\partial G_j}{\partial X_\nu} \lambda_j) \quad \text{with } 0 \le \nu \le n \text{ and } L \in P_R^1.$$

Remark 1.2.1. **(I)** holds if f factors as $S \xrightarrow{g} T \xrightarrow{i} \widetilde{M}$ where g is smooth and i is a regular immersion. In this case $c = \operatorname{codim}_{\widetilde{M}}(T)$.

Remark 1.2.2. In view of (1.22), **(II)** holds if $F_i\mu_i,\ G_j\lambda_j \in W$ for $\forall i, j$ and $J' \subset W$.

Remark 1.2.3. $\mathbf{(IV)}_q$ always holds if $s \le n - r + 1$. Indeed we will see (cf. 1.5.3) that for any $1 \le j_1 < \cdots < j_{n-r} \le s$ and any $\alpha \in |X \cap Y_{j_1} \cap \cdots \cap Y_{j_{n-r}}|$, $A_1(0)/(J' + \mathfrak{m}_\alpha)$ is a k-vector space of dimension $s - 1 - (n - r)$ and $A_1(0)/(J' + \mathfrak{m}_\alpha) = 0$ if $s - 1 \le n - r$.

Remark 1.2.4. $\mathbf{(IV)}_q$ holds if $W = A_1(0)$ and $\delta_{\min} \ge q$ (cf. §1.2.1). In this case the natural map

$$A_1(0) \longrightarrow \bigoplus_{i=0}^{q} A_1(0)/\mathfrak{m}_{\alpha_i} \tag{1.23}$$

is surjective for arbitrary $(q+1)$-points $\alpha_i \in |\mathbb{P}_R^n|$ $(0 \le i \le q)$ lying over a point $x \in |S|$. To see this it suffices to show that

$$P_x^q \longrightarrow \bigoplus_{i=0}^{q} P_x^q/(\mathfrak{m}_{\alpha_i,x} \cap P_x^q) \tag{1.24}$$

is surjective. Let $H_i \in P_x^1$ $(0 \le i \le q)$ be a linear form such that $H_i(\alpha_j) \neq 0$ for $j \neq i$ and $H_i(\alpha_i) = 0$. Then the images of $H_i' := H_0 \cdots \widehat{H_i} \cdots H_q \in P_x^q$ for $0 \le i \le q$ generate the right hand side of (1.24).

Main Theorem. *Fix an integer $q \ge 0$.*

i) Assuming $\mathbf{(IV)}_q$, $\Psi_{U/S}^q$ *is injective.*

ii) Assuming **(I)**, $\mathbf{(II)}_q$, **(III)** *and* $\mathbf{(IV)}_q$, $\Psi_{U/S}^q$ *is an isomorphism.*

In order to clarify the technical conditions of the Main Theorem, we explain in the next section its implications on the image of the regulator map (1.2). The proof of the Main Theorem will be given in the sections following the next.

1.3 Implications of the Main Theorem

Let

$$\Omega_{R/k}^{\bullet,d=0} = \mathrm{Ker}(\Omega_{R/k}^{\bullet} \xrightarrow{d} \Omega_{R/k}^{\bullet+1})$$

be the module of closed differential forms.

Theorem 1.3.1. *Fix an integer $q \geq 0$ and assume* **(I)**, **(II)**, $\mathbf{(III)}_q$ *and* $\mathbf{(IV)}_{q+1}$ *in the Main Theorem. Then the map ψ_ℓ^M (cf. (1.14)) induces an isomorphism*

$$\bigoplus_{\ell=0}^{q} \bigwedge^{\ell+n-r} (G_j) \otimes_{\mathbb{Q}} \Omega_{R/k}^{q-\ell,d=0} \xrightarrow{\cong} \mathrm{Ker}(\nabla_q), \tag{1.25}$$

where $\nabla_q : F^{n-r} H_{\mathrm{dR}}^{n-r}(U/S) \otimes \Omega_{R/k}^q \to F^{n-r-1} H_{\mathrm{dR}}^{n-r}(U/S) \otimes \Omega_{R/k}^{q+1}$.

Proof We first note that $\mathbf{(IV)}_{q+1} \Longrightarrow \mathbf{(IV)}_q$ by definition. Consider the following commutative diagram

$$\begin{array}{ccc}
\bigoplus_{\ell=0}^{q} \bigwedge^{\ell+n-r} (G_j) \otimes_{\mathbb{Q}} \Omega_{R/k}^{q-\ell} & \xrightarrow{\Psi'} & F^{n-r} H_{\mathrm{dR}}^{n-r}(U/S) \otimes \Omega_{R/k}^q \\
\Big\downarrow{\scriptstyle \mathrm{id}\otimes d} & & \Big\downarrow{\scriptstyle \nabla_q} \\
\bigoplus_{\ell=0}^{q+1} \bigwedge^{\ell+n-r} (G_j) \otimes_{\mathbb{Q}} \Omega_{R/k}^{q+1-\ell} & \xrightarrow{\Psi''} & F^{n-r-1} H_{\mathrm{dR}}^{n-r}(U/S) \otimes \Omega_{R/k}^{q+1}.
\end{array}$$

Ψ' is injective and its image is $\ker \overline{\nabla}_q$ by the Main Theorem (ii). Ψ'' is injective by the Main Theorem (i). Thus the assertion follows by diagram chase. □

The second implication of the Main theorem concerns the Hodge filtration on cohomology with coefficients. The Gauss-Manin connection (cf. (1.7)) gives rise to the following complex of Zariski sheaves on S

$$H_{\mathrm{dR}}^{n-r}(U/S) \xrightarrow{\nabla} H_{\mathrm{dR}}^{n-r}(U/S) \otimes \Omega_{S/k}^1 \xrightarrow{\nabla} \cdots \xrightarrow{\nabla} H_{\mathrm{dR}}^{n-r}(U/S) \otimes \Omega_{S/k}^{\dim S}. \tag{1.26}$$

which is denoted by $H_{\mathrm{dR}}^{n-r}(U/S) \otimes \Omega_{S/k}^{\bullet}$. We define the de Rham cohomology with coefficients as the hypercohomology

$$H_{\mathrm{dR}}^q(S, H_{\mathrm{dR}}^{n-r}(U/S)) = H_{\mathrm{Zar}}^q(S, H_{\mathrm{dR}}^{n-r}(U/S) \otimes \Omega_{S/k}^{\bullet}).$$

It is a finite dimensional k-vector space. It follows from the theory of mixed Hodge modules by Morihiko Saito ([17]) that $H_{\mathrm{dR}}^{\bullet}(S, H_{\mathrm{dR}}^{n-r}(U/S))$ carries in a canonical way the Hodge filtration and the weight filtration $W_{\bullet}$ denoted by

$$F^p H_{\mathrm{dR}}^{\bullet}(S, H_{\mathrm{dR}}^{n-r}(U/S)) \quad \text{and} \quad W_p H_{\mathrm{dR}}^{\bullet}(S, H_{\mathrm{dR}}^{n-r}(U/S))$$

respectively. (Arapura [1] has recently given a simpler proof of this fact.) In case $k = \mathbb{C}$ there is the comparison isomorphism between the de Rham cohomology and the Betti cohomology ([9, Thm.6.2])

$$H^q(S^{\mathrm{an}}, R^{n-r}\pi_*^{\mathrm{an}}\mathbb{C}_U) \simeq H^q_{\mathrm{dR}}(S, H^{n-r}_{\mathrm{dR}}(U/S)) \quad (\pi : U \to S) \tag{1.27}$$

which preserves the Hodge and weight filtrations on both sides defined by M. Saito. It endows $H^\bullet(S^{\mathrm{an}}, R^{n-r}\pi_*^{\mathrm{an}}\mathbb{Q}_U)$ with a mixed Hodge structure.

Define the subcomplex G^i of $H^{n-r}_{\mathrm{dR}}(U/S) \otimes \Omega^\bullet_{S/k}$ as

$$\begin{aligned} F^i H^{n-r}_{\mathrm{dR}}(U/S) \to F^{i-1} H^{n-r}_{\mathrm{dR}}(U/S) \otimes \Omega^1_{S/k} \to \cdots \\ \cdots \to F^{i-\dim S} H^{n-r}_{\mathrm{dR}}(U/S) \otimes \Omega^{\dim S}_{S/k} \end{aligned} \tag{1.28}$$

where $F^\bullet H^{n-r}_{\mathrm{dR}}(U/S)$ is the Hodge filtration as in (1.9). If S were proper over k, we would have

$$F^i H^\bullet_{\mathrm{dR}}(S, H^{n-r}_{\mathrm{dR}}(U/S)) = H^\bullet_{\mathrm{Zar}}(S, G^i).$$

When S is not proper, there is in general only a natural injection (cf. [2, Lemma 4.2])

$$F^i H^\bullet_{\mathrm{dR}}(S, H^{n-r}_{\mathrm{dR}}(U/S)) \hookrightarrow H^\bullet_{\mathrm{Zar}}(S, G^i) \quad (\forall i \geq 0). \tag{1.29}$$

The precise description of the Hodge filtration on the de Rham cohomology with coefficients is more complicated in general.

Theorem 1.3.2. *Fix an integer $q \geq 0$. Let $S \subset \overline{S}$ be a smooth compactification with $\partial S := \overline{S} - S$, a normal crossing divisor on $\overline{S}$. Assuming* **(I)**, **(II)**, **(III)**$_q$ *and* **(IV)**$_{q+1}$ *in the main theorem, we have an isomorphism*

$$\bigoplus_{\ell=0}^{q} \bigwedge^{\ell+n-r}(G_j) \otimes_{\mathbb{Q}} \Gamma(\overline{S}, \Omega^{q-\ell}_{\overline{S}/k}(\log \partial S)) \xrightarrow{\cong} F^{n-r+q} H^q_{\mathrm{dR}}(S, H^{n-r}_{\mathrm{dR}}(U/S)). \tag{1.30}$$

Proof We have the following commutative diagram

$$\begin{array}{ccc} \bigoplus_{\ell=0}^q \bigwedge^{\ell+n-r}(G_j) \otimes_{\mathbb{Q}} \Gamma(\overline{S}, \Omega^{q-\ell}_{\overline{S}/k}(\log \partial S)) & \xrightarrow{\phi} & F^{n-r+q} H^q_{\mathrm{dR}}(S, H^{n-r}_{\mathrm{dR}}(U/S)) \\ e \downarrow \cap & & a \downarrow \cap \\ \bigoplus_{\ell=0}^q \bigwedge^{\ell+n-r}(G_j) \otimes_{\mathbb{Q}} \Omega^{q-\ell, d=0}_{R/k} & \xrightarrow{\cong} & H^q(S, G^{n-r+q}) \\ b \downarrow & & \downarrow \\ \bigoplus_{\ell=0}^q \bigwedge^{\ell+n-r}(G_j) \otimes_{\mathbb{Q}} H^{q-\ell}_{\mathrm{dR}}(S/k) & \xrightarrow{c} & H^q_{\mathrm{dR}}(S, H^{n-r}_{\mathrm{dR}}(U/S)) \end{array} \tag{1.31}$$

where by definition

$$H^t_{\mathrm{dR}}(S/k) = H^t(\overline{S}, \Omega^\bullet_{\overline{S}/k}(\log \partial S))$$

and the map b comes from the isomorphism

$$H^t(\overline{S}, \Omega^\bullet_{\overline{S}/k}(\log \partial S)) \simeq H^t(S, \Omega^\bullet_{S/k}) \simeq \Omega^{t,d=0}_{R/k}/d\Omega^{t-1}_{R/k} \tag{1.32}$$

due to [8, II (3.1.11)] and it is surjective. The map a comes from (1.29). The map e comes from [8, II (3.2.14)]. The bijection in the middle row is the composition of the isomorphism in Theorem 1.3.1 and the isomorphism

$$\mathrm{Ker}(\nabla_\ell) \simeq H^\ell(S, G^{\ell+n-r}) \quad \text{for } \forall \ell \geq 0. \tag{1.33}$$

The map c is induced by the composition

$$\begin{aligned}
\bigoplus_{\ell=0}^{q} \bigwedge^{\ell+n-r} (G_j) \otimes_{\mathbb{Q}} \Omega^{q-\ell,d=0}_{R/k} &\longrightarrow \mathrm{Ker}(\nabla_\ell) \otimes_{\mathbb{Q}} H^{q-\ell}_{\mathrm{dR}}(S/k) \\
&\longrightarrow H^\ell(S, H^{n-r}_{\mathrm{dR}}(U/S) \otimes \Omega^\bullet_{S/k}) \otimes H^{q-\ell}(S, \Omega^\bullet_{S/k}) \\
&\longrightarrow H^q(S, H^{n-r}_{\mathrm{dR}}(U/S) \otimes \Omega^\bullet_{S/k})
\end{aligned}$$

where the first map is induced by $\psi^M_{\ell+n-r}$ (1.14) and (1.32), the second by (1.33), and the last by cup product. We claim that there is a map ϕ which makes the upper square of the diagram (1.31) commute. Indeed let V be the source of c. Endowing V with the Hodge filtration defined by

$$F^p\Big(\bigwedge^t(G_j) \otimes_{\mathbb{Q}} H^u_{\mathrm{dR}}(S/k)\Big) = \bigwedge^t(G_j) \otimes_{\mathbb{Q}} F^{p-t} H^u_{\mathrm{dR}}(S/k)$$

with $F^{p-t}H^u_{\mathrm{dR}}(S/k) = H^u(\overline{S}, \Omega^{\geq p-t}_{\overline{S}/k}(\log \partial S))$, c respects the Hodge filtrations. Noting

$$F^{n-r+q}V = \bigoplus_{\ell=0}^{q} \bigwedge^{\ell+n-r} (G_j) \otimes_{\mathbb{Q}} H^0(\overline{S}, \Omega^{q-\ell}_{\overline{S}/k}(\log \partial S)),$$

we see that c induces ϕ as desired. The injectivity of ϕ follows from that of e. To show its surjectivity, note that $\mathrm{Im}(c)$ contains $F^{n-r+q}H^q_{\mathrm{dR}}(S, H^{n-r}_{\mathrm{dR}}(U/S))$ by the diagram. This shows $F^{n-r+q}\,\mathrm{Coker}(c) = 0$. By strictness of the Hodge filtration, we get the surjectivity of ϕ. This completes the proof of the theorem. □

In what follows we assume $k = \mathbb{C}$. Take $S \subset \overline{S}$, a smooth compactification with $\partial S := \overline{S} - S$, a normal crossing divisor on $\overline{S}$. Write for $t \geq 0$

$$H_{\mathbb{Q}}^{t,0}(S) := H^t(S^{\mathrm{an}}, \mathbb{Q}(t)) \cap F^t H^t(S^{\mathrm{an}}, \mathbb{C})$$
$$= H^t(S^{\mathrm{an}}, \mathbb{Q}(t)) \cap H^0(\overline{S}, \Omega^t_{\overline{S}/\mathbb{C}}(\log \partial S))$$

Write $m = n - r = \dim(U/S)$. Let

$$\mathrm{reg}_{U/S}^{m+\ell} : K_{m+\ell}^M(\mathcal{O}(U)) \otimes \mathbb{Q} \to H^\ell(S^{\mathrm{an}}, R^m \pi_*^{\mathrm{an}} \mathbb{Q}(m+\ell)) \cap F^{m+\ell}$$

be as (1.2). It induces for $q \geq 0$

$$\lambda_q : \bigoplus_{\ell=0}^q K_{m+\ell}^M(\mathcal{O}(U)) \otimes_{\mathbb{Q}} H_{\mathbb{Q}}^{q-\ell,0}(S) \to$$
$$\to H^q(S^{\mathrm{an}}, R^m \pi_*^{\mathrm{an}} \mathbb{Q}_U(m+q)) \cap F^{m+q}. \tag{1.34}$$

Theorem 1.1.1 follows from the following corollaries in view of Remarks 1.2.1, 1.2.2, 1.2.3.

Corollary 1.3.3. *Fix an integer $q \geq 1$ and assume* **(I)**, **(II)**, $\mathbf{(III)}_q$ *and* $\mathbf{(IV)}_{q+1}$ *in the Main Theorem. Then the map* (1.2)

$$\mathrm{reg}_{U/S}^{m+q} : K_{m+q}^M(\mathcal{O}(U)) \otimes \mathbb{Q} \to H^q(S^{\mathrm{an}}, R^m \pi_*^{\mathrm{an}} \mathbb{Q}_U(m+q)) \cap F^{m+q}$$

is surjective for $q = 1$. More generally $\mathrm{reg}_{U/S}^{m+q}$ is surjective if the regulator map for S:

$$\mathrm{reg}_S^t : K_t^M(\mathcal{O}(S)) \otimes \mathbb{Q} \to H_{\mathbb{Q}}(S)^{t,0}$$

is surjective for $1 \leq \forall t \leq q$.

Proof The first assertion of 1.3.3 follows from the second in view of the fact that reg_S^1 is surjective, namely $H_{\mathbb{Q}}^{1,0}(S)$ is generated by $\mathrm{dlog}\, \mathcal{O}(S)^*$. The second assertion is a direct consequence of the following isomorphism induced by λ_q:

$$\bigoplus_{\ell=0}^{q} \bigwedge^{\ell+m} (G_j) \otimes_{\mathbb{Q}} H_{\mathbb{Q}}(S)^{q-\ell,0} \xrightarrow{\cong} H^q(S^{\mathrm{an}}, R^m \pi_*^{\mathrm{an}} \mathbb{Q}_U(m+q)) \cap F^{m+q} \tag{1.35}$$

which follows from Theorem 1.3.2. □

Corollary 1.3.4. *Assuming* **(I)**, **(II)**, $\mathbf{(III)}_q$ *and* $\mathbf{(IV)}_{q+1}$ *for $q = 0$, λ_0 induces an isomorphism*

$$\bigwedge^{m} (G_j) \xrightarrow{\cong} H^0(S^{\mathrm{an}}, R^m \pi_*^{\mathrm{an}} \mathbb{Q}_U(m)).$$

Proof Applying (1.35), we have an isomorphism

$$\bigwedge^m(G_j) \xrightarrow{\cong} H^0(S^{\mathrm{an}}, R^m\pi_*^{\mathrm{an}}\mathbb{Q}_U(m)) \cap F^m. \tag{1.36}$$

We need show that the right hand side is equal to $H^0(S^{\mathrm{an}}, R^m\pi_*^{\mathrm{an}}\mathbb{Q}_U)$. It suffices to show that $H^0(S^{\mathrm{an}}, R^m\pi_*^{\mathrm{an}}\mathbb{Q})$ is pure of type (m,m). We need a result from [3, Theorem (III)], which implies that the map

$$\overline{\nabla} : \mathrm{Gr}_F^p\, H_{\mathrm{dR}}^m(U/S) \longrightarrow \mathrm{Gr}_F^{p-1}\, H_{\mathrm{dR}}^m(U/S) \otimes \Omega_{R/\mathbb{C}}^1$$

is injective for all $1 \le p \le m-1$ under the assumption of Corollary 1.3.4. It implies

$$\mathrm{Ker}(\nabla) \cap F^1 H_{\mathrm{dR}}^m(U/S) = \mathrm{Ker}(\nabla) \cap F^m H_{\mathrm{dR}}^m(U/S),$$

where $\nabla : H_{\mathrm{dR}}^m(U/S) \to H_{\mathrm{dR}}^m(U/S) \otimes \Omega_{R/\mathbb{C}}^1$ is the algebraic Gauss-Manin connection. Noting $H^0(S^{\mathrm{an}}, R^m\pi_*^{\mathrm{an}}\mathbb{C}_U) \xrightarrow{\sim} \mathrm{Ker}(\nabla)$ under the comparison isomorphism (1.27), it implies

$$F^1 H^0(S^{\mathrm{an}}, R^m\pi_*^{\mathrm{an}}\mathbb{C}_U) = F^m H^0(S^{\mathrm{an}}, R^m\pi_*^{\mathrm{an}}\mathbb{C}_U). \tag{1.37}$$

Consider the mixed Hodge structure $H := H^0(S^{\mathrm{an}}, R^m\pi_*^{\mathrm{an}}\mathbb{Q}_U)$. By the Hodge symmetry (1.37) implies

$$H^{p,q} := Gr_F^p Gr_{\overline{F}}^q Gr_{p+q}^W H = 0 \quad \text{unless } (p,q) = (m,0), (m,m), (0,m).$$

Hence it suffices to show $H^{m,0} = 0$. Putting $V = F^m H^0(S^{\mathrm{an}}, R^m\pi_{X*}^{\mathrm{an}}\mathbb{C}_X)$ where $\pi_X : X \to S$ is as in (1.6), we have the surjection $V \to H^{m,0}$ while $V = 0$ by Theorem 1.3.2 applied to the case $s = 0$. This completes the proof.

□

1.4 Theory of Generalized Jacobian Rings

We introduce the generalized Jacobian ring. It describes the Hodge cohomology groups $H^\bullet(\Omega_{X/S}^\bullet(\log Z_*))$ of open complete intersections, and enables us to identify the Gauss-Manin connection (cf. (1.11)) with the multiplication of rings. The computational results in this section will play a key role in the proof of the Main Theorem (see §1.5.2).

1.4.1 Fundamental results on generalized Jacobian ring

Recall the notations in §1.2.1. Let

$$A = P_R[\mu_1, \cdots, \mu_r, \lambda_1, \cdots, \lambda_s] = R[X_0, \ldots, X_n, \mu_1, \cdots, \mu_r, \lambda_1, \cdots, \lambda_s]$$

be the polynomial ring over P_R with indeterminants $\mu_1, \cdots, \mu_r, \lambda_1, \cdots, \lambda_s$. For $q \in \mathbb{Z}$ and $\ell \in \mathbb{Z}$, we put

$$A_q(\ell) = \bigoplus_{a_1+\cdots+a_r+b_1+\cdots+b_s=q} P_R^{m(a,b,\ell)} \cdot \mu_1^{a_1} \cdots \mu_r^{a_r} \lambda_1^{b_1} \cdots \lambda_s^{b_s}$$

with $m(a,b,\ell) = \sum_{i=1}^r a_i d_i + \sum_{j=1}^s b_j e_j + \ell$. Here a_i and b_j run over non-negative integers satisfying $a_1 + \cdots + a_r + b_1 + \cdots + b_s = q$. By convention, $A_q(\ell) = 0$ for $q < 0$. Note that the notation in 1.21 is compatible with the above definition.

The Jacobian ideal $J = J(F_1, \cdots, F_r, G_1, \cdots, G_s)$ is defined to be the ideal of A generated by

$$\sum_{i=1}^r \frac{\partial F_i}{\partial X_k}\mu_i + \sum_{j=1}^s \frac{\partial G_j}{\partial X_k}\lambda_j, \quad F_\ell, \quad G_{\ell'}\lambda_{\ell'} \quad (0 \le k \le n, \ 1 \le \ell \le r, \ 1 \le \ell' \le s).$$

The quotient ring $B = A/J$ is called the *generalized Jacobian ring* ([3]). We put

$$J_q(\ell) = A_q(\ell) \cap J \quad \text{and} \quad B_q(\ell) = A_q(\ell)/J_q(\ell).$$

We now recall some fundamental results from [3].

Theorem 1.4.1 ([3], Theorem (I)). *Suppose $n \ge r+1$. For each integer $0 \le p \le n-r$ there is a natural isomorphism*

$$\phi : B_p(\mathbf{d}+\mathbf{e}-n-1) \xrightarrow{\cong} H^p(X, \Omega_{X/S}^{n-r-p}(\log Z_*))$$

and the following diagram is commutative up to a scalar in $R^\times$:

$$\begin{array}{ccc} B_p(\mathbf{d}+\mathbf{e}-n-1) & \xrightarrow{\epsilon} & B_{p+1}(\mathbf{d}+\mathbf{e}-n-1) \otimes A_1(0)^* \\ \downarrow{\scriptstyle\phi} & & \downarrow{\scriptstyle 1\otimes\Theta^*} \\ H^p(X, \Omega_{X/S}^{n-r-p}(\log Z_*)) & \xrightarrow{\overline{\nabla}} & H^{p+1}(X, \Omega_{X/S}^{n-r-p-1}(\log Z_*)) \otimes \Omega_{R/k}^1 \end{array} \tag{1.38}$$

where $\overline{\nabla}$ is induced by the Gauss-Manin connection:

$$\nabla : F^{n-r-p}H_{\mathrm{dR}}^{n-r}(U/S) \longrightarrow F^{n-r-p-1}H_{\mathrm{dR}}^{n-r}(U/S) \otimes \Omega_{R/k}^1$$

and ϵ is induced from the multiplication

$$B_p(\mathbf{d}+\mathbf{e}-n-1) \otimes A_1(0) \to B_{p+1}(\mathbf{d}+\mathbf{e}-n-1)$$

and Θ^* is the dual of the map (1.20)

$$\Theta : T_{R/k} \to A_1(0); \quad \theta \mapsto \sum_{i=1}^{r} \theta(F_i)\mu_i + \sum_{j=1}^{s} \theta(G_j)\lambda_j.$$

The second fundamental result is the duality theorem for generalized Jacobian rings. For an R-module we denote $M^* = \mathrm{Hom}_R(M, R)$.

Theorem 1.4.2 ([3], Theorem (II)). *There is a natural map (called the trace map)*

$$\tau : B_{n-r}(2(\mathbf{d} - n - 1) + \mathbf{e}) \to R.$$

Let

$$h_p : B_p(\mathbf{d} - n - 1) \to B_{n-r-p}(\mathbf{d} + \mathbf{e} - n - 1)^*$$

be the map induced by the following pairing induced by the multiplication

$$B_p(\mathbf{d} - n - 1) \otimes B_{n-r-p}(\mathbf{d} + \mathbf{e} - n - 1) \to B_{n-r}(2(\mathbf{d} - n - 1) + \mathbf{e}) \xrightarrow{\tau} R.$$

Then h_p is bijective if $1 \le p \le n - r$, and surjective if $p = 0$. The kernel of h_0 is a locally free R-module of rank $\binom{s-1}{n-r}$.

1.4.2 Generalized Jacobian rings à la M. Green

We review from [3, §2] a "sheaf theoretic" definition of generalized Jacobian ring. This sophisticated definition originates from M.Green ([10]). It is useful for various computations (cf. §1.4.3 and §1.5.2).

Put

$$\mathbb{E} = \mathbb{E}_0 \bigoplus \mathbb{E}_1 \quad \text{with } \mathbb{E}_0 = \bigoplus_{i=1}^{r} \mathcal{O}(d_i) \text{ and } \mathbb{E}_1 = \bigoplus_{j=1}^{s} \mathcal{O}(e_j)$$

which is a locally free sheaf on $\mathbb{P}^n = \mathbb{P}^n_R$. We consider the projective space bundle

$$\pi : \mathbb{P} := \mathbb{P}(\mathbb{E}) \longrightarrow \mathbb{P}^n.$$

Let $\mathcal{L} = \mathcal{O}_{\mathbb{P}(\mathbb{E})}(1)$ be the tautological line bundle. We have the Euler exact sequence

$$0 \longrightarrow \mathcal{O}_{\mathbb{P}} \longrightarrow \pi^*\mathbb{E}^* \otimes \mathcal{L} \longrightarrow T_{\mathbb{P}/\mathbb{P}^n} \longrightarrow 0. \tag{1.39}$$

We consider the effective divisors

$$\mathbb{Q}_i := \mathbb{P}(\bigoplus_{1\le\alpha\ne i\le r} \mathcal{O}(d_\alpha) \bigoplus \mathbb{E}_1) \hookrightarrow \mathbb{P}(\mathbb{E}) \quad \text{for } 1 \le i \le r,$$

$$\mathbb{P}_j := \mathbb{P}(\mathbb{E}_0 \bigoplus \bigoplus_{1\le\beta\ne j\le r} \mathcal{O}(e_\beta)) \hookrightarrow \mathbb{P}(\mathbb{E}) \quad \text{for } 1 \le j \le s,$$

and let

$$\mu_i \in H^0(\mathbb{P}, \mathcal{L} \otimes \pi^*\mathcal{O}(-d_i)), \quad \lambda_j \in H^0(\mathbb{P}, \mathcal{L} \otimes \pi^*\mathcal{O}(-e_j))$$

be the global sections associated to these. We put

$$\sigma = \sum_{i=1}^{r} F_i\mu_i + \sum_{j=1}^{s} G_j\lambda_j \ \in \Gamma(\mathbb{P}, \mathcal{L}).$$

Let $\Sigma_{\mathcal{L}}$ be the sheaf of differential operators on $\mathcal{L}$ of order ≤ 1, defined as:

$$\begin{aligned}\Sigma_{\mathcal{L}} = \mathcal{D}\mathit{iff}^{\le 1}(\mathcal{L}) &= \{P \in \mathcal{E}nd_k(\mathcal{L}) \ ; \ Pf - fP \text{ is } \mathcal{O}_{\mathbb{P}}\text{-linear for } \forall f \in \mathcal{O}_{\mathbb{P}}\} \\ &\simeq \mathcal{L} \otimes \mathcal{D}\mathit{iff}^{\le 1}(\mathcal{O}_{\mathbb{P}}) \otimes \mathcal{L}^*.\end{aligned}$$

(It might be helpful to mention that $\Sigma_{\mathcal{L}}$ is a prolongation bundle.) By definition it fits into an exact sequence

$$0 \longrightarrow \mathcal{O}_{\mathbb{P}} \longrightarrow \Sigma_{\mathcal{L}} \longrightarrow T_{\mathbb{P}} \longrightarrow 0 \tag{1.40}$$

with extension class

$$-c_1(\mathcal{L}) \in \operatorname{Ext}^1(T_{\mathbb{P}}, \mathcal{O}_{\mathbb{P}}) \simeq \operatorname{Ext}^1(\mathcal{O}_{\mathbb{P}}, \Omega^1_{\mathbb{P}} \otimes \mathcal{O}_{\mathbb{P}}) \simeq H^1(\mathbb{P}, \Omega^1_{\mathbb{P}}).$$

Letting $U \subset \mathbb{P}^n$ be an affine subspace and $x_1, \cdots, x_n$ be its coordinates, $\Gamma(\pi^{-1}(U), \Sigma_{\mathcal{L}})$ is generated by the following sections

$$\frac{\partial}{\partial x_i}, \ \lambda_i\frac{\partial}{\partial \lambda_j}, \ \lambda_i\frac{\partial}{\partial \mu_j}, \ \mu_i\frac{\partial}{\partial \lambda_j}, \ \mu_i\frac{\partial}{\partial \mu_j}, \ \mathcal{O}_{\mathbb{P}}\text{-linear maps.} \tag{1.41}$$

The section σ defines a map

$$j(\sigma) : \Sigma_{\mathcal{L}} \longrightarrow \mathcal{L}, \quad P \longmapsto P(\sigma),$$

which is surjective by the assumption that $X_* + Y_*$ is a simple normal crossing divisor. It gives rise to the exact sequence

$$0 \longrightarrow T_{\mathbb{P}}(-\log \mathcal{Z}) \longrightarrow \Sigma_{\mathcal{L}} \xrightarrow{j(\sigma)} \mathcal{L} \longrightarrow 0, \tag{1.42}$$

where $\mathcal{Z} \subset \mathbb{P}$ is the zero divisor of σ. Put

$$\mathbb{Q}_* = \mathbb{Q}_1 + \cdots + \mathbb{Q}_r \quad \text{and} \quad \mathbb{P}_* = \mathbb{P}_1 + \cdots + \mathbb{P}_s,$$

and define $\Sigma_{\mathcal{L}}(-\log \mathbb{P}_*)$ to be the inverse image of $T_{\mathbb{P}}(-\log \mathbb{P}_*)$ via the map in (1.40). We then have the exact sequence

$$0 \longrightarrow \mathcal{O}_{\mathbb{P}} \longrightarrow \Sigma_{\mathcal{L}}(-\log \mathbb{P}_*) \longrightarrow T_{\mathbb{P}}(-\log \mathbb{P}_*) \longrightarrow 0. \tag{1.43}$$

Moreover (1.42) gives rise to an exact sequence

$$0 \longrightarrow T_{\mathbb{P}}(-\log(\mathcal{Z} + \mathbb{P}_*)) \longrightarrow \Sigma_{\mathcal{L}}(-\log \mathbb{P}_*) \xrightarrow{j(\sigma)} \mathcal{L} \longrightarrow 0. \tag{1.44}$$

Lemma 1.4.3. *For integers k and ℓ, put $A_k(\ell)_\Sigma = H^0(\mathcal{L}^k \otimes \pi^*\mathcal{O}(\ell))$ and*

$$J_k(\ell)_\Sigma = \mathrm{Im}\left(H^0(\Sigma_{\mathcal{L}}(-\log \mathbb{P}_*) \otimes \mathcal{L}^{k-1} \otimes \pi^*\mathcal{O}(\ell)) \xrightarrow{j(\sigma)\otimes 1} H^0(\mathcal{L}^k \otimes \pi^*\mathcal{O}(\ell))\right).$$

Then we have

$$A_k(\ell) = A_k(\ell)_\Sigma, \quad J_k(\ell) = J_k(\ell)_\Sigma.$$

Proof See [3, Lem.(2-2)]. □

Thus we have obtained another definition of the generalized Jacobian ring.

1.4.3 Some computational results

We keep the notations in §1.4.2. In what follows we simply write $\Sigma = \Sigma_{\mathcal{L}}(-\log \mathbb{P}_*)$.

Lemma 1.4.4. *We have*

$$H^w(\mathbb{P}, \bigwedge^p \Sigma^* \otimes \mathcal{L}^\nu \otimes \pi^*\mathcal{O}(\ell)) = 0$$

if one of the following conditions holds:

(i) $p - \nu \leq r + s - 1$ *and* $\nu \leq -1$,
(ii) $p - \nu \geq n + 1$ *and* $\nu \geq -s + 1$.
(iii) $w > 0$, $\nu \geq -s + 1$, $\ell \geq 0$ *and* $(\nu, \ell) \neq (0, 0)$.

Proof See [3, Thm.(4-1)]. □

Proposition 1.4.5. *Let k be an integer.*

(i) $H^\nu(\bigwedge^k \Sigma^*) = 0$ *for any* $k \geq 0$ *and* $\nu \neq 0$, n.
(ii) $H^0(\bigwedge^k \Sigma^*)$ *is a locally free R-module of rank* $\binom{s-1}{k}$.
(iii) $H^n(\bigwedge^k \Sigma^*)$ *is a locally free R-module of rank* $\binom{s-1}{k-n-1}$. *(Note* $\binom{x}{\ell} = 0$ *for* $\ell < 0$ *by convention.)*

The rest of this section is devoted to the proof of Proposition 1.4.5. Recall that there is an exact sequence

$$0 \longrightarrow \Omega^1_{\mathbb{P}}(\log \mathbb{P}_*) \longrightarrow \Sigma^* \longrightarrow \mathcal{O}_{\mathbb{P}} \longrightarrow 0 \tag{1.45}$$

with the extension class

$$c_1(\mathcal{L})|_{\mathbb{P}-\mathbb{P}_*} \in \operatorname{Ext}^1(\mathcal{O}_{\mathbb{P}}, \Omega^1_{\mathbb{P}}(\log \mathbb{P}_*)) = H^1(\Omega^1_{\mathbb{P}}(\log \mathbb{P}_*)).$$

It gives rise to the short exact sequence

$$0 \longrightarrow \Omega^{\bullet}_{\mathbb{P}}(\log \mathbb{P}_*) \longrightarrow \bigwedge^{\bullet}\Sigma^* \longrightarrow \Omega^{\bullet-1}_{\mathbb{P}}(\log \mathbb{P}_*) \longrightarrow 0, \tag{1.46}$$

and we have the long exact sequence

$$\begin{aligned} \cdots H^\nu(\Omega^k_{\mathbb{P}}(\log \mathbb{P}_*)) &\longrightarrow H^\nu(\bigwedge^k\Sigma^*) \\ &\longrightarrow H^\nu(\Omega^{k-1}_{\mathbb{P}}(\log \mathbb{P}_*)) \stackrel{\delta}{\longrightarrow} H^{\nu+1}(\Omega^k_{\mathbb{P}}(\log \mathbb{P}_*)) \cdots, \end{aligned} \tag{1.47}$$

where δ is induced by the cup-product with $c_1(\mathcal{L})|_{\mathbb{P}-\mathbb{P}_*} \in H^1(\Omega^1_{\mathbb{P}}(\log \mathbb{P}_*))$.

The first step is to write down $H^\bullet(\Omega^\bullet_{\mathbb{P}}(\log \mathbb{P}_*))$ and δ explicitly. We prepare some notations. For an integer $k \geq 1$ we put

$$\Delta_k = \{I = (i_1, \cdots, i_k) |\ 1 \leq i_1 < \cdots < i_k \leq s\}.$$

For $I = (i_1, \cdots, i_k) \in \Delta_k$ we write $\mathbb{P}_I = \mathbb{P}_{i_1} \cap \cdots \cap \mathbb{P}_{i_1}$. For $k = 0$ we put $\Delta_0 = \{\varnothing\}$ and $\mathbb{P}_\varnothing = \mathbb{P}$ by convention. To compute $H^\nu(\Omega^k_{\mathbb{P}}(\log \mathbb{P}_*))$ we first note the isomorphisms

$$H^\nu(\Omega^k_{\mathbb{P}}(\log \mathbb{P}_*)) \simeq \operatorname{Gr}^k_F H^{\nu+k}(\mathbb{P} - \mathbb{P}_*) \simeq \operatorname{Gr}^W_{2k} H^{\nu+k}(\mathbb{P} - \mathbb{P}_*),$$

where $H^*(\mathbb{P} - \mathbb{P}_*) = H^*(\mathbb{P}, \Omega^\bullet_{\mathbb{P}/S}(\log \mathbb{P}_*))$, and

$$F^p H^*(\mathbb{P} - \mathbb{P}_*) = H^*(\mathbb{P}, \Omega^{\geq p}_{\mathbb{P}/S}(\log \mathbb{P}_*)) \subset H^*(\mathbb{P}, \Omega^\bullet_{\mathbb{P}/S}(\log \mathbb{P}_*))$$

is the Hodge filtration and $W_p H^*(\mathbb{P} - \mathbb{P}_*) \subset H^*(\mathbb{P} - \mathbb{P}_*)$ denotes the weight filtration induced by the spectral sequence

$$E_2^{pq} = \bigoplus_{I \in \Delta_q} H^p(\mathbb{P}_I) \Longrightarrow H^{p+q}(\mathbb{P} - \mathbb{P}_*) \tag{1.48}$$

where $H^*(\mathbb{P}_I) = H^*(\mathbb{P}_I, \Omega^\bullet_{\mathbb{P}_I/S})$ (cf. [8]). We note $E_2^{p,q} = 0$ unless $0 \leq q \leq s$. Since the spectral sequence (1.48) degenerates at E_3, $H^\nu(\Omega^k_{\mathbb{P}}(\log \mathbb{P}_*))$ is isomorphic to the cohomology at the middle term of the following complex

$$\bigoplus_{I_1 \in \Delta_{k-\nu+1}} H^{2\nu-2}(\mathbb{P}_{I_1}) \longrightarrow \bigoplus_{I_2 \in \Delta_{k-\nu}} H^{2\nu}(\mathbb{P}_{I_2}) \longrightarrow \bigoplus_{I_3 \in \Delta_{k-\nu-1}} H^{2\nu+2}(\mathbb{P}_{I_3}). \tag{1.49}$$

The arrows in (1.49) are described as follows. Let $I_1 = (i_1, \cdots, i_k) \in \Delta_k$ and $I_2 \in \Delta_{k-1}$. If $I_2 \not\subset I_1$, then $H^{2\nu-2}(\mathbb{P}_{I_1}) \to H^{2\nu}(\mathbb{P}_{I_2})$ is the zero map. If $I_2 = (i_1, \cdots, \widehat{i_p}, \cdots, i_k)$, then it is $(-1)^{p-1}\phi_{I_1 I_2}$ where $\phi_{I_1 I_2}$ is the Gysin map. In order to describe it in more convenient way we introduce some notations. Let

$$S_R = R[x, y]$$

be the polynomial ring and S_R^ν be the set of homogeneous polynomials of degree ν. We put

$$Q_I(x,y) = \begin{cases} \prod_{i=1}^{r}(x - d_i y) \cdot \prod_{j \notin I}(x - e_j y) & \text{if } I \neq \varnothing, \\ \prod_{i=1}^{r}(x - d_i y) \cdot \prod_{j=1}^{s}(x - e_j y) & \text{if } I = \varnothing, \end{cases}$$

Lemma 1.4.6. *i) There is an isomorphism of graded rings:*

$$S_R/(Q_I(x,y), y^{n+1}) \xrightarrow{\cong} H^*(\mathbb{P}_I);\ x \mapsto c_1(\mathcal{L})|_{\mathbb{P}_I},\ y \mapsto \pi^* c_1(\mathcal{O}(1))|_{\mathbb{P}_I},$$

where we recall $\pi : \mathbb{P} = \mathbb{P}(\mathbb{E}) \to \mathbb{P}^n$.

ii) For $I \subset I'$ *we have the commutative diagram*

$$\begin{array}{ccc} H^*(\mathbb{P}_I) & \xrightarrow{\cong} & R[x,y]/(Q_I(x,y), y^{n+1}) \\ \downarrow{\scriptstyle \psi_{II'}} & & \downarrow \\ H^*(\mathbb{P}_{I'}) & \xrightarrow{\cong} & R[x,y]/(Q_{I'}(x,y), y^{n+1}) \end{array}$$

where the left vertical map is the restriction map and the right vertical map is the natural surjection.

iii) If $I' = I \cup \{j\}$ *with* $j \notin I$ *we have the commutative diagram*

$$\begin{array}{ccc} H^*(\mathbb{P}_{I'}) & \xrightarrow{\cong} & R[x,y]/(Q_{I'}(x,y), y^{n+1}) \\ \downarrow{\scriptstyle \phi_{II'}} & & \downarrow{\scriptstyle x - e_j y} \\ H^*(\mathbb{P}_I) & \xrightarrow{\cong} & R[x,y]/(Q_I(x,y), y^{n+1}) \end{array}$$

where the left vertical map is the Gysin map and the right vertical map is the multiplication by $x - e_j y$.

Proof The first assertion is well-known, and the second assertion follows immediately from the first. To show the last assertion, we note that $\phi_{II'}$ is the Poincaré dual of $\psi_{II'}$ and the composite $\phi_{II'}\psi_{II'} : H^*(\mathbb{P}_I) \to H^{*+2}(\mathbb{P}_I)$ is the multiplication by the class $c_1(\mathbb{P}_{I'})|_{\mathbb{P}_I}$ of the divisor $\mathbb{P}_{I'}$ in $\mathbb{P}_I$. Hence the assertion follows by noting $c_1(\mathbb{P}_{I'})|_{\mathbb{P}_I} = c_1(\mathbb{P}_j)|_{\mathbb{P}_I} = x - e_j y$. □

For $I=(i_1,\dots,i_\ell)\in\Delta_\ell$ write $\lambda_I=\lambda_{i_1}\wedge\cdots\wedge\lambda_{i_\ell}$ and $\lambda_I=1$ if $I=\varnothing$. Lemma 1.4.6 provides us with an isomorphism

$$\bigoplus_{I\in\Delta_\ell} S_R^p/(Q_I(x,y),y^{n+1})\otimes\lambda_I \xrightarrow{\cong} \bigoplus_{I\in\Delta_\ell} H^{2p}(\mathbb{P}_I) \quad \text{for each } p\geq 0.$$

Under this isomorphism, the arrows in (1.49) are identified with

$$d_\lambda:\xi\otimes\lambda_I\longmapsto\sum_{k=1}^{\ell}(-1)^{k-1}(x-e_{i_k}y)\xi\otimes\lambda_{i_1}\wedge\cdots\wedge\widehat{\lambda_{i_k}}\wedge\cdots\wedge\lambda_{i_\ell}. \tag{1.50}$$

Thus we have obtained the following result.

Lemma 1.4.7. *For any integers k and ν, $H^\nu(\Omega^k_{\mathbb{P}}(\log\mathbb{P}_*))$ is isomorphic to the cohomology at the middle term of the complex*

$$\begin{aligned}\bigoplus_{I_1\in\Delta_{k-\nu+1}} S_R^{\nu-1}/(Q_{I_1}(x,y),y^{n+1})\otimes\lambda_{I_1} &\overset{d_\lambda}{\to} \bigoplus_{I_2\in\Delta_{k-\nu}} S_R^{\nu}/(Q_{I_2}(x,y),y^{n+1})\otimes\lambda_{I_2}\\ &\overset{d_\lambda}{\to} \bigoplus_{I_3\in\Delta_{k-\nu-1}} S_R^{\nu+1}/(Q_{I_3}(x,y),y^{n+1})\otimes\lambda_{I_3}\end{aligned} \tag{1.51}$$

with d_λ defined as in (1.50) *(Note that by convention $\bigoplus_{I\in\Delta_\ell}(\cdots)=0$ unless $0\leq\ell\leq s$).*

In order to calculate the cohomology at the middle term of the complex (1.51), we introduce new symbols $\varepsilon_1,\cdots,\varepsilon_s$ and write $\varepsilon_I=\varepsilon_{i_1}\wedge\cdots\wedge\varepsilon_{i_\ell}$ for $I=(i_1,\dots,i_\ell)\in\Delta_\ell$. Consider the following diagram

$$\begin{array}{ccc}
0 & & 0\\
\downarrow & & \downarrow\\
\bigoplus_{I\in\Delta_\ell} S_R\otimes\varepsilon_I & \xrightarrow{d_\varepsilon} & \bigoplus_{I'\in\Delta_{\ell-1}} S_R\otimes\varepsilon_{I'}\\
{\scriptstyle a_I}\downarrow & & \downarrow{\scriptstyle a_{I'}}\\
\bigoplus_{I\in\Delta_\ell} S_R\otimes\lambda_I\bigoplus S_R\otimes\varepsilon_I & \xrightarrow{d_\lambda+d_\varepsilon} & \bigoplus_{I'\in\Delta_{\ell-1}} S_R\otimes\lambda_{I'}\bigoplus S_R\otimes\varepsilon_{I'}\\
{\scriptstyle b_I}\downarrow & & \downarrow{\scriptstyle b_{I'}}\\
\bigoplus_{I\in\Delta_\ell} S_R\otimes\lambda_I & \xrightarrow{d_\lambda} & \bigoplus_{I'\in\Delta_{\ell-1}} S_R\otimes\lambda_{I'}\\
\downarrow & & \downarrow\\
\bigoplus_{I\in\Delta_\ell} S_R/(Q_I,y^{n+1})\otimes\lambda_I & \xrightarrow{d_\lambda} & \bigoplus_{I'\in\Delta_{\ell-1}} S_R/(Q_{I'},y^{n+1})\otimes\lambda_{I'}\\
\downarrow & & \downarrow\\
0 & & 0
\end{array} \tag{1.52}$$

where d_ε is given by:

$$d_\varepsilon : \xi \otimes \varepsilon_I \longmapsto \sum_{k=1}^{\ell} (-1)^{k-1} \xi \otimes \varepsilon_{i_1} \wedge \cdots \wedge \widehat{\varepsilon_{i_k}} \wedge \cdots \wedge \varepsilon_{i_\ell}, \tag{1.53}$$

a_I and b_I are given by:

$$a_I : \xi \otimes \varepsilon_I \mapsto \xi Q_I \otimes \lambda_I + \xi y^{n+1} \otimes \varepsilon_I \quad \text{and}$$
$$b_I : \xi_1 \otimes \lambda_I + \xi_2 \otimes \varepsilon_I \mapsto (\xi_1 y^{n+1} - \xi_2 Q_I) \otimes \lambda_I.$$

One can easily check that the diagram is commutative and the vertical sequences are exact. Put

$$E^\ell = \operatorname{Ker}\left(\bigoplus_{I \in \Delta_\ell} S_R^0 \otimes \varepsilon_I \xrightarrow{d_\varepsilon} \bigoplus_{I \in \Delta_{\ell-1}} S_R^0 \otimes \varepsilon_I \right). \tag{1.54}$$

Note $E^\ell = 0$ unless $0 \leq \ell \leq s$ by convention.

Claim 1.4.8. *i) The following sequence is exact:*

$$0 \longrightarrow \bigoplus_{I \in \Delta_s} S_R \otimes \varepsilon_I \xrightarrow{d_\varepsilon} \bigoplus_{I \in \Delta_{s-1}} S_R \otimes \varepsilon_I \xrightarrow{d_\varepsilon} \cdots \longrightarrow \bigoplus_{I = \varnothing} S_R \otimes 1 \longrightarrow 0.$$

ii) Assume $s \geq 2$ and $e_i \neq e_j$ for some $i \neq j$. Then, for an integer $\ell \geq 0$, the following sequence is exact:

$$\bigoplus_{I \in \Delta_\ell} S_R^0 \otimes \lambda_I \xrightarrow{d_\lambda} \bigoplus_{I \in \Delta_{\ell-1}} S_R^1 \otimes \lambda_I \xrightarrow{d_\lambda} \cdots \longrightarrow \bigoplus_{I = \varnothing} S_R^\ell \otimes 1 \longrightarrow 0.$$

iii) Assume $e = e_1 = \cdots = e_s$. Then the cohomology at the middle term of the complex

$$\bigoplus_{I_1 \in \Delta_{\ell+1}} S_R^{p-1} \otimes \lambda_{I_1} \xrightarrow{d_\lambda} \bigoplus_{I_2 \in \Delta_\ell} S_R^p \otimes \lambda_{I_2} \xrightarrow{d_\lambda} \bigoplus_{I_3 \in \Delta_{\ell-1}} S_R^{p+1} \otimes \lambda_{I_3} \tag{1.55}$$

is isomorphic to $S_R^p/(x - ey) \otimes E^\ell$.

Proof **(i)** Easy (and well-known).

(ii) Let V_0 be a free R-module with basis $\lambda_1, \cdots, \lambda_s$. Let $c : \mathcal{O}_{\mathbb{P}^1} \otimes V_0 \to \mathcal{O}_{\mathbb{P}^1}(1)$ be the map of locally free sheaves on $\mathbb{P}^1 = \operatorname{Proj}(S_R)$, defined by $\lambda_j \mapsto x - e_j y$. This is surjective by the assumption. It gives rise to the Koszul complex

$$0 \to \mathcal{O}_{\mathbb{P}^1}(\ell - s) \otimes \bigwedge^s V_0 \to \mathcal{O}_{\mathbb{P}^1}(\ell - s + 1) \otimes \bigwedge^{s-1} V_0 \to \cdots$$
$$\cdots \to \mathcal{O}_{\mathbb{P}^1}(\ell - 1) \otimes V_0 \to \mathcal{O}_{\mathbb{P}^1}(\ell) \to 0. \tag{1.56}$$

We decompose (1.56) into the following sequences:

$$0 \to \mathcal{O}_{\mathbb{P}^1}(\ell - s) \otimes \bigwedge^{s} V_0 \to \cdots \to \mathcal{O}_{\mathbb{P}^1}(-1) \otimes \bigwedge^{\ell+1} V_0 \to V_1 \to 0. \tag{1.57}$$

$$0 \to V_1 \longrightarrow \mathcal{O}_{\mathbb{P}^1} \otimes \bigwedge^{\ell} V_0 \to \cdots \to \mathcal{O}_{\mathbb{P}^1}(\ell - 1) \otimes V_0 \to \mathcal{O}_{\mathbb{P}^1}(\ell) \to 0. \tag{1.58}$$

Noting $H^0(\mathcal{O}_{\mathbb{P}^1}(\ell)) = S_R^{\ell}$ and that (1.58) gives an acyclic resolution of V_1, it suffices to show $H^i(\mathbb{P}^1, V_1) = 0$ for $i \geq 1$ which is obvious for $i \geq 2$. To show $H^1(V_1) = 0$ it suffices to prove $H^0(\mathcal{O}(-2) \otimes V_1^*) = 0$ by the Serre duality. By (1.57) there is an injection $\mathcal{O}(-2) \otimes V_1^* \hookrightarrow \mathcal{O}(-1) \otimes \bigwedge^{\ell+1} V_0^*$. The assertion follows from this.

(iii) Since $e = e_1 = \cdots = e_s$, we have the following commutative diagram

$$\begin{array}{ccccc}
\bigoplus_{I_1 \in \Delta_{\ell+1}} S_R^{p-1} \otimes \varepsilon_{I_1} & \xrightarrow{d_\epsilon} & \bigoplus_{I_2 \in \Delta_\ell} S_R^{p-1} \otimes \varepsilon_{I_2} & \xrightarrow{d_\epsilon} & \bigoplus_{I_3 \in \Delta_{\ell-1}} S_R^{p-1} \otimes \varepsilon_{I_3} \\
\downarrow{\scriptstyle \iota_0} & & \downarrow{\scriptstyle \iota_1} & & \downarrow{\scriptstyle \iota_2} \\
\bigoplus_{I_1 \in \Delta_{\ell+1}} S_R^{p-1} \otimes \lambda_{I_1} & \xrightarrow{d_\lambda} & \bigoplus_{I_2 \in \Delta_\ell} S_R^{p} \otimes \lambda_{I_2} & \xrightarrow{d_\lambda} & \bigoplus_{I_3 \in \Delta_{\ell-1}} S_R^{p+1} \otimes \lambda_{I_3}
\end{array}$$

where $\iota_i : \xi \otimes \varepsilon_I \mapsto (x - ey)^i \xi \otimes \lambda_I$. Note that ι_0 is bijective and ι_i are injective for $i > 0$. Due to (i), the cohomology group of the complex (1.55) is isomorphic to

$$\operatorname{Ker}\left(\bigoplus_{I_2 \in \Delta_\ell} S_R^p/(x - ey) \otimes \lambda_{I_2} \xrightarrow{d_\lambda} \bigoplus_{I_3 \in \Delta_{\ell-1}} S_R^{p+1}/(x - ey)^2 \otimes \lambda_{I_3} \right). \tag{1.59}$$

The map $S_R^p/(x-ey) \to S_R^{p+1}/(x-ey)^2$, given by multiplication with $(x-ey)$, is injective. Hence (1.59) is isomorphic to

$$\operatorname{Ker}\left(\bigoplus_{I_2 \in \Delta_\ell} S_R^p/(x - ey) \otimes \varepsilon_{I_2} \xrightarrow{d_\varepsilon} \bigoplus_{I_3 \in \Delta_{\ell-1}} S_R^p/(x - ey) \otimes \varepsilon_{I_3} \right) = S_R^p/(x-ey) \otimes E^\ell.$$

This completes the proof. □

Combining Lemmas 1.4.7, 1.4.8 and (1.52), we get the following explicit description of $H^\nu(\Omega_{\mathbb{P}}^k(\log \mathbb{P}_*))$.

Lemma 1.4.9. *Let $E^\bullet$ be as in (1.54), and put*

$$\Lambda^\ell = \operatorname{Ker}\left(\bigoplus_{I \in \Delta_\ell} S_R^0 \otimes \lambda_I \xrightarrow{d_\lambda} \bigoplus_{I \in \Delta_{\ell-1}} S_R^1 \otimes \lambda_I \right).$$

i) Assume $s \geq 2$ and $e_i \neq e_j$ for some $i \neq j$. Then we have

$$H^\nu(\Omega^k_{\mathbb{P}}(\log \mathbb{P}_*)) \simeq \begin{cases} \Lambda^k & \text{if } \nu = 0 \\ \Lambda^{k-n-1} & \text{if } \nu = n \\ 0 & \text{otherwise} \end{cases}$$

(Note $\Lambda^\ell = 0$ unless $0 \leq \ell \leq s$ by convention.)

ii) Assume $e = e_1 = \cdots = e_s$. Then we have for for all k, $\nu \geq 0$

$$H^\nu(\Omega^k_{\mathbb{P}}(\log \mathbb{P}_*)) \simeq S^\nu_R/(y^{n+1}, x - ey) \otimes E^{k-\nu} \simeq R[y]^\nu/(y^{n+1}) \otimes E^{k-\nu}.$$

In order to complete the proof of Proposition 1.4.5 we need the following Lemma, which gives an explicit description of the map δ in (1.47).

Lemma 1.4.10. *i) Assume $s \geq 2$ and $e_i \neq e_j$ for some $i \neq j$. Then $c_1(\mathcal{L})|_{\mathbb{P}-\mathbb{P}_*} = 0$ and $\delta = 0$.*

ii) Assume $e = e_1 = \cdots = e_s$. Then

$$\pi^* : H^1(\Omega^1_{\mathbb{P}^n}) \longrightarrow H^1(\Omega^1_{\mathbb{P}}(\log \mathbb{P}_*))$$

is injective and $c_1(\mathcal{L})|_{\mathbb{P}-\mathbb{P}_*} = \pi^* c_1(\mathcal{O}(e))$. The map δ in (1.47) is identified with the multiplication by $ey \otimes 1$ under the isomorphisms in Lemma 1.4.7.

Proof (i) Noting $\mathcal{O}(\mathbb{P}_j) \simeq \mathcal{L} \otimes \pi^*\mathcal{O}(-e_j)$ and $c_1(\mathbb{P}_j)|_{\mathbb{P}-\mathbb{P}_*} = 0$, we have

$$c_1(\mathcal{L})|_{\mathbb{P}-\mathbb{P}_*} = \pi^* c_1(\mathcal{O}(e_j)) = e_j \pi^* c_1(\mathcal{O}(1)) \quad \text{for } 1 \leq \forall j \leq s.$$

By the assumption this implies $c_1(\mathcal{L})|_{\mathbb{P}-\mathbb{P}_*} = 0$ and $\pi^* c_1(\mathcal{O}(1)) = 0$.
(ii) The first assertion follows from the existence of an isomorphism $\mathbb{P} \simeq \mathbb{P}^n \times \mathbb{P}^{r+s-1}$ such that $\mathbb{P}_j$ corresponds to $\mathbb{P}^n \times H_j$ with H_j a hyperplane. The second assertion has been already shown in (i). To show the last we first note that the cup product for $H^\nu(\Omega^k_{\mathbb{P}}(\log \mathbb{P}_*))$ is induced by the cup product

$$H^{2i}(\mathbb{P}_I) \otimes H^{2j}(\mathbb{P}_I) \to H^{2(i+j)}(\mathbb{P}_{I+J})$$

when one identifies $H^\nu(\Omega^k_{\mathbb{P}}(\log \mathbb{P}_*))$ with the cohomology at the middle term of the complex (1.49). Here $\mathbb{P}_{I+J} = \mathbb{P}_I \cap \mathbb{P}_J$ if $I \cap J = \varnothing$ and $H^{2(i+j)}(\mathbb{P}_{I+J}) = 0$ otherwise by convention. Under the isomorphisms of Lemma 1.4.6, it is identified with

$$S^i_R/(Q_I(x,y), y^{n+1}) \otimes S^j_R/(Q_J(x,y), y^{n+1}) \to S^{i+j}_R/(Q_{I+J}(x,y), y^{n+1}),$$
$$(f \otimes \lambda_I) \otimes (g \otimes \lambda_J) \mapsto fg \otimes (\lambda_I \bigwedge \lambda_J).$$

Since δ is induced by the cup product with $c_1(\mathcal{L})|_{\mathbb{P}-\mathbb{P}_*} \in H^1(\Omega^1_{\mathbb{P}}(\log \mathbb{P}_*))$, the desired assertion follows by noting $c_1(\mathcal{L})|_{\mathbb{P}-\mathbb{P}_*}$ corresponds to ey

under the isomorphism $H^1(\Omega^1_{\mathbb{P}}(\log \mathbb{P}_*)) \simeq S^1_R/(Q_\varnothing(x,y), y^{n+1})$ due to Lemma 1.4.7. □

Finally we can complete the proof of Proposition 1.4.5. First assume $s \geq 2$ and $e_i \neq e_j$ for some $i \neq j$. The assertion follows from (1.47), Lemma 1.4.10 (i) and Lemma 1.4.9 (i) by noting that the R-module Λ^ℓ is locally free of rank $\binom{s-2}{\ell}$ due to Claim 1.4.8 (ii). (Compare the coefficients of $(1-x)^{s-2} = (1-x)^s \cdot (1-x)^{-2} = (\sum_p (-1)^p \binom{s}{p} x^p) \cdot (\sum_q q x^q)$ to get $\binom{s-2}{\ell} = \binom{s}{\ell} - 2\binom{s}{\ell-1} + \cdots$.) Next assume $e = e_1 = \cdots = e_s$. By (1.47), Lemma 1.4.10 (ii) and Lemma 1.4.9 (ii), we have an exact sequence

$$R[y]^{\nu-1}/(y^{n+1}) \xrightarrow{ey\otimes 1} R[y]^\nu/(y^{n+1}) \otimes E^{k-\nu} \longrightarrow H^\nu(\bigwedge^k \Sigma^*)$$
$$\longrightarrow R[y]^\nu/(y^{n+1}) \otimes E^{k-\nu-1} \xrightarrow{ey\otimes 1} R[y]^{\nu+1}/(y^{n+1}) \otimes E^{k-\nu-1}.$$

The desired assertion follows by noting that the R-module E^ℓ is locally free of rank $\binom{s-1}{\ell}$ due to Claim 1.4.8 (i). This completes the proof of Proposition 1.4.5.

1.5 Proof of the Main Theorem

In this section we prove the Main Theorem stated in §1.2.4.

1.5.1 Proof of (i)

Let

$$\Psi^q_{U/S} : \bigoplus_{\ell=0}^{q} \bigwedge^{\ell+n-r} (G_j)_X \otimes_{\mathbb{Q}} \Omega^{q-\ell}_{R/k} \longrightarrow H^0(\Omega^{n-r}_{X/S}(\log Z_*)) \otimes \Omega^q_{R/k} \tag{1.60}$$

be the map (1.19). We show the stronger assertion that $\Psi^q_{U/S} \otimes_R \kappa(x)$ is injective for any $x \in |S|$ assuming $\mathbf{(IV)}_q$. We fix $x \in |S|$. Without loss of generality we may assume $j_1 = 1, \cdots, j_{n-r} = n-r$ in $\mathbf{(IV)}_q$. We may work in an étale neighbourhood of x to assume that R is a strict henselian local ring with the closed point $x \in \mathrm{Spec}(R)$ and that the 0-dimensional scheme defined by $F_1 = \cdots = F_r = G_1 = \cdots = G_{n-r} = 0$ in $\mathbb{P}^n_R$ is a disjoint union of copies of $\mathrm{Spec}(R)$.

For an integer $\ell \geq 1$ let $\bigwedge^\ell (G_j)_{j\neq 1}$ be the subspace of $\bigwedge^\ell (G_j)$ generated by such $g_{j_0\cdots j_\ell}$ that $j_\nu \neq 1$ for $0 \leq \forall \nu \leq \ell$ (cf. (1.17)). We have the exact sequence

$$0 \to \bigwedge^{\ell}(G_j)_{j\neq 1} \to \bigwedge^{\ell}(G_j) \xrightarrow{\tau} \bigwedge^{\ell-1}(G_j)_{j\neq 1} \to 0,$$

where τ is characterized by the condition that $\tau(g_{1j_1\cdots j_\ell}) = -g_{j_1\cdots j_\ell}$ and that it annihilates $\bigwedge^{\ell}(G_j)_{j\neq 1}$. Put $Z_*^{(1)} = Z_2 + \cdots + Z_s$ where we recall that $Z_j \subset X$ is a smooth hypersurface section defined by G_j. Consider the residue map along Z_1:

$$\mathrm{Res} : \Omega^{n-r}_{X/S}(\log Z_*) \to \Omega^{n-r-1}_{Z_1/S}(\log Z_*^{(1)} \cap Z_1);\ dg_1/g_1 \wedge \omega \mapsto \omega|_{Z_1},$$

where g_1 is a local equation of Z_1. By (1.18) one sees that $\mathrm{Res}\circ\Psi^q_{U/S}$ factors through τ and we get the following commutative diagram:

$$\begin{array}{ccc}
0 & & 0 \\
\downarrow & & \downarrow \\
\bigoplus_{\ell=0}^{q} \bigwedge^{\ell+n-r} (G_j)_{j\neq 1} \otimes_{\mathbb{Q}} \Omega^{q-\ell}_{R/k} & \longrightarrow & H^0(\Omega^{n-r}_{X/S}(\log Z_*^{(1)})) \otimes_R \Omega^q_{R/k} \\
\downarrow & & \downarrow \\
\bigoplus_{\ell=0}^{q} \bigwedge^{\ell+n-r} (G_j) \otimes_{\mathbb{Q}} \Omega^{q-\ell}_{R/k} & \xrightarrow{\Psi^q_{U/S}} & H^0(\Omega^{n-r}_{X/S}(\log Z_*)) \otimes_R \Omega^q_{R/k} \\
\downarrow{\scriptstyle \tau\otimes\mathrm{id}} & & \downarrow{\scriptstyle \mathrm{Res}\otimes\mathrm{id}} \\
\bigoplus_{\ell=0}^{q} \bigwedge^{\ell+n-r-1} (G_j)_{j\neq 1} \otimes_{\mathbb{Q}} \Omega^{q-\ell}_{R/k} & \longrightarrow & H^0(\Omega^{n-r-1}_{Z_1/S}(\log Z_*^{(1)} \cap Z_1)) \otimes_R \Omega^q_{R/k} \\
\downarrow & & \downarrow \\
0 & & 0.
\end{array}$$

By the diagram and induction we are reduced to show the injectivity of $\Psi^q_{U/S} \otimes_R \kappa(x)$ in case $s = 1$ or $n - r = 0$. If $s = 1$, the assertion is clear because $\bigwedge^{\ell}(G_j) = 0$ by convention. We consider the case $n - r = 0$. Then

$$X = \{F_1 = \cdots = F_n = 0\} \subset \mathbb{P}^n_R \quad \text{and} \quad Y_j = \{G_j = 0\} \subset \mathbb{P}^n_R \quad (1 \le j \le s).$$

By the assumption we have $X = \coprod_{\beta\in X(R)} \mathrm{Spec}(R)$ where $X(R)$ is the set of sections of $X \to \mathrm{Spec}(R)$. The map (1.60) becomes

$$\Psi : \bigoplus_{\ell=0}^{q} \bigwedge^{\ell}(G_j) \otimes_{\mathbb{Q}} \Omega^{q-\ell}_{R/k} \longrightarrow H^0(\mathcal{O}_X) \otimes_R \Omega^q_{R/k}.$$

By Nakayama's lemma, condition $\mathbf{(IV)}_q$ in the Main Theorem implies:

(∗) *There are $q+1$ points $\beta_0, \cdots, \beta_q \in X(R)$ such that the map*

$$W \xrightarrow{\Theta} A_1(0)/(J' + \mathfrak{m}_{\beta_0}) \oplus \cdots \oplus A_1(0)/(J' + \mathfrak{m}_{\beta_q}),$$

is surjective, where $\mathfrak{m}_\beta \subset P_R$ denotes the homogeneous ideal defining β in $\mathbb{P}^n_R = \mathrm{Proj}(P_R)$.

We note

$$H^0(\mathcal{O}_X) = \bigoplus_{\beta \in X(R)} R \cdot [\beta]$$

and put

$$H^0(\mathcal{O}_X)' = \bigoplus_{0 \le \nu \le q} R \cdot [\beta_\nu].$$

It suffices to show the injectivity of $\Psi' \otimes_R \kappa(x)$ where

$$\Psi' : \bigoplus_{\ell=0}^{q} \bigwedge^{\ell}(G_j) \otimes_{\mathbb{Q}} \Omega^{q-\ell}_{R/k} \longrightarrow H^0(\mathcal{O}_X)' \otimes_R \Omega^q_{R/k} \tag{1.61}$$

is the composite of Ψ with the projection $H^0(\mathcal{O}_X) \to H^0(\mathcal{O}_X)'$. We have

$$\Psi'(v_{1j_1\cdots j_\ell} \otimes \eta) = \sum_{\nu=0}^{q} [\beta_\nu] \otimes \epsilon_{j_1\nu} \wedge \cdots \wedge \epsilon_{j_\ell\nu} \wedge \eta \quad (\eta \in \Omega^{q-\ell}_{R/k}),$$

$$\text{with} \quad \epsilon_{j\nu} := \mathrm{dlog}((G_j^{e_1}/G_1^{e_j})(\beta_\nu)) \in \Omega^1_{R/k}.$$

Hence the desired assertion follows from the following two lemmas.

Lemma 1.5.1. *The log forms $\epsilon_{j\nu}$ $(2 \le j \le s,\ 0 \le \nu \le q)$ are linearly independent in $\Omega^1_{R/k} \otimes \kappa(x)$.*

Lemma 1.5.2. *Let Ω be a finite dimensional vector space over a field k of characteristic zero. Suppose that $\epsilon_{j\nu} \in \Omega$ $(1 \le j \le s,\ 0 \le \nu \le q)$ are linearly independent. For given $\eta_t \in \bigwedge^t \Omega$ with $0 \le t \le q$, put for $0 \le \nu \le q$ (the product is wedge product)*

$$\left.\begin{aligned} \omega_\nu = \eta_q + \textstyle\sum_{j=1}^{s} \eta_{q-1}\epsilon_{j\nu} + \sum_{1\le j_1<j_2\le s} \eta_{q-2}\epsilon_{j_1\nu}\epsilon_{j_2\nu} + \cdots \\ \cdots + \textstyle\sum_{1\le j_1<\cdots<j_q\le s} \eta_0\epsilon_{j_1\nu}\cdots\epsilon_{j_q\nu} \quad \in \bigwedge^q \Omega. \end{aligned}\right\} \tag{1.62}$$

If $\omega_\nu = 0$ for $0 \le \forall\nu \le q$, then $\eta_t = 0$ for $0 \le \forall t \le q$.

Now we prove the above lemmas. Lemma 1.5.1 follows from the condition (∗) by noting the following:

Lemma 1.5.3. *For $\beta \in X(R)$, $A_1(0)/(J' + \mathfrak{m}_\beta)$ is a free R-module of rank $s-1$, and the dual of $T_{R/k} \to A_1(0)/(J' + \mathfrak{m}_\beta)$ induced by Θ is given by the matrix*

$$(\mathrm{dlog}\, \frac{G_2^{e_1}}{G_1^{e_2}}(\beta), \cdots, \mathrm{dlog}\, \frac{G_s^{e_1}}{G_1^{e_s}}(\beta))$$

for a suitable choice of a basis of $A_1(0)/(J' + \mathfrak{m}_\beta)$.

Proof Giving $\beta = (\beta_0 : \cdots : \beta_n)$ in the homogeneous coordinate of $\mathbb{P}^n_R$,

$$\mathfrak{m}_\beta = (\beta_i X_j - \beta_j X_i)_{0 \le i < j \le n} \subset P_R = R[X_0, \cdots, X_n].$$

We may assume without loss of generality that $\beta_0 = 1$. We write $\partial_\ell := \partial/\partial X_\ell$ for $0 \le \ell \le n$ and $L(\beta) := L(1, \beta_1, \cdots, \beta_n) \in R$ for any homogeneous polynomial L. Then we have an isomorphism

$$A_1(0)/(J' + \mathfrak{m}_\beta) \cong (\bigoplus_{i=1}^{n} R \cdot X_0^{d_i} \mu_i \bigoplus \bigoplus_{j=1}^{s} R \cdot X_0^{e_j} \lambda_j)/J'_\beta$$

$$\text{with} \quad J'_\beta = \sum_{\ell=0}^{n} R \cdot (\sum_{i=1}^{n} \partial_\ell F_i(\beta) X_0^{d_i} \mu_i + \sum_{j=1}^{s} \partial_\ell G_j(\beta) X_0^{e_j} \lambda_j).$$

Using the fact

$$\det \begin{pmatrix} \partial_1 F_1(\beta) & \cdots & \partial_1 F_n(\beta) \\ \vdots & & \vdots \\ \partial_n F_1(\beta) & \cdots & \partial_n F_n(\beta) \end{pmatrix} \in R^*,$$

we get

$$A_1(0)/(J' + \mathfrak{m}_\beta) \cong (\bigoplus_{j=1}^{s} R \cdot X_0^{e_j} \lambda_j)/R \cdot (\sum_{j=1}^{s} e_j G_j(\beta) X_0^{e_j} \lambda_j), \qquad (1.63)$$

which is a free R-module of rank $(s-1)$. To prove the second assertion we first show for $\theta \in T_{R/k}$

$$\Theta(\theta) \equiv \theta(G_1(\beta)) \cdot X_0^{e_1} \lambda_1 + \cdots + \theta(G_s(\beta)) \cdot X_0^{e_s} \lambda_s$$
$$\mod J' + (X_\ell - \beta_\ell X_0)_{1 \le \ell \le n}. \qquad (1.64)$$

In fact

$$\begin{aligned}
\sum_{j=1}^{s}\theta(G_j(\beta))\cdot X_0^{e_j}\lambda_j &= \sum_{j=1}^{s}(\theta G_j)(\beta)\cdot X_0^{e_j}\lambda_j+\sum_{j=1}^{s}\sum_{\ell=0}^{n}\frac{\partial G_j}{\partial X_\ell}(\beta)\cdot\theta(\beta_\ell)\cdot X_0^{e_j}\lambda_j \\
&\equiv \sum_{j=1}^{s}(\theta G_j)(\beta)\cdot X_0^{e_j}\lambda_j-\sum_{i=1}^{n}\sum_{\ell=0}^{n}\frac{\partial F_i}{\partial X_\ell}(\beta)\cdot\theta(\beta_\ell)\cdot X_0^{d_i}\mu_i \mod J' \\
&\underset{(**)}{=} \sum_{j=1}^{s}(\theta G_j)(\beta)\cdot X_0^{e_j}\lambda_j+\sum_{i=1}^{n}(\theta F_i)(\beta)\cdot X_0^{d_i}\mu_i \\
&\equiv \sum_{j=1}^{s}(\theta G_j)\lambda_j+\sum_{i=1}^{n}(\theta F_i)\mu_i \mod (X_\ell-\beta_l X_0)_{1\le\ell\le n} \\
&= \Theta(\theta).
\end{aligned}$$

Here $(**)$ follows from $0=\theta(F_i(\beta))=(\theta F_i)(\beta)+\sum_{\ell=0}^{n}(\partial_\ell F_i)(\beta)\cdot\theta(\beta_\ell)$ (note $F_i(\beta)=0$ since $\beta\in X(R)$). Let $\{(X_0^{e_j}\lambda_j)^*\}_{1\le j\le s}$ be the dual basis of $\bigoplus_{j=1}^{s}R\cdot X_0^{e_j}\lambda_j$. Then

$$\frac{1}{e_jG_j(\beta)}(X_0^{e_j}\lambda_j)^*-\frac{1}{e_1G_1(\beta)}(X_0^{e_1}\lambda_1)^* \quad (2\le j\le s)$$

is a basis of the dual module of the right hand side of (1.63). By (1.64), we see that the dual of $T_{R/k}\to A_1(0)/(J'+\mathfrak{m}_\beta)$ is given by

$$\frac{1}{e_jG_j(\beta)}(X_0^{e_j}\lambda_j)^*-\frac{1}{e_1G_1(\beta)}(X_0^{e_1}\lambda_1)^*\longmapsto\frac{1}{e_1e_j}\operatorname{dlog}\frac{G_j^{e_1}}{G_1^{e_j}}(\beta).$$

This completes the proof of Lemma 1.5.1. □

Finally we prove Lemma 1.5.2. We prove the assertion by induction on $q\ge 0$. If $q=0$, it is clear. Let ϵ_{il}^* $(1\le i\le s,\ 0\le\ell\le q)$ be a linear form on Ω such that $\epsilon_{il}^*(\epsilon_{i'\ell'})=0$ if $(i,\ell)\ne(i',\ell')$ and $\epsilon_{il}^*(\epsilon_{il})=1$. For $1\le\nu\ne\ell\le q$ we have

$$\begin{aligned}
(\omega_\nu,\epsilon_{il}^*) &= (\eta_q,\epsilon_{il}^*)+\sum_{j=1}^{s}(\eta_{q-1},\epsilon_{il}^*)\epsilon_{j\nu}+\cdots+\sum_{1\le j_1<\cdots<j_{q-1}\le s}(\eta_1,\epsilon_{il}^*)\epsilon_{j_1\nu}\cdots\epsilon_{j_{q-1}\nu} \\
&= 0\in\bigwedge^{q-1}\Omega.
\end{aligned}$$

By induction this implies $(\eta_t,\epsilon_{il}^*)=0$ for $0\le\forall t,\forall\ell\le q$ and $1\le\forall i\le s$. Then

$$0=(\omega_\nu,\epsilon_{j_1\nu}^*\cdots\epsilon_{j_{q-t}\nu}^*)=\eta_t.$$

This completes the proof of Lemma 1.5.2.

1.5.2 Proof of (ii) : Case $T_{R/k} \simeq W$

Since we have proved the injectivity of $\Psi^q_{U/S} \otimes_R \kappa(x)$ for $\forall x \in |S|$, it suffices to show that the kernel of $\overline{\nabla}_q$ is a locally free R-module of the same rank as the source of $\Psi^q_{U/S}$. More precisely we want to show the following.

Lemma 1.5.4. *Assuming* **(I)**, **(II)** *and* $\mathbf{(III)}_q$ *in the Main Theorem,* $\mathrm{Ker}(\overline{\nabla}_q)$ *is a locally free R-module of rank*

$$\sum_{k\geq 0} \binom{s-1}{n-r+k} \cdot \operatorname{rank} \bigwedge^{q-k} T_{R/k}. \tag{1.65}$$

In this subsection we show Lemma 1.5.4 assuming $\Theta : T_{R/k} \to W$ is an isomorphism. We note that W is a locally free R-module by **(I)**. By Theorem 1.4.1, $\overline{\nabla}_q$ is identified with the map

$$B_0(\mathbf{d}+\mathbf{e}-n-1) \otimes \bigwedge^{q} W^* \longrightarrow B_1(\mathbf{d}+\mathbf{e}-n-1) \otimes \bigwedge^{q+1} W^*$$

induced by the multiplication $B_q(\ell) \otimes W \to B_{q+1}(\ell)$. By the duality theorem (Theorem 1.4.2), the dual of the map fits into the commutative diagram

$$\begin{array}{ccc} B_1(\mathbf{d}+\mathbf{e}-n-1)^* \otimes \bigwedge^{q+1} W & \longrightarrow & B_0(\mathbf{d}+\mathbf{e}-n-1)^* \otimes \bigwedge^{q} W \\ \cong \uparrow & & \uparrow \iota \\ B_{n-r-1}(\mathbf{d}-n-1) \otimes \bigwedge^{q+1} W & \xrightarrow{\ \Phi\ } & B_{n-r}(\mathbf{d}-n-1) \otimes \bigwedge^{q} W. \end{array} \tag{1.66}$$

The diagram induces an exact sequence

$$0 \to \mathrm{Coker}(\Phi) \to \left(\mathrm{Ker}(\overline{\nabla}_q)\right)^* \to \mathrm{Coker}(\iota) \to 0.$$

Due to Theorem 1.4.2 ι is injective and its cokernel is a locally free R-module of rank $\binom{s-1}{n-r} \cdot \operatorname{rank} \bigwedge^{q} W$. Therefore it suffices to show that the cokernel of the map Φ is a locally free R-module of rank

$$\sum_{k\geq 1} \binom{s-1}{n-r+k} \cdot \operatorname{rank} \bigwedge^{q-k} W. \tag{1.67}$$

In order to show this we recall the notations in §1.4.2 and §1.4.3. For integers k, h and ℓ we put

$$\mathbf{M}_{k,h}(\ell) = \bigwedge^{n+r+s-h} \Sigma^* \otimes \mathcal{L}^{r+k-h} \otimes \pi^*\mathcal{O}(\ell - \mathbf{d} + n + 1)$$

and

$$C_{k,h}(\ell) = H^0(\mathbb{P}, \mathbf{M}_{k,h}(\ell)).$$

Due to Lemma 1.4.3, there is an exact sequence

$$C_{k,1}(\ell) \longrightarrow C_{k,0}(\ell) \longrightarrow B_k(\ell) \longrightarrow 0. \tag{1.68}$$

We now need the following two results Lemmas 1.5.5 and 1.5.6. The first result is a direct consequence of [3, Lem.(7-4)]. It is a generalization of Nori's connectivity [14] to the case of open complete intersections and is the key to the proof of the Main Theorem.

Proposition 1.5.5. *Putting $C_{k,h} = C_{k,h}(\mathbf{d} - n - 1)$, the Koszul complex*

$$C_{n-r+h-1,h} \otimes \bigwedge^{q-h+1} W \to C_{n-r+h,h} \otimes \bigwedge^{q-h} W \to C_{n-r+h+1,h} \otimes \bigwedge^{q-h-1} W \to C_{n-r+h+2,h} \otimes \bigwedge^{q-h-2} W$$

is exact for $\forall h \geq 0$ assuming **(II)** *and* $\mathbf{(III)}_q$ *in the Main Theorem.*

Putting $\mathbf{M}_{k,h} = \mathbf{M}_{k,h}(\mathbf{d} - n - 1)$, (1.44) induces the following exact sequence (cf. [3, Lem.(5-1)])

$$0 \to \mathbf{M}_{n-r+k,n+r+s} \to \mathbf{M}_{n-r+k,n+r+s-1} \to \cdots \to \mathbf{M}_{n-r+k,0} \to 0. \tag{1.69}$$

Lemma 1.5.6. *Let $k \geq 1$ be an integer. Then the complex*

$$0 \to C_{n-r+k,n+r+s} \to C_{n-r+k,n+r+s-1} \to \cdots \to C_{n-r+k,0} \to 0$$

induced by (1.69) is exact except at the term $C_{n-r+k,k-1}$, and the cohomology group at this term is isomorphic to $H^n(\bigwedge^{r+s-k} \Sigma^*)$.

Before proving Lemma 1.5.6, we complete the proof of Lemma 1.5.4 assuming $T_{R/k} \simeq W$. We write $B_\bullet = B_\bullet(\mathbf{d} - n - 1)$. Consider the following commutative diagram:

$$
\begin{array}{ccccccc}
 & & C_{n-r-1,1}\otimes \bigwedge^{q+1} W & \to & C_{n-r-1,0}\otimes \bigwedge^{q+1} W & \xrightarrow{j_1} & B_{n-r-1}\otimes \bigwedge^{q+1} W \\
 & & \downarrow & & \downarrow & & \downarrow \Phi \\
 & & C_{n-r,1}\otimes \bigwedge^{q} W & \to & \boxed{C_{n-r,0}\otimes \bigwedge^{q} W} & \xrightarrow{j_2} & B_{n-r}\otimes \bigwedge^{q} W \\
 & & \downarrow & & \downarrow & & \\
C_{n-r+1,2}\otimes \bigwedge^{q-1} W & \to & \boxed{C_{n-r+1,1}\otimes \bigwedge^{q-1} W} & \to & \boxed{C_{n-r+1,0}\otimes \bigwedge^{q-1} W} & & \\
\downarrow & & \downarrow & & \downarrow & & \\
\boxed{C_{n-r+2,2}\otimes \bigwedge^{q-2} W} & \to & \boxed{C_{n-r+2,1}\otimes \bigwedge^{q-2} W} & \to & \vdots & & \\
\downarrow & & \downarrow & & & & \\
\boxed{C_{n-r+3,2}\otimes \bigwedge^{q-3} W} & \to & \vdots & & & & \\
\downarrow & & & & & & \\
\vdots & & & & & &
\end{array}
$$

Starting from the third row, each horizontal sequence is exact except at the term $C_{n-r+k,k-1}\otimes \bigwedge^{q-k} W$ $(k \geq 1)$ by Lemma 1.5.6. The horizontal sequences in the first and second row are exact, and the maps j_1 and j_2 are surjective (cf. (1.68)). The vertical sequences are exact at the boxed terms by Proposition 1.5.5. A diagram chase now shows that there is a finite decreasing filtration $U^\bullet$ on $\operatorname{Coker}\Phi$ such that $U^0 \operatorname{Coker}\Phi = \operatorname{Coker}\Phi$ and that

$$\operatorname{Gr}_U^{k-1} \operatorname{Coker}\Phi \simeq H^n\Big(\bigwedge^{r+s-k} \Sigma^*\Big) \otimes \bigwedge^{q-k} W \quad (k \geq 1).$$

This shows that the cokernel of Φ is a locally free R-module. Moreover, by Proposition 1.4.5 (iii), we have

$$\operatorname{rank}(\operatorname{Coker}\Phi) = \sum_{k\geq 1} \binom{s-1}{r+s-k-n-1} \cdot \operatorname{rank} \bigwedge^{q-k} W, \tag{1.70}$$

which is equal to (1.67). (Note $\binom{x}{\ell} = 0$ for $\ell < 0$ by convention.) This completes the proof of Lemma 1.5.4 assuming $T_{R/k} \simeq W$.

Now we prove Lemma 1.5.6. Noting $\mathbf{M}_{n-r+k,n+k} = \bigwedge^{r+s-k} \Sigma^*$, we decompose (1.69) into the following exact sequences:

$$0 \to \mathbf{M}_{n-r+k,n+r+s} \to \cdots \to \mathbf{M}_{n-r+k,n+k+1} \to N_1 \to 0, \tag{1.71}$$

$$0 \longrightarrow N_1 \longrightarrow \bigwedge^{r+s-k} \Sigma^* \longrightarrow N_2 \longrightarrow 0, \tag{1.72}$$

$$0 \to N_2 \to \mathbf{M}_{n-r+k,n+k-1} \to \cdots \to \mathbf{M}_{n-r+k,0} \to 0. \tag{1.73}$$

By Lemma 1.4.4 i), $H^w(\mathbf{M}_{n-r+k,h}) = 0$ for $\forall w \geq 0$ and $\forall h \geq n+k+1$. Hence $H^w(N_1) = 0$ and $H^w(\bigwedge^{r+s-k} \Sigma^*) = H^w(N_2)$ for $\forall w \geq 0$. On the other hand, since $H^w(\bigwedge^{\bullet}\Sigma^* \otimes \mathfrak{L}^\nu) = 0$ if $\nu > 0$ and $w > 0$ by Lemma 1.4.4 (iii), we have $H^w(\mathbf{M}_{n-r+k,h}) = 0$ for $\forall w > 0$ and $0 \leq \forall h \leq n+k-1$. This means that (1.73) is a flabby resolution of N_2. Therefore, for $0 \leq h \leq n+k-2$, the cohomology group at $C_{n-r+k,h}$ is isomorphic to $H^{n+k-1-h}(\bigwedge^{r+s-k} \Sigma^*)$. Now the assertion follows from Proposition 1.4.5 (i).

1.5.3 Proof of (ii) : General Case

It remains to show 1.5.4 in case $\Theta : T_{R/k} \to W$ is not necessarily an isomorphism. Setting $I = \mathrm{Ker}(T_{R/k} \to W)$, we get the exact sequence:wAnnual EAGER Conference And Workshop On the Occasion of J. Murre's 75–th Birthday on *Algebraic Cycles and Motives* August 30– September 3, 2004 Leiden University, the Netherlands

$$0 \longrightarrow I \longrightarrow T_{R/k} \longrightarrow W \longrightarrow 0. \tag{1.74}$$

Since W is a locally free R-module, so is I. By the argument in §1.5.2 it suffices to show that the cokernel of the map

$$B_{n-r-1}(\mathbf{d}-n-1) \otimes \bigwedge^{q+1} T_{R/k} \longrightarrow B_{n-r}(\mathbf{d}-n-1) \otimes \bigwedge^{q} T_{R/k} \tag{1.75}$$

is a locally free R-module of rank

$$\sum_{k \geq 1} \binom{s-1}{n-r+k} \cdot \mathrm{rank} \bigwedge^{q-k} T_{R/k}. \tag{1.76}$$

(1.74) gives rise to a filtration $U^\bullet$ on $\bigwedge^q T_{R/k}$ such that $U^i/U^{i+1} = (\bigwedge^{q-i} W) \otimes (\bigwedge^i I)$. Since I annihilates $B_\bullet(\mathbf{d}-n-1)$, the map (1.75) admits a filtration whose graded quotients for $0 \leq i \leq q$ are given by

$$B_{n-r-1}(\mathbf{d}-n-1) \otimes \bigwedge^{q+1-i} W \otimes \bigwedge^i I \longrightarrow B_{n-r}(\mathbf{d}-n-1) \otimes \bigwedge^{q-i} W \otimes \bigwedge^i I. \tag{1.77}$$

By what we have shown in §1.5.2 the cokernel of the map (1.75) is a locally free R-module of rank

$$\sum_{i=0}^{q} \left(\sum_{k \geq 1} \binom{s-1}{n-r+k} \cdot \mathrm{rank}(\bigwedge^{q-i-k} W) \right) \cdot \mathrm{rank}(\bigwedge^i I). \tag{1.78}$$

It is easy to see that the numbers (1.76) and (1.78) are equal (left to the reader).

This completes the proof of the Main Theorem.

References

[1] Arapura, D: The Leray spectral sequence is motivic, Invent. Math. **160**, pp. 567–589 (2005).

[2] Asakura, M: On the K_1-groups of algebraic curves, Invent. Math. **149**, pp. 661–685 (2002).

[3] Asakura, M. and S. Saito: Generalized Jacobian rings for open complete intersections, Math. Nachr. 2004.

[4] Asakura, M. and S. Saito: Noether-Lefschetz locus for Beilinson-Hodge cycles I, Math. Zeit. 2004.

[5] Beilinson, A: Notes on absolute Hodge cohomology, in: *Applications of Algebraic K-theory to Algebraic Geometry and Number theory*, Contemp. Math. **55**, 1986. pp. 35–68, 27–41, Springer, 1987.

[6] Beilinson, A: Height pairings between algebraic cyles, Lecture Notes in Math. 1289, 1–26, Springer, 1987.

[7] Bloch, B.: Algebraic cycles and higher K-theory. Advances in Math. **61**, 1986, 267–304

[8] Deligne, D.: Théorie de Hodge. II, III, Publ. Math. IHES, **40** (1971), 5–58, Publ. Math. IHES, **44** (1975), 5–77.

[9] Deligne, D.: *Equations différentielles à points singuliers réguliers*, Lecture note in Math. **163**, Springer, 1970.

[10] Green, M.: The period map for hypersurface sections of high degree on an arbitrary variety, Compositio Math. **55** (1984), 135–156.

[11] Jannsen, U.: Mixed Motives and Algebraic K-theory, Lecture note in Math. **1400**, Springer, 1990.

[12] Katz, N and T. Oda: On the differentiation of De Rham cohomology classes with respect to parameters, J. Math. Kyoto Univ. **8** (1968), 199–213.

[13] Milnor, J.: *Introduction to Algebraic K-theory*, Ann. of Math Studies **72**, Princeton 1970.

[14] Nori, M. S: Algebraic cycles and Hodge theoretic connectivity, Invent. Math., **111** (1993), 349–373

[15] Quillen, D.: Higher algebraic K-theory. I, 85–147, Lecture Notes in Math. Vol. **341**, Springer, Berlin 1973.

[16] Saito, M.: Modules de Hodge polarisables, Publ. RIMS. Kyoto Univ. **24** (1988), 849-995.

[17] Saito, M.: Mixed Hodge modules, Publ. RIMS. Kyoto Univ. **26** (1990), 221–333.

[18] Schneider, P.: Introduction to the Beilinson conjectures, in: *Beilinson's Conjectures on Special Values of L-functions* (editors: M. Rapoport, N. Schappacher and P. Schneider), Perspectives in Math. **4**, Academic Press

[19] Srinivas, V.: *Algebraic K-theory*, Progress in Math. **90** (1991), Birkhäuser.

[20] Steenbrink, J. and S. Zucker: Variation of mixed Hodge structure, Invent. math. **80** (1985) no. 3, 489–542.

2

On the Splitting of the Bloch-Beilinson Filtration

Arnaud Beauville

Institut Universitaire de France and Laboratoire J.-A. Dieudonné
UMR 6621 du CNRS, Université de Nice
Parc Valrose
F-06108 NICE Cedex 02
beauville@math.unice.fr

Introduction

This paper deals with the *Chow ring* $\mathsf{CH}(X)$ (with rational coefficients) of a smooth projective variety X – that is, the $\mathbb{Q}$-algebra of algebraic cycles on X, modulo rational equivalence. This is a basic invariant of the variety X, which may be thought of as an algebraic counterpart of the cohomology ring of a compact manifold; in fact there is a $\mathbb{Q}$-algebra homomorphism $c_X : \mathsf{CH}(X) \to \mathsf{H}(X, \mathbb{Q})$, the *cycle class map.* But unlike the cohomology ring, the Chow ring, and in particular the kernel of c_X, is poorly understood.

Still some insight into the structure of this ring is provided by the deep conjectures of Bloch and Beilinson. They predict the existence of a functorial ring filtration $(F^j)_{j\geq 0}$ of $\mathsf{CH}(X)$, with $\mathsf{CH}^p(X) = F^0\mathsf{CH}^p(X) \supset \ldots \supset F^{p+1}\mathsf{CH}^p(X) = 0$ and $F^1\mathsf{CH}^p(X) = \operatorname{Ker} c_X$. We refer to [J] for a discussion of the various candidates for such a filtration and the consequences of its existence.

The existence of that filtration is not even known for an abelian variety A. In that case, however, there is a canonical *ring graduation* given by $\mathsf{CH}^p(A) = \bigoplus_s \mathsf{CH}^p_s(A)$, where $\mathsf{CH}^p_s(A)$ is the subspace of elements $\alpha \in \mathsf{CH}^p(A)$ with $k_A^*\alpha = k^{2p-s}\alpha$ for all $k \in \mathbb{Z}$ (k_A denotes the endomorphism $a \mapsto ka$ of A) [B2]. Unfortunately this does not define the required filtration because the vanishing of the terms $\mathsf{CH}^p_s(A)$ for $s < 0$ is not known in general – in fact, this vanishing is essentially equivalent to the existence of the Bloch-Beilinson filtration (the precise relationship is thoroughly analyzed in [Mu]). So if the Bloch-Beilinson filtration indeed exists, it *splits* in the sense that it is the filtration associated to a graduation of $\mathsf{CH}(A)$.

In [B-V] we observed that this also happens for a K3 surface S. Here the filtration is essentially trivial; the fact that it splits means that the image of the intersection product $\mathsf{CH}^1(S) \otimes \mathsf{CH}^1(S) \to \mathsf{CH}^2(S)$ is always one-dimensional – an easy but somewhat surprising property.

The motivation for this paper was to understand whether the splitting of the Bloch-Beilinson filtration for abelian varieties and K3 surfaces is accidental or part of a more general framework. Now asking for a conjectural splitting of a conjectural filtration may look like a rather idle occupation. The point we want to make is that the mere existence of such a splitting has quite concrete consequences, which at least in some cases can be tested. We will restrict for simplicity to the case of regular varieties, that is, varieties X for which $F^1\mathsf{CH}^1(X) = 0$. Then if the filtration comes from a graduation, any product of divisors must have degree 0; therefore, if we denote by $\mathsf{DCH}(X)$ the sub-algebra of $\mathsf{CH}(X)$ spanned by divisor classes, *the cycle class map*

$$c_X : \mathsf{DCH}(X) \longrightarrow \mathsf{H}(X)$$

is injective. In other words, any polynomial relation $P(D_1, \ldots, D_s) = 0$ between divisor classes which holds in cohomology must hold in $\mathsf{CH}(X)$. We will call this property the *weak splitting property.* Despite its name it is rather restrictive: it implies for instance the existence of a class $\xi_X \in \mathsf{CH}^n(X)$, with $n = \dim X$, such that

$$D_1 \cdot \ldots \cdot D_n = \deg(D_1 \cdot \ldots \cdot D_n) \cdot \xi_X \quad \text{in } \mathsf{CH}^n(X)$$

for any divisor classes $D_1, \ldots, D_n$ in $\mathsf{CH}^1(X)$.

What kind of varieties can we expect to have the weak splitting property? A natural class containing abelian varieties and K3 surfaces is that of Calabi-Yau varieties, but that turns out to be too optimistic – it is quite easy to give counter-examples (Example 2.1.5. b)). A more restricted class is that of holomorphic symplectic manifolds – projective manifolds admitting an everywhere non-degenerate holomorphic 2-form. We want to propose the following conjecture:

Conjecture. *A symplectic (projective) manifold satisfies the weak splitting property.*

We have to admit that the evidence we are able to provide is not overwhelming. We will prove that the weak splitting property is invariant under some simple birational transformations called Mukai flops (Proposition 2.2.4). We will also prove that the conjecture holds for the simplest examples of symplectic manifolds, the Hilbert schemes $S^{[2]}$ and $S^{[3]}$ associated to a K3 surface S (Proposition 2.3.1). Already for $S^{[3]}$ the proof is intricate, and makes use of some nontrivial relations in the Chow rings of S^2 and S^3

established in [B-V]. We hope that this might indicate a deep connection between the symplectic structure and the Bloch-Beilinson filtration, but we have not even a conjectural formulation of what this connection could be. *Added in proof*: C. Voisin has recently proved that the weak splitting property (actually a stronger version) holds for the variety of lines in a cubic fourfold, and also for the Hilbert scheme $S^{[n]}$ for $n \leq 48 - 2\rho(S)$, where $\rho(S)$ is the Picard number of S (*On the Chow ring of certain algebraic hyper-Kähler manifolds*, Preprint math.AG/0602400).

2.1 Intersection of divisors

Let X be a projective (complex) manifold. We denote by $\mathsf{CH}(X)$ and $\mathsf{H}(X)$ the Chow and cohomology rings with rational coefficients, and by $\mathsf{CH}(X, \mathbb{C})$ and $\mathsf{H}(X, \mathbb{C})$ the corresponding rings with complex coefficients. We denote by $\mathsf{DCH}(X)$ the sub-algebra of $\mathsf{CH}(X)$ spanned by divisor classes. We will say that X has the *weak splitting property* if the cycle class map $c_X : \mathsf{DCH}(X) \to \mathsf{H}(X)$ is injective.

Remark 2.1.1. The property as stated implies that $\mathsf{CH}^1(X)$ is finite-dimensional, that is, X is regular in the sense that $H^1(X, \mathcal{O}_X) = 0$. For irregular varieties the definition should be adapted, either by considering cycles modulo algebraic equivalence, or by picking up an appropriate subspace of $\mathsf{CH}^1(X)$. We will restrict ourselves to regular varieties in what follows.

Examples 2.1.2. a) A regular surface S satisfies the weak splitting property if and only if the image of the intersection map $\mathsf{CH}^1(S) \otimes \mathsf{CH}^1(S) \to \mathsf{CH}^2(S)$ has rank 1; in other words, there exists a class $\xi_S \in \mathsf{CH}^2(S)$, of degree 1, such that $C \cdot D = \deg(C.D)\, \xi_S$ for all curves C, D on S. This is the case when S is a K3 surface, or also an elliptic surface over $\mathbb{P}^1$ with a section [B-V].

b) Let S be a K3 surface, p a point of S with$[p] \neq \xi_S$ in $\mathsf{CH}^2(S)$. Let $\varepsilon : \widehat{S} \to S$ be the blowing-up of S at p. The space $\mathsf{DCH}^2(\widehat{S})$ is spanned by $\varepsilon^*\xi_S$ and $[q]$, where q is any point of $\widehat{S}$ above p. Since the pushforward map $\varepsilon_* : \mathsf{CH}^2(\widehat{S}) \to \mathsf{CH}^2(S)$ is an isomorphism, theses classes are linearly independent in $\mathsf{CH}^2(\widehat{S})$, so the map $c^2_{\widehat{S}} : \mathsf{DCH}^2(\widehat{S}) \to \mathsf{CH}^2(\widehat{S})$ is not injective.

Observe that we get a family of surfaces parameterized by $p \in S$, for which the weak splitting property fails generically, but holds when p lies in the union of countably many subvarieties of the parameter space.

c) We will give later (2.1.5) examples of Fano and Calabi-Yau threefolds which do not satisfy the weak splitting property.

Proposition 2.1.3. *Let X, Y be two smooth projective* regular *varieties.*

a) We have $\mathsf{DCH}^p(X \times Y) = \bigoplus_{r+s=p} \mathrm{pr}_1^* \mathsf{DCH}^r(X) \otimes \mathrm{pr}_2^* \mathsf{DCH}^s(Y)$. *In particular, $X \times Y$ satisfies the weak splitting property if and only if X and Y do.*

b) Let $f : X \to Y$ be a surjective map. If $c_X^p : \mathsf{DCH}^p(X) \to \mathsf{H}^{2p}(X)$ is injective, then so is $c_Y^p : \mathsf{DCH}^p(Y) \to \mathsf{H}^{2p}(Y)$.

Proof **a)** We have $\mathsf{CH}^1(X \times Y) = \mathrm{pr}_1^* \mathsf{CH}^1(X) \oplus \mathrm{pr}_2^* \mathsf{CH}^1(Y)$ since X and Y are regular; the assertion a) follows at once.
b) Follows from the commutative diagram

$$\begin{array}{ccc} \mathsf{DCH}^p(X) & \xrightarrow{c_X^p} & \mathsf{H}^{2p}(X) \\ {\scriptstyle f^*}\uparrow & & \uparrow{\scriptstyle f^*} \\ \mathsf{DCH}^p(Y) & \xrightarrow{c_Y^p} & \mathsf{H}^{2p}(Y) \end{array}$$

and the injectivity of $f^* : \mathsf{CH}^p(Y) \to \mathsf{CH}^p(X)$ (if h is an ample class in $\mathsf{CH}^1(X)$ and $d = \dim X - \dim Y$, we have $f_*(h^d) = r \cdot 1_Y$, with $r \in \mathbb{Q}^*$, and $f_*(h^d \cdot f^*\xi) = r\,\xi$ for ξ in $\mathsf{CH}(Y)$). $\square$

We now consider the behaviour of the weak splitting property when the variety X is blown up along a smooth subvariety B. We will use the notation summarized in the following diagram:

$$\begin{array}{ccc} E & \xhookrightarrow{i} & \widehat{X} \\ {\scriptstyle \eta}\downarrow & & \downarrow{\scriptstyle \varepsilon} \\ B & \xhookrightarrow{j} & X. \end{array} \tag{2.1}$$

We denote by c the codimension of B in X and by N its normal bundle.

Lemma 2.1.4. *Let p be an integer. Assume*

i) The cycle class map $c_B^q : \mathsf{DCH}^q(B) \to \mathsf{H}^{2q}(B)$ is injective for $p - c < q < p$;
ii) The Chern classes $c_i(N)$ belong to $\mathsf{DCH}(B)$;
iii) The map $c_X^p : \mathsf{CH}^p(X) \to \mathsf{H}^{2p}(X)$ restricted to $\mathsf{DCH}^p(X) + j_*\mathsf{DCH}^{p-c}(B)$ is injective.

Then the cycle class map $c_{\widehat{X}}^p : \mathsf{DCH}^p(\widehat{X}) \to \mathsf{H}^{2p}(\widehat{X})$ is injective.

Proof The projection $p : E \to B$ identifies E to $\mathbb{P}_B(N^\vee)$. Let $h \in \mathsf{CH}^1(E)$ be the class of the tautological bundle $\mathcal{O}_E(1)$; we have $i^*[E] = -h$, and therefore, for $\xi \in \mathsf{CH}(X)$, $[E]^p \cdot \varepsilon^*\xi = i_*(i^*[E]^{p-1} \cdot i^*\varepsilon^*\xi) = (-1)^{p-1} i_*(h^{p-1} \cdot \eta^* j^* \xi)$.

Since $\mathsf{CH}^1(\widehat{X}) = \varepsilon^*\mathsf{CH}^1(X) \oplus \mathbb{Q}[E]$, we get

$$\begin{aligned} \mathsf{DCH}^p(\widehat{X}) &= \varepsilon^*\mathsf{DCH}^p(X) + [E] \cdot \varepsilon^*\mathsf{DCH}^{p-1}(X) + \ldots + \mathbb{Q}[E]^p \\ &\subset \varepsilon^*\mathsf{DCH}^p(X) + i_*\eta^*\mathsf{DCH}^{p-1}(B) + i_*(h \cdot \eta^*\mathsf{DCH}^{p-2}(B)) + \ldots + \mathbb{Q} i_* h^{p-1}. \end{aligned}$$

For $q \geq c$ we have a relation $h^q = h^{c-1} \cdot \eta^* c_{q,c-1} + \ldots + \eta^* c_{q,0}$, where the $c_{i,j}$ are polynomial in the Chern classes of N; by our hypothesis (ii) these classes lie in $\mathsf{DCH}(B)$. Moreover the "key formula" [F, 6.7]

$$i_*(\gamma \cdot \eta^*\xi) = \varepsilon^* j_* \xi \qquad \text{for } \xi \in \mathsf{CH}(B)\ ,$$

with $\gamma = h^{c-1} + h^{c-2} \cdot \eta^* c_1(N) + \ldots + \eta^* c_{c-1}(N)$, implies

$$i_*(h^{c-1} \cdot \eta^*\mathsf{DCH}^{p-c}(B)) \subset \varepsilon^* j_* \mathsf{DCH}^{p-c}(B) + \sum_{k=0}^{c-2} i_*(h^k \cdot \eta^*\mathsf{DCH}^{p-k-1}(B)),$$

so that we finally get

$$\mathsf{DCH}^p(\widehat{X}) \subset \varepsilon^*\big(\mathsf{DCH}^p(X) + j_*\mathsf{DCH}^{p-c}(B)\big) + \sum_{k=0}^{c-2} i_*(h^k \cdot \eta^*\mathsf{DCH}^{p-k-1}(B)).$$

Since the map

$$\begin{aligned} \mathsf{H}^{2p}(X) \oplus \sum_{k=0}^{c-2} \mathsf{H}^{2(p-k-1)}(B) &\longrightarrow \mathsf{H}^{2p}(\widehat{X}) \\ (\alpha\ ;\ \beta_0, \ldots, \beta_{c-2}) &\longmapsto \varepsilon^*\alpha + \sum_k i_*(h^k \cdot \eta^*\beta_k) \end{aligned}$$

is an isomorphism (see for instance [Jo]), our hypotheses (i) and (iii) ensure that $c^p_{\widehat{X}}$ is injective. □

Examples 2.1.5. a) Take $X = \mathbb{P}^3$, and let B be a smooth curve, of degree d and genus g. Let ℓ be the class of a hyperplane in $\mathbb{P}^3$, ℓ_B its pull back to B. The space $\mathsf{DCH}^2(\widehat{X})$ is generated by

$$\varepsilon^*\ell^2 \quad , \quad \varepsilon^*\ell \cdot [E] = i_* p^* \ell_B \quad , \quad [E]^2 = -i_* h = i_*\eta^* c_1(N) - \varepsilon^*[B]$$

We have $c_1(N) = 4\ell_B + K_B$, so $\mathsf{DCH}^2(\widehat{X})$ contains the elements $i_*\eta^*\ell_B$ and $i_*\eta^* K_B$.

The map $i_*\eta^* : \mathsf{CH}^1(B) \to \mathsf{CH}^2(X)$ induces an isomorphism of the subspace of degree 0 divisor classes on B onto the subspace of homologically trivial classes in $\mathsf{CH}^2(X)$. If we choose ℓ_B non proportional to K_B in $\mathsf{CH}^1(B)$, the class $i_*\eta^*(d\,K_B - (2g-2)\ell_B)$ in $\mathsf{DCH}^2(\widehat{X})$ is homologically trivial, but non-trivial. Thus the map $c^2_{\widehat{X}} : \mathsf{DCH}^2(\widehat{X}) \to \mathsf{H}^4(\widehat{X})$ is not injective.

If B is a scheme-theoretical intersection of cubics, $\widehat{X}$ is a Fano variety [M-M] – we can take for instance B of genus 2 and ℓ_B a general divisor class of degree 5 (or B of genus 3 and ℓ_B general of degree 6, or B of genus 5 and $\ell_B \equiv K_B - p$ for p a general point of B). Note that by making the linear system vary we get again families where the general member does not satisfy the weak splitting property, while countably many special members of the family do satisfy it.

b) Going on with the Fano case, let D be a smooth divisor in $|-2K_X|$, and let $V \to X$ be the double covering of X ramified along D. Then by thc above example and Proposition 2.1.3. b), *V is a Calabi-Yau threefold which does not satisfy the weak splitting property.*

2.2 The weak splitting property for symplectic manifolds

By a symplectic manifold we mean here a simply-connected projective manifold which admits a holomorphic, everywhere non-degenerate 2-form. The manifold is said to be *irreducible* if the 2-form is unique up to a scalar; any symplectic manifold admits a canonical decomposition as a product of irreducible ones. In view of Proposition 2.1.3. a), we may restrict ourselves to irreducible symplectic manifolds.

Let X be an irreducible symplectic manifold, of dimension $2r$. Recall that the space $\mathsf{H}^2(X)$ admits a canonical quadratic form q ([B1], [H]) with the following properties:

- every class $x \in \mathsf{H}^2(X,\mathbb{C})$ with $q(x)=0$ satisfies $x^{r+1}=0$;
- there exists $\lambda \in \mathbb{Q}$ such that $\int_X x^{2r} = \lambda\, q(x)^r$ for all $x \in \mathsf{H}^2(X,\mathbb{C})$, where $\int_X$ is the canonical isomorphism $\mathsf{H}^{2r}(X,\mathbb{C}) \xrightarrow{\sim} \mathbb{C}$.

In fact the following more precise statement has been proved by Bogomolov:

Proposition 2.2.1. *Let V be a subspace of $\mathsf{H}^2(X,\mathbb{C})$ such that the restriction of q to V is non-degenerate (for instance $V = \mathsf{H}^2(X,\mathbb{C})$ or $V = \mathsf{CH}^1(X,\mathbb{C})$). The kernel of the map $\operatorname{Sym} V \to \mathsf{H}(X,\mathbb{C})$ is the ideal of $\operatorname{Sym} V$ spanned by the elements x^{r+1} for $x \in V$, $q(x)=0$.*

Proof The case $V = \mathsf{H}^2(X,\mathbb{C})$ is the main result of [Bo], but the proof given there implies the slightly more general statement 2.2.1. Namely, define

$A(V)$ as the quotient of $\operatorname{Sym} V$ by the ideal spanned by the elements x^{r+1} for $x \in V$, $q(x) = 0$. Then Lemma 2.5 in [Bo] says that $A(V)$ is a finite-dimensional graded Gorenstein $\mathbb{C}$- algebra, with socle in degree $2r$ – in other words, $A_{2r}(V)$ is one-dimensional, and the multiplication pairing $A_d(V) \times A_{2r-d}(V) \to A_{2r}(V) \cong \mathbb{C}$ is a perfect duality.

Since any element x of $\mathsf{H}^2(X, \mathbb{C})$ with $q(x) = 0$ satisfies $x^{r+1} = 0$, we get a $\mathbb{C}$- algebra homomorphism $u : A(V) \to \mathsf{H}(X, \mathbb{C})$. The kernel of u is an ideal of $A(V)$; if it is non-zero, it contains the minimal ideal $A_{2r}(V)$ of $A(V)$. But this is impossible because V contains an element h with $q(h) \neq 0$, hence with $h^{2r} \neq 0$. □

Corollary 2.2.2. *The following conditions are equivalent:*

i) The cycle class map $c_X : \mathsf{DCH}(X) \to \mathsf{H}(X)$ *is injective (that is,* X *satisfies the weak splitting property);*
ii) The map $c_X^{r+1} : \mathsf{DCH}^{r+1}(X) \to \mathsf{H}^{2r+2}(X)$ *is injective;*
iii) Every element x *of* $\mathsf{CH}^1(X, \mathbb{C})$ *with* $q(x) = 0$ *satisfies* $x^{r+1} = 0$ *(in* $\mathsf{CH}^{r+1}(X, \mathbb{C})$*).*

Proof Consider the diagram

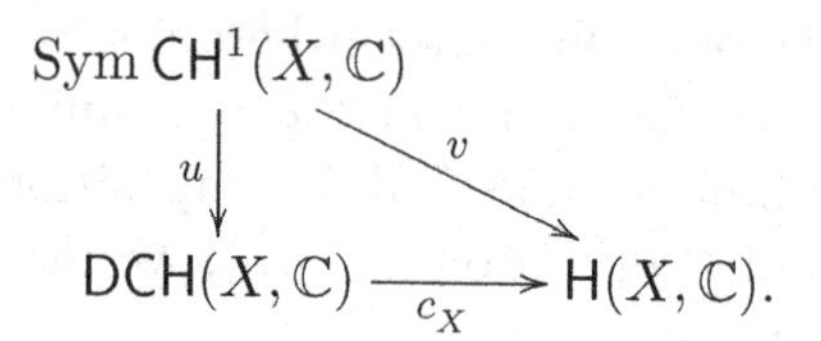

The injectivity of c is equivalent to $\operatorname{Ker} v \subset \operatorname{Ker} u$. In view of the Proposition, this is exactly condition (iii), and it is equivalent to $\operatorname{Ker} v^{r+1} \subset \operatorname{Ker} u^{r+1}$. □

Remark 2.2.3. Assume that there is an element $\alpha \in \mathsf{CH}^1(X)$ with $q(\alpha) = 0$ – this is the case for instance if $\dim_{\mathbb{Q}} \mathsf{CH}^1(X) \geq 5$. Then the set of such elements is Zariski dense in the quadric $q = 0$ of $\mathsf{CH}^1(X, \mathbb{C})$. Thus the conditions of the Corollary are also equivalent to:
(iii′) *Every element* x *of* $\mathsf{CH}^1(X)$ *with* $q(x) = 0$ *satisfies* $x^{r+1} = 0$.

A possible proof of (iii′) could be as follows. It seems plausible that the subset of *nef* classes $x \in \mathsf{CH}^1(X)$ with $q(x) = 0$ is Zariski dense in the quadric $q = 0$ (this holds at least when X is a K3 surface). If this is the case, it would be enough to prove (iii′) for nef classes. Now it is a standard conjecture (see [S]) that a nef class $x \in \mathsf{CH}^1(X)$ with $q(x) = 0$ should be the pull back of the class of a hyperplane in $\mathbb{P}^r$ under a Lagrangian fibration $f : X \to \mathbb{P}^r$, so that $x^{r+1} = f^*(h^{r+1}) = 0$.

We will now consider the behaviour of the weak splitting property under a Mukai flop. Let X be an irreducible symplectic manifold, of dimension $2r$; assume that X contains a subvariety P isomorphic to $\mathbb{P}^r$. Then P is a Lagrangian subvariety, and its normal bundle in X is isomorphic to Ω^1_P. We blow up P in X, getting our standard diagram

$$\begin{array}{ccc} E & \xrightarrow{\ i\ } & \widehat{X} \\ {\scriptstyle\eta}\big\downarrow & & \big\downarrow{\scriptstyle\varepsilon} \\ P & \xrightarrow{\ j\ } & X. \end{array}$$

The exceptional divisor E is the cotangent bundle $\mathbb{P}(T_P)$, which can be identified with the incidence divisor in $P \times P^\vee$, where $P^\vee$ is the projective space dual to P. The projection $\eta^\vee : E \to P^\vee$ identifies E to $\mathbb{P}(T_{P^\vee})$, and E can be blown down to $P^\vee$ by a map $\varphi : \widehat{X} \to X'$, where X' is a smooth algebraic space. To remain in our previous framework we will assume that X' is projective, so that X' is again an irreducible symplectic manifold. The diagram

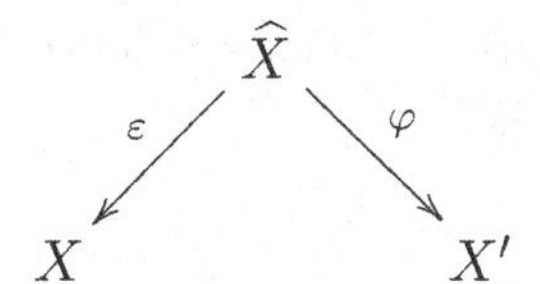

is called a *Mukai flop.* There are many concrete examples of such flops, see [M].

Proposition 2.2.4. *If X satisfies the weak splitting property, so does X'.*

Proof Consider the $\mathbb{Q}$- linear map $\varphi_*\varepsilon^* : \mathsf{CH}^1(X) \to \mathsf{CH}^1(X')$. It is bijective and preserves the canonical quadratic forms (see e.g. [H, Lemma 2.6]. In view of Corollary 2.2.2, the Proposition will follow from

Lemma 2.2.5. *Let $\alpha \in \mathsf{CH}^1(X)$, and $\alpha' := \varphi_*\varepsilon^*\alpha$. Then $\alpha'^{r+1} = \varphi_*\varepsilon^*(\alpha^{r+1})$.*

Proof We have $\varphi^*\alpha' = \varepsilon^*\alpha + m[E]$ for some $m \in \mathbb{Q}$. Let $\ell \in \mathsf{CH}^{2r-1}(\widehat{X})$ be the class of a line contained in a fibre of $\eta^\vee$; we have $\deg([E]\cdot\ell) = -1$, and $\varepsilon_*\ell$ is the class of a line in P. Intersecting the above equality with ℓ gives $m = \deg(\alpha_{|P})$, or equivalently $\alpha_{|P} = mk$ in $\mathsf{CH}^1(P)$, where k is the class of a hyperplane in P. Then

$$\varphi^*\alpha'^{r+1} = (\varepsilon^*\alpha + m[E])^{r+1} = \sum_{p=0}^{r+1} \binom{r+1}{p} m^{r+1-p}\, \varepsilon^*\alpha^p \cdot [E]^{r+1-p}.$$

As in (2.1.4), let $h \in \mathsf{CH}^1(E)$ be the class of $\mathcal{O}_E(1)$. For $p \leq r$ we have

$$\varepsilon^*\alpha^p \cdot [E]^{r+1-p} = (-1)^{r-p} i_*(h^{r-p} \cdot i^*\varepsilon^*\alpha^p) = (-1)^{r-p} i_*(h^{r-p} \cdot \eta^*\alpha^p_{|P}),$$

Thus $$\varphi^*\alpha'^{r+1} = \varepsilon^*\alpha^{r+1} + m^{r+1} i_*\Big(\sum_{p=0}^{r} \binom{r+1}{p} (-1)^{r-p} h^{r-p}\, \eta^* k^p\Big).$$

Now since the total Chern class of T_P is $(1+k)^{r+1}$ we have in $\mathsf{CH}^r(E)$

$$\sum_{p=0}^{r} \binom{r+1}{p} (-1)^p h^{r-p}\, \eta^* k^p = \sum_{p=0}^{r} (-1)^p h^{r-p}\, \eta^* c_p(T_P) = 0,$$

hence $\varphi^*\alpha'^{r+1} = \varepsilon^*\alpha^{r+1}$. Applying φ_* gives the lemma, hence the Proposition.

□

Corollary 2.2.6. *Let X, X' be birationally equivalent projective symplectic fourfolds. Then X satisfies the weak splitting property if and only if X' does.*

Indeed any birational map between projective symplectic fourfolds is a composition of Mukai flops [W].

2.3 The weak splitting property for $S^{[2]}$ and $S^{[3]}$

The simplest symplectic manifolds are K3 surfaces, for which we have already seen that the weak splitting property holds (Example 2.1.2). More precisely [B-V], let S be a K3 surface and o a point of S lying on a (singular) rational curve R. The class of o in $\mathsf{CH}^2(S)$ is independent of the choice of R, and we have, for every $\alpha, \beta \in \mathsf{CH}^1(S)$,

$$\alpha \cdot \beta = \deg(\alpha \cdot \beta)\,[o] \quad \text{in} \quad \mathsf{CH}^2(S).$$

Let $\Delta : S \hookrightarrow S \times S$ be the diagonal embedding. For $\alpha \in \mathsf{CH}^1(S)$, we have in $\mathsf{CH}^3(S \times S)$ ([B-V, Prop. 1.6],)

$$\Delta_*\alpha = \mathrm{pr}_1^*\,\alpha \cdot \mathrm{pr}_2^*[o] + \mathrm{pr}_1^*[o] \cdot \mathrm{pr}_2^*\,\alpha. \tag{2.2}$$

K3 surfaces are the first instance of a famous series of symplectic manifolds, the *Hilbert schemes* $S^{[r]}$ parameterizing finite subschemes of length r on the K3 surface S.

Proposition 2.3.1. *Let S be a K3 surface. The symplectic varieties $S^{[2]}$ and $S^{[3]}$ satisfy the weak splitting property.*

Proof – Let us warm up with the easy case of $S^{[2]}$. Let $S^{\{2\}}$ be the variety obtained by blowing up the diagonal of $S \times S$. The Hilbert scheme $S^{[2]}$ is the quotient of $S^{\{2\}}$ by the involution which exchanges the factors. In view of Corollary 2.2.1 and Proposition 2.1.3. b) it suffices to prove that the cycle class map $c^3_{S^{\{2\}}} : \mathsf{DCH}^3(S^{\{2\}}) \to \mathsf{H}^6(S^{\{2\}})$ is injective. We will check that the hypotheses of Lemma 2.1.4 are satisfied. Condition (i) is the weak splitting property for S. The normal bundle to the diagonal in $S \times S$ is T_S, so (ii) means that the class $c_2(T_S)$ belongs to $\mathsf{DCH}^2(S)$; this is proved in ([B-V, thm. 1 c]). Formula (2.2) implies $\Delta_*\mathsf{CH}^1(S) \subset \mathsf{DCH}^3(S \times S)$, so condition (iii) reduces to the injectivity of $c^3_{S\times S}$, which follows from Proposition 2.1.3. a) and the corresponding result for S.

– Let us pass to the more difficult case of $S^{[3]}$. The Hilbert scheme $S^{[3]}$ is dominated by the *nested Hilbert scheme* $S^{[2,3]}$ which parameterizes pairs $(Z, Z') \in S^{[2]} \times S^{[3]}$ with $Z \subset Z'$; it is isomorphic to the blow-up of $S \times S^{[2]}$ along the incidence subvariety $\mathcal{I} = \{(x, Z) \mid x \in Z\}$. Let $\pi : S^{\{2\}} \to S^{[2]}$ be the quotient map, and $p : S^{\{2\}} \to S$ the first projection. Then the map $j = (p, \pi) : S^{\{2\}} \hookrightarrow S \times S^{[2]}$ induces an isomorphism of $S^{\{2\}}$ onto $\mathcal{I}$ (see for instance [L, 1.2]).

To prove the theorem, it suffices, by Corollary 2.2.2 and Proposition 2.1.3. b), to prove that the cycle class map $\mathsf{DCH}^4(S^{[2,3]}) \to \mathsf{H}^8(S^{[2,3]})$ is injective. We will again check that the hypotheses of Lemma 2.1.4 are satisfied. Condition (i) is the injectivity of the cycle class map $c^3_{S^{\{2\}}} : \mathsf{DCH}^3(S^{\{2\}}) \to \mathsf{H}^6(S^{\{2\}})$, which has just been proved. Let N be the normal bundle to the embedding $j : S^{\{2\}} \hookrightarrow S \times S^{[2]}$, and $E \subset S^{\{2\}}$ the exceptional divisor, which is the ramification locus of π. From the exact sequences

$$
\begin{gathered}
0 \to N^\vee \longrightarrow p^*\Omega^1_S \oplus \pi^*\Omega^1_{S^{[2]}} \longrightarrow \Omega^1_{S^{\{2\}}} \to 0 \\
0 \to \pi^*\Omega^1_{S^{[2]}} \longrightarrow \Omega^1_{S^{\{2\}}} \longrightarrow \mathcal{O}_E(-E) \to 0 \\
0 \to \mathcal{O}_{S^{\{2\}}}(-2E) \longrightarrow \mathcal{O}_{S^{\{2\}}}(-E) \longrightarrow \mathcal{O}_E(-E) \to 0
\end{gathered}
$$

we obtain the equality in K-theory $[N^\vee] = [p^*\Omega^1_S] + [\mathcal{O}_{S^{\{2\}}}(-2E)] - [\mathcal{O}_{S^{\{2\}}}(-E)]$. We conclude that $c_2(N) = c_2(N^\vee)$ belongs to $\mathsf{DCH}^2(S^{\{2\}})$, so that condition (ii) holds.

– The rest of the proof will be devoted to check condition (iii), namely the injectivity of

$$
\mathsf{DCH}^4(S \times S^{[2]}) + j_*\mathsf{DCH}^2(S^{\{2\}}) \longrightarrow \mathsf{H}^8(S \times S^{[2]}).
$$

Let us fix some notation. We will use our standard diagram (2.1)

$$\begin{array}{ccc} E & \stackrel{i}{\hookrightarrow} & S^{\{2\}} \\ \eta \downarrow & & \downarrow \varepsilon \\ S & \stackrel{\Delta}{\hookrightarrow} & S^2. \end{array}$$

We denote by p and q the two projections of $S^{\{2\}}$ onto S. We define an injective $\mathbb{Q}$-linear map $\iota : \mathsf{CH}(S) \to \mathsf{CH}(S^{[2]})$ by $\iota(\xi) := \pi_* p^*\xi$; we will use the same notation for cohomology classes. We have $\pi^*\iota(\xi) = p^*\xi + q^*\xi$ for ξ in $\mathsf{CH}(S)$ or $\mathsf{H}(S)$. Finally if $\alpha \in \mathsf{CH}(S)$ and $\xi \in \mathsf{CH}(S^{[2]})$ we put $\alpha \boxtimes \xi := \mathrm{pr}_1^*\alpha \otimes \mathrm{pr}_2^*\xi$.

We have $\mathsf{CH}^1(S^{\{2\}}) = p^*\mathsf{CH}^1(S) \oplus q^*\mathsf{CH}^1(S) \oplus \mathbb{Q}[E]$. In $\mathsf{CH}^2(S^{\{2\}})$ we have $[E]\cdot \varepsilon^*\alpha = i_*\eta^*\Delta^*\alpha$ for $\alpha \in \mathsf{CH}^1(S^2)$, and $[E]^2 = -\varepsilon^*[\Delta(S)]$. Therefore:

$$\begin{aligned} \mathsf{DCH}^2(S^{\{2\}}) = \mathbb{Q}\,p^*[o] + \mathbb{Q}\,q^*[o] \ + \ & p^*\mathsf{CH}^1(S) \otimes q^*\mathsf{CH}^1(S) \\ & + i_*\eta^*\mathsf{CH}^1(S) + \mathbb{Q}\,\varepsilon^*[\Delta(S)]. \end{aligned}$$

We want to describe the space $j_*\mathsf{DCH}^2(S) + \mathsf{DCH}^4(S \times S^{\{2\}})$.

Lemma 2.3.2. *Let $\alpha, \beta \in \mathsf{CH}^1(S)$. The classes $j_*p^*[o]$, $j_*(\,p^*\alpha \cdot q^*\beta)$, and $j_*i_*\eta^*\alpha$ belong to $\mathsf{DCH}^4(S \times S^{\{2\}}) + \mathbb{Q}\,([o] \boxtimes \iota([o]))$.*

Proof Let $j' : S^{\{2\}} \hookrightarrow S \times S^{\{2\}}$ be the embedding given by $j'(z) = (p(z), z)$, so that $j = (1,\pi)\circ j'$. From the cartesian diagram

$$(2.3.2)\qquad \begin{array}{ccc} S^{\{2\}} & \stackrel{j'}{\hookrightarrow} & S \times S^{\{2\}} \\ p \downarrow & & \downarrow (1,p) \\ S & \stackrel{\Delta}{\hookrightarrow} & S \times S \end{array}$$

we obtain $j'_*p^*[o] = (1,p)^*\Delta_*[o] = [o] \boxtimes p^*[o]$, hence $j_*p^*[o] = [o] \boxtimes \iota([o])$. In the same way we have $j'_*p^*\alpha = (1,p)^*\Delta_*\alpha$, hence, using (2.2),

$$j'_*p^*\alpha = \alpha \boxtimes p^*[o] + [o] \boxtimes p^*\alpha.$$

Multiplying by $\mathrm{pr}_2^*\,q^*\beta$ and using $\mathrm{pr}_2 \circ j' = \mathrm{Id}$ we obtain

$$j'_*(p^*\alpha \cdot q^*\beta) = \alpha \boxtimes (p^*[o] \cdot q^*\beta) + [o] \boxtimes (p^*\alpha \cdot q^*\beta),$$

hence $$j_*(p^*\alpha \cdot q^*\beta) = \alpha \boxtimes \pi_*(p^*[o] \cdot q^*\beta) + [o] \boxtimes \pi_*(p^*\alpha \cdot q^*\beta).$$

For $\alpha, \beta \in \mathsf{CH}^1(S)$, put $\langle \alpha, \beta \rangle := \deg(\alpha \cdot \beta)$. Then

$$\begin{aligned}\pi^*\pi_*(p^*\alpha \cdot q^*\beta) &= p^*\alpha \cdot q^*\beta + p^*\beta \cdot q^*\alpha \\ &= (p^*\alpha + q^*\alpha)(p^*\beta + q^*\beta) - \langle \alpha, \beta \rangle (p^*[o] + q^*[o]) \\ &= \pi^*\big(\iota(\alpha)\iota(\beta) - \langle \alpha, \beta \rangle \iota([o])\big);\end{aligned}$$

we find similarly $\pi^*\pi_*(p^*[o] \cdot q^*\beta) = \pi^*\iota([o])\iota(\beta)$, and finally

$$j_*(p^*\alpha \cdot q^*\beta) = \alpha \boxtimes \iota([o])\iota([\beta]) \ + \ [o] \boxtimes \big(\iota(\alpha)\iota(\beta) - \langle \alpha, \beta \rangle \, \iota([o])\big).$$

Let $\gamma \in \mathsf{CH}^1(S)$. We have $\iota(\beta)^2 \cdot \iota(\gamma) = \langle \beta^2 \rangle \, \iota([o])\iota(\gamma) + 2\langle \beta \cdot \gamma \rangle \, \iota([o])\iota(\beta)$ (this is easily checked by applying π^* as above). If $\langle \beta^2 \rangle \neq 0$ we conclude by taking $\gamma = \beta$ that $\iota([o])\iota(\beta)$ is proportional to $\iota(\beta)^3$. If $\langle \beta^2 \rangle = 0$ we can choose γ so that $(\beta \cdot \gamma) \neq 0$; then $\iota([o])\iota(\beta)$ is proportional to $\iota(\beta)^2\iota(\gamma)$. In each case we see that $\iota([o])\iota(\beta)$ belongs to $\mathsf{DCH}^3(S^{[2]})$, hence the assertion of the lemma about $j_*(\, p^*\alpha \cdot q^*\beta)$.

Consider finally the cartesian diagram

$$\begin{array}{ccc} E & \xhookrightarrow{\ k\ } & S \times E \\ {\scriptstyle \eta}\downarrow & & \downarrow{\scriptstyle (1,\eta)} \\ S & \xhookrightarrow{\ \Delta\ } & S \times S \end{array}$$

with $k(e) = (\eta(e), e)$. Using again (2.2) we get

$$k_*\eta^*\alpha = (1, \eta)^*\Delta_*\alpha = \alpha \boxtimes \eta^*[o] + [o] \boxtimes \eta^*\alpha.$$

Pushing forward in $S \times S^{[2]}$ we obtain $j_*i_*\eta^*\alpha = \alpha \boxtimes i'_*\eta^*[o] + [o] \boxtimes i'_*\eta^*\alpha$, where $i' = \pi \circ i$ is the embedding of E in $S^{[2]}$.

To avoid confusion let us denote by $\bar{E}$ the image of E in $S^{[2]}$, so that $\pi^*[\bar{E}] = 2[E]$. We have $i'_*\eta^*\alpha = \pi_*([E] \cdot p^*\alpha) = \frac{1}{2}[\bar{E}] \cdot \iota(\alpha) \in \mathsf{DCH}^2(S^{[2]})$. Likewise $[E]^3 = i_*h^2 = -24 i_*\eta^*[o]$, hence $i'_*\eta^*[o] = -\frac{1}{96}\,[\bar{E}]^3 \in \mathsf{DCH}^2(S^{[2]})$. This finishes the proof of the lemma. □

The lemma and the formula for $\mathsf{DCH}^2(S^{\{2\}})$ show that $j_*\mathsf{DCH}^2(S^{\{2\}})$ is spanned modulo $\mathsf{DCH}^4(S \times S^{[2]})$ by the classes

$$[o] \boxtimes \iota([o]), \quad j_*q^*[o], \quad j_*\varepsilon^*[\Delta(S)].$$

In fact there is one more relation, much more subtle, between these classes modulo $\mathsf{DCH}^4(S \times S^{[2]})$.

Lemma 2.3.3. *We have*

$$2[o] \boxtimes \iota([o]) - 2j_*q^*[o] + j_*\varepsilon^*[\Delta(S)] \ \in \ \mathsf{DCH}^4(S \times S^{[2]}).$$

Proof We start from a relation in $\mathsf{CH}^4(S^3)$, proved in [B-V, Prop. 3.2]. For $1 \le i < j \le 3$, let us denote by $p_{ij} : S^3 \to S^2$ the projection onto the i- th and j- th factors, and by $p_i : S^3 \to S$ the projection onto the i- th factor. We will write simply Δ for the diagonal $\Delta(S) \subset S^2$, and $\delta \subset S^3$ for the small diagonal, that is, the subvariety of triples (x,x,x) for $x \in S$. Then:

$$[\delta] - \sum_{i<j, k\neq i,j} p_{ij}^*[\Delta]\cdot p_k^*[o] + \sum_{i<j} p_i^*[o]\cdot p_j^*[o] = 0.$$

Pull back this relation by the map $\varepsilon_S = (1_S, \varepsilon) : S \times S^{\{2\}} \to S \times S^2$. Since

$$p_1 \circ \varepsilon_S = \mathrm{pr}_1, \quad p_2 \circ \varepsilon_S = p \circ \mathrm{pr}_2, \quad p_3 \circ \varepsilon_S = q \circ \mathrm{pr}_2, \quad p_{23} \circ \varepsilon_S = \varepsilon,$$

we obtain $\varepsilon_S^*[\delta] = j'_*\varepsilon^*[\Delta], \qquad \varepsilon_S^*(p_1^*[o]\cdot p_{23}^*[\Delta]) = [o]\boxtimes \varepsilon^*[\Delta],$

$$\varepsilon_S^*(p_2^*[o]\cdot p_3^*[o]) = 1 \boxtimes p^*[o]\cdot q^*[o], \quad \varepsilon_S^*(p_1^*[o]\cdot p_2^*[o]) = [o]\boxtimes p^*[o],$$

$$\varepsilon_S^*(p_1^*[o]\cdot p_3^*[o]) = [o]\boxtimes q^*[o].$$

We have $p_{12} \circ \varepsilon_S = (1,p)$, hence $\varepsilon_S^* p_{12}^*[\Delta] = j'_*1$ (see diagram 2.3.2) and $\varepsilon_S^*(p_3^*[o]\cdot p_{12}^*[\Delta]) = j'_*q^*[o]$. Let $j'' = (q,1) : S^{\{2\}} \to S \times S^{\{2\}}$; the same argument gives $\varepsilon_S^* p_{13}^*[\Delta] = j''_*1$ and $\varepsilon_S^*(p_2^*[o]\cdot p_{13}^*[\Delta]) = j''_*p^*[o]$. Finally we have $\mathrm{k}j'_*q^*[o] + j''_*p^*[o] = \pi_S^* j_* q^*[o]$. Pushing forward by π_S we obtain in $\mathsf{CH}^4(S \times S^{[2]})$:

$$j_*\varepsilon^*[\Delta] - 2j_*q^*[o] - [o]\boxtimes \pi_*\varepsilon^*[\Delta] + 2[o]\boxtimes \iota([o]) + 1 \boxtimes \iota([o])^2 = 0.$$

It remains to observe that $[o]\boxtimes \pi_*\varepsilon^*[\Delta]$ and $1\boxtimes \iota([o])^2$ belong to $\mathsf{DCH}^4(S \times S^{[2]})$. Indeed from $[E]^2 = -\varepsilon^*[\Delta]$ we deduce $\pi_*\varepsilon^*[\Delta] = -\frac{1}{2}[\bar{E}]^2 \in \mathsf{DCH}^2(S^{[2]})$. And if h is any element of $\mathsf{CH}^1(S)$ with $h^2 = d \neq 0$, we have

$$\pi^*\iota(h)^4 = 6d^2 p^*[o]\cdot q^*[o] = 3d^2\pi^*\iota([o])^2, \quad \text{hence} \quad \iota([o])^2 \in \mathsf{DCH}^4(S^{[2]}).$$

□

For a smooth projective variety X, let us denote by $\mathsf{DH}(X)$ the (graded) subspace of $\mathsf{H}(X)$ spanned by intersection of divisor classes – that is, the image of $\mathsf{DCH}(X)$ by the cycle class map. It remains to prove that the cycle class map $c^4_{S\times S^{[2]}}$ is injective on $\mathsf{DCH}^4(S\times S^{[2]}) + \mathbb{Q}\,([o]\boxtimes\iota([o])) + \mathbb{Q}\,j_*q^*[o]$. Since we know by (2.3) and (2.1.3. a)) that it is injective on $\mathsf{DCH}^4(S\times S^{[2]})$, this amounts to:

Lemma 2.3.4. *There is no non-trivial relation*

$$a\,[o]\boxtimes\iota([o]) + b\,j_*q^*[o] \in \mathsf{DH}^8(S\times S^{[2]}),$$

with $a,b\in\mathbb{Q}$.

To prove this, suppose that such a relation holds. Let ω be a non-zero class in $\mathsf{H}^{2,0}(S)$; for any class ξ in $\mathsf{H}^8(S\times S^{[2]},\mathbb{C})$ put $h(\xi) := (\mathrm{pr}_2)_*(\mathrm{pr}_1^*\,\omega\cdot\xi)$. Since the product of ω with any algebraic class in $\mathsf{H}^2(S)$ is zero, h is zero on $\mathsf{DH}^8(S\times S^{[2]})$. Clearly $h([o]\boxtimes\iota([o]))=0$, while $h(j_*q^*[o]) = \pi_*(p^*\omega\cdot q^*[o]) = \iota(\omega)\iota([o])$. This class is nonzero, for instance because $\langle\iota(\omega)\iota([o]),\iota(\bar\omega)\rangle = \langle\omega,\bar\omega\rangle > 0$.

Thus $b=0$, and our relation reduces to $[o]\boxtimes\iota([o])\in\mathsf{DH}^8(S\times S^{[2]})$. Since $\mathsf{DH}^8(S\times S^{[2]}) = \bigoplus_{i+j=8}\mathsf{DH}^i(S)\boxtimes\mathsf{DH}^j(S^{[2]})$ (see Proposition 2.1.3. a)), this is equivalent to $\iota([o])\in\mathsf{DH}^4(S^{[2]})$. Thus the proof reduces to the following assertion:

Lemma 2.3.5. *The class $\iota([o])$ does not belong to $\mathsf{DH}^4(S^{[2]})$.*

Proof We have

$$\begin{aligned}\mathsf{H}^4(S^{\{2\}}) &= \varepsilon^*\mathsf{H}^4(S^2)\oplus i_*\eta^*\mathsf{H}^2(S)\\ &= \mathbb{Q}p^*[o]\oplus\mathbb{Q}q^*[o]\oplus(p^*\mathsf{H}^2(S)\otimes q^*\mathsf{H}^2(S))\oplus i_*\eta^*\mathsf{H}^2(S).\end{aligned}$$

Taking the invariants under the involution of $S^{\{2\}}$ which exchanges the factors, we find

$$\mathsf{H}^4(S^{[2]}) = \mathbb{Q}\,\iota([o])\oplus\mathrm{Sym}^2\,\mathsf{H}^2(S)\oplus i'_*\eta^*\mathsf{H}^2(S),$$

where $\mathrm{Sym}^2\,\mathsf{H}^2(S)$ is identified with a subspace of $\mathsf{H}^4(S^{[2]})$ by $\alpha\cdot\beta\mapsto\pi_*(p^*\alpha\cdot q^*\beta)$, and $i' := \pi\circ i$ is the natural embedding of E in $S^{[2]}$. Since $\mathsf{CH}^1(S^{[2]}) = \iota(\mathsf{CH}^1(S))\oplus\mathbb{Q}\cdot[E]$, the subspace $\mathsf{DH}^4(S^{[2]})$ is spanned by the classes

$$\iota(\alpha)\iota(\beta),\quad \iota(\alpha)\cdot[E] = 2i'_*\eta^*\alpha,\quad [E]^2 = -2\pi_*\varepsilon^*[\Delta]\quad\text{for }\ \alpha,\beta\in\mathsf{CH}^1(S).$$

Suppose that we have a relation

$$\iota([o]) = \sum_{i<j} m_{ij}\,\iota(\alpha_i)\iota(\alpha_j) + i'_*\eta^*\gamma + m\,\pi_*\varepsilon^*[\Delta]\quad\text{in}\quad \mathsf{H}^4(S^{[2]}),$$

where (α_i) is a basis of $\mathsf{CH}^1(S)$. This gives in $\mathsf{H}^4(S^{\{2\}})$:

$$p^*[o]+q^*[o] = \sum_{i<j} m_{ij}\,(p^*\alpha_i+q^*\alpha_i)(p^*\alpha_j+q^*\alpha_j) + 2i_*\eta^*\gamma + 2m\,\varepsilon^*[\Delta].$$

Projecting onto the direct summand $i_*\eta^*\mathsf{H}^2(S)$ of $\mathsf{H}^4(S^{\{2\}})$ we find $i_*\eta^*\gamma = 0$. Multiplying by $p^*\omega$ and pushing forward by q we find as in the proof of (2.3.4) that all terms but $\varepsilon^*[\Delta]$ give 0, so $m=0$. Finally the equality

$$p^*[o]+q^*[o] = \sum_{i<j} m_{ij}\,(p^*\alpha_i+q^*\alpha_i)(p^*\alpha_j+q^*\alpha_j)$$

projected onto $\mathrm{Sym}^2 \mathsf{H}^2(S)$ gives $m_{ij} = 0$ for all i, j. This achieves the proof of the lemma and therefore of the Proposition. □

Comments. A variation of this method can be used to prove that the *generalized Kummer variety* K_2 associated to an abelian surface A [B1] has the weak splitting property; one must replace $\mathsf{CH}^1(A)$ by the subspace of *symmetric* divisor classes. We leave the details to the reader.

We should point out, however, that even among symplectic fourfolds these examples are quite particular. Indeed for each integer $g \geq 2$, the projective K3 surfaces of genus g (that is, embedded in $\mathbb{P}^g$ with degree $2g-2$) form an irreducible 19-dimensional family; the corresponding family of Hilbert schemes $S^{[2]}$ is contained in a 20-dimensional irreducible family of projective symplectic manifolds (see [B1]). Since the weak splitting property is not invariant under deformation, we do not know whether it holds for the general member of such a family. It would be interesting, in particular, to check whether the property holds for the variety of lines contained in a smooth cubic hypersurface in $\mathbb{P}^5$.

References

[B1] Beauville, A.: Variétés kählériennes dont la première classe de Chern est nulle, J. of Diff. Geometry **18**, 755–782 (1983).

[B2] Beauville, A.: Sur l'anneau de Chow d'une variété abélienne, Math. Annalen **273**, 647–651 (1986).

[B-V] Beauville, A and C. Voisin: On the Chow ring of a K3 surface, J. Algebraic Geom. **13** (2004), 417–426.

[Bo] Bogomolov, F.: On the cohomology ring of a simple hyper-Kähler manifold (on the results of Verbitsky), Geom. Funct. Anal. **6** (1996), 612–618.

[F] Fulton, W.: *Intersection theory*, Ergebnisse der Mathematik und ihrer Grenzgebiete (3) **2**, Springer-Verlag, Berlin (1984).

[J] Jannsen, U.: Motivic sheaves and filtrations on Chow groups, in *Motives (Seattle, 1991)*, 245–302, Proc. Sympos. Pure Math. **55** (Part 1), Amer. Math. Soc., Providence, RI, 1994.

[Jo] Jouanoulou, J.-P.: Cohomologie de quelques schémas classiques et théorie cohomologique des classes de Chern, SGA 5 (Cohomologie l- adique et fonctions L), Lecture Notes in Math. **589**, 282–350, Springer-Verlag, Berlin (1977).

[H] Huybrechts, D.: Compact hyper-Kähler manifolds: basic results, Invent. Math. **135** (1999), 63–113.

[L] Lehn, M.: Chern classes of tautological sheaves on Hilbert schemes of points on surfaces, Invent. Math. **136** (1999), 157–207.

[M] Mukai, S.: Symplectic structure of the moduli space of sheaves on an abelian or $K3$ surface, Invent. Math. **77** (1984), no. 1, 101–116.

[M-M] Mori, S. and S. Mukai: Classification of Fano 3-folds with $B_2 \geq 2$. Manuscripta Math. **36** (1981/82), 147–162.

[Mu] Murre, J. P.: On a conjectural filtration on the Chow groups of an algebraic variety, I: The general conjectures and some examples, Indag. Math. (N.S.) **4** (1993), 177–188.

[S] Sawon, J.: Abelian fibred holomorphic symplectic manifolds, Turk. J. Math. **27** (2003), 197–230.

[W] Wierzba, J.: Birational Geometry of Symplectic 4-folds, Preprint (2002).

3

Künneth Projectors for Open Varieties

Spencer Bloch
Dept. of Mathematics, University of Chicago,
Chicago, IL 60637, USA
bloch@math.uchicago.edu

Hélène Esnault
Mathematik, Universität Duisburg-Essen
FB6, Mathematik, 45117 Essen, Germany
esnault@uni-essen.de

To Jacob Murre

Abstract

We consider correspondences on smooth quasiprojective varieties U. An algebraic cycle inducing the Künneth projector onto $H^1(U)$ is constructed. Assuming normal crossings at infinity, the existence of relative motivic cohomology is shown to imply the independence of ℓ for traces of open correspondences.

3.1 Introduction

Let X be a smooth, projective algebraic variety over an algebraically closed field k, and let $H^*(X)$ denote a Weil cohomology theory. The existence of algebraic cycles on $X \times X$ inducing as correspondences the various Künneth projectors $\pi^i : H^*(X) \to H^i(X)$ is one of the standard conjectures of Grothendieck, [10, 11]. It is known in general only for the cases $i = 0, 1, 2d-1, 2d$ where $d = \dim X$. The purpose of this note is to consider correspondences on smooth quasi-projective varieties U. In the first section we prove the existence of an "algebraic" Künneth projector $\pi^1 : H^*(U) \to H^1(U)$ assuming that U admits a smooth, projective completion X. The word algebraic is placed in quotes here because in fact the algebraic cycle on $X \times U$ inducing π^1 is not, as one might imagine, trivialized on $(X - U) \times U$. It is only partially trivialized. This partial trivialization is sufficient to define a class in $H_c^{2d-1}(U) \otimes H^1(U)$ giving the desired projection. Of course, our cycle on $X \times U$ will be trivialized on $(X - U) \times V$ for $V \subset U$ suitably small nonempty open, but our method does not in any obvious way yield a full trivialization on $(X - U) \times U$. We finish this first section with some comments on π^i for $i > 1$ and some speculation, mostly coming from discussions with A. Beilinson, on how these ideas might be applied to study the Milnor conjecture that the Galois cohomology ring of the function field $H^*(k(X), \mathbb{Z}/n\mathbb{Z})$ is generated by H^1.

In the last section, we use the existence of relative motivic cohomology [12] to prove an integrality and independence of ℓ result for the trace of an algebraic correspondence Γ on $U \times U$. We are indebted to G. Laumon for pointing out that one may endeavor to prove this using results already in the literature([5, 14, 15], and [7]) by reduction mod p and composition with a high power of Frobenius. Our objective in what follows is to show how techniques in motivic cohomology can apply to such questions, at least when the divisor at infinity $D = X - U$ is a normal crossings divisor.

When the Zariski closure of the correspondence stabilizes the various strata D_I at infinity (e.g. when the correspondence is the graph of Frobenius) then the trace on $H^*(U)$ is realized as an alternating sum of traces on $H^*(D_I)$. When in addition all the intersections with the diagonals are transverse, the contribution to the alternating sum coming from points lying off U cancels, and the trace on $H^*(U)$ is just the sum of the fixed points on U.

We would like to acknowledge helpful correspondence with A. Beilinson, M. Levine, J. Murre, T. Saito, and V. Srinivas. We thank G. Laumon and L. Lafforgue for explaining to us [7], and the referee for very useful comments and advice.

3.2 The first Künneth component

Let k be an algebraically closed field. We work in the category of algebraic varieties over k. $H^*(X)$ will denote étale cohomology with $\mathbb{Q}_\ell$-coefficients for some ℓ prime to the characteristic of k. If $k = \mathbb{C}$, we take Betti cohomology with $\mathbb{Q}$-coefficients.

Let C be a smooth, complete curve over k, and let $\delta \subset C$ be a nonempty finite set of reduced points. Let $J(C)$ be the Jacobian of C, and let $J(C, \delta)$ be the semiabelian variety which represents the functor

$$X \mapsto \{(\mathcal{L}, \phi) \mid \mathcal{L} \text{ line bundle on } C \times X, \ \deg \ \mathcal{L}|_{C \times k(X)} = 0 \\ \phi : \mathcal{L}|_{\delta \times X} \cong \mathcal{O}_{\delta \times X}\} / \cong, \tag{3.1}$$

where the equivalence relation $\cong$ consists of the line bundle isomorphisms commuting with ϕ. There is an exact sequence

$$0 \to T \to J(C, \delta) \to J(C) \to 0 \tag{3.2}$$

where T is the torus $\Gamma(\delta, \mathcal{O}^\times)/\Gamma(C, \mathcal{O}^\times)$. By abuse of notation we shall write $J(C, \delta)$ rather than $J(C, \delta)(k)$. We can identify the character group $\mathrm{Hom}(T, \mathbb{G}_m)$ with $\mathrm{Div}^0_\delta(C)$, the group of 0-cycles of degree 0 supported on δ. A split subgroup $\Delta \subset \mathrm{Div}^0_\delta(C)$ corresponds to a quotient $T \twoheadrightarrow T_\Delta = T/\ker \Delta$, where $\ker \Delta \subset T$ is the subtorus killed by all characters in Δ. We may push out (3.2) and define $J(C, \Delta) := J(C, \delta)/\ker \Delta$:

$$0 \to T_\Delta \to J(C, \Delta) \to J(C) \to 0. \tag{3.3}$$

The functor represented by $J(C, \Delta)$ is the following quotient of (3.1)

$$\begin{aligned} X \mapsto \{(\mathcal{L}, \phi) \mid \mathcal{L} \text{ line bundle on } X \times C, \ \deg \mathcal{L}|_{k(X)\times C} = 0 \\ \phi : \otimes_i \mathcal{L}^{\otimes n_i}|_{X\times\{c_i\}} \cong \mathcal{O}_X \text{ for all } \textstyle\sum n_i c_i \in \Delta\}/\cong . \end{aligned} \tag{3.4}$$

These trivializations should be compatible in an evident way with the group law on Δ.

Lemma 3.2.1. *We write $H^1(C, \delta) = H^1_c(C - \delta)$. Define*

$$H^1(C, \Delta) := (H^1(C, \delta)/\Delta^\perp) \otimes \mathbb{Q}_\ell,$$

where

$$\Delta^\perp \otimes \mathbb{Q}_\ell \subset \mathbb{Q}_\ell[\delta]/\mathbb{Q}_\ell \subset H^1(C, \delta; \mathbb{Q}_\ell)$$

is perpendicular to $\Delta \subset \mathrm{Div}^0_\delta(C)$ under the evident coordinatewise duality. There is a well defined first Chern class $c_1(\mathcal{L}_\Delta)$ of the universal Poincaré bundle $\mathcal{L}_\Delta$ on $J(C, \Delta) \times C$ which lies in $H^1(J(C, \Delta))(1) \otimes H^1(C, \Delta)$.

Proof Let $I_\delta \subset \mathcal{O}_{J(C,\Delta)\times C}$ be the ideal of $J(C, \Delta) \times \delta$. Let $\pi : C \to C'$ be the singular curve obtained from C by gluing all the points of δ to a single point $\delta' \in C'$. Define $M_\Delta \subset (1 \times \pi)_*(\mathcal{O}^\times_{J(C,\Delta)\times C})/k^\times$ to be the pullback as indicated:

$$\begin{array}{ccccccc} 0 \to (1\times\pi)_*(1+I_\delta) & \to & M_\Delta & \to & (\ker\Delta)_{J(C,\Delta)\times\delta'} & \to 0 \\ \| & & \downarrow & & \downarrow & \\ 0 \to (1\times\pi)_*(1+I_\delta) & \to & \frac{(1\times\pi)_*\mathcal{O}^\times_{J(C,\Delta)\times C}}{k^\times} & \to & \frac{(1\times\pi)_*\mathcal{O}^\times_{J(C,\Delta)\times\delta}}{k^\times} & \to 0. \end{array} \tag{3.5}$$

Using that $R^1(1 \times \pi)_*\mathbb{G}_m = (0)$, it is straightforward to check that pairs $(\mathcal{L}, \phi)$ in (3.4) with $X = J(C, \Delta)$ correspond to M_Δ torsors on $J(C, \Delta) \times C'$. In particular, we have a class $[\mathcal{L}_\Delta] \in H^1(J(C, \Delta) \times C', M_\Delta)$ corresponding to the Poincaré bundle.

One gets a diagram of Kummer sequences of sheaves on $J(C,\Delta) \times C'$ (Here $j : C - \delta \hookrightarrow C$)

$$\begin{array}{ccccccc} & & 0 & & 0 & & 0 & \\ & & \downarrow & & \downarrow & & \downarrow & \\ 0 \to & (1\times\pi)_*(1\times j)_!\mu_{\ell^n} & \to & (1\times\pi)_*(1+I_\delta) & \xrightarrow{\ell^n} & (1\times\pi)_*(1+I_\delta) & \to 0 \\ & \downarrow & & \downarrow & & \downarrow & \\ 0 \to & M_{\Delta,\ell^n} & \to & M_\Delta & \xrightarrow{\ell^n} & M_\Delta & \to 0 \\ & \downarrow & & \downarrow & & \downarrow & \\ 0 \to & ((\ker\Delta)_{J(C,\Delta)\times\delta'})_{\ell^n} & \to & (\ker\Delta)_{J(C,\Delta)\times\delta'} & \xrightarrow{\ell^n} & (\ker\Delta)_{J(C,\Delta)\times\delta'} & \to 0 \\ & \downarrow & & \downarrow & & \downarrow & \\ & 0 & & 0 & & 0 & \end{array} \tag{3.6}$$

We have $[\mathcal{L}_\Delta] \in H^1(J(C,\Delta) \times C', M_\Delta)$ and so by the Kummer coboundary, $c_1(\mathcal{L}_\Delta) \in \varprojlim_n H^2(J(C,\Delta) \times C', M_{\Delta,\ell^n})$. But $M_{\Delta,\ell^n} \cong \mathbb{Z}/\ell^n_{J(C,\Delta)} \boxtimes \psi_{\ell^n}$, where ψ_{ℓ^n} fits into an exact sequence of sheaves on C'

$$0 \to \pi_* j_! \mu_{\ell^n} \to \psi_{\ell^n} \to (\ker\Delta)_{\delta',\ell^n} \to 0. \tag{3.7}$$

We can identify $\Delta^\perp \otimes \mu_{\ell^n}$ with $(\ker\Delta)_{\ell^n}$. the exact cohomology sequence from (3.7) yields

$$(\ker\Delta)_{\mu_{\ell^n}} \to H^1(C,\delta;\mu_{\ell^n}) \to H^1(C',\psi_{\ell^n}) \to 0. \tag{3.8}$$

Passing to the limit over n, it now follows that we may define $c_1(\mathcal{L}_\Delta) \in H^1(J(C,\Delta)) \otimes H^1(C,\Delta)(1)$ as in the statement of the lemma. (Note that $\mathcal{L}_\Delta$ is trivial on $(0) \times C$. Further we are free to replace $\mathcal{L}_\Delta$ by $\mathcal{L}_\Delta \otimes (\mathcal{M} \boxtimes \mathcal{O}_C)$ for $\mathcal{M}$ a line bundle on $J(C,\Delta)$. We may therefore assume the Künneth components of $c_1(\mathcal{L}_\Delta)$ in degrees $(2,0)$ and $(0,2)$ vanish.) □

Lemma 3.2.2. *Suppose given a morphism $\rho : X \to J(C)$. Let Ξ be a Cartier divisor on $C \times X$ representing ρ. We assume Ξ is flat over C so we may define a correspondence $\Xi_* : \mathrm{Div}(C) \to \mathrm{Div}(X)$. Let $U \subset X$ be nonempty open in X. Then there exists a lifting $\rho_{U,\Delta} : U \to J(C,\Delta)$ of ρ if and only if $(\Xi|_{C\times U})_*(\Delta) \subset \mathrm{Div}(U)$ consists of principal divisors. The set of such liftings is a torsor under $\mathrm{Hom}(\Delta, \Gamma(U, \mathcal{O}_U^\times))$.*

Proof Choose a basis $z_i = \sum_j n_{ij}c_j$ for the free abelian group Δ. Write $\mathcal{O}_{C\times X}(\Xi)_{z_i\times X} := \otimes_j \mathcal{O}_{C\times X}(\Xi)^{\otimes n_{ij}}|_{\{c_j\}\times X}$. The assumption that $\Xi_*(\Delta)$ consists of principal divisors is precisely the assumption that all the line bundles $\mathcal{O}_{C\times X}(\Xi)_{z_i\times X}|_U$ are trivial. The choice of the trivializations for a basis of Δ yields the choice of the desired lifting $\rho_{U,\Delta}$. □

Lemma 3.2.3. *Assume X is a smooth variety, and let $\rho : X \to J(C)$ be as above. Suppose $U \subset X$ is a dense open set such that $\rho|_U$ admits a lifting $\rho_{U,\Delta} : U \to J(C,\Delta)$. Let $\mathrm{Div}^0_{X-U}(X)$ be the free abelian group on Cartier divisors supported on $X - U$ which are homologous to 0 on X. Then we get a commutative diagram on cohomology*

$$\begin{array}{ccccccccc} 0 & \to & H^1(J(C)) & \to & H^1(J(C,\Delta)) & \to & \Delta\otimes\mathbb{Q}_\ell(-1) & \to & 0 \\ & & \downarrow{\scriptstyle \rho^*} & & \downarrow{\scriptstyle \rho^*_{U,\Delta}} & & \downarrow{\scriptstyle a} & & \\ 0 & \to & H^1(X) & \to & H^1(U) & \to & \mathrm{Div}^0_{X-U}(X)\otimes\mathbb{Q}_\ell(-1) & \to & 0. \end{array} \tag{3.9}$$

Proof The left hand square is commutative by functoriality. That the cokernels on the top and bottom row are as indicated follows on the top row from the Leray spectral sequence for the projection $\pi : J(C,\Delta) \to J(C)$ and on the bottom from the localization sequence which may be written

$$0 \to H^1(X) \to H^1(U) \to H^2_{X-U}(X) \to H^2(X). \tag{3.10}$$

The identification $H^2_{X-U}(X) \cong \mathrm{Div}_{X-U}(X)\otimes\mathbb{Q}_\ell(-1)$ is saying that by purity, the Gysin homomorphism is an isomorphism. □

Remark 3.2.4. (i) Fixing $\rho_{U,\Delta}$ amounts to fixing trivializations of the restriction $\mathcal{O}_{C\times X}(\Xi)_{z_i\times X}|_U$ as above. Such trivializations exhibit

$$\mathcal{O}_{C\times X}(\Xi)_{z_i\times X} \cong \mathcal{O}_X(D_i)$$

for some divisor D_i with support on $X - U$. The map labeled a in (3.9) sends $z_i \mapsto D_i$.
(ii) The diagram

$$\begin{array}{ccc} \Delta & \longrightarrow & J(C) \\ \downarrow{\scriptstyle a} & & \downarrow{\scriptstyle \rho^*} \\ \mathrm{Div}^0_{X-U}(X) & \longrightarrow & \mathrm{Pic}^0(X) \end{array} \tag{3.11}$$

is commutative, where the horizontal arrows are cycle classes. Indeed, both a and ρ^* are defined by the divisor on $C\times X$. Note that a depends on the choice of $\rho_{U,\Delta}$ but only up to rational equivalence.

Now suppose X is smooth, projective, of dimension d. Let $U \subset X$ be a dense open subset. Write $X - U = D \cup Z$ where $D \subset X$ is a divisor and $\mathrm{codim}(Z \subset X) \geq 2$. We have $H^1(X - D) \cong H^1(U)$. Since we are interested in $H^1(U)$, we may assume $U = X - D$ is the complement of a divisor.

Let $i : C \hookrightarrow X$ be a general linear space section of dimension 1, and let $\delta = C \cap D$. We may choose $\rho : X \to J(C)$ such that the composition

$$\mathrm{Pic}^0(X) \xrightarrow{i^*} J(C) \xrightarrow{\rho^*} \mathrm{Pic}^0(X) \tag{3.12}$$

is multiplication by an integer $N \neq 0$. Indeed, let H be a very ample line bundle so that C is the $(d-1)$-fold product of general sections of H. Intersection with H yields an isogeny $\mathrm{Pic}^0(X) \to \mathrm{Alb}(X)$, which defines an inverse isogeny $\mathrm{Alb}(X) \to \mathrm{Pic}^0(X)$ of degree N. We pull back the Poincaré bundle from $J(C) \times J(C)$ to $C \times X$ via the composite map $C \times X \to J(C) \times \mathrm{Alb}(X) \to J(C) \times \mathrm{Pic}^0(X) \to J(C) \times J(C)$, where the first map is the cycle map, the second one is $1 \times$isogeny, the third one is $1\times$ restriction. We define $\mathcal{O}_{C\times X}(\Xi)$ to be the inverse image of the Poincaré bundle. The morphism $\rho : X \to J(C)$ is the correspondence $x \mapsto \mathcal{O}_{C\times X}(\Xi)|_{C\times\{x\}}$ and does not depend on the choice of the section Ξ.

Consider the diagram

$$\begin{array}{ccccccc} 0 \to & \mathbb{Q}_\ell[\delta]/\mathbb{Q}_\ell & \to & H^1_c(C-\delta) & \to & H^1(C) & \to 0 \\ & \downarrow{\scriptstyle b} & & \downarrow{\scriptstyle i_*} & & \downarrow{\scriptstyle i_*} & \\ 0 \to & \frac{H^{2d-2}(D)}{H^{2d-2}(X)}(d-1) & \to & H^{2d-1}_c(U)(d-1) & \to & H^{2d-1}(X)(d-1). & \end{array} \tag{3.13}$$

Here, the rows are long exact sequences associated to restriction to closed subsets, and the vertical arrows are Gysin maps. The map b can be described as follows. The $\mathbb{Q}_\ell$-vector space $H^{2d-2}(D)(d-1)$ has basis the irreducible components of D, and $b(x)$ is the basis element $[D_x]$ associated to the unique component D_x of D containing x. We have dual exact sequences (defining $\mathrm{Div}^0_D(X)$)

$$0 \to \mathrm{Div}^0_D(X) \to H_{2d-2}(D)(1-d) \to H^2(X)(1) \tag{3.14}$$

$$H^{2d-2}(X)(d-1) \to H^{2d-2}(D)(d-1) \to \frac{H^{2d-2}(D)}{H^{2d-2}(X)}(d-1) \to 0.$$

If we view $\mathbb{Q}_\ell[\delta]$ and $H^{2d-2}(D)(d-1)$ as endowed with symmetric pairings with orthonormal bases the points $x \in \delta$ and the cohomology classes of irreducible components $D_i \subset D$, then b is adjoint to the map $D_i \mapsto D_i \cdot \delta$. We conclude

Lemma 3.2.5. *Define* $\mathrm{Div}^0_D(X)$ *to be the* $\mathbb{Q}_\ell$*-vector space spanned by divisors on* X *supported on* D *and homologous to* 0 *on* X. *Define* $\Delta \subset \mathrm{Div}^0_\delta(C)$ *to be the image of* $\mathrm{Div}^0_D(X)$ *under pullback* i^*. *Then there is a commutative diagram*

$$\begin{array}{ccccccc} 0 \to & \mathbb{Q}_\ell[\delta]/\Delta^\perp & \to & H^1_c(C,\Delta) & \to & H^1(C) & \to 0 \\ & \downarrow b & & \downarrow i_* & & \downarrow i_* & \\ 0 \to & \frac{H^{2d-2}(D)}{H^{2d-2}(X)}(d-1) & \to & H^{2d-1}_c(U)(d-1) & \to & H^{2d-1}(X)(d-1). & \end{array} \tag{3.15}$$

Proof The map b is dual to the restriction map $\mathrm{Div}^0_D(X) \xrightarrow{i^*} \mathrm{Div}^0_\delta(C)$. By definition $\Delta^\perp$ is orthogonal to the image of i^*, i.e. $\Delta^\perp = \ker b$. □

Lemma 3.2.6. *Let* $\Delta = i^*(\mathrm{Div}^0_D(X)) \subset \mathrm{Div}^0_\delta(C)$ *be as in Lemma 3.2.5. Then* ρ *defined in* (3.12) *lifts to some* $\rho_{U,\Delta} : U \to J(C,\Delta)$.

Proof The correspondence defined by $\mathcal{O}_{C\times X}(\Xi)$ in (3.12) carries $\mathcal{O}_C(z)$ for $z \in \Delta = i^*(\mathrm{Div}^0_D(X))$ to line bundles in $\mathrm{Pic}^0(X)$, the classes of which fall in the image of $\rho^* i^*(\mathrm{Div}^0_D(X)) \equiv N \cdot \mathrm{Div}^0_D(X)$ in $\mathrm{Pic}^0(X)$. To be more precise, let D_p be a basis for $\mathrm{Div}^0_D(X)$, and set $z_p = i^* D_p$. This is a basis of Δ. Then $\mathcal{O}_{C\times X}(\Xi)|_{z_p \times X} = \mathcal{O}_X(D_p)$. Thus choose $\rho_{U,\Delta}$ in Lemma 3.2.2 using this trivialization on U. □

Using the Lemmas 3.2.1, 3.2.5, 3.2.6 together with (3.9), we pull back

$$c_1(\mathcal{L}_\Delta) \in H^1(C,\Delta) \otimes H^1(J(C,\Delta))(1) \xrightarrow{i_* \otimes \rho^*_{U,\Delta}} H^{2d-1}_c(U)(d) \otimes H^1(U) \cong H^1(U)^\vee \otimes H^1(U) \tag{3.16}$$

and define a correspondence $\Phi : H^1(U) \to H^1(U)$.

Lemma 3.2.7. *The map* Φ *is the multiplication by* N.

Proof We consider Φ. It acts on $H^1(U)$, compatibly with the exact sequence

$$0 \to H^1(X) \to H^1(U) \to \mathrm{Div}^0_D(X)(-1) \to 0. \tag{3.17}$$

By definition of $\rho_{U,\Delta}$, it is equal to $N \cdot \mathrm{Id}$ on $H^1(X)$ and on $\mathrm{Div}^0_D(X)(-1)$. Thus $\Phi - N \cdot \mathrm{Id}$ is a correspondence from $\mathrm{Div}^0_D(X)(-1)$ to $H^1(X)$. We use purity in the sense of Deligne. There is no nontrivial correspondence $\mathrm{Div}^0_D(X)(-1) \to H^1(X)$. If $k = \mathbb{C}$ and we consider Betti cohomology, $\mathrm{Div}^0_D(X)(-1)$ is pure of weight 2 while $H^1(X)$ is pure of weight 1. If k is the algebraic closure of a finite field, we have the same conclusion. Otherwise, all the objects used are defined over a finitely generated field k over a finite field k_0. By Čebotarev's theorem, the Galois group of k/k_0 is generated by the Frobenius maps, so we may make sense of the notion of weight for $H^1(U)$. We conclude as in the complex case.

□

We now express in terms of cycles the trivialization of $\mathcal{O}_{C\times X}(\Xi)_{p\times X} = \mathcal{O}_X(D_p)$ used in the proof of Lemma 3.2.6.

Theorem 3.2.8. *With notation as above, there exists a cycle Γ on $X \times U$ of dimension $d = \dim X$ together with rational functions f_μ on X for each divisor μ homologically equivalent to 0 on X and supported on $D = X_U$ such that $pr_{2*}(\Gamma \cdot (\mu \times U)) = (f_\mu)$. The data $(\Gamma, \{f_\mu\})$ define a class in $H_c^{2d-1}(U) \otimes H^1(U)$ which gives the identity map on $H^1(U)$.*

We close this section with a comment about Künneth projectors $\pi^i : H^*(U) \to H^i(U)$ for $i > 1$. We consider the somewhat weaker question of the existence of an algebraic projector when we localize at the generic point of the target, i.e. we consider $H^*(U) \to H^i(U) \to \varinjlim_{V\subset U} H^i(V)$. We assume $U = X - D$ with X smooth, projective, and D a Cartier divisor.

Proposition 3.2.9. *Let $n < \dim X$ be an integer. Let $Y \subset X$ be a multiple hyperplane section of dimension n which is general with respect to D. Write $\delta = Y \cap D$. Then the restriction map*

$$H_D^{n+1}(X) \to H_\delta^{n+1}(Y) \tag{3.18}$$

is injective.

Proof Let $d = \dim(X)$. By duality, we have to show surjectivity of the Gysin map $H^{n-1}(\delta) \to H^{2d-(n+1)}(D)(d-n)$. More generally, one has

Theorem 3.2.10 (P. Deligne). *Let $\mathcal{F}$ be a ℓ-adic sheaf on $\mathbb{P}^N$. Then there exists a non-empty open set $U \subset (\mathbb{P}^N)^\vee$ such that for $\iota : A \hookrightarrow \mathbb{P}^N$ a hyperplane section corresponding to a point of U, the Gysin homomorphism*

$$H^{i-2}(A, \iota^*\mathcal{F})(-1) \to H^i_A(\mathbb{P}^N, \mathcal{F}),$$

is an isomorphism for all i. In particular, if $V \subset \mathbb{P}^N$ is a projective variety, then the Gysin homomorphism $H^i(A\cap V) \to H^{i+2}(V)(1)$ is an isomorphism for $i > \dim(A\cap V)$ and surjective for $i = \dim(A\cap V)$ for a non-empty open set of A.

The proof of the general theorem is written in [6, Theorem 2.1], Applied to $\mathcal{F} = a_*\mathbb{Q}_\ell$, where $a : V \to \mathbb{P}^N$ is the projective embedding, it shows that the Gysin isomorphism $H^i(A\cap V) \to H^{i+2}_{A\cap V}(V)(1)$ is an isomorphism. Then the application follows from Artin's vanishing theorem $H^i(V-(A\cap V))=0$ for $i>\dim(V)$. $\square$

Let L be the Lefschetz operator on $H^*(X)$. One of the standard conjectures ($B(X,L)$ in [11]) is the existence of an algebraic correspondence Λ which is a "weak inverse" to L. Assume now that this standard conjecture B is true for X and for all smooth linear space sections $Y\subset X$. The strong Lefschetz theorem implies that $L^{d-n}: H^n(X) \xrightarrow{\cong} H^{2d-n}(X)(d-n)$. Assuming $B(X,L)$, $\Lambda^{d-n}=(L^{d-n})^{-1}: H^{2d-n}(X)(d-n)\cong H^n(X)$. Write $P=\Lambda^{d-n}|_{Y\times X}$. It is easy to check that the composition

$$H^n(X) \xrightarrow{i^*} H^n(Y) \xrightarrow{P} H^n(X) \tag{3.19}$$

is the identity, so $H^n(Y)=\mathrm{Image}(i^*)\oplus\ker(P)$. Consider the diagram

$$\begin{array}{ccccc} H^n(X) & \to & H^n(U) & \xrightarrow{a} & H^{n+1}_D(X) \\ \downarrow{\scriptstyle i^*} & & \downarrow & & \downarrow{\scriptstyle b} \\ H^n(Y) & \xrightarrow{d} & H^n(Y-\delta) & \xrightarrow{c} & H^{n+1}_\delta(Y). \end{array} \tag{3.20}$$

Define

$$H^n(Y-\delta)^0=\{x\in H^n(Y-\delta)\mid c(x)\in \mathrm{Im}(b\circ a)\}. \tag{3.21}$$

As a consequence of proposition 3.2.9 and (3.20) we see that $H^n(U)\twoheadrightarrow H^n(Y-\delta)^0/d(\ker(P))$, and the kernel of this map is the image in $H^n(U)$ of elements $x\in H^n(X)$ such that $i^*x\in\ker(P)\oplus\mathrm{Image}(H^n_\delta(Y)\to H^n(Y))$. For such an x, it will necessarily be the case that $x=P(i^*x)$ is supported on a proper closed subset of X. In particular, for some $V\subset U$ open dense, P will induce a map

$$P_U: H^n(Y-\delta)^0\to H^n(V). \tag{3.22}$$

The map $i^*: H^n(U)\to H^n(Y-\delta)^0$ dualises to $i_*: (H^n(Y-\delta)^0)^\vee \to H^{2d-n}_c(U)$, so we may define

$$(i_*\otimes P_U): H^n(Y-\delta)^0)^\vee\otimes H^n(Y-\delta)^0\to H^{2d-n}_c(U)\otimes H^n(V). \tag{3.23}$$

Let $T\subset H^n(Y-\delta)^0$ be the subgroup of cohomology classes supported in codimension 1. Assuming inductively that we are able to define an algebraic correspondence on Y which carries a class

$$\gamma\in (H^n(Y-\delta)^0)^\vee\otimes(H^n(Y-\delta)^0/T) \tag{3.24}$$

corresponding to the evident map $H^n(Y-\delta)^0\twoheadrightarrow H^n(Y-\delta)^0/T$, it would follow since $P_U(T) \subset \ker(H^n(V) \to \varinjlim_{V\subset U} H^n(V))$ that we could view

$$(i_* \otimes P_U)(\gamma) \in H_c^{2d-n}(U) \otimes \varinjlim_{V\subset U} H^n(V). \tag{3.25}$$

This correspondence would have the desired properties.

One interest in pursuing this line of investigation concerns the Milnor conjecture that the Galois cohomology with $\mathbb{Z}/\ell\mathbb{Z}$-coefficients prime to the residue characteristic is generated as an algebra by H^1. There is a geometric proof of this result in top degree [2], so, for example, elements in $H^n(Y-\delta)$ lie in the subalgebra generated by H^1 after localization. If P_U exists as an algebraic correspondence, then using the existence of a norm in Milnor K-theory, one could show that the Milnor conjecture was true for $H^*(k(X), \mathbb{Z}/\ell\mathbb{Z})$ for almost all ℓ. (The condition on ℓ arises because the standard conjectures only make sense after tensoring with $\mathbb{Q}$.) Here, the idea that cohomology classes in degree n might come by correspondence from an algebraic variety of dimension $\leq n$ was suggested to us by Alexander Beilinson.

3.3 Open correspondences

The aim of this section is to give a simple motivic proof of the independence of ℓ or of a complex embedding of a ground field k of the trace for open correspondences. If we assume that k is finite, then, as conjectured by Deligne, high Frobenius power twists move the correspondence to a general position correspondence and the local factors have been computed in [5], [14], [15], [7]. Surely in this case the simple observations which follow are weaker.

We consider open correspondences. This means the following. Let X be a smooth projective variety of dim d over an algebraically closed field k, and let $U \subset X$ be a nontrivial open subvariety, with complement $D = X - U$. One considers codim d cycles $\Gamma \subset U \times U$ which have the property that they induce a correspondence $\Gamma_* : H^i(U) \to H^i(U)$ or equivalently $H_c^i(U) \to H_c^i(U)$ for all i. Here cohomology is étale $\mathbb{Q}_\ell$ cohomology or Betti cohomology if $k = \mathbb{C}$ and we denote by $p_i : X \times X \to X$ the two projections. We write

$$\Gamma = \sum n_j \Gamma_j \tag{3.26}$$

where Γ_j is irreducible, $n_j \in \mathbb{Z}$ and define

$$\bar{\Gamma} := \sum n_j \bar{\Gamma}_j, \ \Gamma_j \subset U \times U \tag{3.27}$$

where $\bar{}$ is the Zariski closure in $X \times X$. We will use the following facts ([4], Théorème 2.9).

Proposition 3.3.1. *Let $q : \Gamma \to U$ be a proper map of quasi-projective varieties. Then one has a pull-back map*

$$q^* : H^i_c(U) \to H^i_c(\Gamma). \tag{3.28}$$

Let

$$\begin{array}{ccc} \Gamma & \xrightarrow{\iota} & \Gamma' \\ {\scriptstyle p}\downarrow & & {\scriptstyle p'}\downarrow \\ U & \xrightarrow{=} & U \end{array} \tag{3.29}$$

be a commutative diagram of quasi-projective varieties of dimension d, with ι an open embedding, p' proper and U smooth. Then there are push-down maps p_*, p'_* making the following diagram commutative

$$\begin{array}{ccc} H^i_c(\Gamma) & \xrightarrow{\iota^*} & H^i_c(\Gamma') \\ {\scriptstyle p_*}\downarrow & & {\scriptstyle p'_*}\downarrow \\ H^i_c(U) & \xrightarrow{=} & H^i_c(U) \end{array} \tag{3.30}$$

Definition 3.3.2. If $p_2|_{(\Gamma)_j} : \Gamma_j \to U$ is proper for all j, or equivalently $\Gamma_j \subset X \times U$ is Zariski closed or equivalently if

$$\bar{\Gamma}_j \cap (D \times X) \subset X \times D \ \forall\, j, \tag{3.31}$$

then one defines

$$(\Gamma_j)_* : H^i_c(U) \xrightarrow{p_2^*} H^i_c(\Gamma_j) \xrightarrow{(p_1)_*} H^i_c(U), \tag{3.32}$$

and call it the open correspondence defined by Γ_j. The correspondence defined by Γ is then by definition $\Gamma_* = \sum n_j(\Gamma_j)_*$.

(The ordering (p_1, p_2) here is chosen as in [9].)

Remark 3.3.3. If we compare this condition to the one yielding to Künneth correspondences in section 2, then it is much stronger. Indeed, Theorem 3.2.8 yields a cycle Γ which meets physically $D \times U$, but cohomologically it washes out, while in this section we handle the case where there is physically no intersection.

Remark 3.3.4. We have $\Gamma_j = \bar{\Gamma}_j - (X \times D)$ and we set $\Gamma'_j = \bar{\Gamma}_j - (D \times X)$. Thus $p_1|_{\Gamma'_j} : \Gamma'_j \to U$ is projective. One has the following commutative diagram

$$\begin{array}{ccccccc}
H^i_c(U) & \xrightarrow{p_2^*} & H^i_c(\Gamma_j) & \xrightarrow{\iota^*} & H^i_c(\Gamma'_j) & \xrightarrow{(p_1)_*} & H^i_c(U) \\
{=}\downarrow & & \downarrow & & \downarrow & & {=}\downarrow \\
H^i_c(U) & \to & H^{i+2d}_{c,\Gamma_j}(X \times U, d) & & H^{i+2d}_{c,\Gamma'_j}(U \times X, d) & \to & H^i_c(U) \\
{=}\downarrow & & {=}\downarrow & & \downarrow & & {=}\downarrow \\
H^i_c(U) & \xrightarrow{p_2^*} & H^{i+2d}_{c,\Gamma_j}(U \times U, d) & \xrightarrow{\iota^*} & H^{i+2d}_{c,\Gamma'_j}(U \times X, d) & \xrightarrow{(p_1)_*} & H^i_c(U)
\end{array} \tag{3.33}$$

with factorization of the upper horizontal composition of maps, which is $\Gamma_{j,*}$, through the lower horizontal composition of maps. So from far left to far right on the bottom line, this is the correspondence $\Gamma_{j,*}$.

We assume now that we have the assumption as in Definition 3.3.2 and we wish to give conditions under which one can compute the trace of Γ_* which is defined by

$$\mathrm{Tr}(\Gamma_*) := \sum_{i=0}^{2d} (-1)^i \, \mathrm{Tr}(\Gamma_*|_{H^i_c(U)}). \tag{3.34}$$

As it stands, the trace of Γ_* depends a priori on ℓ, or, for varieties defined over a field k of characteristic 0, and Betti cohomology taken with respect to a complex embedding $\iota : k \to \mathbb{C}$, it depends on ι. One has

Theorem 3.3.5. *Let X be a smooth projective variety of dimension d defined over a field k, together with a strict normal crossings divisor $D \subset X$ of open complement $U = X - D$. Let $\Gamma \subset U \times U$ be a dimension d cycle defining an open correspondence Γ_* on ℓ-adic cohomology or Betti cohomology as in Definition 3.3.2. Then $\mathrm{Tr}(\Gamma_*)$ does not depend on ℓ in ℓ-adic cohomology or on the complex embedding of k in Betti cohomology.*

Proof We use the relative motivic cohomology $H^{2d}_M(X \times U, D \times U, \mathbb{Z}(d))$, as defined in [12], chapter 4, 2.2 and p. 209. The group $H^m_M(X \times U, D \times U, \mathbb{Z}(n))$ is the homology $H_{2n-m}(\mathcal{Z}^n(X \times U, D \times U, *))$, where $\mathcal{Z}^n(X \times U, D \times U, *)$ is the single complex associated to the double higher Chow cycle complex

$$\begin{array}{ccccc} \cdots & & \cdots & & \cdots \\ \partial\big\downarrow & & \partial\big\downarrow & & \big\downarrow \\ \mathcal{Z}^n(X\times U,1) & \xrightarrow{\text{rest}} & \mathcal{Z}^n(D^{(1)}\times U,1) & \xrightarrow{\text{rest}} & \mathcal{Z}^n(D^{(2)}\times U,1) \\ \partial\big\downarrow & & \partial\big\downarrow & & \big\downarrow \\ \mathcal{Z}^n(X\times U,0) & \xrightarrow{\text{rest}} & \mathcal{Z}^n(D^{(1)}\times U,0) & \xrightarrow{\text{rest}} & \mathcal{Z}^n(D^{(2)}\times U,0). \end{array} \tag{3.35}$$

Here $D^{(a)}$ is the normalization of all the strata of codimension a, $\mathcal{Z}^n(D^{(a)}\times U,b)$ is a group of cycles on $D^{(a)}\times U\times S^b$ where $S^\bullet$ is the cosimplicial scheme $S^n = \mathrm{Spec}(k[t_0,\ldots,t_n]/(\sum_{i=0}^n t_i - 1))$ with face maps $S^n \hookrightarrow S^{n+1}$ defined by $t_i = 0$. More precisely, $\mathcal{Z}^n(D^{(a)}\times U,b)$ is generated by the codimension n subvarieties $Z\subset D^{(a)}\times U\times S^b$ such that, for each face F of S^b, and each irreducible component $F'\subset D^{(a)}$ of the strata of D we have $\mathrm{codim}_{F'\times U\times F}(Z\cap(F'\times U\times F))\geq n$. The horizontal restriction maps are the intersection with the smaller strata, the vertical ∂'s are the boundary maps.

This relative motivic cohomology acts as correspondences on $H_c^*(U)$, where $H_c^*(U)$ is ℓ-adic or Betti cohomology ([3], section 4). Let us write $\Gamma = \sum n_j\Gamma_j$. By Definition 3.3.2, one has $\Gamma_j\subset X\times U$ closed with $\Gamma_j\cap(D\times U) = \varnothing$, thus in particular, $\Gamma\in\mathcal{Z}^d(X\times U,0)$ with $\mathrm{rest}(\Gamma) = 0$ in $\mathcal{Z}^d(D^{(1)}\times U,0)$, thus it defines a class

$$[\Gamma]\in H^{2d}_M(X\times U, D\times U, d). \tag{3.36}$$

Similarly, we consider the restriction $\Delta_U\subset U\times X$ of the diagonal $\Delta\subset X\times X$. This defines a class in $\mathcal{Z}^d(U\times X,0)$. As $\mathrm{rest}(\Gamma_U) = 0$ in $\mathcal{Z}^d(U\times D^{(1)},0)$, it defines a class

$$[\Delta_U]\in H^{2d}_M(U\times X, U\times D, d). \tag{3.37}$$

We want to pair $[\Gamma]$ with $[\Delta_U]$. We argue using M. Levine's work. Let Y be a N-dimensional smooth projective variety defined over k, with two strict normal crossings divisors A, B so that $A+B$ is a strict normal crossings divisor. By [12], Chapter IV, lemma 2.3.5 and lemma 2.3.6, the motive $M(Y-A, B-B\cap A)$ is dual to the motive $M(Y-B, A-B\cap A)$. It yields a cup product

$$H^a_M(Y-A, B-B\cap A, b)\times H^{2N-a}_M(Y-B, A-A\cap B, N-b)\to\mathbb{Z}. \tag{3.38}$$

This cup product is compatible with the cup product in ℓ-adic or Betti cohomology. We apply this to

$$Y = X \times X, \ A = D \times X, \ B = X \times D \tag{3.39}$$

so we can cup $[\Delta_U] \in H^{2d}_M(Y - A, B - B \cap A, d)$ and $[\Gamma] \in H^{2d}_M(Y - B, A - A \cap B, d)$

$$[\Delta_U] \cup [\Gamma] \in \mathbb{Z}. \tag{3.40}$$

The theorem is then the consequence of the following proposition.

Proposition 3.3.6.

$$\mathrm{Tr}(\Gamma_*) = [\Delta_U] \cup [\Gamma] \in \mathbb{Z}.$$

Proof By the compatibility of the cup product (3.38) with cohomology, we just have to prove the proposition with $[\Delta_U]$ and $[\Gamma]$ replaced by their classes $\mathrm{cl}(\Delta_U) \in H^{2d}(U \times X, U \times D, d)$ and $\mathrm{cl}(\Gamma) \in H^{2d}(X \times U, D \times U, d)$ in cohomology. We may assume that $\Gamma \subset X \times U$ is irreducible. On the other hand, by (3.33), the map

$$H^i_c(\Gamma) \xrightarrow{p_{1*}} H^i_c(U) \tag{3.41}$$

factors through

$$H^i_c(\Gamma) \xrightarrow{\text{Gysin}} H^{i+2d}_{\Gamma,c}(X \times U, d) = H^{i+2d}_{\Gamma,c}(U \times U, d) \tag{3.42}$$
$$\to H^{i+2d}_c(U \times U, d) \to H^{i+2d}_c(U \times X, d) \xrightarrow{p_{1*}} H^i_c(U)$$

so the correspondence

$$\Gamma_* : H^i_c(U) \xrightarrow{p_2^*} H^i_c(\Gamma) \xrightarrow{p_{1*}} H^i_c(U) \tag{3.43}$$

is just defined on $\alpha \in H^i_c(U)$ as follows. As the cup product from $H^i_c(U)$ with $H^j(U)$ is well defined with values in $H^{i+j}_c(U)$, so a fortiori in $H^{i+j}(X)$, and $\mathrm{cl}(\Gamma) \in H^{2d}(X \times U, D \times U, d) = \oplus_{i=0}^{2d} H^i(X, D) \otimes H^{2d-i}(U)(d)$, one has a well defined cup-product

$$p_2^*(\alpha) \cup \mathrm{cl}(\Gamma) \in H^{i+2d}(X \times X, D \times X, d) = H^{i+2d}_c(U \times X, d). \tag{3.44}$$

Then one applies p_{1*} to the value of this cup-product. Now we argue as in the classical case. Let e^i_a be a basis of $H^i_c(U)$, and $(e^i_a)^\vee$ be its dual basis in $H^{2d-i}(U)(d)$. Write $\mathrm{cl}(\Gamma) = \sum_i \sum_a f^i_a \otimes (e^i_a)^\vee, f^i_a = \sum f^i_{ab} e^i_b \in H^i_c(U)$. So $\Gamma_*(e^i_a) = \sum_b f^i_{ab}$, and $\mathrm{Tr}(\Gamma_*) = \sum(-1)^i \sum_a \sum_b f^i_{ab}$. On the other hand, one has $\mathrm{cl}(\Delta_U) = \sum_i \sum_a (e^i_a)^\vee \otimes (e^i_a)$. Thus $\mathrm{cl}(\Delta_U \cup \Gamma) = \sum(-1)^i \sum_a \sum_b f^i_{ab}$. □

The proposition finishes the proof of the theorem. □

The rest of the section is devoted to giving a concrete expression for (3.34) under stronger geometric assumptions on Γ. We define

$$X^o = X - \bigcup_{i<j}(D_i \cap D_j), \tag{3.45}$$

$$\Gamma^o = \bar{\Gamma} \cap (X^o \times X^o),\ (\Gamma')^o = \Gamma' \cap (X^o \times X^o),\ D^o = X^o \cap D$$

and similarly for the components $_j$. One has the following compatibility.

Lemma 3.3.7. *One has a commutative diagram*

$$\begin{array}{ccc} H^i_c(U) & \to & H^i(X^o) \\ {\scriptstyle p_2^*}\downarrow & & {\scriptstyle p_2^*}\downarrow \\ H^i_c(\Gamma_j) & \to & H^i(\Gamma^o_j) \\ {\scriptstyle \iota_*}\downarrow & & {\scriptstyle \iota^*}\downarrow \\ H^i_c((\Gamma_j)') & \to & H^i((\Gamma')^o_j) \\ {\scriptstyle p_{1*}}\downarrow & & {\scriptstyle p_{1*}}\downarrow \\ H^i_c(U) & \to & H^i(X^o) \end{array}$$

where the composition of the left vertical arrows is the correspondence $(\Gamma_j)_*$. We deem $(\Gamma^o_j)_*$ the composition of the right vertical arrows.

Proof Given (3.30) the lemma follows directly from Definition (3.27). □

We remark that the Grothendieck-Lefschetz trace formula allows to compute the trace of the correspondence $\bar{\Gamma}_*$ on X

$$\mathrm{Tr}(\bar{\Gamma}_*) = \deg(\bar{\Gamma} \cdot \Delta_X). \tag{3.46}$$

Thus, in the corollary, we would like to complete the commutative diagram in an exact sequence of commutative diagrams, so that we can apply the trace formula on all the terms but the one we seek. In the sequel, we give a strong geometric condition under which it is possible.

Definition 3.3.8. We assume that D is a strict normal crossings divisor. The dim d cycle $\Gamma \subset U \times U$ is said to be in good position with respect to $D \times X$ if the following two conditions are fulfilled.

i) Each $\bar{\Gamma}_j$ cuts each stratum $D_I \times X$ in codim $\geq d$, where $D_I = D_{i_1} \cap \ldots \cap D_{i_r}$ for $I = \{i_1, \ldots, i_r\}$ with $|I| = r$.

ii)

$$\bar{\Gamma}_j \cap (D_I \times X) \subset D_I \times D_I$$

set theoretically.

In this case, for all I we define the cycles

$$Z_{jI} = \bar{\Gamma}_j \cdot (D_I \times X) \subset D_I \times D_I \tag{3.47}$$

Let us be more precise. We denote motivic cohomology by $H^a_M(b)$. We drop the subscript $_j$, thus $\Gamma = \Gamma_j$. This defines

$$Z_I = \sum m_{I,a} Z_{I,a} \tag{3.48}$$

as in (3.39), where the $Z_{I,a}$ are the reduced irreducible components of Z_I. One has the Gysin isomorphisms

$$\oplus_a \mathbb{Q}_\ell[Z_{I,a}] \xrightarrow{\cong} H^{2(d-r)}_{Z_I}(D_I \times D_I, d-r) \xrightarrow{\cong} H^{2d}_{Z_I}(D_I \times X, d) \tag{3.49}$$

This yields the commutative diagram

$$\begin{array}{ccc}
\oplus_a \mathbb{Q} \cdot [Z_{I,a}] & \xrightarrow{\otimes_{\mathbb{Q}} \mathbb{Q}_\ell} & \oplus_a \mathbb{Q}_\ell \cdot [Z_{I,a}] \\
\cong \downarrow & & \downarrow \cong \\
H^{2(d-r)}_{M,Z_I}(D_I \times D_I, d-r) & \xrightarrow{\otimes_{\mathbb{Q}} \mathbb{Q}_\ell} & H^{2(d-r)}_{Z_I}(D_I \times D_I, d-r) \\
\cong \downarrow & & \downarrow \cong \\
H^{2d}_{M,Z_I}(D_I \times X, d) & \xrightarrow{\otimes_{\mathbb{Q}} \mathbb{Q}_\ell} & H^{2d}_{Z_I}(D_I \times X, d).
\end{array} \tag{3.50}$$

So we conclude that Z_I is a well defined cycle

$$Z_I \in H^{2(d-r)}_{M,Z_I}(D_I \times D_I, d-r) \to H^{2(d-r)}_M(D_I \times D_I, d-r), \tag{3.51}$$

which defines a correspondence

$$(Z_I^o)_* = \sum_a m_{I,a}(Z_{I,a})_* : H^i_c(D_I - \bigcup_{i \notin I} D_{I,i}) \to H^i_c(D_I - \bigcup_{i \notin I} D_{I,i}) \tag{3.52}$$

$$Z_I^o = Z_I \cap (D_I - \bigcup_{i \notin I} D_{I,i} \times D_I - \bigcup_{i \notin I} D_{I,i})$$

by the Definition 3.3.2. We use the notations from (3.45), setting $\Gamma = \Gamma_j$, together with

$$Z = \bigcup_i \Gamma \cap (D_i^o \times D_i^o). \tag{3.53}$$

Lemma 3.3.9. *One has a commutative diagram*

$$\begin{array}{ccc} H_c^{i-1}(D^o) & \to & H_c^i(U) \\ \downarrow{=} & & \downarrow{p_2^*} \\ H_c^{i-1}(D^o) & \to & H_c^i(\Gamma) \\ \downarrow{p_2^*} & & \downarrow{\iota^*} \\ H_c^{i-1}(Z) & \to & H_c^i(\Gamma') \\ \downarrow{(p_1)_*} & & \downarrow{(p_1)_*} \\ H_c^{i-1}(D^o) & \to & H_c^i(U) \end{array}$$

Proof Indeed, by assumption, one has $\Gamma' \subset X^o \times U$, thus, with the notations of (3.45), one has $\Gamma' = \Gamma^o - Z$. □

In order to have a unified notation, we set for $|I| = 0$

$$Z_I = \bar{\Gamma},\ \Delta_I = \Delta_X. \tag{3.54}$$

One has

Theorem 3.3.10. *Let X be a smooth projective variety over an algebraically closed field k, $D \subset X$ be a strict normal crossing divisor, with complement $U = X - D$, and $\Gamma \subset U \times U$ be a dim d correspondence, with $p_2|_\Gamma : \Gamma \to U$ proper, and in good position with respect to $D \times X$ in the sense of Definition 3.3.2. Then one has*

$$\mathrm{Tr}(\Gamma_*) = \sum_{r=0}^{d} (-1)^r \sum_{|I|=r} \deg(Z_I \cdot \Delta_I).$$

Proof We consider the long exact sequence

$$\ldots \to H_c^{i-1}(D^o) \to H_c^i(U) \to H_c^i(X^o) \to \ldots \tag{3.55}$$

and apply Lemma 3.3.7 and Lemma 3.3.9. We find that the trace on U is the sum of the traces on X^o and on D^o. If D was smooth, this would finish the proof applying Grothendieck's formula (3.46). If there are higher codimensional strata, we argue as follows. We know the trace on D_i^o by induction on the dimension. We have to understand the trace on X^o. We define $X^{oo} = X - \cup_{i<j<k}(D_i \cap D_j \cap D_k)$ and redo Lemma 3.3.7 with U replaced by X^o and X^o replaced by X^{oo}. Then we redo Lemma 3.3.9 with U replaced by X^o and $D^o = X^o - U$ replaced by $X^{oo} - X^o$. Then we redo (3.55) with D^o replaced

by $X^{oo} - X^o$, U replaced by X^o and X^o replaced by X^{oo}. Again this shifts the trace computation to X^{oo}. We continue like this till the highest codimensional strata and finish with the trace on X, for which we of course apply Grothendieck formula (3.46). □

We now give a scheme-theoretic condition under which the expression given in Theorem 3.3.10 depends only on local contributions in U. This condition is inspired by [9], Lemma 2.3.1.

Definition 3.3.11. We assume that D is a strict normal crossings divisor. The dim d cycle $\Gamma \subset U \times U$ is said to be in scheme theoretic good position with respect to $D \times X$ if it is in good position in the sense of Definition 3.3.8 and ii) is replaced by

$$\bar{\Gamma}_j \cap (D_I \times X) \subset D_I \times D_I$$

scheme theoretically, that is all the intersections multiplicities are equal to 1.

Proposition 3.3.12. *Let X be a smooth projective variety over an algebraically closed field k, $D \subset X$ be a strict normal crossing divisor, with complement $U = X - D$, and $\Gamma \subset U \times U$ be a dim d correspondence, with $p_2|_\Gamma : \Gamma \to U$ proper, and in scheme theoretic good position with respect to $D \times X$ in the sense of Definition 3.3.11. We assume moreover that $\bar{\Gamma}_j$ and $\Delta|_U$ cut transversally. Then* $\mathrm{Tr}(\Gamma_*) = \deg(\Delta_U \cdot \Gamma)$.

Proof Due to the good position assumption, all intersection multiplicities are 1 and the contributions lying on $(D \times X) \cup (X \times D)$ cancel in Theorem 3.3.10. □

Example 3.3.13. One case where the conditions of Proposition 3.3.12 hold is in characteristic p when Γ is the graph of Frobenius. In this case, of course, the result is known by other methods.

Example 3.3.14. This example is inspired by [9, Remark 2.3.6]. We take $X = \mathbb{P}^1, D = \{\infty\}, U = \mathbb{A}^1, \Gamma = \Gamma_{pq} = \{x^p - y^q = 0\} \subset \mathbb{A}^1 \times \mathbb{A}^1$. Then Γ defines an open correspondence with $\mathrm{Tr}(\Gamma_*) = p$. On the other hand, one has

$$\deg(\Gamma_{pq} \cdot \Delta_U) = \dim k[t]/(t^p - t^q) = \begin{cases} \max(p,q) & \text{if } p \neq q \\ \infty & \text{if } p = q \end{cases}$$

and one has

$$\deg(\bar{\Gamma}_{pq} \cdot (\infty \times \mathbb{P}^1)) = q, \ \deg(\bar{\Gamma}_{pq} \cdot (\mathbb{P}^1 \times \infty)) = p. \tag{3.56}$$

Thus Γ_{pq} is in scheme theoretic good position with respect to $\infty \times \mathbb{P}^1$ if and only if $p > q$, and is always in good position with respect to $\mathbb{P}^1 \times \infty$. Since $\deg(\Delta_{\mathbb{P}^1} \cdot \bar{\Gamma}_{pq}) = \deg(\mathcal{O}(1,1) \cdot \mathcal{O}(p,q)) = p+q$, we see exactly how the formula of Theorem 3.3.10 works both in Theorem 3.3.10 and in Proposition 3.3.12.

References

[1] Beilinson, A.: Higher regulators and values of L-functions, J. Soviet Math. **30** (1985), 2036–2070.

[2] Bloch, S.: *Lecture on Algebraic cycles*, Duke University Mathematics Series, IV, (1980).

[3] Bloch, S., H. Esnault and M. Levine: Decomposition of the diagonal and eigenvalues of Frobenius for Fano hypersurfaces, Am. J. of Mathematics, **127** no1 (2005), 193-207.

[4] Deligne, P.: La formule de dualité globale, SGA 4, XVIII, Lecture Notes in Mathematics **305**, Springer Verlag.

[5] Deligne, P.: Rapport sur la formule des traces, SGA 4,5, Lecture Notes in Mathematics **569**, Springer Verlag.

[6] Esnault, H.: Eigenvalues of Frobenius acting on the ℓ-adic cohomology of complete intersections of low degree, C. R. Acad. Sci. Paris, Ser. I **337** (2003), 317–320.

[7] Fujiwara, K.: Rigid geometry, Lefschetz-Verdier trace formula and Deligne's conjecture.

[8] Jannsen, U.: Motivic Sheaves and Filtrations on Chow Groups, in *Motives*, Proc. Symp. Pure Math., Vol. **55** (1994), 245–302.

[9] Kato, K., Saito, T.: Ramification theory for varieties over a perfect field, preprint 59 pages, 2004.

[10] Kleiman, S: Algebraic Cycles and the Weil Conjectures, in *Dix Exposés sur la Cohomologie des Schémas*, North-Holland, Amsterdam, 1968, 359–386.

[11] Kleiman, S: The Standard Conjectures, in *Motives*, Proc. Symp. Pure Math., Vol. **55** (1994), 13–20.

[12] Levine, M.: *Mixed Motives*, Mathematical Surveys and Monographs **57** (1998), American Mathematical Society.

[13] Murre, J. P.: On a conjectural filtration on the Chow groups of an algebraic variety, (I and II), Indag. Math. New Series Vol. **4** (1993), 177–201.

[14] Pink, R.: On the calculation of local terms in the Lefschetz-Verdier trace formula and its application to a conjecture of Deligne, Ann. of Math. **135** (1992), 483–525.

[15] Zink, T.: The Lefschetz trace formula for open surfaces, In: *Automorphic Forms, Shimura Varieties and L-functions, vol. II*, Prospectives in Math. (1989), 337–376.

4

The Brill-Noether Curve of a stable Vector Bundle on a Genus Two Curve

Sonia Brivio

Dipartimento di Matematica, Universitá di Pavia,
via Ferrata 1, 27100 Pavia,
brivio@dimat.unipv.it

Alessandro Verra†

Dipartimento di Matematica, Terza Universitá di Roma,
Largo S.Murialdo, 00146 Roma,
verra@mat.uniroma3.it

4.1 Introduction

In this note we deal with the moduli space $\mathcal{U}_r$ of semistable vector bundle of rank r and degree $r(g-1)$ over a smooth, irreducible complex projective curve of genus $g \geq 2$. $\mathcal{U}_r$ contains the Brill-Noether locus

$$\Theta_r := \{[E] \in \mathcal{U}_r \mid h^0(E) \geq 1\}$$

which is an integral Cartier divisor and it is known as the generalized theta divisor of $\mathcal{U}_r$, see [6], [5]. Moreover the tensor product defines a morphism

$$f : \mathcal{U}_r \times \mathrm{Pic}^0(C) \to \mathcal{U}_r$$

and we can consider the pull-back $f^*\Theta_r$ of Θ_r. Let $[E] \in \mathcal{U}_r$ be the moduli point of the vector bundle E and let $\det E \cong M^{\otimes r}$. It is well known that then

$$\mathcal{O}_{[E]\times \mathrm{Pic}^0(C)}(f^*\Theta_r) \cong \mathcal{O}_{\mathrm{Pic}^0(C)}(r\Theta_M),$$

where

$$\Theta_M := \{N \in \mathrm{Pic}^0(C) \mid h^0(M \otimes N) \geq 1\}.$$

Note that M is a line bundle of degree $g-1$ and that Θ_M is a theta divisor on $\mathrm{Pic}^0(C)$. We define

$$\Theta_E := f^*\Theta_r \cdot [E] \times \mathrm{Pic}^0(C)$$

† Partially supported by the research programs *'Moduli Spaces and Lie Theory'* and *'Geometry on Algebraic Varieties'* of the Italian Ministry of Education.

‡ 2000 Mathematics Subject Classification: 14H60

if the intersection is proper. In this case we will say that Θ_E *is the theta divisor of* E. The construction of Θ_E allows us to define a rational map as follows. Consider

$$\mathcal{T}_r := \bigcup_{M \in \mathrm{Pic}^{g-1}(C)} |r\Theta_M|.$$

It is a standard fact that $\mathcal{T}_r$ has a natural structure of projective bundle over $\mathrm{Pic}^{r(g-1)}(C)$. So we omit its construction; we only mention that the corresponding projection

$$p : \mathcal{T}_r \to \mathrm{Pic}^{r(g-1)}(C)$$

is defined as follows: $p(D) = M^{\otimes r}$ if and only if $D \in |r\Theta_M|$.

Note that the only elements of multiplicity r in $|r\Theta_M|$ are exactly the divisors $r\Theta_{M\otimes\eta}$, where η varies in the set of the elements of order r in $\mathrm{Pic}^0(C)$. Therefore the map p is well defined. In the following we will study the rational map

$$\theta_r : \mathcal{U}_r \to \mathcal{T}_r$$

which associates to a general $[E] \in \mathcal{U}_r$ the corresponding theta divisor $\Theta_E \in \mathcal{T}_r$. Let

$$\det : \mathcal{U}_r \to \mathrm{Pic}^{r(g-1)}(C)$$

be the determinant map. It is well known that $\mathcal{T}_r$ is the projectivization of $\det_* \mathcal{O}_{\mathcal{U}_r}(\Theta_r)^*$ and that θ_r is the induced tautological map. In particular it follows that $p \cdot \theta_r = \det$. We will say that θ_r is *the theta map*.

Too many questions are still unsettled about the theta map, except if $r \leq 2$: see e.g. [4] for a general survey. This situation is probably related to the fact that the following basic question is still largely unsolved.

Question. Is θ_r generically finite onto its image?

Actually the main difficulty here is that in most of the cases θ_r is not a morphism [9]. Thus, in spite of the ampleness of Θ_r, it is not a priori granted that θ_r is generically finite onto its image.

In this paper we give a natural geometric interpretation of the fibres of the map θ_r for a curve C of genus two. A very special feature in this case is that

$$\dim\ \mathcal{U}_r = \dim \mathcal{T}_r = r^2 + 1,$$

so the generic finiteness of θ_r is even more expected. Applying our description of the fibres we prove the generic finiteness of θ_r.

Such a result is not new: Beauville recently proved it using a different, relatively simple method, see [3]. We believe that our description has some interest in itself and we hope to use it for further applications, in particular to compute the degree of θ_r.

Our approach relies on Brill-Noether theory for curves contained in a genus two Jacobian. Let $D \in \theta_r(\mathcal{U}_r)$ be a sufficiently general element, then D is a smooth curve of genus r^2+1 in $\mathrm{Pic}^0(C)$ (§ 4.2). Consider the Brill-Noether locus

$$W^{r-1}_{r^2}(D) \;=\; \{L \in \mathrm{Pic}^{r^2}(D) \mid h^0(L) \geq r\}$$

and observe that its expected dimension is one, i.e. the Brill-Noether number $\rho(r-1, r^2, r^2+1)$ is one. Our main result can be summarized as follows:

Theorem. *Each point* $[E] \in \theta_r^{-1}(D)$ *defines an irreducible component*

$$C_E \subset W^{r-1}_{r^2}(D)$$

biregular to C*. Let* Z *be the set of all irreducible components of* $W^{r-1}_{r^2}(D)$ *and let*

$$i_D : \theta_r^{-1}(D) \to Z$$

be the map sending $[E]$ *to* C_E*. Then* i_D *is injective.*

The statement clearly implies that θ_r is generically finite. We define C_E as the *Brill-Noether curve of* E. Fixing a Poincaré bundle $\mathcal{P}$ on $D \times \mathrm{Pic}^{r^2}(D)$ in an appropriate way, it turns out that E is the restriction of $\nu_*\mathcal{P}$ to C_E, where ν is the projection onto $\mathrm{Pic}^{r^2}(D)$. In particular the family of the fibres of E is just the family of the spaces $H^0(L)$, $L \in C_E$. Note that the choice of $\mathcal{P}$, hence of $\det E$, depends on the embedding $D \subset \mathrm{Pic}^0(C)$. This is essentially explained in the final part of this note.

To have a typical example of what happens, the reader can consider the case $r = 2$. In this case D is a curve of genus 5 endowed with a fixed point free involution which is induced by the -1 multiplication of $\mathrm{Pic}^0(C)$. Since $r = 2$, the Brill-Noether locus $W^1_4(D)$ is exactly the singular locus of the theta divisor of $\mathrm{Pic}^4(D)$. It follows from the theory of Prym varieties that $W^1_4(D)$ is the union of two irreducible curves: one of them has genus 4, the other one is just a copy of C (see also [10]). This is the Brill-Noether curve of a stable rank two vector bundle E such that $\theta_2([E]) = D$. In higher rank the general theory of Prym-Tjurin varieties can certainly provide further information on $W^{r-1}_{r^2}(D)$ and hence on the fibres of θ_r. However, in order to reconstruct them, a very explicit description is needed for the Prym-Tjurin realizations of a genus two Jacobian.

On Jacobians of higher genus several extensions of the above constructions are possible and perhaps deserve to be considered in the study of the theta maps. We hope to have underlined with this note the manifold links between moduli of vector bundles on a curve C, Prym-Tjurin realizations of its Jacobian JC and Brill-Noether theory for curves in JC.

We wish to thank the referee for some helpful comments.
The second author wishes to express his gratitude to Jacob Murre, on the occasion of the Proceedings of the Conference in his honour, for the friendship and for the scientific attention received during three decades.

4.2 Notation and preliminary results

From now on C is a smooth, irreducible, complex projective curve of genus 2. Let $C^{(2)}$ be the 2-symmetric product of C. A point of such a surface is a divisor $x + y$ with $x, y \in C$. We consider the map

$$a : C^{(2)} \to \mathrm{Pic}^0(C)$$

sending $x + y \in C^{(2)}$ to $\omega_C(-x - y)$. Of course a is the composition of the Abel map defined by ω_C with -1 multiplication on $\mathrm{Pic}^0(C)$. Therefore $a = -\sigma$, where $\sigma : C^{(2)} \to \mathrm{Pic}^0(C)$ is the blowing up of the zero point. For each fibre $|r\Theta_M|$ of the projective bundle $\mathcal{T}_r$ we have the linear isomorphism

$$a_M^* : |r\Theta_M| \to |a^* r\Theta_M|$$

defined by the pull-back. With Θ_E the theta divisor of $[E]$, we will keep the following notation

$$D_E := a^*\Theta_E.$$

D_E is an effective divisor in $C^{(2)}$ which is supported on the set

$$\{x + y \in C^{(2)} \mid h^0(E \otimes \omega_C(-x - y)) \geq 1\}.$$

D_E is biregular to Θ_E if the zero point is not in Θ_E, otherwise D_E is the union of the projective line $|\omega_C|$ and of a curve birational to Θ_E. We find it more convenient to deal with D_E than with Θ_E.

First we want to prove that a general D_E is smooth. To do so, we need Laszlo's singularity theorem.

Theorem 4.2.1 ([8]). *The multiplicity of Θ_r at its stable point $[E]$ is $h^0(E)$.*

Next, we show:

Proposition 4.2.2. *Let $[E]$ be a general stable point of $\mathcal{U}_r$ then*

$$h^0(E \otimes \omega_C(-x - y)) \leq 1, \quad \forall x + y \in C^{(2)}.$$

Proof By induction on r. Let $r = 1$ then $D_E = C$ and E is a general line bundle of degree 1, in particular $|E \otimes \omega_C|$ is a base-point-free pencil and this implies the statement. Let $r \geq 2$. We can assume by induction that there exist general $[B] \in \mathcal{U}_{r-1}$ and $[A] \in \mathcal{U}_1 = \mathrm{Pic}^1(C)$ satisfying the statement. We consider the exact sequence

$$0 \to B \to E \to A \to 0$$

defined by the vector $e \in \mathrm{Ext}^1(A, B)$. Tensoring such a sequence by $\omega_C(-x-y)$ and passing to the long exact sequence we obtain the coboundary map

$$e_{x+y} : H^0(\omega_C \otimes A(-x-y)) \to H^1(\omega_C \otimes B(-x-y)).$$

Claim The statement holds for E if and only if e_{x+y} has maximal rank for every $x + y$.

Proof We have $h^0(\omega_C \otimes A(-x-y)) \leq 1$ and

$$h^0(\omega_C \otimes B(-x-y)) = h^1(\omega_C \otimes B(-x-y)) \leq 1.$$

Then the statement follows from the above mentioned long exact sequence. Finally it is obvious that e_{x+y} has maximal rank except possibly for points $x+y$ with $h^0(\omega_C \otimes A(-x-y)) = h^0(\omega_C \otimes B(-x-y)) = 1$. The set of these points is $D_A \cap D_B$. Since A is general we can assume that $D_A \cap D_B$ is finite. Let $x+y \in D_A \cap D_B$ then e_{x+y} has not maximal rank if and only if it is the zero map. It is a standard property that, in the present case, the locus

$$H_{x+y} = \{e \in \mathrm{Ext}^1(A, B) \mid e_{x+y} \text{ is the zero map}\}$$

is a hyperplane. Let $H := \bigcup H_{x+y}$, $x + y \in D_A \cap D_B$, then a general $e \in \mathrm{Ext}^1(A, B) - H$ defines a semistable E satisfying the condition of the statement. Since this condition is open on $\mathcal{U}_r$ the result follows. □

Corollary 4.2.3. *Let $[E]$ be a general stable point of $\mathcal{U}_r$, then D_E is smooth.*

Proof Let $f : \mathcal{U}_r \times \mathrm{Pic}^0(C) \to \mathcal{U}_r$ be the map defined via tensor product. Recall that

$$\Theta_E = f^*\Theta_r \cdot [E] \times \mathrm{Pic}^0(C) \subset \mathcal{U}_r \times \mathrm{Pic}^0(C).$$

Therefore Θ_E is the fibre of the projection $q : f^*\Theta_r \to \mathcal{U}_r$. Then, by generic smoothness, a general Θ_E is smooth if $\Theta_E \cap \mathrm{Sing}\, f^*\Theta_r = \varnothing$. On the other hand f is smooth, with fibres biregular to $\mathrm{Pic}^0(C)$. The smoothness of f implies that $\mathrm{Sing}\, f^*\Theta_r = f^*\,\mathrm{Sing}\,\Theta_r$. Therefore, by Laszlo's singularity theorem (Theorem 4.2.1) and the definition of f, we have

$$\mathrm{Sing}\, f^*\Theta_r = \{([E], \xi) \in \mathcal{U}_r \times \mathrm{Pic}^0(C) \mid h^0(E(\xi)) \geq 2\}.$$

But the previous proposition implies that $h^0(E(\xi)) \leq 1$, for all $\xi \in \mathrm{Pic}^0(C)$. Then it follows that $\Theta_E \cap \mathrm{Sing}\, f^*\Theta_r = \varnothing$ and hence a general Θ_E is smooth. The same holds for D_E. □

4.3 The tautological model P_E

Now we want to see that the above curve D_E appears as the singular locus of some natural tautological model of $\mathbb{P}E^*$ in $\mathbb{P}^{2r-1}$.

Proposition 4.3.1. *Let E be any semistable bundle of degree r then*

1) $h^1(\omega_C \otimes E) = 0$ *and* $h^0(\omega_C \otimes E) = 2r$.
2) $\omega_C \otimes E$ *is globally generated unless E is not stable and* $\mathrm{Hom}(E, \mathcal{O}_C(x))$ *is non zero for some point $x \in C$.*

Proof **1)** By Serre duality $h^1(\omega_C \otimes E) = h^0(E^*)$. Since E^* is semistable of slope -1 it follows $h^0(E^*) = 0$. Then we have $h^0(\omega_C \otimes E) = 2r$ by Riemann-Roch.
2) By 1) E is globally generated if and only if $h^0(\omega_C \otimes E(-x)) = r$, $\forall\, x \in C$. By Serre duality this is equivalent to $\mathrm{Hom}(E, \mathcal{O}_C(x)) = 0$, $\forall\, x \in C$. Note also that $\mathrm{Hom}(E, \mathcal{O}_C(x)) \neq 0$ implies that E is not stable. This completes the proof. □

In this section we assume that $[E] \in \mathcal{U}_r$ has the following properties (satisfied by a general $[E]$):

- $\omega_C \otimes E$ is globally generated,
- D_E exists i.e. $[E]$ is not in the indeterminacy locus of $\theta_r : \mathcal{U}_r \to \mathcal{T}_r$,
- D_E is smooth,

To simplify notation we put

$$F := \omega_C \otimes E \quad \text{and} \quad \mathbb{P}_E := \mathbb{P}F^*.$$

Lemma 4.3.2. *Let F be general and let $\overline{F}$ be defined by the standard exact sequence*

$$0 \to F^* \to H^0(F)^* \otimes \mathcal{O}_C \to \overline{F} \to 0$$

induced by the evaluation map. Then $\overline{F}$ is stable. In particular the map

$$j : \mathcal{U}_r \to \mathcal{U}_r$$

sending $[\omega_C^{-1} \otimes F]$ to $[\omega_C^{-1} \otimes \overline{F}]$ is a birational involution.

Proof First of all we claim that $\overline{F}$ is semistable for F general enough. Let $F_o = L^r$, where $L \in \mathrm{Pic}^3(C)$ is globally generated, then $h^0(F_o) = 2r$ and $\overline{F}_o = F_o$. Up to a base change there exist an integral variety T and a vector bundle $\mathcal{F}$ over $T \times C$ such that the family of vector bundles $\{F_t := \mathcal{F} \otimes \mathcal{O}_{t\times C},\ t \in T\}$ dominates $\mathcal{U}_r$ and contains F_o. By semicontinuity we can assume, up to replacing T by a non empty open subset, that $h^0(F_t) = h^0(F_o) = 2r$ and F_t is globally generated. So it is standard to construct from $\mathcal{F}$ a vector bundle $\overline{\mathcal{F}}$ on $T \times C$ with the following property: $\overline{\mathcal{F}} \otimes \mathcal{O}_{t\times C} = \overline{F}_t$, for each $t \in T$. Since $\overline{\mathcal{F}} \otimes \mathcal{O}_{o\times C} = L^r$ is semistable, the same holds for a general vector bundle $\overline{\mathcal{F}} \otimes \mathcal{O}_{t\times C}$. Hence the claim follows. Let F be a general stable bundle: $\overline{F}$ is semistable, by lemma 4.3.1 $h^0(\overline{F}) = 2r$, moreover since $h^0(F^*) = 0$, we have $H^0(F)^* \simeq H^0(\overline{F})$ and $\overline{F}$ is globally generated. So j is defined at $\overline{F}$, actually $j(\overline{F}) = F$. This implies that j is a birational involution and $\overline{F}$ is stable too. □

Since F is globally generated the map defined by $\mathcal{O}_{\mathbb{P}_E}(1)$ is a morphism

$$u_E : \mathbb{P}_E \to \mathbb{P}^{2r-1} := \mathbb{P}H^0(F)^*.$$

In particular the restriction of u_E to any fibre $\mathbb{P}_{E,x}$ of $\mathbb{P}_E$ is a linear embedding

$$u_{E,x} : \mathbb{P}_{E,x} \to \mathbb{P}^{2r-1}.$$

Definition 4.3.3. The image of u_E, (of $u_{E,x}$), will be denoted P_E,$(P_{E,x})$.

For any $d \in C^{(2)}$, $F_d := F \otimes \mathcal{O}_d$ can be naturally seen as a rank r vector bundle over d. Note that its projectivization is p^*d, where $p : \mathbb{P}_E \to C$ is the projection map. In particular the evaluation map $e_d : H^0(F_d) \otimes \mathcal{O}_d \to F_d$ defines an embedding

$$i_d : p^*d \to \mathbb{P}H^0(F_d)^*.$$

We have $\mathbb{P}H^0(F_d)^* = \mathbb{P}^{2r-1}$ and moreover $i_d(p^*d)$ is the union of two disjoint linear spaces of dimension $r-1$ if d is smooth. The next lemma is therefore elementary.

Lemma 4.3.4. *Let $o \in \mathbb{P}H^0(F_d)^*$ be a point not in $i_d(p^*d)$, then there exists exactly one line L containing o and such that $Z := i_d^*L$ is a 0-dimensional scheme of length two. Moreover with λ the linear projection from o, the scheme Z is the unique 0-dimensional scheme of length two on which $\lambda \cdot i_d$ is not an embedding.*

The central arrow in the long exact sequence

$$0 \to H^0(F(-d)) \to H^0(F) \to H^0(F_d) \to H^1(F(-d)) \to 0 \tag{4.1}$$

defines a linear map

$$\lambda_d : \mathbb{P}H^0(F_d)^* \to \mathbb{P}H^0(F)^* = \mathbb{P}^{2r-1},$$

and from the construction clearly $\lambda_d \circ i_d$ is the map

$$u_E|_{p^*d} : p^*d \to \mathbb{P}^{2r-1}.$$

Proposition 4.3.5. *$u_E|_{p^*d}$ is not an embedding if and only if $d \in D_E$.*

Proof From the previous remarks and lemma 4.3.4 it follows that $u_E|_{p^*d}$ is not an embedding if and only if λ_d is not an isomorphism. By the long exact sequence (4.1) λ_d is not an isomorphism if and only if $h^0(E(-d)) \geq 1$, that is, if $d \in D_E$. □

We want to use the previous results to study the singular locus of P_E. Let $\mathrm{Hilb}_2(\mathbb{P}_E)$ be the Hilbert scheme of 0-dimensional schemes $Z \subset \mathbb{P}_E$ of length two. Consider

$$\Delta = \{Z \in \mathrm{Hilb}_2(\mathbb{P}_E) \mid u_E|_Z \text{ is not an embedding}\}.$$

Let $Z \in \Delta$ then $Z \subset p^*d$, where $d := p_*Z$ belongs to D_E. So we have a morphism

$$p_* : \Delta \to D_E$$

sending Z to d.

Proposition 4.3.6. *Let E be general then $p_* : \Delta \to D_E$ is biregular.*

Proof Let $Z \in \Delta$ and let $p_*Z = d$, then Z is embedded in p^*d. Since $d \in D_E$ we have $h^0(F(-d)) = 1$, see prop. 4.2.2. This implies that the linear map

$$\lambda_d : \mathbb{P}H^0(F_d)^* \to \mathbb{P}^{2r-1}$$

is the projection from a point onto a hyperplane in $\mathbb{P}^{2r-1}$. Then, by lemma 4.3.4, Z is the unique element of Δ which is contained in p^*d. Hence p_* is injective. Conversely let $d \in D_E$, then $\lambda_d \circ i_d$ is not an embedding on exactly one 0-dimensional scheme $Z \subset p^*d$ of length two. Since $\lambda_d \circ i_d = u_E|_{p^*d}$, it follows that Z is in Δ and that p_* is surjective. Since D_E is a smooth curve, p_* is biregular. □

Proposition 4.3.7. *Assume $r \geq 2$ and E general, then $u_E : \mathbb{P}_E \to P_E$ is the normalization map and* Sing *P_E is an irreducible curve.*

Proof Let $\tilde{D} \subset \mathbb{P}_E$ be the image of the curve

$$\tilde{\Delta} = \{(Z, q) \in \Delta \times \mathbb{P}_E \mid q \in \mathrm{Supp}\ Z\}$$

under the projection $\Delta \times \mathbb{P}_E \to \mathbb{P}_E$. The set $\tilde{D}$ is the locus of points where u_E is not an embedding: it is a proper closed set as soon as $r \geq 2$. Hence $u_E : \mathbb{P}_E \to P_E$ is a morphism of degree one if $r \geq 2$. Since $\mathbb{P}_E$ is smooth, u_E is the normalization map if each of its fibres is finite. Assume u_E contracts an irreducible curve B to a point o. B cannot be in a fibre $\mathbb{P}_{E,p}$. Otherwise D_E would contain the curve $\{z+p, z \in C\}$ and would be reducible. Hence $o \in \cap P_{E,x}$, $x \in C$. Let $x+y \in C^{(2)}$ with $x \neq y$, then $P_{E,x} \cup P_{E,y}$ is contained in a hyperplane and hence $h^0(E \otimes \omega_C(-x-y)) \geq 1$. This implies $D_E = C^{(2)}$: a contradiction. It remains to show that Sing P_E is an irreducible curve: this is clear because $\operatorname{Sing} P_E = u_E(\tilde{D})$. □

4.4 The line bundle H_E

We will keep the generality assumptions and the notation of the previous section. Recall that $d \in D_E$ uniquely defines a 0-dimensional scheme $Z_d \subset p^*d$ of length two such that $u_E|Z_d$ is not an embedding, in particular $u_E(Z_d)$ is a point.

Definition 4.4.1. $h_E : D_E \to \mathbb{P}^{2r-1}$ is the morphism sending d to $u_E(Z_d)$, moreover

$$H_E := h_E^* \mathcal{O}_{\mathbb{P}^{2r-1}}(1).$$

Remarks 4.4.2. 1) Let $F := \omega_C \otimes E$ and let $q_1, q_2 : C \times C \to C$ be the projections. Note that $q_1^*F \oplus q_2^*F$ descends to a vector bundle $F^{(2)}$ on $C^{(2)}$ via the quotient map $C \times C \to C^{(2)}$. Moreover the evaluation $H^0(F) \to F_x \oplus F_y$ induces a natural map

$$e : H^0(F) \otimes \mathcal{O}_{C^{(2)}} \to F^{(2)}.$$

Then D_E is the degeneracy locus of e and H_E is its cokernel. This implies that the sheaf H_E can be defined for every curve D_E and that H_E is a line bundle if and only if $h^0(\omega_C \otimes E(-x-y)) = 1$ for each $x + y \in D_E$.

2) A very simple geometric definition of h_E can be given as follows: let $d = x + y \in D_E$ with $x \neq y$ then

$$h_E(d) = P_{E,x} \cap P_{E,y} = u_E(Z_d).$$

Proposition 4.4.3. *$h_E : D_E \to \mathbb{P}^{2r-1}$ is generically injective if E is general and $r \geq 2$.*

Proof Let $U = \{x + y \in D_E \mid x \neq y\}$, assume that $d_1, d_2 \in U$ and $h_E(d_1) = h_E(d_2) = o$: then $o \in \bigcap_{i=1\ldots4} P_{E,x_i}$, where $\Sigma\, x_i = d_1 + d_2$. We consider the standard exact sequence

$$0 \to F^* \to H^0(F)^* \otimes \mathcal{O}_C \to \overline{F} \to 0,$$

where $F = \omega_C \otimes E$. The long exact sequence identifies $H^0(F)^*$ with a subspace of $H^0(\overline{F})$. Hence o is a 1-dimensional space generated by some $s \in H^0(\overline{F})$. It is standard to verify that $o \in \bigcap_{i=1\dots4} P_{E_{x_i}}$ if and only if s is zero on $d_1 + d_2 - d$, where d is the gcd of d_1, d_2. By Lemma 4.3.4 $\overline{F}$ is stable, hence we must have deg $d \geq 2$, that is, $d_1 = d_2 = d$. □

Proposition 4.4.4. *H_E has degree $r^2 + 2r$.*

Proof Set theoretically we have $h_E(D_E) = \operatorname{Sing} P_E$, hence $\operatorname{Sing} P_E$ is an irreducible curve. Let $o = h_E(x+y)$ be general then $x \neq y$ and moreover $u_E^*(o)$ is supported on two closed points o' and o'': this follows because h_E is generically injective.

Claim: *The tangent map du_E is injective at o' and o''.*

Let $\widetilde{D(u_E)}$ be the double point scheme of u_E, defined as in [7, p. 166]. $\widetilde{D(u_E)}$ is contained in $\widetilde{\mathbb{P}_E \times \mathbb{P}_E}$, where $\pi : \widetilde{\mathbb{P}_E \times \mathbb{P}_E} \to \mathbb{P}_E \times \mathbb{P}_E$ is the blowing up of the diagonal Δ. A point of $\widetilde{D(u_E)}$ is either the inverse image by π of a pair (o', o'') in $\mathbb{P}_E \times \mathbb{P}_E - \Delta$ such that $u_E(o') = u_E(o'')$ or it is a point in $\pi^{-1}(\Delta)$ parametrizing a 1 dimensional space of tangent vectors to $\mathbb{P}_E$ on which du_E is zero. In the former case we have also $p(o') \neq p(o'')$ because u_E is injective on each fibre of p. On the other hand it is clear that, in our situation,

$$q^{-1}(D_E) = (p \times p) \cdot \pi(\widetilde{D(u_E)})$$

where $q : C \times C \to C^{(2)}$ is the quotient map. Thus du_E is not injective at most along fibres $\mathbb{P}_{E,z}$ such that $2z \in D_E$. D_E cannot be the diagonal of $C^{(2)}$ because $D_E^2 = 2r^2$. Hence D_E contains finitely many points $2z$ and we can choose the above point $o = h_E(x+y)$ so that $2x$ and $2y$ are not in D_E. This implies our claim.

Let T be the tangent space to P_E at o and let $T', T'' \subset T$ be the images of du_E at o', o''. Since du_E is injective at o', o'' and $u_E^{-1}(o) = \{o', o''\}$, it follows that $T' \cup T''$ spans T and that $T' \cap T''$ is the tangent space to $\operatorname{Sing} P_E$ at o. We have $\dim T' \cap T'' \geq 1$ because $\operatorname{Sing} P_E$ is a curve. On the other hand $P_{E,x} \cap P_{E,y} = o$ implies dim $T \geq 2r - 1$. Since dim $T' =$ dim $T'' = r$, we deduce that dim $T' \cap T'' = 1$. Hence, as a scheme defined by the Jacobian ideal of P_E, $\operatorname{Sing} P_E$ is reduced. Finally the degree of $\operatorname{Sing} P_E$ can be obtained via double point formula, see [7, 9.3], as follows:

$$V \cdot V - c_{r-1}(\mathcal{N}_{V|\mathbb{P}^{2r-2}}) = 2\deg(\operatorname{Sing} P_E)$$

where $V \subset \mathbb{P}^{2r-2}$ is a general hyperplane section of P_E, corresponding to a global section $\sigma \in H^0(\mathcal{O}_{\mathbb{P}P_E}(1)) \simeq H^0(\omega_C \otimes E)$, c_{r-1} denotes the $(r-1)$-st Chern class of the normal bundle $\mathcal{N}_{V|\mathbb{P}^{2r-2}}$. Note that $V = P_{E'}$, with E', the vector bundle of rank $r-1$ defined by the section σ as follows:

$$0 \to \mathcal{O}_C \to E \otimes \omega_C \to E' \otimes \omega_C \to 0,$$

and $\mathcal{O}_{\mathbb{P}_{E'}}(1) = \mathcal{O}_{\mathbb{P}_E}(1)_{|\mathbb{P}_{E'}}$. Let $H = [\mathcal{O}_{\mathbb{P}_{E'}}(1)]$ and f be the class of a fibre of $\mathbb{P}_{E'}$, then by computing the total Chern class of the normal bundle, we find

$$c_{r-1}(\mathcal{N}_{V|\mathbb{P}^{2r-2}}) = rH^{r-1} + fH^{r-2}[4r^2 - 4r] = 7r^2 - 4r.$$

Finally, we have $\deg \operatorname{Sing} P_E = \deg\ H_E = r^2 + 2r$. □

The genus of D_E is r^2+1 and H_E has degree r^2+2r, hence $h^0(H_E) \geq 2r$. In fact:

Proposition 4.4.5. *For a general E the line bundle H_E is non special, that is,*

$$h^0(H_E) = 2r.$$

Proof By induction on r. Let $r = 1$ then $A := \omega_C \otimes E$ is a general line bundle of degree 3, $\mathbb{P}_E = C$ and $u_E : \mathbb{P}_E \to \mathbb{P}^1$ is the triple covering defined by A. Moreover D_E is the family of divisors $x + y$ which are contained in a fibre of u_E. It is easy to see that D_E is a copy of C and that $h_E = u_E$. Then $H_E = A$ and hence $h^0(H_E) = 2$.

Let $r \geq 2$ and let $[E_{r-1}] \in \mathcal{U}_{r-1}$ and $E_1 \in \operatorname{Pic}^1(C)$ be general points satisfying the statement, then their corresponding curves D_{r-1} and D_1 are smooth and transversal. Taking a general semistable extension

$$0 \to E_{r-1} \to E \to E_1 \to 0 \tag{4.2}$$

we have $h^0(E \otimes \omega_C(-x-y)) \leq 1$ for any $x+y$, (see 4.2.2 and its proof). Observe also that $D_E = D_1 \cup D_{r-1}$ and that h_E is a morphism. The restrictions of h_E to D_1 and D_{r-1} can be described as follows:

a) Let $F := \omega_C \otimes E$ and let $A := \omega_C \otimes E_1$: tensoring (4.2) by ω_C and passing to the long exact sequence, we obtain a surjective map $H^0(F) \to H^0(A)$. Its dual is a linear embedding $i : \mathbb{P}H^0(A)^* \to \mathbb{P}^{2r-1}$. On the other hand we already know that h_{E_1} is the triple cover of $\mathbb{P}H^0(A)^*$ defined by A. It is easy to conclude that $h_E|_{D_1} = i \cdot h_{E_1}$. In particular $h_E(D_1)$ is a line ℓ in $\mathbb{P}^{2r-1}$ which is triple for $h_E(D_E)$.

b) Let $B := \omega_C \otimes E_{r-1}$: tensoring (4.2)) by ω_C and passing to the long exact sequence we get an injection $H^0(B) \to H^0(F)$. Its dual induces a projection $p : \mathbb{P}^{2r-1} \to \mathbb{P}H^0(B)^*$ of centre ℓ. It is again easy to conclude that $p \cdot h_E E|_{D_{r-1}} = h_{E_{r-1}}$.

It follows from the remarks in (b) that $H_E \otimes \mathcal{O}_{D_{r-1}} = H_{E_{r-1}}(a)$, where $a := (h_E|_{D_{r-1}})^* \ell = D_1 \cdot D_{r-1}$. On the other hand (a) implies that $H_E \otimes \mathcal{O}_{D_1} = H_{E_1} = A$. Finally, tensoring the Mayer-Vietoris exact sequence

$$0 \to \mathcal{O}_{D_E} \to \mathcal{O}_{D_{r-1}} \oplus \mathcal{O}_{D_1} \to \mathcal{O}_a \to 0,$$

by H_E we obtain

$$0 \to H_E \to H_{E_{r-1}}(a) \oplus A \to \mathcal{O}_a \otimes H_E \to 0.$$

By induction $h^1(H_{E_{r-1}}) = 0$, hence $h^1(H_{E_{r-1}}(a)) = 0$. Moreover $h^1(A) = 0$. Passing to the long exact sequence, the vanishing of $H^1(H_E)$ follows if the restriction $\rho : H^0(H_{E_{r-1}}(a)) \to \mathcal{O}_a(a) \otimes H_{E_{r-1}}$ is surjective. Since $h^1(H_{E_{r-1}}) = 0$, this follows from the long exact sequence of

$$0 \to H_{E_{r-1}} \to H_{E_{r-1}}(a) \to \mathcal{O}_a(a) \otimes H_{E_{r-1}} \to 0.$$

The vanishing of $h^1(H_E)$ extends by semicontinuity to a general point of $\mathcal{U}_r$.

□

4.5 The Brill-Noether curve of E

In the following we will set for simplicity: $D := D_E$. D is an abstract curve endowed with an embedding $D \subset C^{(2)}$. These data are in general not sufficient to reconstruct the vector bundle E. As we will see the additional datum of H_E makes such a reconstruction possible.

The embedding $D \subset C^{(2)}$ uniquely defines the family of divisors

$$b_x := C_x \cdot D$$

where $x \in C$ and $C_x := \{x + y \mid y \in C\}$. b_x fits in the standard exact sequence

$$0 \to H^0(\omega_C \otimes E(-x)) \otimes \mathcal{O}_C \to \omega_C \otimes E(-x) \to \mathcal{O}_{b_x} \to 0$$

and its degree is $2r$. The determinant of E can be reconstructed from the family $\{b_x \mid x \in C\}$. Indeed let $x + x' \in |\omega_C|$, then the previous exact sequence implies

$$\det E \cong \mathcal{O}_C(b_x - rx').$$

Let $t + \Theta_E$ be the translate of Θ_E by $t \in \mathrm{Div}^0(C)$: $D_{E(-t)} = a^*(t + \Theta_E)$. Thus, upon replacing E by $E(-t)$, D is transversal to C_x and b_x is smooth for a general x. Mainly we will consider b_x as a divisor on D. Let $d \in D$, it is clear that: $d \in \mathrm{Supp}\ b_x \iff d = x + y \iff h_E(d) \in P_{E,x}$. This implies that

$$b_x = h_E{}^* P_{E,x}$$

for each $x \in C$. The line bundles $H_E(-b_x)$ have degree r^2. Since D has genus $r^2 + 1$ they define a family of points in the theta divisor of $\mathrm{Pic}^{r^2}(D)$. We can say more:

Proposition 4.5.1. *Let E be general, then for every point $x \in C$ we have $h^0(H_E(-b_x)) = r$.*

Proof We know from Prop. 4.4.5 that $h^0(H_E) = 2r$. We also know that $b_x = h_E{}^*(P_{E,x})$. Since the space $P_{E,x}$ has dimension $r - 1$, it follows $h^0(H_E(-b_x)) \geq r$. Moreover the equality holds if the set $h_E(\mathrm{Supp}\ b_x)$ spans $P_{E,x}$. We prove this by induction on r. Let $r = 1$ then $P_{E,x}$ is a point: since h_E is a morphism $h_E(\mathrm{Supp}\ b_x) = P_{E,x}$. Let $r \geq 2$, as in the proof of 4.4.5 we consider a general extension

$$0 \to E_{r-1} \to E \to E_1 \to 0 \tag{4.3}$$

with $[E_{r-1}] \in \mathfrak{U}_{r-1}$ and $E_1 \in \mathrm{Pic}^1(C)$ general. We use the same assumptions and notations of the proof of 4.4.5 which is similar to this proof. In particular the curves $D_{E_{r-1}}$ and D_{E_1} are smooth and transversal, moreover the exact sequence

$$0 \to B \to F \to A \to 0$$

just denotes the above exact sequence (4.3) tensored by ω_C. Such a sequence induces a linear embedding $i : \mathbb{P}H^0(A)^* \to \mathbb{P}H^0(F)^*$. The image of i is the line ℓ considered in 4.4.5 and it holds the equality proved there: $h_E|D_{E_1} = i \circ u_{E_1}$. Then it turns out that

$$\ell \cap P_{E,x} = h_E(C_x \cap D_{E_1}) = \text{ one point } o_x$$

for each $x \in C$. On the other hand let $p : P_E \to \mathbb{P}H^0(B)^*$ be the projection of centre ℓ, then the latter exact sequence implies that $p(P_{E,x}) = P_{E_{r-1},x}$. Moreover we also know from the proof of 4.4.5 that $h_{E_{r-1}} = p \circ h_E$. By induction $P_{E_{r-1},x}$ is spanned by

$$h_{E_{r-1}}(C_x \cap D_{E_{r-1}}) = p(h_E(C_x \cap D_{E_{r-1}})).$$

Hence the linear span of $h_E(C_x \cap D_{E_{r-1}})$ is a space $L \subset P_{E,x}$ of dimension $\geq r - 2$ and such that $p(L) = P_{E_{r-1},x}$. If $o_x \in P_{E,x} - L$ then $P_{E,x}$ is spanned by $h_E(C_x \cap D)$. If $o_x \in L$ then $\dim L = r-1$ and $L = P_{E,x}$. In both cases the statement follows. □

For each $\ell \in \mathrm{Pic}^1(C)$ we consider the curve

$$B_\ell := \{x+y \in |\ell(z)|, \ z \in C\}.$$

B_ℓ is biregular to C unless $\ell = \mathcal{O}_C(x)$ for some $x \in C$. In the latter case B_ℓ is $C_x \cup |\omega_C|$. We define

$$b_\ell := D \cdot B_\ell.$$

Note that $b_\ell = b_x$ if $B_\ell = C_x \cup |\omega_C|$. The reason is that we are assuming E general, then $h^0(E) = 0$ and hence $D \cap |\omega_C| = \varnothing$.

Lemma 4.5.2. *The morphism $b : \mathrm{Pic}^1(C) \to \mathrm{Pic}^{2r}(D)$ sending ℓ to b_ℓ is an embedding. We will denote its image by J_D:*

$$J_D := \{\mathcal{O}_D(b_\ell) \mid \ell \in \mathrm{Pic}^1(C)\}.$$

In particular J_D contains the canonical theta divisor

$$C_D := \{\mathcal{O}_D(b_x) \mid x \in C\}. \tag{4.4}$$

Proof Up to shifting degrees, b is a morphism between the complex tori $\mathrm{Pic}^0(C)$ and $\mathrm{Pic}^0(D)$. Hence it is an isogeny up to translations, so b is an embedding if it is injective. Let $\ell_1, \ell_2 \in \mathrm{Pic}^1(C)$ and set $\mathcal{L} := \mathcal{O}_D(b_{\ell_1} - b_{\ell_2})$. $\mathcal{L}$ is defined by the standard exact sequence:

$$0 \to \mathcal{O}_{C^{(2)}}(-D + B_{\ell_1} - B_{\ell_2}) \to \mathcal{O}_{C^{(2)}}(B_{\ell_1} - B_{\ell_2}) \to \mathcal{L} \to 0.$$

$D + B_{\ell_2} - B_{\ell_1}$ is the pull-back by the Abel map $a\colon C^{(2)} \to \mathrm{Pic}^0(C)$ of a divisor homologous to $r\Theta$, where Θ is a theta divisor in $\mathrm{Pic}^0(C)$. Since $r\Theta$ is ample, it follows: $h^0(-D + B_{\ell_1} - B_{\ell_2}) = h^1(-D + B_{\ell_1} - B_{\ell_2}) = 0$. So the associated long exact sequence gives:

$$h^0(\mathcal{L}) = h^0(\mathcal{O}_{C^{(2)}}(B_{\ell_1} - B_{\ell_2})).$$

Moreover, it is easy to see that if $\ell_1 \neq \ell_2$, then $h^0(\mathcal{O}_{C^{(2)}}(B_{\ell_1} - B_{\ell_2})) = 0$ hence b_{ℓ_1} and b_{ℓ_2} are not linearly equivalent. □

As an immediate consequence of the lemma, the map

$$b_0\colon \mathrm{Pic}^0(C) \to \mathrm{Pic}^0(D)$$

sending $\mathcal{O}_C(x-y) \to \mathcal{O}_D(b_x - b_y)$ is an embedding too.

As we already pointed out, the knowledge of C_D is not sufficient to reconstruct E. Instead we need another curve:

Definition 4.5.3. The *Brill-Noether curve of E* is the curve

$$C_E := \{H_E(-b_x),\ x \in C\}.$$

C_E is a copy of C embedded in $\mathrm{Pic}^{r^2}(D)$. Since $h^0(H_E(-b_x)) = r$, each point of C_E is a point of multiplicity r for the theta divisor

$$\Theta_D := \{L \in \mathrm{Pic}^{r^2}(D) \mid h^0(L) \geq 1\}.$$

In particular C_E is contained in the Brill-Noether locus

$$W^{r-1}_{r^2}(D) := \{L \in \mathrm{Pic}^{r^2}(D) \mid h^0(L) \geq r\}.$$

The Brill-Noether number $\rho(r-1, r^2, r^2+1)$ yields the expected dimension of $W^{r-1}_{r^2}(D)$. We have $\rho(r-1, r^2, r^2+1) = 1$ for each r, so we expect that C_E is an irreducible component of $W^{r-1}_{r^2}(D)$, see [1, ch. V]. Of course D is not a general curve of genus r^2+1, so the latter property is not a priori granted.

Remark 4.5.4. E can uniquely be reconstructed from the pair (C_D, H_E) as follows. Consider the correspondence

$$B = \{(x, y+z) \in C \times D \mid x \in \{y, z\}\},$$

with $B \cdot (x \times D) = b_x$. Let $p_1 : C \times D \to C$ and $p_2 : C \times D \to D$ be the projection maps, then we apply the functor p_{1*} to the exact sequence

$$0 \to p_2^* H_E(-B) \to p_2^* H_E \to p_2^* H_E \otimes \mathcal{O}_B \to 0.$$

This yields the exact sequence

$$0 \to \overline{F}^* \to H^0(H_E) \otimes \mathcal{O}_C \to p_{1*}\mathcal{O}_B \otimes p_2^* H_E \to R^1 p_{1*} p_2^* H_E(-B) \to 0,$$

where $\overline{F}^* := p_{1*} p_2^* H_E(-B)$. With $F = \omega_C \otimes E$, we have the natural identities

$$\overline{F_x}^* = H^0(H_E(-b_x)) = H^0(F(-x)).$$

The left one is immediate. Let $\mathcal{I}$ be the ideal of $P_{E,x}$, then we have $H^0(H_E(-b_x)) = H^0(\mathcal{I}(1))$ by prop. 4.5.1. Hence the right equality follows from the identity $H^0(\mathcal{I}(1)) = H^0(F(-x))$. The above identities, together with $H^0(H_E) = H^0(F)$, imply that

$$F = H^0(H_E) \otimes \mathcal{O}_C / \overline{F}^*.$$

As an immediate consequence of the above construction we have:

Proposition 4.5.5. *Let $[E_1], [E_2]$ be general points of $\mathcal{U}_r$. Assume that $\theta_r([E_1]) = \theta_r([E_2]) = D$ and $H_{E_1} = H_{E_2}$. Then $[E_1] = [E_2]$.*

Remark 4.5.6. The previous construction also defines the vector bundles

$$\overline{E}: = \overline{F} \otimes \omega_C^{-1} \quad , \quad \widetilde{E}: = \omega_C^{-1} \otimes (R^1 p_{1*} p_2^* H_E(-B)).$$

We already know from 4.3.2 that the assignment $[E] \to [\overline{E}]$ defines a birational involution $j : \mathcal{U}_r \to \mathcal{U}_r$. Note also that $\widetilde{E}$ is semistable for E general: to prove this it suffices to produce one semistable E_o such that $\widetilde{E}_o$ is semistable. The existence of E_o follows by induction on r: this is obvious for $r = 1$. Let $r \geq 2$ and let E_o be defined by a semistable extension $e \in \mathrm{Ext}^1(E_1, E_{r-1})$ where $[\mathcal{U}]_{r-1} \in \mathcal{U}_{r-1}$ and $E_1 \in \mathcal{U}_1$. It is easy to show that $\widetilde{E}_o$ is defined by some $\widetilde{e} \in \mathrm{Ext}^1(\widetilde{E}_1, \widetilde{E}_{r-1})$: we leave the details to the reader. Hence $\widetilde{E}_o$ is semistable. Due to this property we can define a rational map κ : $\mathcal{U}_r \to \mathcal{U}_r$ sending $[E]$ to $[\widetilde{E}]$.

Proposition 4.5.7. *The map $\kappa : \mathcal{U}_r \to \mathcal{U}_r$ sending $[E]$ to $[\widetilde{E}]$ is birational: its inverse is $j \circ k \circ j$.*

Proof Let $T := \mathcal{O}_D(b_x + b_{i(x)})$, where $i : C \to C$ is the hyperelliptic involution. T does not depend on x because the family of divisors $\{b_x + b_{i(x)},\ x + i(x) \in |\omega_C|\}$ is rational. Then we define the line bundle of degree $r^2 + 2r$

$$\widetilde{H}_E := \omega_D \otimes T \otimes H_E^{-1}.$$

First, we note that $h^1(\widetilde{H}_E) = 0$ for E general. Indeed $\omega_D \otimes \widetilde{H}_E^{-1}$ is $H_E(-b_x - b_{i(x)})$ and hence $h^1(\widetilde{H}_E) = h^0(H_E(-b_x - b_{i(x)})$ by Serre duality. Since $h^0(H_E(-b_x - b_{i(x)}) = h^0(\omega_C \otimes E(-x - i(x)) = h^0(E)$, it follows $h^1(\widetilde{H}_E) = 0$ for each $[E] \in \mathcal{U}_r - \Theta_r$. Secondly we note that, with the previous notations, Serre duality yields a natural identification

$$R^1 p_{1*} p_2^* H_E(-B)_x = H^0(\widetilde{H}_E(-b_{i(x)}))^*,\ \forall x \in C.$$

It is then easy to deduce that $\omega_C \otimes \widetilde{E} = R^1 p_{1*} p_2^* H_E(-B) \cong p_{1*} p_2^* \widetilde{H}_E(-B)^*$.

Starting from $\widetilde{H}_E$ it is clear that one obtains H_E and with the same construction $\overline{E} = \omega_C^{-1} \otimes p_{1*} p_2^* H_E$. Note also that $\widetilde{H}_E$ is the line bundle $H_{\overline{\widetilde{E}}}$ defined by the vector bundle $\overline{\widetilde{E}}$. This implies that $k^{-1} = j \circ k \circ j$ and hence that k is birational. □

4.6 The fibres of the theta map

We want to see that E can also be uniquely reconstructed from the pair (D, C_E). For this we consider, more generally, any smooth curve $D \subset C^{(2)}$ such that $a_* D \in \mathcal{T}_r$:

Definition 4.6.1. A *Brill-Noether curve of D* is a copy

$$C' \subset \mathrm{Pic}^{r^2}(D)$$

of C satisfying the following property: there exists $H \in \mathrm{Pic}^{r^2+2r}(D)$ such that

$$C' = \{H(-b_y), y \in C\},$$

moreover H is non special and $h^0(H(-b_x)) = r$ for every point $x \in C$. The set of the Brill-Noether curves of D will be denoted by S_D.

Let $D = \theta_r([E])$ then C_E is a Brill-Noether curve of D. Let $r = 1$ then $D = C$ and the canonical theta divisor of $\mathrm{Pic}^1(C)$ is the unique Brill-Noether curve of D.

Lemma 4.6.2. *Let H be as in the previous definition. Then H is unique.*

Proof Assume that $C' = \{H'(-b_x), x \in C\}$ for a second H'. Then there exists an automorphism $u : C \to C$ which defined by setting $u(x) = y$ precisely if $H'(-b_x) = H(-b_y)$. Let $\gamma : C \times C \to \mathrm{Pic}^0(D)$ be the map sending (x, y) to $H' \otimes H^{-1}(b_x - b_y)$: we claim that the image of γ is the copy $J_D \subset \mathrm{Pic}^0(D)$ of $\mathrm{Pic}^0(C)$ and that $\gamma : C \times C \to \mathrm{Pic}^0(C)$ is the difference map. To see this recall that $C_D = \{b_x,\ x \in C\}$ is the theta divisor of J_D, see (4.4). The map $\tilde{\gamma}: C \times C \to \mathrm{Pic}^0(D)$ sending $(x, y) \to \mathcal{O}_D(b_x - b_y)$ factors through the isomorphism $t: C \times C \to C_D \times C_D$, sending $(x, y) \to (b_x, b_y)$, and the difference map. Moreover, we have the following commutative diagram

$$\begin{array}{ccc} C \times C & \xrightarrow{t} & C_D \times C_D \\ \downarrow & & \downarrow \\ \mathrm{Pic}^0(C) & \xrightarrow{b_0} & \mathrm{Pic}^0(D) \end{array}$$

where the vertical arrows are difference maps and $b_0\colon \mathrm{Pic}^0(C) \to \mathrm{Pic}^0(D)$ sending $\mathcal{O}_C(x - y) \to \mathcal{O}_D(b_x - b_y)$ is an embedding, see 4.5.2. This implies that $\tilde{\gamma}$ is a difference map and γ too. The graph of u is obviously contracted by γ, on the other hand the only curve contracted by the difference map is the diagonal of $C \times C$. Then u is the identity and $H(-b_x) = H'(-b_x)$ for each $x \in C$. Hence $H = H'$. □

Proposition 4.6.3. *Let $[E_1], [E_2]$ be general points in U_r, assume that $C_{E_1} = C_{E_2}$ and that $\theta_r([E_1]) = \theta_r([E_2]) = D$. Then $[E_1] = [E_2]$.*

Proof By the previous lemma $H_{E_1} = H_{E_2}$ and by 4.5.5 this implies $[E_1] = [E_2]$. □

Theorem 4.6.4. *The theta map $\theta_r : \mathcal{U}_r \to \mathcal{T}_r$ is generically finite.*

Proof It suffices to show that $\theta_r|_U : U \to \theta_r(U)$ is generically finite for a suitable dense open set U. Hence we can assume that $D \in \theta_r(U)$ is a smooth curve and that the points of $(\theta_r|_U)^{-1}(D) = \theta_r^{-1}(D) \cap U$ are sufficiently general in $\mathcal{U}_r$. Let

$$i : (\theta_r|_U)^{-1}(D) \to S_D$$

be the map sending $[E]$ to C_E. By 4.6.3 E can uniquely be reconstructed from (D, C_E), hence it follows that i is injective. On the other hand recall that C_E is contained in the Brill-Noether locus $W^{r-1}_{r^2}(D)$. Since the Brill-Noether number $\rho(r-1, r^2, r^2+1)$ is one, each irreducible component of $W^{r-1}_{r^2}(D)$ has dimension ≥ 1. This implies that $\theta_r|_U$ is finite if C_E is an irreducible component of $W^{r-1}_{r^2}(D)$. This property is proved in the next theorem. Hence the statement follows. □

Lemma 4.6.5. *Let $D = \theta_r([E])$ for a general $[E] \in \mathcal{U}_r$ and let $a = D \cdot D_1$ for a general $D_1 \in \mathcal{T}_1$, then the line bundle $H_E(a - b_x)$ is non special.*

Proof Let $D_1 = \theta_1([E_1]) \subset C^{(2)}$ with $E_1 = \mathcal{O}_C(x)$, then we have: $D_1 = x \times C \cup |\omega_C| \subset C^{(2)}$. Note that $a = D \cdot D_1 = b_x$ if E is general. Hence $H_E(a - b_x) = H_E$ is non special. By semicontinuity, the same is true for a general D_1. □

Theorem 4.6.6. *For a general $[E] \in \mathcal{U}_r$ the Brill-Noether curve C_E is an irreducible component of $W^{r-1}_{r^2}(D)$, $D = \theta_r([E])$.*

Proof Let $H := H_E$, it is sufficient to show the injectivity of the Petri map

$$\mu : H^0(H(-b_x)) \otimes H^0(\omega_D \otimes H^{-1}(b_x)) \to H^0(\omega_D)$$

for a general $x \in C$. This implies that the tangent space to $W^{r-1}_{r^2}(D)$ at its point $H(-b_x)$ is 1-dimensional, see [1, Ch. V]. We proceed by induction on r. Let $r = 1$, then $D = C$ and $C_E = \{\mathcal{O}_C(x), x \in C\}$. Hence the injectivity of μ is immediate. For $r \geq 2$ we borrow once more the notations and the method from the proof of proposition 4.4.5. So we specialize E to the semistable vector bundle defined by the exact sequence

$$0 \to E_{r-1} \to E \to E_1 \to 0.$$

Then D is the transversal union of the curves $D_{r-1} = \theta_{r-1}([E_{r-1}])$ and $D_1 = \theta_1([E_1])$, h_E is a morphism and H is the line bundle $h_E^* \mathcal{O}_{\mathbb{P}^{2r-1}}(1)$. Let

$a = D_{r-1} \cdot D_1$: from the proof of 4.4.5 we have $D_1 = C$ and $H_{E_1} = H_1$ and moreover

$$H \otimes \mathcal{O}_{D_{r-1}} = H_{E_{r-1}}(a) \text{ and } H \otimes \mathcal{O}_{D_1} = H_1.$$

Since x is general we can assume $\operatorname{Supp} b_x \cap \operatorname{Sing} D = \varnothing$ so that $\mathcal{O}_D(b_x)$ is a line bundle. Let $\mathcal{I}$ be the ideal sheaf of D_1 in D: at first we show that $\mu | I \otimes W$ is injective, where

$$I := H^0(\mathcal{I} \otimes H(-b_x)) \text{ and } W := H^0(\omega_D \otimes H^{-1}(b_x)).$$

Since $\mathcal{I} \otimes \mathcal{O}_{D_{r-1}} = \mathcal{O}_{D_{r-1}}(-a)$ and $\omega_D \otimes \mathcal{O}_{D_{r-1}} = \omega_{D_{r-1}}(a)$, we have the restriction maps $\rho_I : I \to H^0(H_{E_{r-1}}(-b_{x,r-1}))$ and $\rho_W : W \to H^0(\omega_{D_{r-1}} \otimes H^{-1}_{E_{r-1}}(b_{x,r-1}))$ with $b_{x,r-1} = b_x \cdot D_{r-1}$.

Claim. *ρ_I is an isomorphism and ρ_W is surjective.*

Assuming the claim, we shall complete the proof. First of all, $\rho := \rho_I \otimes \rho_W$ is surjective and defines the exact sequence

$$0 \to \ker \rho \to I \otimes W \to H^0(H_{E_{r-1}}(-b_{x,r-1})) \otimes H^0(\omega_{D_{r-1}} \otimes H^{-1}_{E_{r-1}}(b_{x,r-1})) \to 0.$$

In particular it follows $\dim \ker\ \rho = r - 1$. By induction on r the Petri map on the tensor product at the right side is injective. Therefore $\mu | I \otimes W$ is injective $\iff$ $\mu | \ker \rho$ is injective. But our claim implies $\dim \ker \rho_W = 1$ and $\ker \rho = I \otimes \langle w \rangle$, where w generates $\ker \rho_W$. Hence $\mu | \ker \rho$ is injective as well as $\mu | I \otimes W$. Let $V := H^0(H(-b_x))$ and consider the exact sequence

$$0 \to I \otimes W \to V \otimes W \to (V/I) \otimes W \to 0.$$

The map μ induces a multiplication

$$\nu : (V/I) \otimes W \to H^0(\omega_D)/\mu(I \otimes W).$$

The injectivity of $\mu | I \otimes W$ implies that μ is injective if and only if ν is injective. On the other hand, ρ_I is an isomorphism, hence $\dim\ I = r-1$ and $\dim\ V/I = 1$. Let $v \in V - I$. Then ν is injective $\iff$ $vW \cap \mu(I \otimes W) = (0)$ $\iff$ no $w \in W - (0)$ vanishes on D_1. This is equivalent to the injectivity of the restriction map

$$u : H^0(\omega_D \otimes H^{-1}(b_x)) \to H^0(\omega_C(a - x));$$

in fact $W = H^0(\omega_D \otimes H^{-1}(b_x))$ and $\omega_D \otimes H^{-1}(b_x) \otimes \mathcal{O}_{D_1} = \omega_C(a - x)$. To prove that u is injective consider the Mayer-Vietoris sequence

$$0 \to W \to H^0(\omega_{D_{r-1}} \otimes H^{-1}_{E_{r-1}}(b_{x,r-1})) \oplus H^0(\omega_C(a - x)) \to \mathcal{O}_a \to \dots$$

The left non zero arrow followed by the projection onto $H^0(\omega_C(a - x))$ is exactly u. This implies that $\ker\ u$, via the restriction map, injects in

$H^0(\omega_{D_{r-1}} \otimes H^{-1}_{E_{r-1}}(b_{x,r-1} - a))$. So u is injective if the latter space is zero, that is, if $H^1(H_{E_{r-1}}(a - b_{x,r-1})) = 0$: this has been shown in lemma 4.6.5. Hence μ is injective. Then, by semicontinuity, the same property is true for a general $[E] \in \mathcal{U}_r$ and the statement follows.

It remains to show the above claim.

- Let $h : D \to \mathbb{P}^{2r-1}$ be the map defined by H. As in 4.4.3 $h(D_1)$ is a line ℓ and $P_{E,x} \cap \ell$ is a point. Moreover $P_{E,x}$ is spanned by $h(b_x)$. Hence we have $I = H^0(\mathcal{J}(1))$ and dim $I = r - 1$, $\mathcal{J}$ being the ideal of $P_{E,x} \cup \ell$. In particular ρ_ℓ is the pull-back $(h|_{D_{r-1}})^*$ restricted to a space of linear forms vanishing on $h(D_1)$. Since $h(D)$ is non degenerate ρ_ℓ is injective. Then, for dimension reasons, ρ_ℓ is an isomorphism.

- As in 4.4.5 the projection $p : P_{E,x} \to P_{E_{r-1},x}$ from $P_{E,x} \cap \ell$ is surjective. Equivalently the restriction $H^0(E \otimes \omega_C(-x)) \to H^0(E_{r-1} \otimes \omega_C(-x))$ is surjective. So this property holds for general E, E_{r-1}. Let $\widetilde{E}_r$, $\widetilde{E}_{r-1}$ be defined from E_r, E_{r-1} as in Remark 4.5.6. By 4.5.7 they are general. Hence the restriction $H^0(\widetilde{E} \otimes \omega_C(-x)) \to H^0(\widetilde{E}_{r-1} \otimes \omega_C(-x))$ is surjective: this map is just ρ_W. □

Our partial geometric description of the theta map can be summarized as follows.

Theorem 4.6.7. *Let $D \in \mathcal{T}_r$ be general and smooth, then there exists a natural injective map i_D between the fibre of θ_r at D and the set of the Brill-Noether curves of D. Namely the map*

$$i_D : \theta_r{}^{-1}(D) \to S_D$$

associates to $[E] \in \theta_r^{-1}(D)$ its Brill-Noether curve $C_E \in S_D$.

Proof Since θ_r is generically finite, each point $[E] \in \theta^{-1}(D)$ is sufficiently general in $\mathcal{U}_r$. The injectivity then follows from corollary 4.6.4. □

Remark 4.6.8. Each Brill-Noether curve $C \in S_D$ uniquely defines a vector bundle E_C of rank r and degree r: to construct E_C it suffices to take the line bundle H appearing in the definition 4.6.1 of Brill-Noether curve. Applying the reconstruction given in remark 4.5.4 to the pair (C_D, H) we indeed obtain such a vector bundle E_C. If E_C is semistable it turns out that $\theta_r([E_C]) = D$ and that $i_D([E_C]) = C$. In particular i_D is bijective if each $C \in S_D$ defines a semistable E_C. This property seems very plausible for a general D, however we do not have a rigorous proof.

References

[1] Arbarello, E., M. Cornalba, P. A. Griffiths and J. Harris: *Geometry of Algebraic Curves*, vol. I, (1984), Springer-Verlag, New York, Berlin, Heidelberg, Tokyo.

[2] Beauville, A.: Fibrés de rang 2 sur les courbes, fibre determinant et fonctions thêta, Bull. Soc. Math. France **116** (1988), 431–448.

[3] Beauville, A.: Vector bundles and theta functions on curves of genus 2 and 3, Preprint math.AG/0406030,(2004), to appear.

[4] Beauville, A.: Vector bundles on curves and theta functions, Preprint math. AG/0502179, (2005).

[5] Beauville, A., M. S. Narasimhan and S. Ramanan:, Spectral curves and the generalised theta divisor, J.Reine Angew. Math. **398** (1989), 169–179.

[6] Drezet, J. M. and M. S. Narasimhan: Groupe de Picard des variétés de modules de fibrés semi-stables sur les courbes algébriques, Invent. math. **97** (1989), 53–94.

[7] Fulton, W.: *Intersection Theory*, (1984), Springer Verlag Berlin.

[8] Laszlo, Y.: Un théorème de Riemann pour le diviseur Thêta generalisé sur les espaces de modules de fibrés stables sur une courbe, Duke Math. J. **64** (1991), 333–347.

[9] Raynaud, M.: Sections des fibrés vectoriels sur une courbe, Bull. Soc. Math. France **110** (1982), 103–125

[10] Teixidor, M.: For which Jacobi varieties is Sing Θ reducible? J. f. reine u. angew. Math. **354** (1985), 141–149.

5

On Tannaka duality for vector bundles on p-adic curves

Christopher Deninger

Mathematisches Institut
Einsteinstr. 62
48149 Münster, Germany

Annette Werner

Fachbereich Mathematik
Pfaffenwaldring 57
70569 Stuttgart, Germany
werner@mathematik.uni-stuttgart.de

Dedicated to Jacob Murre

5.1 Introduction

In our paper [DW2] we have introduced a certain category $\mathcal{B}^{ps}$ of degree zero bundles with "potentially strongly semistable reduction" on a p-adic curve. For these bundles it was possible to establish a partial p-adic analogue of the classical Narasimhan–Seshadri theory for semistable vector bundles of degree zero on compact Riemann surfaces. One of the main open questions of [DW2] is whether our category is abelian. The first main result of the present note, Corollary 5.3.4, asserts that this is indeed the case. It follows that $\mathcal{B}^{ps}$ and also the subcategory $\mathcal{B}^{ps}_{red}$ of all polystable bundles in $\mathcal{B}^{ps}$ is a neutral Tannakian category. In the second main result, theorem 5.4.6, we calculate the group of connected components of the Tannaka dual group of $\mathcal{B}^{ps}_{red}$. This uses a result of A. Weil characterizing vector bundles that become trivial on a finite étale covering as the ones satisfying a "polynomial equation" over the integers.

Besides, in section 5.3 we give a short review of [DW2], and in section 5.2 we discuss the "strongly semistable reduction" condition.

Finally we would like to draw the reader's attention to the paper of Faltings [F] where a non-abelian p-adic Hodge theory is developped.

It is a pleasure for us to thank Uwe Jannsen for a helpful discussion.

5.2 Vector bundles in characteristic p

Throughout this paper, we call a purely one-dimensional separated scheme of finite type over a field k a curve over k.

Let C be a connected smooth, projective curve over k. For a vector bundle E on C we denote by $\mu(E) = \frac{\deg(E)}{\mathrm{rk}(E)}$ the slope of E. Then E is called semistable (respectively stable), if for all proper non-trivial subbundles F of E the inequality $\mu(F) \leq \mu(E)$ (respectively $\mu(F) < \mu(E)$) holds.

Lemma 5.2.1. *If $\pi : C' \to C$ is a finite separable morphism of connected smooth projective k-curves, then semistability of E is equivalent to semistability of $\pi^* E$.*

Proof See [Gie2], 1.1. □

If $\mathrm{char}(k) = 0$, then by lemma 5.2.1 any finite morphism of smooth connected projective curves preserves semistability. However in the case $\mathrm{char}(k) = p$, there exist vector bundles which are destabilized by the Frobenius map, see [Gie1], Theorem 1. Assume that $\mathrm{char}(k) = p$, and let $F : C \to C$ be the absolute Frobenius morphism, defined by the p-power map on the structure sheaf.

Definition 5.2.2. A vector bundle E on C is called strongly semistable of degree zero if $\deg(E) = 0$ and if $F^{n*}E$ is semistable on C for all $n \geq 0$.

Now we also consider non-smooth curves over k. Let Z be a proper curve over k. By $C_1, \ldots, C_r$ we denote the irreducible components of Z endowed with their reduced induced structures. Let $\tilde{C}_i$ be the normalization of C_i, and write $\alpha_i : \tilde{C}_i \to C_i \to Z$ for the canonical map. Note that the curves $\tilde{C}_i$ are smooth irreducible and projective over k.

Definition 5.2.3. A vector bundle E on the proper k-curve Z is called strongly semistable of degree zero, if all $\alpha_i^* E$ are strongly semistable of degree zero.

A alternative characterization of this property is given by the following well-known result.

Proposition 5.2.4. *A vector bundle E on Z is strongly semistable of degree zero if and only if for any k-morphism $\pi : C \to Z$, where C is a smooth connected projective curve over k, the pullback $\pi^* E$ is semistable of degree zero on C.*

Note that in [DM], (2.34) bundles with this property are called semistable of degree zero.

Proof Let X be a scheme over k. The absolute Frobenius F sits in a commutative diagram

$$\begin{array}{ccc} X & \xrightarrow{F} & X \\ \downarrow & & \downarrow \\ \operatorname{spec} k & \xrightarrow{F} & \operatorname{spec} k \end{array}$$

If we denote for all $r \geq 1$ by $X^{(r)}$ the scheme X together with the structure map $X \to \operatorname{spec} k \xrightarrow{F^r} \operatorname{spec} k$, then $F^r : X^{(r)} \to X$ is a morphism over $\operatorname{spec} k$. Assume that E and Z have the property in the claim. Applying it to the smooth projective curves $\tilde{C}_i$ and the k-morphisms

$$\tilde{C}_i^{(r)} \xrightarrow{F^r} \tilde{C}_i \xrightarrow{\alpha_i} Z,$$

we find that all $\alpha_i^* E$ are strongly semistable of degree zero, i.e. that E is strongly semistable of degree zero in the sense of definition 5.2.3.

Conversely, assume that all $\alpha_i^* E$ are strongly semistable of degree zero. Let $\pi : C \to Z$ be a k-morphism from a smooth connected projective curve C to Z. Then π factors through one of the C_i. If π is constant, then $\pi^* E$ is trivial, hence semistable of degree zero. Hence we can assume that $\pi(C) = C_i$. Since C is smooth, π also factors through the normalization $\tilde{C}_i$, i.e. there is a morphism $\pi_i : C \to \tilde{C}_i$ satisfying $\alpha_i \circ \pi_i = \pi$. Since π is dominant, it is finite and hence π_i is the composition of a separable map and a power of Frobenius, see e.g. [Ha], IV, 2.5. Hence there exists a smooth projective curve D over k and a finite separable morphism $f : D \to \tilde{C}_i$ such that $C \xrightarrow{\sim} D^{(r)}$ for some $r \geq 1$ and π_i factors as

$$\pi_i : C \xrightarrow{\sim} D^{(r)} \xrightarrow{F^r} D \xrightarrow{f} \tilde{C}_i.$$

Write $E_i = \alpha_i^* E$. Then we have to show that $\pi^* E = \pi_i^* E_i$ is semistable of degree 0 on C. By assumption, E_i is strongly semistable of degree 0 on $\tilde{C}_i$. Using Lemma 5.2.1 and the fact that F commutes with all morphisms in characteristic p, we find that the pullback $f^* E_i$ under the finite, separable map f is strongly semistable of degree 0 on D. Hence $F^{r*} f^* E_i$ is semistable, which implies that $\pi_i^* E_i$ is semistable of degree zero. □

Generalizing a result by Lange and Stuhler in [LS], one can show

Proposition 5.2.5. *If $k = \mathbb{F}_q$ is a finite field, then a vector bundle E on the proper k-curve Z is strongly semistable of degree zero, if and only if there exists a finite surjective morphism*

$$\varphi : Y \to Z$$

of proper k-curves such that $\varphi^* E$ is trivial. In fact, one can take φ to be the composition

$$\varphi : Y \xrightarrow{\mathrm{Fr}_q^s} Y \xrightarrow{\pi} Z$$

of a power of the k-linear Frobenius morphism Fr_q (defined by the q-th power map on $\mathcal{O}_Y$) and a finite, étale and surjective morphism π.

Proof See [DW2], Theorem 18. □

Note that every semistable vector bundle E of degree zero on a smooth geometrically connected projective curve C of genus $g \leq 1$ is strongly semistable. Namely, for $g = 0$, every semistable vector bundle of degree zero is in fact trivial. For $g = 1$, the claim follows from Atiyah's classification [At]: let $E = \bigoplus_i E_i$ be the decomposition of E into indecomposable components. Since E is semistable of degree zero, all E_i have degree zero since they are subbundles and quotients. Therefore by [At] Theorem 5, we have $E_i \simeq L \otimes G$, where L is a line bundle of degree zero and G is an iterated extension of trivial line bundles. The pullback of E_i under some Frobenius power is also of this form. Since the category of semistable vector bundles of degree 0 on C is closed under extensions and contains all line bundles of degree zero, we conclude that E_i is indeed strongly semistable of degree 0. This proves the following fact:

Lemma 5.2.6. *Let Z be a proper k-curve such that the normalizations $\tilde{C}_i$ of all irreducible components C_i are geometrically connected of genus $g(\tilde{C}_i) \leq 1$. Consider a vector bundle E on Z. If all restrictions $E\,|_{\tilde{C}_i}$ are semistable of degree zero, then E is strongly semistable of degree zero.*

By [Giel], for every genus ≥ 2 there are examples of semistable vector bundles of degree zero which are not strongly semistable.

On the other hand, there are results indicating that there are "a lot of" strongly semistable vector bundles of degree zero. In [LP], Laszlo and Pauly show that for an ordinary smooth projective curve C of genus two over an algebraically closed field k of characteristic two, the set of strongly semistable rank two bundles is Zariski dense in the coarse moduli space of all semistable rank two bundles with trivial determinant. See [JRXY] for generalizations to higher genus.

In [Du], Ducrohet investigates the case of a supersingular smooth projective curve of genus two over an algebraically closed field k with $\mathrm{char}(k) = 2$. It turns out that in this case all equivalence classes of semistable bundles with trivial determinant but one are in fact strongly semistable.

5.3 Vector bundles on p-adic curves

Before discussing the p-adic case, let us recall some results in the complex case, i.e. regarding vector bundles on a compact Riemann surface X. Let $x \in X$ be a base point and denote by $\pi : \tilde{X} \to X$ the universal covering of X. Every representation $\rho : \pi_1(X, x) \to \mathrm{GL}_r(\mathbb{C})$ gives rise to a flat vector bundle E_ρ on X, which is defined as the quotient of the trivial bundle $\tilde{X} \times \mathbb{C}^r$ by the $\pi_1(X, x)$-action given by combining the natural action of $\pi_1(X, x)$ on the first factor with the action induced by ρ on the second factor. It is easily seen that every flat vector bundle on X is isomorphic to some E_ρ. Regarding E_ρ as a holomorphic bundle on X, a theorem of Weil [W] says that a holomorphic bundle E on X is isomorphic to some E_ρ (i.e. E comes from a representation of $\pi_1(X, x)$) if and only if in the decomposition $E = \bigoplus_{i=1}^r E_i$ of E into indecomposable subbundles all E_i have degree zero. A famous result by Narasimhan and Seshadri [NS] says that a holomorphic vector bundle E of degree 0 on X is stable if and only if E is isomorphic to E_ρ for some irreducible unitary representation ρ. Hence a holomorphic vector bundle comes from a unitary representation ρ if and only if it is of the form $E = \bigoplus_{i=1}^r E_i$ for stable (and hence indecomposable) subbundles of degree zero.

Now let us turn to the p-adic case. Let X be a connected smooth projective curve over the algebraic closure $\overline{\mathbb{Q}}_p$ of $\mathbb{Q}_p$ and put $X_{\mathbb{C}_p} = X \otimes_{\overline{\mathbb{Q}}_p} \mathbb{C}_p$. We want to look at p-adic representations of the algebraic fundamental group $\pi_1(X, x)$ where $x \in X(\mathbb{C}_p)$ is a base point. It is defined as follows. Denote by F_x the functor from the category of finite étale coverings X' of X to the category of finite sets which maps X' to the set of $\mathbb{C}_p$-valued points of X' lying over x.

For $x, x' \in X(\mathbb{C}_p)$ we call any isomorphism $F_x \xrightarrow{\sim} F_{x'}$ of fibre functors an étale path from x to x'. (Note that any topological path on a Riemann surface induces naturally such an isomorphism of fibre functors.) Then the étale fundamental group $\pi_1(X, x)$ is defined as

$$\pi_1(X, x) = \mathrm{Aut}(F_x).$$

The goal of our papers [DW1] and [DW2] is to associate p-adic representations of the étale fundamental group $\pi_1(X, x)$ to certain vector bundles on $X_{\mathbb{C}_p}$. Let us briefly describe the main result. We call any finitely presented, proper and flat scheme $\mathfrak{X}$ over the integral closure $\overline{\mathbb{Z}}_p$ of $\mathbb{Z}_p$ in $\overline{\mathbb{Q}}_p$ with generic fibre X a model of X. By $\mathfrak{o}$ we denote the ring of integers in $\mathbb{C}_p$, and by $k = \overline{\mathbb{F}}_p$ the residue field of $\overline{\mathbb{Z}}_p$ and $\mathfrak{o}$. We write $\mathfrak{X}_\mathfrak{o} = \mathfrak{X} \otimes_{\overline{\mathbb{Z}}_p} \mathfrak{o}$ and $\mathfrak{X}_k = \mathfrak{X} \otimes_{\overline{\mathbb{Z}}_p} k$.

Definition 5.3.1. We say that a vector bundle E on $X_{\mathbb{C}_p}$ has strongly semistable reduction of degree zero if E is isomorphic to the generic fibre of a vector bundle $\mathcal{E}$ on $\mathfrak{X}_{\mathfrak{o}}$ for some model $\mathfrak{X}$ of X, such that the special fibre $\mathcal{E}_k$ is a strongly semistable vector bundle of degree zero on the proper k-curve $\mathfrak{X}_k$.

E has potentially strongly semistable reduction of degree zero if there is a finite étale morphism $\alpha : Y \to X$ of connected smooth projective curves over $\overline{\mathbb{Q}}_p$ such that $\alpha^*_{\mathbb{C}_p} E$ has strongly semistable reduction of degree zero on $Y_{\mathbb{C}_p}$.

By $\mathcal{B}^s$ (respectively $\mathcal{B}^{ps}$) we denote the full subcategory of the category of vector bundles on $X_{\mathbb{C}_p}$ consisting of all E with strongly semistable (respectively potentially strongly semistable) reduction of degree zero. Besides, for every divisor D on $X_{\mathbb{C}_p}$ we define $\mathcal{B}_{X_{\mathbb{C}_p},D}$ to be the full subcategory of those vector bundles E on $X_{\mathbb{C}_p}$ which can be extended to a vector bundle $\mathcal{E}$ on $\mathfrak{X}_{\mathfrak{o}}$ for some model $\mathfrak{X}$ of X, such that there exists a finitely presented proper $\overline{\mathbb{Z}}_p$-morphism

$$\pi : \mathcal{Y} \longrightarrow \mathfrak{X}$$

satisfying the following two properties:

i) The generic fibre of π is finite and étale outside D
ii) The pullback $\pi_k^* \mathcal{E}_k$ of the special fibre of $\mathcal{E}$ is trivial on $\mathcal{Y}_k$ (c.f. [DW2], definition 6 and theorem 16).

Then we show in [DW2], Theorem 17:

$$\mathcal{B}^s = \bigcup_D \mathcal{B}_{X_{\mathbb{C}_p},D},$$

where D runs through all divisors on $X_{\mathbb{C}_p}$. By [DW2], Theorem 13, every bundle in $\mathcal{B}_{X_{\mathbb{C}_p},D}$ is semistable of degree zero, so that $\mathcal{B}^s$ and also $\mathcal{B}^{ps}$ are full subcategories of the category $\mathfrak{T}^{ss}$ of semistable bundles of degree zero on $X_{\mathbb{C}_p}$. Line bundles of degree zero lie in $\mathcal{B}^{ps}$ by [DW2] Theorem 12 a.

The main result in [DW2] is the following (c.f. [DW2], theorem 36):

Theorem 5.3.2. *Let E be a bundle in $\mathcal{B}^{ps}$. For every étale path γ from x to y in $X(\mathbb{C}_p)$ there is an isomorphism*

$$\rho_E(\gamma) : E_x \xrightarrow{\sim} E_y$$

of "parallel transport", which behaves functorially in γ. The association $E \mapsto \rho_E(\gamma)$ is compatible with tensor products, duals and internal homs of vector bundles in the obvious way. It is also compatible with $\mathrm{Gal}(\overline{\mathbb{Q}}_p/\mathbb{Q}_p)$*-conjugation. Besides, if $\alpha : X \to X'$ is a morphism of smooth projective* *curves over* $\overline{\mathbb{Q}}_p$ *and* E' *a bundle in* $\mathcal{B}^{ps}_{X'_{\mathbb{C}_p}}$, *then* $\rho_{\alpha^* E'}(\gamma)$ *and* $\rho_{E'}(\alpha_*\gamma)$ *coincide, where* $\alpha_*\gamma$ *is the induced étale path on* X'. *For every* $x \in X(\mathbb{C}_p)$ *the fibre functor*

$$\mathcal{B}^{ps} \longrightarrow \mathrm{Vec}_{\mathbb{C}_p},$$

mapping E *to the fibre* E_x *in the category* $\mathrm{Vec}_{\mathbb{C}_p}$ *of* $\mathbb{C}_p$*-vector spaces, is faithful.*

In particular one obtains a continuous representation $\rho_{E,x} : \pi_1(X,x) \to \mathrm{GL}_r(E_x)$. The functor $E \mapsto \rho_{E,x}$ is compatible with tensor products, duals, internal homs, pullbacks of vector bundles and $\mathrm{Gal}(\overline{\mathbb{Q}}_p/\mathbb{Q}_p)$-conjugation.

Let us look at two special cases of this representation: For line bundles on a curve X with good reduction, ρ induces a homomorphism

$$\mathrm{Pic}^0_X(\mathbb{C}_p) \longrightarrow \mathrm{Hom}_{\mathrm{cont}}(\pi_1(X,x), \mathbb{C}_p^*)$$

mapping L to $\rho_{L,x}$. As shown in [DW1] this map coincides with the map defined by Tate in [Ta] § 4 on an open subgroup of $\mathrm{Pic}^0_X(\mathbb{C}_p)$. Secondly, applying ρ to bundles E in $H^1(X_{\mathbb{C}_p}, \mathcal{O}) = \mathrm{Ext}^1_{X_{\mathbb{C}_p}}(\mathcal{O}, \mathcal{O})$, one recovers the Hodge–Tate map to $H^1(X_{\text{ét}}, \mathbb{Q}_p) \otimes \mathbb{C}_p = \mathrm{Ext}^1_{\pi_1(X,x)}(\mathbb{C}_p, \mathbb{C}_p)$, see [DW1], corollary 8.

It follows from [DW2], Proposition 9 and Theorem 11 that the categories $\mathcal{B}^s$ and $\mathcal{B}^{ps}$ are closed under tensor products, duals, internal homs and extensions. We will now prove another important property of those categories.

Theorem 5.3.3. *If a vector bundle* E *on* $X_{\mathbb{C}_p}$ *is contained in* $\mathcal{B}^s$ *(respectively* $\mathcal{B}^{ps}$*), then every quotient bundle of degree zero and every subbundle of degree zero of* E *is also contained in* $\mathcal{B}^s$ *(respectively* $\mathcal{B}^{ps}$*).*

Proof It suffices to show this property for the category $\mathcal{B}^s$. By duality, it suffices to treat quotient bundles. So let $\tilde{\mathcal{E}}$ be a vector bundle with strongly semistable reduction of degree zero on $\mathfrak{X}_\mathfrak{o}$, where $\mathfrak{X}$ is a model of X. Denote by E the generic fibre of $\tilde{\mathcal{E}}$, and let

$$0 \to E' \to E \to E'' \to 0 \tag{5.1}$$

be an exact sequence of vector bundles $X_{\mathbb{C}_p}$, where E'' has degree zero. By [DW2], Theorem 5, E' can be extended to a vector bundle $\mathcal{F}'$ on $\mathcal{Y}_\mathfrak{o}$, where $\mathcal{Y}$ is a model of X such that there is a morphism $\varphi : \mathcal{Y} \to \mathfrak{X}$ inducing an isomorphism on the generic fibres. Since $\mathrm{Hom}(\mathcal{F}', \varphi_\mathfrak{o}^*\tilde{\mathcal{E}}) \otimes_\mathfrak{o} \mathbb{C}_p = \mathrm{Hom}(E', E)$, we may assume that the embedding $E' \to E$ can be extended to a $\mathcal{O}_{\mathcal{Y}_\mathfrak{o}}$-module homomorphism $\mathcal{F}' \to \varphi_\mathfrak{o}^*\tilde{\mathcal{E}}$ after changing the morphisms in the diagram (5.1).

Let $\mathcal{F}''$ be the quasi-coherent sheaf on $\mathcal{Y}_{\mathfrak{o}}$ such that $\mathcal{F}' \to \varphi_{\mathfrak{o}}^* \tilde{\mathcal{E}} \to \mathcal{F}'' \to 0$ is exact. Then $\mathcal{F}''$ is of finite presentation. Note that the generic fibre of this sequence is isomorphic to the sequence (5.1).

Let r be the rank of E''. The same argument as in the proof of [DW2], Theorem 5 shows that the blowing-up

$$\psi_{\mathfrak{o}} : \mathfrak{Z}_{\mathfrak{o}} \to \mathcal{Y}_{\mathfrak{o}}$$

of the r-th Fitting ideal of $\mathcal{F}''$ descends to a finitely presented morphism $\psi : \mathfrak{Z} \to \mathcal{Y}$ inducing an isomorphism on the generic fibres. Besides, if $\mathcal{A}$ denotes the annihilator of the r-th Fitting ideal of $\psi_{\mathfrak{o}}^* \mathcal{F}''$, the sheaf $\psi_{\mathfrak{o}}^* \mathcal{F}'' / \mathcal{A}$ is locally free by [RG] (5.4.3). Hence it gives rise to a vector bundle $\mathcal{E}''$ on $\mathfrak{Z}_{\mathfrak{o}}$ with generic fibre E''. Let us write $\mathcal{E} = \psi_{\mathfrak{o}}^* \varphi_{\mathfrak{o}}^* \tilde{\mathcal{E}}$. Then we have a natural surjective homomorphism of vector bundles $\mathcal{E} \to \mathcal{E}''$ on $\mathfrak{Z}_{\mathfrak{o}}$ extending the quotient map $E \to E''$ on the generic fibre $X_{\mathbb{C}_p}$.

Let K be a finite extension of $\mathbb{Q}_p$ such that $\mathfrak{Z}$ descends to a proper and flat scheme $\mathfrak{Z}_{\mathfrak{o}_K}$ over the ring of integers $\mathfrak{o}_K$. We can choose K big enough so that all irreducible components of the special fibre of $\mathfrak{Z}$ are defined over the residue field of K. Let X_K be the generic fibre of $\mathfrak{Z}_{\mathfrak{o}_K}$. Then $X_K \otimes_K \overline{\mathbb{Q}}_p \simeq X$. The scheme $\mathfrak{Z}_{\mathfrak{o}}$ is the projective limit of all $\mathfrak{Z}_A = \mathfrak{Z}_{\mathfrak{o}_K} \otimes_{\mathfrak{o}_K} A$, where A runs over the finitely generated $\mathfrak{o}_K$-subalgebras of $\mathfrak{o}$.

By [EGAIV], (8.5.2), (8.5.5), (8.5.7), (11.2.6) there exists a finitely generated $\mathfrak{o}_K$-subalgebra A of $\mathfrak{o}$ with quotient field $Q \subset \mathbb{C}_p$ such that $\mathcal{E} \to \mathcal{E}''$ descends to a surjective homomorphism $\mathcal{E}_A \to \mathcal{E}''_A$ of vector bundles on $\mathfrak{Z}_A$.

Let $x \in \operatorname{spec} A$ be the point corresponding to the prime ideal $A \cap \mathfrak{m}$ in A, where $\mathfrak{m} \subset \mathfrak{o}$ is the valuation ideal. If π_K is a prime element in $\mathfrak{o}_K$, we have $\mathfrak{o}_K/(\pi_K) \subset A/A \cap \mathfrak{m} \subset \mathfrak{o}/\mathfrak{m} = k$, so that $A \cap \mathfrak{m}$ is a maximal ideal in A. Hence x is a closed point with residue field $\kappa = \kappa(x)$ which is a finite extension of $\mathfrak{o}_K/(\pi_K)$ in k.

By assumption, the vector bundle $\tilde{\mathcal{E}}_k$ on the special fibre $\mathfrak{X}_k$ of $\mathfrak{X}$ is strongly semistable of degree zero. By Proposition 5.2.4, strong semistability is preserved under pullbacks via k-morphisms, so that $\mathcal{E}_k = (\varphi \circ \psi)_k^* \tilde{\mathcal{E}}_k$ is also strongly semistable of degree zero.

The bundle $\mathcal{E}_\kappa = \mathcal{E}_A \otimes_A \kappa$ satisfies $\mathcal{E}_\kappa \otimes_\kappa k \simeq \mathcal{E}_k$, hence it is strongly semistable of degree zero on $\mathfrak{Z}_\kappa = \mathfrak{Z}_A \otimes_A \kappa$.

Let $C_1, \ldots, C_r$ be the irreducible components of $\mathfrak{Z}_\kappa$ with normalizations $\tilde{C}_1, \ldots, \tilde{C}_r$ and denote by $\alpha_i : \tilde{C}_i \to C_i \to \mathfrak{Z}_\kappa$ the natural map. Since the Euler characteristics are locally constant in the fibres of the flat and proper A-scheme $\mathfrak{Z}_A$, we find $\deg \mathcal{E}''_\kappa = \deg E''_Q = 0$.

By the degree formula in [BLR], 9.1, Proposition 5, $\deg(\mathcal{E}''_\kappa)$ is a linear combination of the $\deg(\alpha_i^* \mathcal{E}''_\kappa)$'s with positive coefficients. Since $\alpha_i^* \mathcal{E}''_\kappa$ is a quotient bundle of the semistable degree zero vector bundle $\alpha_i^* \mathcal{E}_\kappa$ on $\tilde{C}_i$, it has degree ≥ 0. Hence for all i we find $\deg(\alpha_i^* \mathcal{E}''_\kappa) = 0$.

Now let $\mathcal{F}$ be a vector bundle on the smooth projective curve $\tilde{C}_i$ which is a quotient of $\alpha_i^* \mathcal{E}''_\kappa$. Then $\mathcal{F}$ is also a quotient of the semistable degree zero bundle $\alpha_i^* \mathcal{E}_\kappa$, which implies $\deg(\mathcal{F}) \geq \deg \alpha_i^* \mathcal{E}_\kappa = 0$. This shows that $\alpha_i^* \mathcal{E}''_\kappa$ is semistable of degree 0 on $\tilde{C}_i$. The same argument applies to all Frobenius pullbacks of $\alpha_i^* \mathcal{E}''_\kappa$, so that $\alpha_i^* \mathcal{E}''_\kappa$ is strongly semistable of degree 0. Hence $\mathcal{E}''_\kappa$ is strongly semistable of degree zero on $\mathfrak{Z}_\kappa$. By [HL], 1.3.8 the base change $\mathcal{E}''_k = \mathcal{E}''_\kappa \otimes_\kappa k$ is also strongly semistable of degree zero. Since $\mathcal{E}''$ has generic fibre E'', it follows that E'' is indeed contained in $\mathcal{B}^s$. □

Corollary 5.3.4. *$\mathcal{B}^s$ and $\mathcal{B}^{ps}$ are abelian categories.*

Proof Recall that $\mathcal{B}^s$ and $\mathcal{B}^{ps}$ are full subcategories of the abelian category $\mathcal{T}^{ss}$ of semistable vector bundles of degree zero on $X_{\mathbb{C}_p}$. Since the trivial bundle is contained in $\mathcal{B}^s$ and $\mathcal{B}^{ps}$, and both categories are closed under direct sums by [DW1], Proposition 9, they are additive.

By the theorem, $\mathcal{B}^s$ and $\mathcal{B}^{ps}$ are also closed under kernels and cokernels, hence they are abelian categories. □

5.4 Tannakian categories of vector bundles

In this section we look at several categories of semistable vector bundles from a Tannakian point of view. Useful references in this context are [DM] and [S] for example.

As before, let X be a smooth projective curve over $\overline{\mathbb{Q}}_p$ with a base point $x \in X(\mathbb{C}_p)$. We call a vector bundle on $X_{\mathbb{C}_p}$ polystable of degree zero if it is isomorphic to the direct sum of stable vector bundles of degree zero. Let $\mathcal{T}^{ss}_{\text{red}}$ be the strictly full subcategory of vector bundles on $X_{\mathbb{C}_p}$ consisting of polystable bundles of degree zero and set $\mathcal{B}^{ps}_{\text{red}} = \mathcal{B}^{ps} \cap \mathcal{T}^{ss}_{\text{red}}$. Then we have the following diagram of fully faithful embeddings

$$\begin{array}{ccc} \mathcal{B}^{ps}_{\text{red}} & \subset & \mathcal{B}^{ps} \\ \cap & & \cap \\ \mathcal{T}^{ss}_{\text{red}} & \subset & \mathcal{T}^{ss}. \end{array} \tag{5.2}$$

Note that because of theorem 5.3.3, every vector bundle E in $\mathcal{B}^{ps}_{\text{red}}$ is the direct sum of stable vector bundles of degree zero contained in $\mathcal{B}^{ps}$.

Lemma 5.4.1. *The categories $\mathcal{T}^{ss}_{\text{red}}$ and $\mathcal{B}^{ps}_{\text{red}}$ are closed under taking subquotients in $\mathcal{T}^{ss}$.*

Proof Since $\mathcal{B}^{ps}$ is closed under subquotients in $\mathcal{T}^{ss}$ by theorem 5.3.3, it suffices to consider $\mathcal{T}^{ss}_{\text{red}}$. A bundle E in $\mathcal{T}^{ss}_{\text{red}}$ can be written as $E = \bigoplus_i E_i$

with E_i stable of degree zero. Let $\varphi : E \to E''$ be a surjective map in $\mathcal{T}^{ss}$. Then the images $\varphi(E_i)$ lie in $\mathcal{T}^{ss}$ and the surjective map $\varphi|_{E_i} : E_i \to \varphi(E_i)$ is an isomorphism or the zero map since E_i is stable. Hence we have $E'' = \sum_j E''_j$ for stable degree zero vector bundles E''_j (the nonzero $\varphi(E_i)$). For every j we have

$$E''_j \cap \sum_{k \neq j} E''_k = E''_j \quad \text{or } = 0$$

since the bundle E''_j being stable is a simple object of $\mathcal{T}_{ss}$. It follows that E'' is the direct sum of suitably chosen E''_j's and hence lies in $\mathcal{T}^{ss}_{\text{red}}$. The case of subobjects follows by duality. □

Consider the fibre functor ω_x on $\mathcal{T}^{ss}$ defined by $\omega_x(E) = E_x$ and $\omega_x(f) = f_x$. It induces fibre functors on the other categories as well.

Theorem 5.4.2. a *The categories $\mathcal{B}^{ps}_{\text{red}}, \mathcal{B}^{ps}, \mathcal{T}^{ss}_{\text{red}}$ and $\mathcal{T}^{ss}$ with the fibre functor ω_x are neutral Tannakian categories over $\mathbb{C}_p$.*
b *The categories $\mathcal{B}^{ps}_{\text{red}}$ and $\mathcal{T}^{ss}_{\text{red}}$ are semisimple. Every object in $\mathcal{B}^{ps}$ (resp. $\mathcal{T}^{ss}$) is a successive extension of objects of $\mathcal{B}^{ps}_{\text{red}}$ (resp. $\mathcal{T}^{ss}_{\text{red}}$).*
c *The natural inclusion $\mathcal{B}^{ps} \subset \mathcal{T}^{ss}$ is an equivalence of categories if and only if $\mathcal{B}^{ps}_{\text{red}} \subset \mathcal{T}^{ss}_{\text{red}}$ is an equivalence of categories.*

Proof **a** For $\mathcal{T}^{ss}$ and $\mathcal{T}^{ss}_{\text{red}}$ this is well known, see e.g. [Si], p. 29. The categories $\mathcal{B}^{ps}$ and $\mathcal{B}^{ps}_{\text{red}}$ are abelian by corollary 5.3.4 and lemma 5.4.1. It was shown in [DW2] that $\mathcal{B}^{ps}$ is closed under tensor products and duals. The same follows for $\mathcal{B}^{ps}_{\text{red}} = \mathcal{B}^{ps} \cap \mathcal{T}^{ss}_{\text{red}}$. Faithfulness of ω_x on $\mathcal{B}^{ps}$ and $\mathcal{B}^{ps}_{\text{red}}$ follows because ω_x is faithful on $\mathcal{T}^{ss}$. Alternatively a direct proof was given in [DW2] Theorem 36.
b Every object in $\mathcal{B}^{ps}_{\text{red}}$ and $\mathcal{T}^{ss}_{\text{red}}$ is the direct sum of simple objects since stable bundles are simple. It is well known that objects of $\mathcal{T}^{ss}$ are successive extensions of stable bundles of degree zero. Since subquotients in $\mathcal{T}^{ss}$ of objects in $\mathcal{B}^{ps}$ lie in $\mathcal{B}^{ps}$ by theorem 5.3.3, the corresponding assertion for $\mathcal{B}^{ps}$ follows.
c This is a consequence of **b** because both $\mathcal{T}^{ss}$ and $\mathcal{B}^{ps}$ are closed under extensions, c.f. [DW2]. □

Let

$$\begin{array}{ccc} G^{ps}_{\text{red}} & \longleftarrow & G^{ps} \\ \uparrow & & \uparrow \\ G^{ss}_{\text{red}} & \longleftarrow & G^{ss} \end{array} \tag{5.3}$$

be the diagram of affine group schemes over $\mathbb{C}_p$ corresponding to diagram (5.2) by Tannakian duality.

Proposition 5.4.3. *All morphisms in (5.3) are faithfully flat. The connected components of G^{ps}_{red} and G^{ss}_{red} are pro-reductive.*

Proof The following is known [DM] Proposition 2.21:
A fully faithful $\otimes$-functor $F : \mathcal{C} \to \mathcal{D}$ of neutral Tannakian categories over a field k of characteristic zero induces a faithfully flat morphism $F^* : G_{\mathcal{D}} \to G_{\mathcal{C}}$ of the Tannakian duals if and only if we have: Every subobject in $\mathcal{D}$ of an object $F(C)$ for some C in $\mathcal{C}$ is isomorphic to $F(C')$ for a subobject C' of C.

This criterion can be verified immediately for the functors in (5.2) by using either theorem 5.3.3 or lemma 5.4.1. The second assertion of the proposition follows from [DM] Proposition 2.23 and Remark 2.28. □

Let $\mathcal{T}_{\mathrm{fin}}$ be the category of vector bundles on $X_{\mathbb{C}_p}$ which are trivialized by a finite étale covering of $X_{\mathbb{C}_p}$. Since $\pi_1(X_{\mathbb{C}_p}, x) \xrightarrow{\sim} \pi_1(X, x)$ is an isomorphism it follows that $\mathcal{T}_{\mathrm{fin}}$ is equivalent to the category of representations of $\pi_1(X, x)$ with open kernels on finite dimensional $\mathbb{C}_p$-vector spaces V, c.f. [LS], 1.2. Such a representation factors over a finite quotient G of $\pi_1(X, x)$ and the corresponding bundle in $\mathcal{T}_{\mathrm{fin}}$ is $E = X'_{\mathbb{C}_p} \times^G \mathbf{V}$. Here $\alpha : X' \to X$ is the Galois covering corresponding to the quotient $\pi_1(X, x) \to G$ and $\mathbf{V}$ is the affine space over $\mathbb{C}_p$ corresponding to V. With the fibre functor ω_x the category $\mathcal{T}_{\mathrm{fin}}$ is neutral Tannakian over $\mathbb{C}_p$ with Tannaka dual

$$\pi_1(X, x)_{/\mathbb{C}_p} = \varprojlim_N (\pi_1(X, x)/N)_{/\mathbb{C}_p}.$$

Here N runs over the open normal subgroups of $\pi_1(X, x)$ and for a finite (abstract) group H we denote by $H_{/\mathbb{C}_p}$ the corresponding constant group scheme. Using Maschke's theorem it follows that $\mathcal{T}_{\mathrm{fin}}$ is semisimple.

Proposition 5.4.4. *The category $\mathcal{T}_{\mathrm{fin}}$ is a full subcategory of $\mathcal{B}^{ps}_{\mathrm{red}}$. The induced morphism $G^{ps}_{\mathrm{red}} \twoheadrightarrow \pi_1(X, x)_{/\mathbb{C}_p}$ is faithfully flat.*

Proof Obviously, $\mathcal{T}_{\mathrm{fin}}$ is a full subcategory of $\mathcal{B}^{ps}$. Let E be a vector bundle in $\mathcal{T}_{\mathrm{fin}}$. Since a finite étale pullback of E is trivial and hence polystable, E is also polystable by [HL], Lemma 3.2.3. Hence E is an object of $\mathcal{T}^{ss}_{\mathrm{red}}$. It follows that $\mathcal{T}_{\mathrm{fin}}$ is a full subcategory of $\mathcal{B}^{ps}_{\mathrm{red}} = \mathcal{B}^{ps} \cap \mathcal{T}^{ss}_{\mathrm{red}}$. The next assertion follows from fully faithfulness of $\mathcal{T}_{\mathrm{fin}} \hookrightarrow \mathcal{B}^{ps}_{\mathrm{red}}$ since $\mathcal{B}^{ps}_{\mathrm{red}}$ is semisimple, c.f. [DM], Remark 2.29. □

Consider a Galois covering $\alpha : X' \to X$ with group $\mathrm{Gal}(X'/X)$ of smooth projective curves over $\overline{\mathbb{Q}}_p$ and choose a point $x' \in X'(\mathbb{C}_p)$ above $x \in X(\mathbb{C}_p)$. Let us write $\mathcal{C}_X$ for any of the categories $\mathcal{B}^{ps}_{\mathrm{red}}, \mathcal{B}^{ps}, \mathcal{T}^{ss}_{\mathrm{red}}$ and $\mathcal{T}^{ss}$ of vector bundles on $X_{\mathbb{C}_p}$. The pullback functor $\alpha^* : \mathcal{C}_X \to \mathcal{C}_{X'}$ is a morphism of neutral Tannakian categories over $\mathbb{C}_p$ commuting with the fibre functors ω_x and $\omega_{x'}$. Let $i : G_{X'} \to G_X$ be the morphism of Tannaka duals induced by α^*. We also need the faithfully flat homomorphism obtained by composition:

$$q : G_X \twoheadrightarrow \pi_1(X, x)_{/\mathbb{C}_p} \twoheadrightarrow \mathrm{Gal}(X'/X)_{/\mathbb{C}_p}.$$

Here the second arrow is determined by our choice of x'. Note that every σ in $\mathrm{Gal}(X'/X)$ induces an automorphism σ^* of $\mathcal{C}_{X'}$ and hence an automorphism $\sigma : G_{X'} \to G_{X'}$ of group schemes over $\mathbb{C}_p$.

Lemma 5.4.5. *There is a natural exact sequence of affine group schemes over* $\mathbb{C}_p$

$$1 \to G_{X'} \xrightarrow{i} G_X \xrightarrow{q} \mathrm{Gal}(X'/X)_{/\mathbb{C}_p} \to 1.$$

Proof Every bundle E' in $\mathcal{C}_{X'}$ is isomorphic to a subquotient of $\alpha^*(E)$ for some bundle E in $\mathcal{C}_X$. Namely, thinking of E' as a locally free sheaf, the sheaf $E = \alpha_* E'$ is locally free again and we have $\alpha^* E \cong \bigoplus_\sigma \sigma^* E'$ where σ runs over $\mathrm{Gal}(X'/X)$. Incidentally, E lies in $\mathcal{C}_X$ because $\alpha^* E \cong \bigoplus_\sigma \sigma^* E'$ lies in $\mathcal{C}_{X'}$. This is clear for $\mathcal{C} = \mathcal{T}^{ss}$ or $\mathcal{B}^{ps}$. For $\mathcal{T}^{ss}_{\mathrm{red}}$ and hence for $\mathcal{B}^{ps}_{\mathrm{red}}$ it follows from [HL] Lemma 3.2.3.

It follows from [DM] Proposition 2.21 (b) that i is a closed immersion. By descent the category $\mathcal{C}_X$ is equivalent to the category of bundles in $\mathcal{C}_{X'}$ equipped with a $\mathrm{Gal}(X'/X)$-operation covering the one on X'. In other words, the category of representations of G_X is equivalent to the category of representations of $G_{X'}$ together with a $\mathrm{Gal}(X'/X)$-action, i.e. a transitive system of isomorphisms $\sigma^*\rho = \rho \circ \sigma \to \rho$ for all σ in $\mathrm{Gal}(X'/X)$. Hence, G_X is an extension of $G_{X'}$ by $\mathrm{Gal}(X'/X)_{/\mathbb{C}_p}$ inducing the above $\mathrm{Gal}(X'/X)$-action on $G_{X'}$. (Because such an extension has the same $\otimes$-category of representations.) In particular, the sequence in the lemma is exact. □

For a commutative diagram of Galois coverings

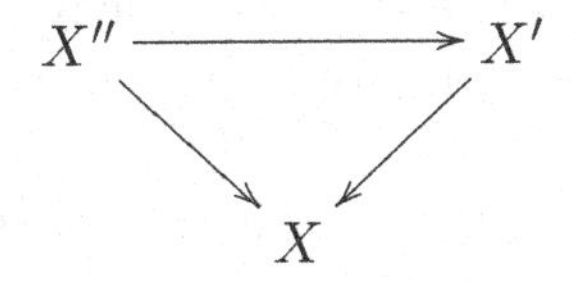

and the choice of points $x' \in X'(\mathbb{C}_p)$ and $x'' \in X''(\mathbb{C}_p)$ over x we get a commutative diagram of affine group schemes over $\mathbb{C}_p$:

$$\begin{array}{ccccccccc} 1 & \longrightarrow & G_{X''} & \longrightarrow & G_X & \longrightarrow & \mathrm{Gal}(X''/X)_{/\mathbb{C}_p} & \longrightarrow & 0 \\ & & \downarrow & & \| & & \downarrow & & \\ 1 & \longrightarrow & G_{X'} & \longrightarrow & G_X & \longrightarrow & \mathrm{Gal}(X'/X)_{/\mathbb{C}_p} & \longrightarrow & 0. \end{array}$$

Passing to the projective limit, we get an exact sequence

$$1 \longrightarrow \varprojlim_{X'} G_{X'} \longrightarrow G_X \longrightarrow \pi_1(X,x)_{/\mathbb{C}_p} \longrightarrow 1. \tag{5.4}$$

Right exactness follows from propositions 5.4.3 and 5.4.4.

If G is a group scheme, we denote by G^0 its connected component of identity.

Theorem 5.4.6. *We have a commutative diagram*

$$\begin{array}{ccccccccc} 1 & \longrightarrow & (G^{ss}_{\mathrm{red}})^0 & \longrightarrow & G^{ss}_{\mathrm{red}} & \longrightarrow & \pi_1(X,x)_{/\mathbb{C}_p} & \longrightarrow & 1 \\ & & \downarrow & & \downarrow & & \| & & \\ 1 & \longrightarrow & (G^{ps}_{\mathrm{red}})^0 & \longrightarrow & G^{ps}_{\mathrm{red}} & \longrightarrow & \pi_1(X,x)_{/\mathbb{C}_p} & \longrightarrow & 1\,. \end{array}$$

In particular $\pi_1(X,x)_{/\mathbb{C}_p}$ *is the common group scheme of connected components of both* G^{ss}_{red} *and* G^{ps}_{red}. *Moreover we have:*

$$(G^{ss}_{\mathrm{red}})^0 = \varprojlim_{X'} G^{ss}_{\mathrm{red},X'} \quad \text{and} \quad (G^{ps}_{\mathrm{red}})^0 = \varprojlim_{X'} G^{ps}_{\mathrm{red},X'}.$$

Here X'/X *runs over a cofinal system of pointed Galois covers of* (X,x).

Proof Let $\mathcal{C}_X$ denote either $\mathcal{B}^{ps}_{\mathrm{red},X}$ or $\mathcal{T}^{ss}_{\mathrm{red},X}$ and let G_X be its Tannaka dual. The exact sequence (5.4) implies that $G^0_X \subset \varprojlim_{X'} G_{X'}$. Hence it suffices to show that $\varprojlim_{X'} G_{X'}$ is connected. The category of finite dimensional representations of $\varprojlim_{X'} G_{X'}$ on $\mathbb{C}_p$-vector spaces is $\varinjlim_{X'} \mathcal{C}_{X'}$. In order to show that $\varprojlim_{X'} G_{X'}$ is connected, by [DM] Corollary 2.22 we have to prove the following:

Claim Let A be an object of $\varinjlim_{X'} \mathcal{C}_{X'}$. Then the strictly full subcategory $[[A]]$ of $\varinjlim_{X'} \mathcal{C}_{X'}$ whose objects are isomorphic to subquotients of $A^N, N \geq 0$ is not stable under $\otimes$ unless A is isomorphic to a trivial bundle.

Proof Let $[[A]]$ be stable under $\otimes$. The category $\varinjlim_{X'} \mathcal{C}_{X'}$ is semisimple since $(\varprojlim_{X'} G_{X'})^0 = \varprojlim_{X'} G^0_{X'}$ is pro-reductive. Hence we may decompose A into simple objects $A = A_1 \oplus \ldots \oplus A_s$. By assumption, for every $j \geq 1$ the object $A_1^{\otimes j}$ is isomorphic to a subquotient of A^N for some $N = N(j)$. The same argument as in the proof of Lemma 5.4.1 shows that up to isomorphism the subquotients of $NA := A^N$ have the form $m_1 A_1 \oplus \ldots \oplus m_s A_s$ for integers $m_i \geq 0$. Hence we get isomorphisms where $\sum$ means "direct sum":

$$A_1^{\otimes j} \cong \sum_{i=1}^{s} m_{ij} A_i \quad \text{for } 1 \leq j \leq r.$$

Here $M = (m_{ij})$ is an $s \times r$-matrix over $\mathbb{Z}$. Fixing some $r > s$ there is a relation with integers c_j, not all zero:

$$\sum_{j=1}^{r} c_j (m_{1j}, \ldots, m_{sj})^t = 0.$$

This gives the relation

$$\sum_{j=1}^{r} c_j^+ (m_{1j}, \ldots, m_{sj})^t = \sum_{j=1}^{r} c_j^- (m_{1j}, \ldots, m_{sj})^t$$

where $c_j^+ = \max\{c_j, 0\}$ and $c_j^- = -\min\{c_j, 0\}$. "Left multiplication" with $(A_1, \ldots, A_s)$ gives isomorphisms

$$\sum_{j=1}^{r} c_j^+ \sum_{i=1}^{s} m_{ij} A_i \cong \sum_{j=1}^{r} c_j^- \sum_{i=1}^{s} m_{ij} A_i,$$

and hence

$$\sum_{j=1}^{r} c_j^+ A_1^{\otimes j} \cong \sum_{j=1}^{r} c_j^- A_1^{\otimes j}.$$

For the polynomials $P^{\pm}(T) = \sum_{j=1}^{r} c_j^{\pm} T^j$ with coefficients in $\mathbb{Z}^{\geq 0}$ we have $P^+ \neq P^-$ and:

$$P^+(A_1) \cong P^-(A_1).$$

Let E_1 be a bundle in $\mathcal{C}_{X'}$ representing A_1 in $\varinjlim_{X'} \mathcal{C}_{X'}$. Then we have an isomorphism

$$P^+(\beta^* E_1) \cong P^-(\beta^* E_1)$$

of vector bundles on a suitable Galois cover $\beta : X'' \to X'$. A theorem of Weil, c.f. [W] Ch. III or [N], now implies that $\beta^* E_1$ and hence E_1 is trivialized by a finite étale covering of X'. Hence A_1, the class of E_1 is isomorphic in $\varinjlim_{X'} \mathcal{C}_{X'}$ to a trivial bundle. The same argument applies to $A_2, A_3, \ldots$ Hence A is isomorphic to a trivial bundle as well. This proves the claim and hence the theorem. □

We now determine the structure of G^{ab} for $G = G^{ps}_{\mathrm{red}}$ and $G = G^{ss}_{\mathrm{red}}$. In either case, this group is pro-reductive and abelian, hence diagonalizable and therefore determined by its character group

$$X(G^{\mathrm{ab}}) = \mathrm{Mor}_{\mathbb{C}_p}(G^{\mathrm{ab}}, \mathbb{G}_m).$$

The characters of G^{ab} correspond to isomorphism classes of one-dimensional representations of G i.e. to isomorphism classes of degree zero line bundles in $\mathcal{B}^{ps}_{\mathrm{red}}$ resp. $\mathcal{T}^{ss}_{\mathrm{red}}$. Since both categories contain all degree zero line bundles we get

$$X(G^{\mathrm{ab}}) = \mathrm{Pic}^0_X(\mathbb{C}_p)$$

and hence

$$\begin{aligned} G^{\mathrm{ab}} &= \mathrm{Hom}(\mathrm{Pic}^0_X(\mathbb{C}_p), \mathbb{G}_{m,\mathbb{C}_p}) \\ &= \varprojlim_{A} \mathrm{Hom}(A, \mathbb{G}_{m,\mathbb{C}_p}). \end{aligned}$$

Here A runs over the finitely generated subgroups of $\mathrm{Pic}^0_X(\mathbb{C}_p)$. A similar argument using the fact that $(G^{ps}_{\mathrm{red}})^0$ resp. $(G^{ss}_{\mathrm{red}})^0$ is the Tannaka dual of $\varinjlim_{X'} \mathcal{B}^{ps}_{\mathrm{red},X'}$ resp. $\varinjlim_{X'} \mathcal{T}^{ss}_{\mathrm{red},X'}$ shows the following: For G as above, the group $(G^0)^{\mathrm{ab}}$ is diagonalizable with character group

$$X((G^0)^{\mathrm{ab}}) = \varinjlim_{X'} \mathrm{Pic}^0_{X'}(\mathbb{C}_p).$$

Note that the right hand group is torsionfree because line bundles of finite order become trivial in suitable finite étale coverings. This corresponds to the fact that $(G^0)^{\mathrm{ab}}$ is connected. We can therefore write as well:

$$X((G^0)^{\mathrm{ab}}) = \varinjlim_{X'} (\mathrm{Pic}^0_{X'}(\mathbb{C}_p)/\mathrm{tors})$$

and hence

$$(G^0)^{\mathrm{ab}} = \varprojlim_{X'} \mathrm{Hom}(\mathrm{Pic}^0_{X'}(\mathbb{C}_p)/\mathrm{tors}, \mathbb{G}_{m,\mathbb{C}_p}).$$

This is a pro-torus over $\mathbb{C}_p$.

In particular we have seen that $(G^{ps}_{\mathrm{red}})^0$ and $(G^{ss}_{\mathrm{red}})^0$ have the same maximal abelian quotient. Incidentally we may compare (G, G^0) with (G^0, G^0): There is an exact sequence:

$$1 \longrightarrow (G, G^0)/(G^0, G^0) \longrightarrow (G^0)^{\mathrm{ab}} \longrightarrow (G^{\mathrm{ab}})^0 \longrightarrow 1.$$

The sequence of character groups is

$$1 \longrightarrow \mathrm{Pic}^0_X(\mathbb{C}_p)/\mathrm{tors} \longrightarrow \varinjlim_{X'} (\mathrm{Pic}^0_{X'}(\mathbb{C}_p)/\mathrm{tors}) \longrightarrow X((G, G^0)/(G^0, G^0)) \longrightarrow 1.$$

If X is a curve of genus $g(X) \geq 2$ then by the Riemann Hurwitz formula, the middle group is infinite dimensional and in particular $(G, G^0)/(G^0, G^0)$ contains a non-trivial pro-torus.

If X is an elliptic curve, it follows from Atiyah's classification in [At] that every stable bundle of degree 0 on X is in fact a line bundle. Hence $\mathfrak{T}^{ss}_{\text{red}} = \mathcal{B}^{ps}_{\text{red}}$ is the full subcategory of all vector bundles which can be decomposed as a direct sum of line bundles of degree 0. This implies that the corresponding Tannaka group G is abelian and diagonalizable with character group $\text{Pic}^0_X(\mathbb{C}_p)$, which fits in the picture above.

We proceed with some remarks on the structure of G^0 which follow from the general theory of reductive groups. Let C be the neutral component of the center of G^0. Then we have

$$G^0 = C \cdot (G^0, G^0).$$

Here C is a pro-torus and (G^0, G^0) is pro-semisimple. The projection $C \rightarrow (G^0)^{\text{ab}}$ is faithfully flat and its kernel is a commutative pro-finite group scheme H. In the exact sequence:

$$1 \longrightarrow X((G^0)^{\text{ab}}) \longrightarrow X(C) \longrightarrow X(H) \longrightarrow 1$$

the group $X((G^0)^{\text{ab}})$ is divisible because $\text{Pic}^0_{X'}(\mathbb{C}_p)$ is divisible. Since $X((G^0)^{\text{ab}})$ is also torsionfree it is a $\mathbb{Q}$-vector space. The group $X(H)$ being torsion it follows that $X(C) \otimes \mathbb{Q} = X((G^0)^{\text{ab}})$ canonically and $X(C) \cong X((G^0)^{\text{ab}}) \oplus X(H)$ non-canonically. It would be interesting to determine $X(H)$ for both $G = G^{ps}_{\text{red}}$ and G^{ss}_{red}.

We end with a remark on the Tannaka dual G_E of the Tannaka subcategory of $\mathcal{B}^{ps}_{\text{red}}$ generated by a vector bundle E in $\mathcal{B}^{ps}_{\text{red}}$. The group G_E is a subgroup of $\mathbf{GL}_{E_x}$ the linear group over $\mathbb{C}_p$ of the $\mathbb{C}_p$-vector space E_x. It can be characterized as follows: The group $G_E(\mathbb{C}_p)$ consists of all g in $\text{GL}\,(E_x)$ with $g(s_x) = s_x$ for all $n, m \geq 0$ and all sections s in

$$\Gamma(X_{\mathbb{C}_p}, (E^*)^{\otimes n} \otimes E^{\otimes m}) = \text{Hom}_{X_{\mathbb{C}_p}}(E^{\otimes n}, E^{\otimes m}).$$

Here $g(s_x)$ means the extension of g to an automorphism g of $(E^*_x)^{\otimes n} \otimes E^{\otimes m}_x$ applied to s_x.

Consider the representation attached to E by theorem 5.3.2:

$$\rho_{E,x} : \pi_1(X, x) \longrightarrow \text{GL}\,(E_x).$$

Its image is contained in $G_E(\mathbb{C}_p)$ because the functor $F \mapsto \rho_{F,x}$ on $\mathcal{B}^{ps}_{\text{red}}$ is compatible with tensor products and duals and maps the trivial line bundle to the trivial representation. Hence G_E contains the Zariski closure of $\text{Im}\,\rho_{E,x}$ in $\mathbf{GL}_{E_x}$. It follows from a result by Faltings [F] that the faithful functor $F \mapsto \rho_{F,x}$ is in fact fully faithful. If $\rho_{E,x}$ is also semisimple then a standard argument shows that G_E is actually equal to the Zariski closure of $\operatorname{Im} \rho_{E,x}$.

References

[At] Atiyah, M.F.: Vector bundles over an elliptic curve, Proc. London Math. Soc. (3) **7** (1957), 414–452

[BLR] Bosch, S., W. Lütkebohmert and M. Raynaud: *Néron models*. Springer 1990

[DM] Deligne, P. and J.S. Milne: *Tannakian categories*. LNM 900. Springer 1982

[DW1] Deninger, C. and A. Werner: Line bundles and p-adic characters. In: G. van der Geer, B. Moonen, R. Schoof (eds.): *Number Fields and Function Fields - Two Parallel Worlds*. Birkhäuser 2005, 101–131.

[DW2] Deninger, C. and A. Werner: Vector bundles on p-adic curves and parallel transport. Ann. Scient. Éc. Norm. Sup. **38** (2005), 553–597

[Du] Ducrohet, L.: The action of the Frobenius map on rank 2 vector bundles over a supersingular genus 2 curve in characteristic 2. Preprint 2005. `http://www.arXiv.math.AG/0504500`

[EGAIV] Grothendieck, A. and J. Dieudonné: Éléments de Géométrie Algébrique IV, Publ. Math. IHES **20** (1964), **24** (1965), **28** (1966), **32** (1967)

[F] Faltings, G.: A p-adic Simpson correspondence. Adv. Math. **198** (2005), 847–862

[Gie1] Giesecker, D.: Stable vector bundles and the Frobenius morphism. Ann. Scient. Éc. Norm. Sup. **6** (1973), 95–101

[Gie2] Giesecker, D.: On a theorem of Bogomolov on Chern classes of stable bundles., Am. J. Math. **101** (1979), 79–85

[Ha] Hartshorne, R.: *Algebraic Geometry*, Springer 1977

[HL] Huybrechts, D. and M. Lehn: *The geometry of moduli spaces of sheaves* Viehweg 1997

[JRXY] Joshi, K., S. Ramanan, E. Z. Xia and J.-K. Yu: On vector bundles destabilized by Frobenius pull-back. Comp. Math. **142** (2006), 616–630

[LP] Laszlo,Y. and C. Pauly: The action of the Frobenius map on rank 2 vector bundles in characteristic 2, preprint 2004 `http://www.arXiv. math.AG/0005044`

[LS] Lange, H. and U. Stuhler: Vektorbündel auf Kurven und Darstellungen der algebraischen Fundamentalgruppe, Math. Z. **156** (1977), 73–83

[Liu] Q. Liu: *Algebraic Geometry and Arithmetic Curves*. Oxford University Press 2002

[N] Nori, M.V.: On the representations of the fundamental group, Comp. Math. **33** (1976), 29–41

[NS] Narasimhan, M.S. and C.S. Seshadri: Stable and unitary vector bundles on a compact Riemann surface. Ann. Math. **82** (1965), 540–567

[RG] Raynaud, M. and L. Gruson: Crières de platitude et de projectivité. Invent. Math. **13** (1971), 1–89

[S] Serre, J.-P: Propriétés conjecturales des groupes de Galois motiviques et des représentations l-adiques. In: *U. Jannsen et al. (eds.): Motives*. Proc. Symp. Pure Math. **55** vol. 1. AMS 1994, 377–400.

[Si] Simpson, C.: Higgs bundles and local systems. Publ. Math. IHES **75** (1992), 5–92

[Ta] Tate, J.: p-divisible groups, in *Proceedings of a Conference on local fields, Driebergen 1966*, 158–183

[W] Weil, A.: Généralisation des fonctions abéliennes, J. de Math. P. et App., (IX) **17** (1938), 47–87

6

On finite-dimensional motives and Murre's conjecture

Uwe Jannsen

NWF I - Mathematik
Universität Regensburg
93040 Regensburg
GERMANY
uwe.jannsen@mathematik.uni-regensburg.de

To Jacob Murre

6.1 Introduction

The conjectures of Bloch, Beilinson, and Murre predict the existence of a certain functorial filtration on the Chow groups (with $\mathbb{Q}$-coefficients) of all smooth projective varieties, whose graded quotients only depend on cycles modulo homological equivalence. This filtration would offer a rather good understanding of these Chow groups, and would allow to prove several other conjectures, like Bloch's conjecture on surfaces of geometric genus 0. In Murre's formulation (cf. 6.5.1 below) one can check the validity of the conjecture for particular smooth projective varieties, and in fact, a slightly weaker form of the conjecture has been proved for several cases, e.g., for surfaces [Mu1] and several threefolds [GM] (proving parts (A), (B) and (D) of the conjecture, and giving evidence for (C)). But to my knowledge, there are few results for higher-dimensional varieties, and the strongest form of Murre's conjecture (including part (C)) is only known for curves, rational surfaces, and, trivially, for Brauer-Severi varieties.

The first aim of this paper is to exhibit some cases, where the full Murre conjecture can be shown. The positive aspect is that we get this for some non-trivial cases of varieties of higher (in fact arbitrarily high) dimension, the negative aspect is that we get this just for some special varieties and special ground fields. In particular, not over some universal domain. As a sample, we get the following:

Theorem 6.1.1. *Let k be a rational or elliptic function field (in one variable) over a finite field $\mathbb{F}$. Let X_0 be an arbitrary product of rational and elliptic curves over $\mathbb{F}$, and let $X = X_0 \times_{\mathbb{F}} k$. Then Murre's conjecture holds for X.*

One ingredient is the notion of "finite-dimensionality" of motives, as introduced independently by Kimura [Ki] and O'Sullivan [OSu]. Up to now it is only known that the Chow motive of a smooth projective variety is finite-dimensional, if it lies in the tensor category generated by the motives of abelian varieties. But I believe that this notion will be fundamental for further progress on Chow groups, and motivic cohomology in general, in view of the nilpotence properties it implies.

Therefore a second aim of this paper is to investigate this property in several directions. We add some observations to the existing results (cf. [An1] for a survey) which may be interesting in their own right, but also bear on our investigation of Murre's conjecture. In particular, we are interested in Chow endomorphisms and nilpotence results: For a smooth projective variety X of pure dimension d let $\mathsf{CH}^d(X \times X)_{\mathbb{Q}}$ be the ring of Chow self-correspondences, i.e., the endomorphism ring $\mathrm{End}(h_{\mathrm{rat}}(X))$ of the Chow motive $h_{\mathrm{rat}}(X)$ associated to X, and let $J(X) = \mathsf{CH}^d(X \times X)_{\mathbb{Q},\mathrm{hom}}$ be the ideal of homologically trivial correspondences. Then we get the following (a similar result appeared in [DP]):

Theorem 6.1.2. *Let X be a smooth projective variety over a field k, and let π_+^X be the projector onto the even degree part of the cohomology. Then the following properties are equivalent.*

i) π_+^X is algebraic and $J(X^N)$ is nilpotent for all $N > 0$.
ii) π_+^X is algebraic and $J(X^N)$ is a nil ideal for all $N > 0$.
iii) X is finite-dimensional (i.e., $h(X)$, the motive of X, is finite-dimensional).

The implication from (iii) to (ii) is Kimura's nilpotence theorem. The other implications give a certain converse. The following again sharpens results of Kimura.

Theorem 6.1.3. *Let M be a motive modulo rational equivalence over a field k, and assume that M is either oddly or evenly finite-dimensional. For an endomorphism $f \in \mathrm{End}(M)$ let $P(t) = \det(f \mid H^*(M, \mathbb{Q}_\ell))$ be the characteristic polynomial of f acting on the ℓ-adic cohomology of M ($\ell \neq \mathrm{char}(k)$). Then $P(f) = 0$ in $\mathrm{End}(M)$.*

This fact (which in the even case was also proved by O'Sullivan) is somewhat surprising, because a priori the equality $P(f) = 0$ only holds modulo homological equivalence.

Coming back to Murre's conjecture, it is well-known by now [GP] that part (A) (the Chow-Künneth decomposition) follows from the standard conjecture on algebraicity of the Kuenneth components, and finite-dimensionality. In particular, this applies to abelian varieties over arbitrary ground fields. In this paper, we explore which additional ingredients can give the remaining part of Murre's conjecture.

We start with the case of a finite ground field $\mathbb{F}$. Here it is known by work of Geisser [Gei] and Kahn [Ka] that the conjunction of finite-dimensionality and Tate's conjecture "implies everything". In particular it implies that rational and numerical equivalence agree (with $\mathbb{Q}$-coefficients). Evidently, the latter also implies Murre's conjecture. However, we are interested in getting some unconditional theorems, and hence we take some pain to single out the minimal conditions to get such results. In particular, we don't want to argue with Tate's conjecture for all varieties, but want to get by with conditions just on the given variety X (For this, we have to rectify some statements in the literature.) As a sample, we get:

Theorem 6.1.4. *Let X be a smooth projective variety over the finite field $\mathbb{F}$. Assume that the ideal $J(X)$ is nilpotent (e.g., assume that $h(X)$ is finite-dimensional). Fix an integer $j \geq 0$ and assume that the Tate conjecture holds for $H^{2j}(X \times_F \bar{F}, \mathbb{Q}_\ell(j))$, and that the Frobenius eigenvalue 1 is semi-simple on $H^{2j}(X \times_F \bar{F}, \mathbb{Q}_\ell(j))$. Then the cycle map induces an isomorphism*

$$\mathsf{CH}^j(X)_{\mathbb{Q}} \otimes_{\mathbb{Q}} \mathbb{Q}_\ell \xrightarrow{\sim} H^{2j}(X \times_F \overline{F}, \mathbb{Q}_\ell(j))^{\mathrm{Gal}(\overline{F}/F)},$$

and the motivic cohomology $H^i_{\mathcal{M}}(X, \mathbb{Z}(j))$ is of finite exponent for all $i \neq 2j$.

Corollary 6.1.5. *Let X be a smooth projective variety of pure dimension d over the finite field F, and assume that $h(X)$ is finite-dimensional. Then $H^i_{\mathcal{M}}(X, \mathbb{Z}(d))$ has finite exponent for all $i \neq 2d$.*

This generalizes results of Soulé [So1]. The theorems on function fields over F (like Theorem 6.1.1) are obtained by considering arbitrary (not necessarily smooth or projective) varieties over F and passing to certain limits of ℓ-adic cohomology, as in [Ja1].

The results in this paper were mainly obtained during a stay at the University of Tokyo during the academic year 2003/2004, and it is my pleasure to thank the department and my host Takeshi Saito for the invitation and the hospitality. I thank the referee for suggesting a more elegant version and proof of Theorem 6.4.3(a).

6.2 Weil cohomology theories, motives, tensor categories

In this section, we recall some notions and properties needed later. We fix a base field k and consider the category SP_k of smooth projective varieties over k. For X in SP_k, we denote by $Z^j(X)$ the group of algebraic cycles of codimension j on X, *with* $\mathbb{Q}$*-rational coefficients*. The following definition is equivalent to the one in [Kl].

Definition 6.2.1. Let E be a field of characteristic 0. An E-linear Weil cohomology theory H is a contravariant functor $X \mapsto H(X)$, $(f : X \to Y) \mapsto f^* := H(f) : H(Y) \to H(X)$ from SP_k to the category of graded commutative E-algebras ($a.b = (-1)^{i+j} b.a$ for $a \in H^i(X), b \in H^j(X)$), together with the following data and properties:

i) $\dim_E H(X) < \infty$, and $H^i(X) = 0$ for $i < 0$ or $i > 2 \cdot \dim X$.
ii) To each morphism $f : X \to Y$ there is functorially associated an E-linear map $f_* : H(X) \to H(Y)$ which is of degree $\dim Y - \dim X$ if X and Y are irreducible.
iii) (projection formula) $f_*(f^* y \cdot x) = y \cdot f_* x$ for $f : X \to Y$, $x \in H(X)$ and $y \in H(Y)$.
iv) (Künneth formula) The association $a \otimes b \mapsto a \times b := pr_X^* a \cdot pr_Y^* b$ gives an isomorphism $H(X) \otimes_E H(Y) \to H(X \times Y)$.
v) (Poincaré duality) $H(\text{Spec } k) = H^0(\text{Spec } k) \cong E$, and the bilinear pairing $H(X) \times H(X) \xrightarrow{\cdot} H(X) \xrightarrow{f_*} H(\text{Spec} k) \cong E$, $(a, b) \mapsto \langle a, b \rangle := f_*(a.b)$, is non-degenerate.
vi) (cycle map) There are cycle class maps $c\ell^j : Z^j(X) \to H^{2j}(X)$ compatible with products, pull-backs f^* and push-forwards f_*, whenever these operations are defined on the algebraic cycles.

Remarks 6.2.2. a) It follows that f_* is the transpose of f^* under Poincaré duality.

b) Let X be a smooth projective variety of pure dimension d, and denote, as usual, by Δ_X the cycle in $Z^d(X \times X)$ corresponding to the diagonal $X \hookrightarrow X \times X$, and also the associated cycles class in $H^{2d}(X \times X)$ for a given Weil cohomology theory H. The Künneth components of the diagonal, $\pi_i \in H^{2d-i}(X) \otimes H^i(X)$ $(i = 1, \ldots, 2d)$ are defined by decomposing $\Delta_X = \sum_{i=0}^{2d} \pi_i$ according to the Künneth isomorphism (6.2.1)(iv). The standard conjecture $C(X)$ predicts that the π_i are algebraic, i.e., again classes of algebraic cycles.

c) Let X and $Y \in SP_k$. Any correspondence from X to Y, i.e., any element $\alpha \in Z^*(X \times Y)$, induces an E-linear map, again denoted α, from $H(Y)$ to $H(X)$ by defining $\alpha(x) = (p_Y)_*((p_X)^*(x).c\ell(\alpha))$ for $x \in H(Y)$, where $p_X : X \times Y \to X$ and $p_Y : X \times Y \to Y$ are the projections. The same already holds for any cohomology class $\alpha' \in H^*(X \times Y)$ in place of $c\ell(\alpha)$. Via this interpretation, the element

π_i, as cohomological correspondence from X to X, is the identity on $H^i(X)$, and zero on $H^j(X)$ for $j \neq i$.

Examples 6.2.3. The following are examples of Weil cohomology theories:

a) If $k = \mathbb{C}$: singular cohomology $H(X) = H^*(X(\mathbb{C}), \mathbb{Q})$ $(E = \mathbb{Q})$.
b) For arbitrary k: ℓ-adic cohomology $H^*_{et}(X \times_k \overline{k}, \mathbb{Q}_\ell)$, for $\ell \neq$ char k $(E = \mathbb{Q}_\ell)$.
c) If char $k = 0$: de Rham cohomology $H^*_{dR}(X/k)$, $(E = k)$.
d) If k is a perfect field of characteristic $p > 0$: crystalline cohomology $H^*_{\text{crys}}(X/B(k)) := H^*_{\text{crys}}(X/W(k)) \otimes_{W(k)} B(k)$ $(E = B(k) := \text{Frac}(W(k))$.

6.2.4. Let $\sim$ be an adequate equivalence relation on algebraic cycles, i.e., an equivalence relation on all cycle groups $Z^j(X)$ for all X in SP_k such that product, push-forward and pull-back of cycles is well-defined on the cycle groups $A^j_\sim(X) := Z^j(X)/\sim$ [Ja3]. We recall that we have the adequate equivalence relations rational, algebraic, homological and numerical equivalence with the relationship

$$\alpha \sim_{\text{rat}} 0 \;\Rightarrow\; \alpha \sim_{\text{alg}} 0 \;\Rightarrow\; \alpha \sim_{\text{hom}} 0 \;\Rightarrow\; \alpha \sim_{\text{num}} 0$$

The category $\mathcal{M}_\sim(k)$ of ($\mathbb{Q}$-rational) motives modulo $\sim$ over k can be defined as follows. For $X, Y \in SP_k$ the group of correspondences (modulo $\sim$) of degree n from X to Y is defined as $\text{Corr}^n_\sim(X,Y) = \oplus_i A_\sim^{\dim(X_i)+n}(X_i \times Y)$, where the X_i are the irreducible components of X. The composition of correspondences $f \in \text{Corr}^m_\sim(X,Y)$ and $g \in \text{Corr}^m_\sim(Y,Z)$ is defined as $g \circ f = (p_{XZ})_*(p^*_{XY}(f).p^*_{YZ}(g)) \in \text{Corr}^{m+n}(X,Z)$, where p_{XZ}, p_{XY} and p_{YZ} are the projections from $X \times Y \times Z$ to $X \times Z$, $X \times Y$ and $Y \times Z$, respectively. Then the objects of $\mathcal{M}_\sim(k)$ can be described as triples (X, p, m), with $X \in SP_k$, $p \in \text{Corr}^0_\sim(X,X)$ an idempotent and $m \in \mathbb{Z}$, and one has $\text{Hom}((X,p,m),(Y,q,n)) = q\,\text{Corr}^{n-m}(X,Y)p$, with composition given by the above composition of correspondences. The Tate objects are defined by $1(n) = (\text{Spec}\, k, id, n)$ for $n \in \mathbb{Z}$.

6.2.5. The precise definition of a tensor category can be found in [DM]. Let us just recall that it is a category with a bifunctor $(A, B) \mapsto A \otimes B$ together with asssociativity constraints $\psi_{A,B,C} : A \otimes (B \otimes C) \cong (A \otimes B) \otimes C$, commutativity constraints $\phi_{A,B} : A \otimes B \cong B \otimes A$ and an unity constraint $l : 1 \otimes A \cong A$, satisfying certain compatibilities modelled after the situation of the tensor product of vector spaces. A tensor category is called rigid, if it has internal Homs $\underline{\text{Hom}}(A, B)$ (characterized by $\text{Hom}(A \otimes B, C) =$

$\mathrm{Hom}(A, \underline{\mathrm{Hom}}(B, C)))$ satisfying some reasonable properties [DM, 1.7]. In this case, the dual of an object is defined as $A^\vee = \underline{\mathrm{Hom}}(A, 1)$.

Examples 6.2.6. a) In particular, let E be a field. Then the category Vec_E of finite-dimensional E-vector spaces is a rigid E-linear tensor category, with the usual tensor product and the obvious constraints.

b) The category GrVec_E of finite-dimensional ($\mathbb{Z}$-) graded E-vector spaces V^* is a rigid E-linear tensor category by defining $(V \otimes W)^r = \oplus_{i+j=r} V^i \otimes_E W^j$, taking the associativity constraints from Vec_E, defining $1 = E$ placed in degree 0, and defining $\phi_{A,B}(a \otimes b) = (-1)^{i+j} b \otimes a$ for $a \in A^i$ and $b \in B^j$.

The relationship between the objects introduced in 6.2.2, 6.2.3 and 6.2.4 is as follows.

6.2.7. For any adequate equivalence relation $\sim$, the category $\mathcal{M}_\sim(k)$ of motives modulo $\sim$ becomes a rigid $\mathbb{Q}$-rational tensor category by defining $(X, p, m) \otimes (Y, q, n) = (X \times Y, p \times q, m + n)$, and taking the obvious associativity constraint, the unit object $1 = (\mathrm{Spec}(k), id, 0)$, and the commutativity constraints induced by the transpositions $\tau_{X,Y} : X \times Y \cong Y \times X$.

Recall that a tensor functor $\Phi : \mathcal{A} \to \mathcal{B}$ between tensor categories is a functor together with functorial isomorphisms $\alpha_{A,B} : \Phi(A) \otimes \Phi(B) \cong \Phi(A \otimes B)$, satisfying some obvious compatibilities with respect to the constraints [DM, 1.8]. Then one has:

Lemma 6.2.8. *Let E be a field. Giving an E-linear Weil cohomology theory H is the same as giving a tensor functor $\Phi : \mathcal{M}_{\mathrm{rat}}(k) \longrightarrow \mathrm{GrVec}_F^f$ with $\Phi(1(-1))$ of degree 2.*

This is well-known, and the proof is straightforward (cf. [An2, 4.1.8.1]): Given a Weil cohomology theory H we can extend it to a (covariant) functor on $\mathcal{M}_{\mathrm{rat}}(k)$ by defining $H^*(X, p, m) = pH^{*+m}(X)$. Here we have used the fact that correspondences act on the cohomology, cf. 6.2.2 (ii), and this also gives the functoriality of this association. Conversely, given a tensor functor Φ, we can compose it with the functor $SP_k \to \mathcal{M}_{\mathrm{rat}}(k)$ to obtain a Weil cohomology theory. Here, for a morphism $f : X \to Y$, f_* is induced by $\Phi(\Gamma_f)$, where Γ_f is the graph of f, and f^* is induced by $\Phi(\Gamma_f^t)$, where Γ_f^t is the transpose of Γ_f.

6.3 Finite-dimensional motives

For any object M in a tensor category $\mathcal{C}$, and every natural number N, the symmetric group S_N acts on the N-fold tensor product $M^{\otimes N}$ as follows. For an elementary transposition $(i, i+1)$, $1 \leq i < N$, the induced isomorphism $(i, i+1)_*$ of $M^{\otimes N}$ is induced by applying the commutativity constraint between the i-th and $(i+1)$-st place, i.e., we have $(i, i+1)_* = id_{M^{\otimes(i-1)}} \otimes \phi_{M,M} \otimes id_{M^{\otimes(N-i-1)}}$. One can check that one obtains a well defined action of S_N by decomposing each element σ as a product $\sigma = \prod \tau_\nu$ of such elementary transpositions, and defining $\sigma_* = \prod(\tau_\nu)_*$. Now let $\mathcal{C}$ be a $\mathbb{Q}$-linear pseudo-abelian tensor category. Then, by linearity, the group ring $\mathbb{Q}[S_N]$ acts on $M^{\otimes N}$, and we can define the symmetric product as $\mathrm{Sym}^N M = e_{\mathrm{sym}} M^{\otimes N}$ and the exterior product as $\bigwedge^N M = e_{\mathrm{alt}} M^{\otimes N}$, where $e_{\mathrm{sym}} = \frac{1}{N!} \sum \sigma$ and $e_{\mathrm{alt}} = \frac{1}{N!} \sum \mathrm{sign}(\sigma)\sigma$, with the sum taken over all $\sigma \in S_N$. Note that these are idempotents in $\mathbb{Q}[S_N]$, and that by the very definition of pseudo-abelian categories, every projector has an image in $\mathcal{C}$.

The following definition goes back to Kimura [Ki] and, independently, to O'Sullivan [OSu] (with a different terminology).

Definition 6.3.1. An object M in a $\mathbb{Q}$-linear pseudo-abelian tensor category $\mathcal{C}$ is called

i) evenly finite-dimensional, if there is some $N > 0$ with $\bigwedge^N M = 0$,
ii) oddly finite-dimensional, if there is some $N < 0$ with $\mathrm{Sym}^N M = 0$,
iii) finite-dimensional, if $M = M_+ \oplus M_-$ with M_+ evenly and M_- oddly finite-dimensional, respectively.

For such objects one has the following notion of dimension.

Definition 6.3.2. If M is finite-dimensional, and if M_+ and M_- are as in definition 6.3.1 (iii), define the (Kimura-) dimension of M as $\dim M = \dim_+ M + \dim_- M$, where $\dim_+ M := \max\{r \mid \bigwedge^r M_+ \neq 0\}$ and $\dim_- M := \max\{s \mid \mathrm{Sym}^s M_- \neq 0\}$.

This is well-defined, because M_+ and M_- are unique up to (non unique) isomorphism (loc.cit.).

Examples 6.3.3. a) If E is a field of characteristic zero, then any object V in Vec_E is evenly finite-dimensional. In fact, $\bigwedge^n V$ is the usual alternating power $\bigwedge^n_E V$, and this is zero for $n > \dim_E V$. One has $\dim V = \dim_E V$.

b) With E as above, every object in the tensor category GrVec_E (cf. 6.2.6 (b) for our conventions) is finite-dimensional: For $V = \oplus_{i \in \mathbb{Z}} V^i$ let $V_+ = \oplus_{i\,\mathrm{even}} V^i$ and $V_- = \oplus_{i\,\mathrm{odd}} V^i$. Then one has $\bigwedge^N_{\mathrm{Gr}} V_+ =$

$\bigwedge_E^N V_+$ and $\mathrm{Sym}^N_{\mathrm{Gr}} V_- = \bigwedge_E^N V_-$, by the sign rule for our commutation constraints. Thus $\dim V = \dim_E V$ is the usual dimension of V as an E-vector space (This should not be confused with the rank of V (cf. [DM, p.113]) with respect to the structure of GrVec_E as a rigid tensor category, which would be $\mathrm{rank} V = \dim_E V_+ - \dim_E V_-$).

Fix a Weil cohomology theory H^*. For a smooth projective variety X let $\pi_i = \pi_i^X$ be the Künneth components of the diagonal, and let

$$\pi_+ = \pi_+^X = \pi_0 + \pi_2 + \pi_4 + \dots \,, \qquad \pi_- = \pi_-^X = \pi_1 + \pi_3 + \pi_5 + \dots$$

be the projectors onto the even and odd degree part of the cohomology, respectively. Then we have the 'sign conjecture'

Conjecture $\mathbf{S(X)}$ The projectors π_+^X and π_-^X are algebraic.

It is implied by standard conjecture $C(X)$ (cf. 6.2.2 (b)), and hence known for curves, surfaces and abelian varieties, and in general over finite fields. Moreover $S(X)$ and $S(Y)$ imply $S(X \times Y)$ (because $\pi^+_{X\times Y} = \pi^+_X \times \pi^+_Y + \pi^-_Y \times \pi^-_Y$).

If $S(X)$ holds (for the given H), then the motive $h_{\mathrm{hom}}(X)$ modulo homological equivalence (for the given H) is finite-dimensional. In fact, decompose $h_{\mathrm{hom}}(X) = M_+ \oplus M_-$, with $M_\pm = (h_{\mathrm{hom}}(X), \pi_\pm)$, and let

$$b_\pm(X) = \dim H^*(M_\pm) = \dim H^*(X)_\pm,$$

where $H^*(X)_\pm$ is defined as in 6.3.3(b). Then

$$\bigwedge^{b_+(X)+1} M_+ = 0 = \mathrm{Sym}^{b_-(X)+1} M_-.$$

because $H^* : \mathcal{M}_{\mathrm{hom}}(k) \to \mathrm{GrVec}_F$ is a faithful tensor functor).

In particular, conjecture $S(X)$ implies that also the motive $h_{\mathrm{num}}(X)$ modulo numerical equivalence is finite-dimensional. However, the following conjecture is much deeper:

Conjecture 6.3.4. (Kimura, O'Sullivan) Every motive modulo rational equivalence is finite-dimensional.

Remark 6.3.5. If $M = (X, p, m)$ is a motive modulo the equivalence relation $\sim$, then $\mathrm{Sym}^n M = (X^n, e_{\mathrm{sym}} \circ p^{\otimes n}, n \cdot m)$. So this is 0 if and only if $e_{\mathrm{sym}} \circ p^{\otimes n} \sim 0$. Here $p^{\otimes n} = p \times \dots \times p$ on $(X \times X)^n \cong X^n \times X^n$, and $e_{\mathrm{sym}} \circ p^{\otimes n} = p^{\otimes n} \circ e_{\mathrm{sym}}$, so this is again an idempotent. In fact, for every endomorphism f of $h(X)$ and every $\sigma \in S_n$ obviously $\sigma \circ f^{\otimes n} = f^{\otimes n} \circ \sigma$. Similarly for $\bigwedge^n$ and e_{alt}.

Now let C be a smooth projective curve over k, and let $x \in C$ be a closed point of degree m. Then we have a decomposition $h_{\mathrm{rat}}(C) = 1 \oplus h^1_{\mathrm{rat}}(C) \oplus 1(-1)$, with $h^1_{\mathrm{rat}}(C) := (C, \Delta_C - \frac{1}{m} x \times C - \frac{1}{m} C \times x, 0)$. As for the notation, note that $\tilde{\pi}_o := \frac{1}{m} x \times C$ and $\tilde{\pi}_2 := \frac{1}{m} C \times x$ are orthogonal idempotents lifting the Künneth components π_0 and π_2, respectively. Hence $\Delta_C - \tilde{\pi}_0 - \tilde{\pi}_2$ is an idempotent lifting the Künneth component $\pi_1 = \Delta_C - \pi_0 - \pi_2$.

Now 1 and $1(r)$, for every $r \in \mathbb{Z}$, are evenly finite-dimensional, since S_2 acts trivially on $1 \otimes 1$ and $1(r) \otimes 1(r)$. The following results thus show that $h(C)$ is finite-dimensional.

Theorem 6.3.6. *([Ki, 4.2]) The motive $h^1_{\mathrm{rat}}(C)$ is oddly finite-dimensional. More precisely, one has $Sym^{2g+1} h^1_{\mathrm{rat}}(C) = 0$ where g is the genus of C.*

Proposition 6.3.7. *([Ki]) Let M and N be objects in a $\mathbb{Q}$-linear pseudo-abelian tensor category.*

a) *If M and N are finite-dimensional, then $M \oplus N$ is finite-dimensional, with* $\dim M \oplus N \leq \dim M + \dim N$.
b) *If M and N are finite-dimensional, then $M \otimes N$ is finite-dimensional, with* $\dim M \otimes N \leq \dim M . \dim N$.
c) *If M is finite-dimensional, then also every direct factor of M.*
d) *$M = 0$ if and only if M is finite-dimensional with* $\dim M = 0$.

6.4 Nilpotence and finite-dimensionality

For each smooth projective variety X over k, let $J(X) \subseteq \mathrm{Corr}^0_{\mathrm{rat}}(X, X)$ be the ideal of correspondences which are numerically equivalent to zero. Recall the following conjecture.

Conjecture $\mathbf{N(X)}$ $J(X)$ is a nilpotent ideal.

A remarkable consequence of this conjecture would be that there is no phantom motive, i.e., no non-trivial motive which becomes zero after passing to numerical equivalence, and that every idempotent modulo numerical or homological equivalence can be lifted to an idempotent modulo rational equivalence. In fact, for any motive M modulo rational equivalence let $J(M) \subseteq \mathrm{End}(M)$ be the ideal of numerically trivial endomorphisms (so that $J(X) = J(h(X))$ for X in SP_k). Then we have

Lemma 6.4.1. *Assume that $J(M)$ is a nil ideal. Let M_{num} and M_{hom} be the images of M in $\mathcal{M}_{\mathrm{num}}(k)$ and $\mathcal{M}_{\mathrm{hom}}(k)$ (with respect to a given Weil cohomology), respectively. Then the following holds.*

i) If $M_{\mathrm{num}} = 0$ (e.g., if $H^*(M) = 0$ for a Weil cohomology theory), then $M = 0$.

ii) Any idempotent in $\mathrm{End}(M_{\mathrm{num}})$ or $\mathrm{End}(M_{\mathrm{hom}})$ can be lifted to an idempotent in $\mathrm{End}(M)$, and any two such liftings are conjugate by a unit of $\mathrm{End}(M_{\mathrm{hom}})$ lying above the identity of $\mathrm{End}(M_{num})$.

iii) If the image of $f \in \mathrm{End}(M)$ in $\mathrm{End}(M_{\mathrm{num}})$ is invertible, then so is f.

Proof (i): If id_M maps to zero in $\mathrm{End}(M_{num})$, it is nilpotent, hence zero. (ii) and (iii): These properties holds for any surjection $A \twoheadrightarrow \overline{A} = A/I$ where A is a (not necessarily commutative) ring with unit, and I is a (two-sided) nil ideal. For (iii), it suffices to assume that the element $a \in A$ maps to $1 \in \overline{A}$. But then a is unipotent, hence invertible. As for (ii), if $\overline{e}$ is idempotent in $\overline{A}$ and a is any lift in A, then $(a - a^2)^N = 0$ for some $N > 0$, and it follows easily that $\tilde{e} = (1 - (1 - a)^N)^N$ is an idempotent lifting $\overline{e}$ ([Ki, 7.8]). If e and e' are idempotents of A lying above $\overline{e}$, then $u = e'e + (1 - e')(1 - e)$ lies above $1 \in \overline{A}$. Thus u is invertible, and the equality $e'u = e'e = ue$ shows that $e' = ueu^{-1}$. □

Conjecture $N(X)$ would follow from the existence of the Bloch-Beilinson filtration [Ja2], or Murre's conjecture [Mu1], or the following conjecture of Voevodsky:

Conjecture 6.4.2. ([Voe]) If an algebraic cycle z is numerically trivial, it is smash nilpotent, i.e., there is an $n > 0$ such that $z^{\times n} = 0$.

In fact, as is observed in loc. cit., a smash nilpotent correspondence from X to X is nilpotent; more precisely, $z^{\times n} = 0$ implies $z^n = 0$ in $\mathrm{Corr}(X, X)$. The following result gives (in part (b)) another criterion for nilpotence. Here we may consider motives modulo any (fixed) adequate equivalence relation $\sim$. Recall that, for a motive $M = (X, p, m)$ and an endomorphism f of M, the trace of f is defined as $\mathrm{tr}(f) = \langle f.p^t \rangle$, where $\langle \alpha.\beta \rangle$ is the intersection number of two cycles α and β. This coincides with the trace coming from the rigid tensor category structure of $\mathcal{M}_\sim(k)$.

Theorem 6.4.3. *Let $f : M \to M$ be an endomorphism of a motive.*

a) If $\wedge^{d+1} M = 0$ (resp. $\mathrm{Sym}^{d+1} M = 0$), then $\sum_{i=0}^{d} (-1)^{d-i} \mathrm{tr}(\wedge^{d-i} f) f^i = 0$ (resp. $\sum_{i=0}^{d} \mathrm{tr}(\mathrm{Sym}^{d-i} f) f^i = 0$).

b) In particular, if M is either evenly finite-dimensional or oddly finite-dimensional, and $d = \dim M$, then there is a monic polynomial $G(t) \in \mathbb{Q}[t]$ of degree d with $G(f) = 0$. If f is numerically equivalent to 0, then f is nilpotent, viz., $f^d = 0$.

This was originally proved by Kimura [Ki] in a slightly weaker form – giving (b) with $d+1$ instead of d, and not giving the description (a) of $G(t)$. The following corollaries already follow from this original form, except for 6.4.7.

Corollary 6.4.4. *If M is a finite dimensional motive, then the ideal $J_M \subseteq End(M)$ of numerically trivial endomorphisms is nilpotent.*

In fact, by decomposing $M = M_+ \oplus M_-$, it is shown in [Ki] that $J(M)$ is a nil ideal, with degree of nilpotence bounded by $n = (\dim_+ M . \dim_- M + 1).\max(\dim_+ M, \dim_- M) + 1$. By a result of Nagata-Higman (cf. [AK, 7.2.8)] it follows that $J(M)$ is in fact a nilpotent ideal, of nilpotence degree $\leq 2^n - 1$ (since we assume $\mathbb{Q}$-coefficients).

Corollary 6.4.5. *If M is a finite-dimensional motive, and H is any F-rational Weil cohomology theory, then $\dim M = \sum_{i\in\mathbb{Z}} \dim_F H^i(M)$. In particular, the right hand side is independent of H.*

Proof (cf. [Ki, 3.9 and 7.4]) We may assume that M is either evenly or oddly finite-dimensional. Obviously, the dimension decreases under any tensor functor, so $\dim M \geq \dim_F H(M)$. On the other hand, by the nilpotence result (together with 6.3.7 and 6.4.1), $\bigwedge^r M = 0$ if $\bigwedge^r H(M) = H(\bigwedge^r M) = 0$, similarly for Sym^r. Thus $\dim M \leq \dim_F H(M)$. □

Corollary 6.4.6. *(compare 6.3.7) If M and N are finite-dimensional motives, then $\dim(M \oplus N) = \dim M + \dim N$ and $\dim M \otimes N = \dim M \cdot \dim N$.*

In fact, this holds for $\sum\limits_{i\in\mathbb{Z}} \dim_F H^i(-)$. For $d = 1$ Theorem 6.4.3 implies:

Corollary 6.4.7. *If M is a finite-dimensional motive with $\dim M = 1$, then $J(M) = 0$, i.e., on $End(M)$ numerical and rational equivalence coincide. Moreover, $End(M) = \mathbb{Q}$.*

Examples 6.4.8. a) If M is an evenly (resp. oddly) finite-dimensional motive of dimension d, then $\bigwedge^d M$ (resp. $\mathrm{Sym}^d M$) is one-dimensional. In fact, by 6.4.5 it suffices to show this after applying some Weil cohomology theory, and then it holds, again by 6.4.5.

b) If C is a curve of genus g, then $h^1_{\mathrm{rat}}(C)$ is oddly finite-dimensional with $\dim h^1_{\mathrm{rat}}(C) = 2g$: This follows from 6.4.6, 6.4.5, and the fact that for the ℓ-adic cohomology, $\ell \neq \mathrm{char}\, k$, one has $\dim_{\mathbb{Q}_\ell} H^1(C \times_k \overline{k}, \mathbb{Q}_\ell) = 2g$. Moreover $\mathrm{Sym}^{2g} h^1_{\mathrm{rat}}(c) \cong 1(-g)$: First of all, $\mathrm{Sym}^{2g} h^1_{\mathrm{rat}}(C)$ is one-dimensional, by (a). Then, by 6.4.4 and 6.4.1, we only have to show this isomorphism modulo (some) homological equivalence. But one knows that $h^1_{\hom}(C) \cong h^1_{\hom}(\mathrm{Jac}(C))$, where $\mathrm{Jac}(C)$ is the Jacobian of C, and that $\bigwedge^{2g} h^1_{\mathrm{rat}}(\mathrm{Jac}(C)) \cong h^{2g}_{\mathrm{rat}}(\mathrm{Jac}(C)) \cong 1(-g)$. Here we have used the fact that $\mathrm{Jac}(C)$ is an abelian variety of dimension g, that for an abelian variety A the Künneth components π_i of the diagonal are algebraic, and that for $h^i_{\hom}(A) := h_{\hom}(A, \pi_i)$ one has a canonical isomorphism $h^i_{hom}(A) \cong \bigwedge^i h^1_{\hom}(A)$.

We can deduce a certain converse of Theorem 6.4.3. Consider the following, a priori weaker variant of conjecture $N(X)$ (for a smooth projective variety X).

Conjecture ($\mathbf{N'(X)}$). $J(X)$ is a nil ideal.

Corollary 6.4.9. *Let X be a smooth projective variety X, and let H^* be any Weil cohomology theory. Then the following statements are equivalent, where $S(X)$ is meant with respect to H^*:*

a) *$h(X)$ is finite-dimensional.*
b) *$S(X)$ holds, and $N(X^n)$ holds for all $n \geq 1$.*
c) *$S(X)$ holds, and $N'(X^n)$ holds for all $n \geq 1$.*

Proof If $M = h(X) = M_+ \oplus M_-$, where $M_+ = h(X, p_+)$ (resp. $M_- = h(X, p_-)$) is evenly (resp. oddly) finite-dimensional, then $H^*(M_+)$ (resp. $H^*(M_-)$) is the even (resp. odd) degree part of $H^*(M)$), because $H^* : \mathcal{M}_{rat}(k) \to GrVec_E$ is a tensor functor. Therefore, modulo homological equivalence, $p_\pm = \pi^X_\pm$. Therefore (a) implies $S(X)$, and by 6.4.4, it also implies $N(X)$. Since (a) also implies finite-dimensionality of $h(X^n) = h(X)^{\otimes n}$, for all $n \geq 1$ (by 2.7), (a) implies (b).
(b) $\Rightarrow$ (c) is trivial.
(c) $\Rightarrow$ (a): If $S(X)$ holds, the $\pi_\pm$ are algebraic projectors modulo homological equivalence, and if $N'(X)$ holds, these lift to orthogonal projectors $\tilde{\pi}_+$ and $\tilde{\pi}_-$ modulo rational equivalence with $\tilde{\pi}_+ + \tilde{\pi}_- = id$ by 3.1 (lift π_+ to a projector $\tilde{\pi}_+$ and let $\tilde{\pi}_- = id - \tilde{\pi}_+$). Let $M_\pm = (X, \tilde{\pi}_\pm, 0)$ modulo rational equivalence. Then $M = M_+ \oplus M_-$, and for $b_\pm = \dim H^*(M_\pm)$ one has $\wedge^{b_+ +1} M_+ = 0 = \mathrm{Sym}^{b_- +1} M_-$ modulo homological equivalence. By 3.1 and $N'(X^n)$, for $n = b_+ + 1$ and $n = b_- + 1$, one concludes that this vanishing also holds modulo rational equivalence, i.e., we obtain (a). □

Corollary 6.4.10. *Voevodsky's nilpotence conjecture (cf. 6.4.2) implies the conjecture of Kimura and O'Sullivan (cf. 6.3.4).*

Proof It implies the standard conjecture $D(X)$ (postulating $\sim_{\mathrm{hom}} = \sim_{\mathrm{num}}$), hence $B(X)$, hence $C(X)$, hence $S(X)$. Moreover, it implies $N'(X)$ (cf. the lines after 6.4.2). □

Remark 6.4.11. O'Sullivan (cf. [An1, Th. 3.33]) has proved: Let $\mathcal{C}$ be a rigid tensor subcategory of $\mathcal{M}(k)$. If every motive in $\mathcal{C}$ is finite-dimensional, and if every tensor functor

$$\omega : \mathcal{C} \longrightarrow s\mathrm{Vec}_F$$

(where F is a field of characteristic 0 and sVec_F is the category of finite-dimensional super (i.e., $\mathbb{Z}/2$-graded) vector spaces over K) factors through numerical equivalence, then Voevodsky's conjecture holds for $\mathcal{C}$.

So far we have only applied the nilpotence result in 6.4.3 (b). Theorem 6.4.3 (a) gives the following Cayley-Hamilton theorem.

Theorem 6.4.12. *Let f be an endomorphism of a motive M, and assume that M is either evenly finite-dimensional or oddly dimensional. Let H^* be a Weil cohomology theory, and let $P(t) = \det(t - f \mid H^*(M))$ be the characteristic polynomial of f on $H^*(M)$. Then $P(t)$ is independent of the chosen Weil cohomology theory, and one has $P(g) = 0$.*

Proof If M is an evenly finite-dimensional motive, its cohomology is even, and by the trace formula [Kl, 1.3.6 c] one has $\mathrm{tr}(f) = \mathrm{tr}(f, H^*(M))$. Therefore one has

$$\sum_{i=0}^{d} (-1)^{d-i}\mathrm{tr}(\wedge^{d-i} f)t^i = \sum_{i=0}^{d} (-1)^{d-i}\mathrm{tr}(\wedge^{n-i} f \mid \bigwedge^{n-i} H^*(M))t^i .$$

On the other hand, it is known that the right hand side is the charcteristic polynomial $P(t)$. For an oddly finite-dimensional motive its cohomology is odd and one has $tr(f) = -tr(f, H^*(M))$, so that

$$\sum_{i=0}^{d} \mathrm{tr}(\mathrm{Sym}^{d-i} f)t^i = \sum_{i=0}^{d} (-1)^{d-i}\mathrm{tr}(\wedge^{n-i} f \mid \bigwedge^{n-i} H^*(M))t^i$$

is again equal to $P(t)$. Therefore the claim follows with 6.4.3 (a). □

We now come to the proof of Theorem 6.4.3. It is straightforward to prove the following two lemmas. (Note that for $n = 3$, Lemma 6.4.13 is just the definition of composition of correspondences.)

Lemma 6.4.13. *Let $p_{ij} : X^n \to X \times X$ be the projection onto the i-th and j-th factor $((x_1, \ldots, x_n) \mapsto (x_i, x_j))$. Consider algebraic cycles $f_1, \ldots, f_{n-1}$ on $X \times X$, regarded as correspondence from X to X. Then one has*

$$(p_{1,n})_*(p_{1,2}^* f_1 \cdot p_{2,3}^* f_2 \cdots p_{n-2,n-1}^* f_{n-2} \cdot p_{n-1,n}^* f_{n-1}) = f_{n-1} \circ f_{n-2} \circ \ldots \circ f_2 \circ f_1$$

(composition of correspondences on the right hand side).

Lemma 6.4.14. *Consider morphisms $f : V \to M$, $g : W \to N$ of smooth, projective varieties, and the diagram*

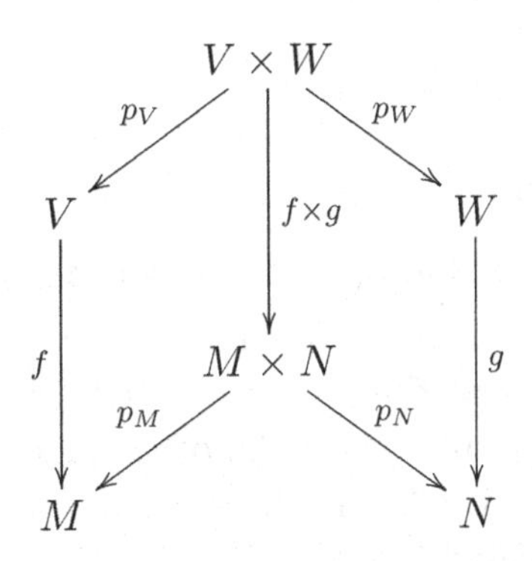

where p_V, p_W, p_M and p_N are the projections. Then, for algebraic cycles α on V and β on W one has

$$(f \times g)_*(p_V^*\alpha \cdot p_W^*\beta) = p_M^* f_*\alpha \cdot p_N^* g_*\beta,$$

i.e., $(f \times g)_*(\alpha \times \beta) = f_*\alpha \times g_*\beta$ for the exterior products.

Proof of Theorem 6.4.3 Let us consider the case where M is evenly finite-dimensional, with $\bigwedge^n M = 0$ (The odd case is similar). If $M = (X, p, m)$, then f is a cycle on $X \times X$ such that $pf = f = fp$. Then, by assumption, we have

$$\sum_{\sigma \in S_n} \operatorname{sgn}(\sigma)\ \sigma \circ p \times \ \ldots \ \times p = 0$$

(where we have n factors p) since this endomorphism factors through $\bigwedge^n M$. This means

$$\sum_{\sigma \in S_n} \operatorname{sign}(\sigma)\ p_{1,n+\sigma(1)}^* p \cdot p_{2,n+\sigma(2)}^* p \ \cdots \ p_{n,n+\sigma(n)}^* p = 0$$

since $p \times \ldots \times p = p_{1,n+1}^* p \cdot p_{2,n+2}^* p \ \cdots \ p_{n,2n}^* p$ on X^{2n}, where $p_{ij} : X^{2n} \to X \times X$ is the projection onto the i-th and j-th factor as in 6.4.13. In particular, we have

$$\sum_{\sigma \in S_n} \operatorname{sign}(\sigma)\ (p_{1,n+1})_*(p_{1,n+\sigma(1)}^* p \cdot \ \cdots \ \cdot p_{n,n+\sigma(n)}^* p \cdot p_{2,n+2}^* f^t$$
$$\cdot\, p_{3,n+3}^* f^t \cdot \ \cdots \ \cdot p_{n,2n}^* f^t) = 0.$$

Let $o_1(\sigma) = \{1, \sigma(1), \sigma^2(1), \ldots, \sigma^{s-1}(1)\}$ (with $\sigma^s(1) = 1$) be the orbit of $1 \in \{1, \ldots, n\}$ under $\sigma \in S_n$. Then σ is the product

$$\sigma = (1\, \sigma(1)\, \sigma^2(1)\, \ldots\, \sigma^{s-1}(1)) \cdot \sigma'$$

of the s-cycle $\sigma_1 : 1 \mapsto \sigma(1) \mapsto \sigma^2(1) \mapsto \ldots \mapsto \sigma^{s-1}(1) \mapsto 1$ and a product σ' of cycles which are disjoint from σ_1. Thus, in the above sum, the summand corresponding to σ is

$$\begin{aligned}\operatorname{sgn}(\sigma)\,(p_{1,n+1})_*(p^*_{1,n+\sigma(1)}p \cdot p^*_{n+\sigma(1),\sigma(1)}g \cdot p_{\sigma(1),n+\sigma^2(1)}p \cdot p_{n+\sigma^2(1),\sigma^2(1)}g \\ \cdots\, p^*_{n+\sigma^{s-1}(1),\sigma^{s-1}(1)}g \cdot p^*_{\sigma^{s-1},n+1}p \cdot \beta)\end{aligned}$$

where β is the product of the $2(n-s)$ factors $p^*_{i,n+\sigma(i)}p$ and $p^*_{i,n+i}f^t$ with $i \in \overline{o_1(\sigma)} = \{1,\ldots,n\} \setminus o_1(\sigma)$. Writing $X^{2n} = V \times W$, with V being the product of the $2s$ factors at the places i or $n+i$ for $i \in o_1(\sigma)$, and W the product over the $2(n-s)$ factors at the other places, it follows from 6.4.11 (applied to $p_{1,n+1} : V \to X \times X$ and the structural morphism $W \to \operatorname{Spec}(k)$) and 6.4.10 that the summand is

$$\operatorname{sign}(\sigma) \cdot (f)^{s-1} \deg(\beta')$$

where β' is a zero cycle on W and $(f)^{s-1}$ is the $(s-1)$-fold self product in $\operatorname{End}(M)$. If σ is an n-cycle, then $\operatorname{sign}(\sigma) = (-1)^{n-1}$, $s = n$, $\beta = \beta' = 1$ and $\deg(\beta') = 1$, so that the summand is $(-1)^{n-1}(f)^{n-1}$. If σ is not an n-cycle, then $s < n$.

This shows that we get a polynomial equation for f with leading term $(-1)^{n-1}(n-1)!\,(f)^{n-1}$. If f is numerically equivalent to zero, then so is β' for $s < n$, so that $\deg(\beta') = 0$ unless σ is an n-cycle. This proves 6.4.3 (a). For 3.3 (b) we note that, by choosing a bijection $\rho : \{1,\ldots,n-s\} \to \overline{o_1(\sigma)}$, we may identify W with $X^{n-s} \times X^{n-s}$ and β' with

$$\begin{aligned}\beta'' &= p^*_{1,n-s+\sigma''(1)}p \cdot \cdots \cdot p^*_{n-s,n-s+\sigma''(n-s)}p \cdot p^*_{1,n-s+1}f^t \cdot \cdots \cdot p^*_{n-s,2(n-s)}f^t \\ &= (\sigma'' \circ p \times \cdots \times p).(f \times \cdots \times f)^t,\end{aligned}$$

where $\sigma'' = \rho^{-1}\sigma'\rho \in S_{n-s}$. Thus

$$\deg(\beta') = tr(\sigma'' \circ p \times \cdots \times p \circ f \times \cdots \times f) = tr(\sigma'' \circ f \times \cdots \times f).$$

Summing over all $\sigma \in S_n$ with fixed σ_1, and keeping the bijection ρ, we thus get

$$(-1)^{s-1} \sum_{\tau \in S_{n-s}} \operatorname{sign}(\tau) tr(\tau \circ f \times \ldots \times f) = (-1)^{s-1}(n-s)!tr(\wedge^{n-s} f) f^{s-1}.$$

After summing over all $\sigma \in S_n$ we then see that the coefficient of f^{s-1} is

$$(-1)^{n-1}(n-1)!\,(-1)^{n-s} tr(\wedge^{n-s} f),$$

because there are $(n-1)!/(n-s)!$ cycles σ_1 containing 1 of length s. This proves 3.3 (b).

6.5 Finite fields

Let us recall Murre's conjecture. Let X be a purely d-dimensional smooth projective variety over a field k, fix a Weil cohomology theory, and assume the standard conjecture $C(X)$, i.e., that the Künneth components $\pi_i = \pi_i^X \in H^{2d}(X \times X)$ of the diagonal Δ_X are algebraic.

Conjecture 6.5.1 (Murre, [Mu1]). A) X has a Chow-Künneth decomposition, i.e., the π_i^X lift to an orthogonal set of idempotents $\{\tilde{\pi}_i\}$ with $\sum_i \tilde{\pi}_i = \Delta_X$ in $\mathsf{CH}^d(X \times X)$.

B) The correspondences $\tilde{\pi}_{2j+1}, \ldots, \tilde{\pi}_{2d}$ act as zero on $\mathsf{CH}^j(X)$.

C) Let $F^\nu\mathsf{CH}^j(X) = \operatorname{Ker} \tilde{\pi}_{2j} \cap \operatorname{Ker} \tilde{\pi}_{2j-1} \cap \ldots \cap \operatorname{Ker} \tilde{\pi}_{2j-\nu+1} \subseteq \mathsf{CH}^j(X)$. Then the descending filtration $F^\cdot$ is independent of the choice of the $\tilde{\pi}_i$.

D) $F^1\mathsf{CH}^j(X) = \mathsf{CH}^j(X)_{\mathrm{hom}} := \{z \in \mathsf{CH}^j(X) | z \sim_{\mathrm{hom}} 0\}$.

It is known [Ja2] that this conjecture, taken for all smooth projective varieties, is equivalent to the conjecture of Bloch-Beilinson on a certain functorial filtration on Chow groups, and that this Bloch-Beilinson filtration would be equal to the filtration $F^\cdot$ defined above. The advantage of Murre's conjecture is that it can be formulated and proved for specific varieties, and that will be used below.

Remarks 6.5.2. a) The condition in 6.5.1(A) is called the Chow-Künneth decomposition, because it amounts to saying that the Künneth decomposition $h_{\mathrm{hom}}(X) = \oplus_{i=0}^{2d} h^i(X)$ in $\mathcal{M}_{\mathrm{hom}}(k)$, with $h^i(X) = (X, \pi_i^X)$, can be lifted to a decomposition $h_{\mathrm{rat}}(X) = \oplus_{i=0}^{2d} \tilde{h}^i(X)$ in the category $\mathcal{M}_{\mathrm{rat}}(k)$ of Chow motives, via $\tilde{h}^i(X) = (X, \tilde{\pi}_i)$.

b) Conjecture 6.5.1 (A) would follow from the nilpotence of $J(X)$, i.e., from conjecture $N(X)$ (cf. 6.4.1), and hence from finite-dimensionality of $h_{\mathrm{rat}}(X)$. On the other hand it is known that the Bloch-Beilinson-Murre conjecture would imply the conjecture (6.3.4) of Kimura-O'Sullivan ([AK], [An1]). Let us note here that, more precisely, Murre's conjecture for X, $X \times X$ and $X \times X \times X$ implies $N(X)$ ([Ja2, pp. 294, 295]), so that Murre's conjecture for all sufficient high powers X^N implies finite-dimensionality of X (6.4.9; note that we have assumed $C(X)$, hence $S(X)$).

Now let k be a finite field. Then the standard conjecture $C(X)$ holds for every smooth projective variety X over k [KM]. It is furthermore known (cf. [Ja3, 4.17]) that the conjecture of Bloch-Beilinson-Murre over k is equivalent to the equality $\sim_{\mathrm{rat}} = \sim_{\mathrm{hom}}$ where homological equivalence is taken with respect to any Weil cohomology theory H satisfying weak Lefschetz (cf. [Kl,

p. 368] or [KM, p. 74]) (e.g., the ℓ-adic cohomology (6.2.3 (b)) for any fixed $\ell \neq \mathrm{char}(k)$). Again we want to make this more precise.

Theorem 6.5.3. *Let k be a finite field. The equality $\sim_{\mathrm{rat}} = \sim_{\mathrm{hom}}$ on $X \times X$ implies Murre's conjecture for X. Conversely, Murre's conjecture for X and $X \times X$ implies the equality $\sim_{\mathrm{rat}} = \sim_{\mathrm{hom}}$ for X.*

Proof The first claim is trivial. For the second claim we use a result of Soulé:

Proposition 6.5.4 (Prop. 2, [So1]). *Let X be smooth projective over k. The k-linear Frobenius $F : X \to X$ acts on $\mathsf{CH}^j(X)$ as the multiplication by $q =$ cardinality of k.*

Given this, assume Murre's conjecture for X and $X \times X$. We may assume that X is irreducible of dimension d. Let $\tilde{\pi}_0, \ldots, \tilde{\pi}_{2d}$ be orthogonal idempotents lifting the Künneth components $\pi_0^X, \ldots, \pi_{2d}^X$ of the diagonal, and define $\tilde{h}^i(X) = (X, \tilde{\pi}_i)$ in the category $\mathcal{M}_{\mathrm{rat}}(k)$ of Chow motives. By Murre's conjecture for $X \times X$,

$$\mathsf{CH}^d(X \times X)_{\mathrm{hom}} = \bigoplus_{r<2d} \widetilde{\pi_r^{X \times X}}\, \mathsf{CH}^d(X \times X)$$

where $\widetilde{\pi_r^{X \times X}} = \sum_{\mu+\nu=r} (\tilde{\pi}_{2d-\mu})^t \times \tilde{\pi}_\nu$ lifts the Künneth component $\pi_r^{X \times X} = \sum_{\mu+\nu=r} \pi_\mu^X \times \pi_\nu^X$ of $X \times X$ (note that $(\pi_{2d-\mu}^X)^t = \pi_\mu^X$). But

$$((\tilde{\pi}_{2d-\mu})^t \times \tilde{\pi}_\nu)\, \mathsf{CH}^d(X \times X) \;=\; \tilde{\pi}_\nu) \circ \mathsf{CH}^d(X \times X) \circ \tilde{\pi}_{2d-\mu},$$

and for $\alpha \in \mathsf{CH}^d(X \times X)$ we have

$$\tilde{\pi}_\nu \alpha \tilde{\pi}_{2d-\mu} \tilde{\pi}_i \mathsf{CH}^j(X) = 0 \quad \text{for} \quad i \neq 2d - \mu.$$

On the other hand, for $i = 2d - \mu$ and $\mu + \nu < 2d$ we have

$$\tilde{\pi}_\nu \alpha \tilde{\pi}_{2d-\mu} \tilde{\pi}_i \mathsf{CH}^j(X) \subseteq \tilde{\pi}_\nu \mathsf{CH}^j(X)$$

with $\nu < i$. This shows that $\mathsf{CH}^d(X \times X)_{\mathrm{hom}}$ acts trivially on $\mathrm{Gr}_F^i \mathsf{CH}^j(X)$ for Murre's filtration, because

$$F^i \mathsf{CH}^j(X) = \sum_{m \leq 2j-i} \tilde{\pi}_m \mathsf{CH}^j(X)$$

by 6.5.1 (B) and (C). In other words, the correspondences in $\mathsf{CH}^d(X \times X)$ act on $\mathrm{Gr}_F^i \mathsf{CH}^j(X)$ modulo homological equivalence, and then this quotient just depends on the motive modulo homological equivalence $h^{2j-i}(X) =$

(X, π^X_{2j-i}). Let $P_i(t) = \det(t - F^* \mid H^*(X))$ be the characteristic polynomial of the k-linear Frobenius $F : X \to X$ acting on the cohomology. It is known from [KM] that

$$P_i(t) = \det(t - F^* \mid H^i(X \times_k \overline{k}, \mathbb{Q}_\ell))$$

for any $\ell \neq \mathrm{char}(k)$ and hence, by Deligne's proof of the Weil conjectures, that $P_i(t)$ is in $\mathbb{Z}[t]$, and has zeros with complex absolute values $q^{i/2}$.

By the Cayley-Hamilton theorem, $P_i(F)$ acts as zero on $H^i(X)$, hence $P_{2j-\nu}(F)$ acts as zero on $\mathrm{Gr}^\nu_F \mathsf{CH}^j(X)$. Since $F = q^j$ on $\mathsf{CH}^j(X)$ by Soulé's result, and $P_{2j-\nu}(q^i) \neq$ for $\nu \neq 0$, we deduce $\mathrm{Gr}^\nu_F \mathsf{CH}^j(X) = 0$ for $\nu \geq 1$. q.e.d. □

One can prove part of Murre's conjecture from finite-dimensionality, by applying ideas of Soulé [So1], Geisser [Gei], and Kahn [Ka].

Theorem 6.5.5. *Let k be a finite field, and let X/k be a smooth projective variety such that $J(X)$ is a nil ideal (e.g., assume that $h_{rat}(X)$ is finite-dimensional). Then there is a unique Chow-Künneth decomposition $h_{rat}(X) = \oplus_{i=0}^{2d} \widetilde{h}^i(X)$, and one has*

$$CH^j(\widetilde{h}^i(X)) = 0 \textit{ for } i \neq 2j.$$

In particular, parts (A), (B) and (C) of Murre's conjecture hold for X and, moreover, $\tilde{\pi}_i$ acts as zero on $CH^j(X)$ for all $i \neq 2j$, so that $F^\nu = 0$ for all $\nu \geq 1$.

Proof The existence of the Chow-Künneth decomposition was noted in 6.5.2. Let $P_i(t) = \det(t - F \mid H^i(X))$ be as above. By Cayley-Hamilton we have $P_i(F) = 0$ in $\mathrm{End}(h^i_{\mathrm{hom}}(X))$, so that $P_i(F)^r = 0$ in $\mathrm{End}(\widetilde{h}^i(X))$ for some $r \geq 1$, by assumption. Therefore

$$0 = P_i(F)^r \cdot \mathsf{CH}^j(\widetilde{h}^i(X)) = P_i(q^j)^r \cdot \mathsf{CH}^j(\widetilde{h}^i(X)),$$

but $P_i(q^j) \neq 0$ for $j \neq 2i$, by Deligne's proof of the Weil conjecture. The claimed consequences for Murre's conjecture are now immediate. Finally, the uniqueness of the Chow-Künneth decomposition is seen as follows. Let $P(t) = \prod P_i(t)$. Then $P(F)$ is homologically trivial, so that $P(F)^r = 0$ for some $r \geq 0$ in $\mathrm{End}(h_{\mathrm{rat}}(X))$. Again by Deligne, the polynomials $P_\nu(t)$ are also pairwise coprime, so that, for each $i \in \{0, \ldots, 2d\}$ there are polynomials $a_i(t)$ and $b_i(t)$ in $\mathbb{Q}(t)$ with

$$a_i(t)P_i(t)^r + b_i(t)Q_i(t)^r = 1,$$

where $Q_i(t) = \prod_{j \neq i} P_j(t)$. Then the elements $\tilde{\pi}_i = b_i(F)Q_i(F)^r$ are pairwise orthogonal idempotents in $\mathrm{End}(h_{\mathrm{rat}}(X))$ summing up to 1, and $\tilde{\pi}_i$ is a lift of the i-th Künneth projector π_i^X, as follows from Cayley-Hamilton. Finally, by 6.4.1 every other idempotent lifting π_i^X is of the form $(1+a)\tilde{\pi}_i(1+a)^{-1}$ with $a \in J(X)$, cf. [Ja2, 5.4]. But every endomorphism of $h_{rat}(X)$ commutes with F (cf. [So1, Prop. 2 ii]), hence with $\tilde{\pi}_i$, so that we obtain $\tilde{\pi}_i$ again. This shows the uniqueness of the Chow-Künneth decomposition. □

Remarks 6.5.6. a) The above proof, together with the fact that the Frobenius F commutes with all morphisms of Chow motives ([So1, Prop. 2 ii)]), shows that the full (tensor) subcategory $\mathcal{M}_{\mathrm{rat}}^{fin}(k) \subset \mathcal{M}_{\mathrm{rat}}(k)$ consisting of the finite-dimensional motives, possesses a unique weight grading in the sense of [Ja3, 4.11], i.e., a grading lifting the weight grading of $\mathcal{M}_{\mathrm{hom}}(k)$: Every motive M has a unique decomposition $M = \oplus_i M^i$ with $H^i(M) = H^i(M^i)$ and $H^j(M^i) = 0$ for $j \neq i$, and this filtration is respected by any morphism.

b) By 2.6 and 2.7, the category $\mathcal{M}_{\mathrm{rat}}^{fin}(k)$ contains $\mathcal{M}_{\mathrm{rat}}^{av}(k)$, the full rigid pseudo-abelian tensor subcategory of $\mathcal{M}_{\mathrm{rat}}(k)$ generated by the motives of abelian varieties. Hence the assumptions of Theorem 6.5.5 hold for (products of) curves, abelian varieties, Fermat hypersurfaces of degree m invertible in k, and Kummer or Enriques surfaces.

As in loc. cit., one can extend the above result to higher algebraic K-theory, or rather motivic cohomology with rational coefficients. Recall that, for any smooth variety V over a field L its $\mathbb{Q}$-rational motivic cohomology can be defined by $H^i_{\mathcal{M}}(V, \mathbb{Q}(j)) = K_{2j-i}(V)^{(j)}$, where

$$K_m(V)^{(j)} := \{\, x \in K_m(V)_{\mathbb{Q}} \mid \psi_n(x) = n^j x \text{ for all } n \geq 1 \,\}.$$

Here ψ_n is the n-th Adams operator, acting on the algebraic K-group $K_m(V)$. One has $H^{2j}_{\mathcal{M}}(V, \mathbb{Q}(j)) = K_0(V)^{(j)} \cong \mathrm{CH}^j(V)$ by the Riemann-Roch theorem for Chow groups. It is known (and follows from the fact just recalled) that algebraic correspondences modulo rational coefficients act on motivic cohomology, so that motivic cohomology extends to a covariant functor on $\mathcal{M}_{\mathrm{rat}}(k)$ by defining

$$H^i_{\mathcal{M}}(M, \mathbb{Q}(j)) = pH^{i+2n}_{\mathcal{M}}(X, \mathbb{Q}(j+n))$$

for $M = (X, p, n)$.

Theorem 6.5.7. *Under the assumptions of theorem 6.5.5, one has*

$$H^i_{\mathcal{M}}(\tilde{h}^\nu(X), \mathbb{Q}(j)) = 0 \textit{ for } \nu \neq 2j.$$

In particular, $H^i_{\mathcal{M}}(X, \mathbb{Q}(j)) = 0$ *for* $j > d = \dim(X)$.

Proof This follows as above, by using that F acts on $H^i(X, \mathbb{Q}(j))$ as q^j, because $F = \psi_q$ on $K_m(X)$ [So2, 6.1], while $P_\nu(q^j) \neq 0$ for $\nu \neq 2j$. □

It remains to investigate part (D) of Murre's conjecture, i.e., the equality $F^1\mathsf{CH}^j(X) = \mathsf{CH}^j(X)_{\mathrm{hom}}$. Recall that the Tate conjecture for $H^{2j}(\overline{X}, \mathbb{Q}_\ell)$ states the surjectivity of the cycle map

$$\mathsf{CH}^j(X) \otimes_{\mathbb{Q}} \mathbb{Q}_\ell \longrightarrow H^{2j}(\overline{X}, \mathbb{Q}_\ell(j))^\Gamma ,$$

where $\Gamma = \mathrm{Gal}(k^{sep}/k)$ is the absolute Galois group of k.

Theorem 6.5.8. *Let X be smooth, projective of pure dimension d. Assume that*

i) $J(X)$ is a nil ideal (e.g., assume that X is finite-dimensional),
ii) the Tate conjecture holds for $H^{2j}(\overline{X}, \mathbb{Q}_\ell)$ and $H^{2(d-j)}(\overline{X}, \mathbb{Q}_\ell)$, and
iii) the eigenvalue 1 of F is semi-simple on $H^{2j}(\overline{X}, \mathbb{Q}_\ell(j))$.

Then the following holds.

a) $\sim_{\mathrm{rat}} = \sim_{\mathrm{num}}$ on $\mathsf{CH}^\nu(X)$ (i.e., $\mathsf{CH}^\nu(X)_{\mathrm{num}} = 0$), for $\nu = j,\ d-j$.
b) $H^i_{\mathcal{M}}(X, \mathbb{Q}(\nu)) = 0$ for all $i \neq 2\nu$, for $\nu = j,\ d-j$.

Proof This follows from results of Geisser [Gei] and Kahn [Ka]. Let us give a brief argument, for avoiding a little problem with the arguments given in [Ka], and for getting a statement used below.

By Poincaré duality, (iii) also holds for $d-j$, so it suffices to consider $\nu = j$. Then, by theorems 6.5.5 and 6.5.7, it suffices to consider $\tilde{h}^{2j}(X)$ instead of X in the statements. Now it is well-known (cf. [Ta, (2.9)]) that the assumptions on the Tate conjecture and the semi-simplicity of F imply that $\sim_{\mathrm{num}} = \sim_{\mathrm{hom}}$ on $\mathsf{CH}^\nu(X)$ for $\nu = j$ and $d-j$, and that

$$A^\nu_{\mathrm{num}}(X) \otimes_{\mathbb{Q}} \mathbb{Q}_\ell \xrightarrow{\sim} H^{2j}(\overline{X}, \mathbb{Q}_\ell(j))^\Gamma \text{ for } \nu = j,\ d-j$$

via the cycle map. Let $P_{2j}(t) = \det(t - F \mid H^{2j}(X))$ where H is the Weil cohomology theory given by ℓ-adic cohomology, $\ell \neq p = \mathrm{char}(k)$. Write $P_{2j}(t) = Q(t)(t-q^j)^\rho$ with $(t-q^j) \nmid Q(t)$ and some $\rho \geq 0$. By assumption, there is an integer $r > 0$ with $P_{2j}(F)^r = 0$ in $\mathrm{End}(\tilde{h}^{2j}(X))$. Now we have $1 = q(t)Q(t)^r + r(t)(t-q^j)^{\rho r}$ with polynomials $q(t)$ and $r(t)$ in $\mathbb{Q}$. This shows that $Q'(F) = q(F)Q(F)^r$ and $P'(F) = r(F)P(F)^r$, with $P(t) = (t-q^j)^\rho$, are orthogonal idempotents in $\mathrm{End}(\tilde{h}^{2j}(X))$ with $Q'(F) + P'(F) = 1$. Let $M_1 = P'(F)\tilde{h}^{2j}(X)$ and $M_2 = Q'(F)\tilde{h}^{2j}(X)$. Then $M = M_1 \oplus M_2$ and $\mathsf{CH}^j(M_1) = 0$ as in the proof of 6.5.5, because $Q(F)^r M_1 = 0$ and $Q(q^j) \neq 0$, and similarly $H^i_{\mathcal{M}}(M_1, \mathbb{Q}(j)) = 0$.

We now claim that $M_2 \cong 1(-j)^\rho$. Then the claims follow, because it is clear that $\sim_{\text{rat}} = \sim_{\text{hom}}$ on $1(-j)$, and well-known (by work of Quillen) that $H^i_{\mathcal{M}}(1(-j), \mathbb{Q}(j)) = H^{i-2j}(\operatorname{Spec} k, \mathbb{Q}(0)) = 0$ for $i \neq 2j$ if k is a finite field.

The characteristic polynomial of F on $P'(F)H^{2j}(\overline{X}, \mathbb{Q}_\ell(j))$ is $(t - q^j)^\rho$. Hence

$$H(M_2(j)) = Q'(F)H^{2j}(\overline{X}, \mathbb{Q}_\ell(j)) \cong H^{2j}(\overline{X}, \mathbb{Q}_\ell(j))^{F=1} \cong \mathbb{Q}_\ell^\rho$$

as a Galois module, by semi-simplicity (iii). By Tate's conjecture (ii), this cohomology has a basis given by algebraic cycles. Using the equality $A^j_{\text{hom}}(X) = \operatorname{Hom}(1, h_{\text{hom}}(X)(j)) = \operatorname{Hom}(1, (M_2)_{\text{hom}}(j))$, and the identification of the composition map

$$\operatorname{Hom}(1, h_{\text{hom}}(X)(j)) \times \operatorname{Hom}(h_{\text{hom}}(X)(j), 1) \to \mathcal{H}om(1,1) = \mathbb{Q}$$

sending (α, β) to $\beta \circ \alpha$ with the intersection number pairing

$$A^j_{\text{hom}}(X) \times A^{d-j}_{\text{hom}}(X) \to \mathbb{Q}$$

sending (α, β) to $\langle \alpha \,.\, \beta \rangle$ we now get two maps $1^\rho \xrightarrow{\varphi} M_2(-j) \xrightarrow{\psi} 1^\rho$ whose composition is the identity. (Note that the above intersection number pairing is non-degenerate, because $A^\nu_{\text{hom}}(X) = A^\nu_{\text{num}}(X)$ for $\nu = j$ and $d - j$, as remarked above.) Therefore 1^ρ becomes a direct factor of $(M_2)_{\text{hom}}$, and we conclude that $\varphi : 1^\rho \cong (M_2)_{\text{hom}}$ is an isomorphism with inverse ψ, because $H(M_2(j)) \cong \mathbb{Q}_\ell^\rho$ as was shown above. But this implies that one also has an isomorphism $1^\rho \cong M_2(j)$ in the category of Chow motives, because $J(M_2)$ is a nil ideal, and $J(1) = 0$. □

Corollary 6.5.9. *If $J(X)$ is a nil ideal (e.g., if X is finite-dimensional), then $H^i_{\mathcal{M}}(X, \mathbb{Q}(d)) = 0$ for $i \neq 2d$.*

Proof In fact, it is clear that the Tate conjecture holds in degrees 0 and $2d$, and that the corresponding cohomology groups are semi-simple Galois representations. We remark that the bijectivity of $c\ell^d : \mathsf{CH}^d(X) \otimes \mathbb{Q}_\ell \to H^{2d}(\overline{X}, \mathbb{Q}_\ell(d))$ is known without the assumption on X, by higher class field theory (note that we have $\mathbb{Q}$-coefficients). □

Corollary 6.5.10. *Assume that $J(X)$ is a nil ideal, the Tate conjecture holds for X (i.e., for all cohomology groups of X), and the eigenvalue 1 is semi-simple on all groups $H^{2j}(\overline{X}, \mathbb{Q}_\ell(j)$. Then*

a) $\mathsf{CH}^j(X)_{\text{num}} = 0$, *and* $\mathsf{CH}^j(X) \otimes_{\mathbb{Q}} \mathbb{Q}_\ell \xrightarrow{\sim} H^{2j}(\overline{X}, \mathbb{Q}_\ell(j)^\Gamma$ *via the cycle map (strong Tate conjecture) for all $j \geq 0$,*

b) $K_m(X) \otimes \mathbb{Q} = 0$ *for all $m \neq 0$ (Parshin conjecture).*

Remarks 6.5.11. a) The problems with the arguments in [Ka] concern the meaning of the statement that rational and numerical equivalence agree on X. In this paper, the meaning is that $\sim_{\mathrm{rat}} = \sim_{\mathrm{num}}$ on $\mathsf{CH}^j(X)$ for all j, and this would also fit with the assumptions in [Ka]. It does not imply that one can identify $h_{\mathrm{rat}}(X)$ and $h_{\mathrm{num}}(X)$ as written in the parenthesis following loc.cit. Cor. 2.2, because that would rather mean that rational and numerical equivalence agree on $X \times X$. Similarly, the reference in [Ka, 2.2] to [Gei, th. 3.3] has to be completed, because in the latter reference the argument is by assuming $\sim_{\mathrm{rat}} = \sim_{\mathrm{num}}$ for all varieties, and deducing an action of $\mathrm{End}(h_{\mathrm{num}}(X))$ on $K_a(X)^{(j)}$, which again requires $\sim_{\mathrm{rat}} = \sim_{\mathrm{num}}$ on $X \times X$. Finally, in the proof of [Ka, Théorème 1.10], the reference to [Mi, th. 2.6] has to be taken with similar care, because again, in that reference the (strong) Tate conjecture is assumed for all varieties, and in principle used for a product of two varieties when deducing semi-simplicity of the category $\mathcal{M}_{\mathrm{hom}}(k)$ and considering the question of isomorphy of two motives. The final conclusion is that the stated results in [Ka] remain correct, while the proofs have to be modified - basically by noting that in the considered cases it suffices to consider morphisms between Tate objects $1(j)$ and $h(X)$ instead of endomorphisms of $h(X)$.

b) In principle, the proof given in [An1, 4.2] is correct, but the short formulation might disguise the fact that, to my knowledge, it does not suffice to assume the Tate conjecture and 1-semi-simplicity just for $H^{2j}(\overline{X}, \mathbb{Q}_\ell(j))$ if one wants to get the results for $\mathsf{CH}^j(X)$.

c) In view of 6.5.6 (b), the assumptions of Corollary 6.5.10 hold, e.g., for arbitrary products of elliptic curves [Sp], for abelian varieties of dimension ≤ 3 [So1, Th. 4], Fermat surfaces of degree m invertible in k and dimension ≤ 3 (loc. cit.), for rational, Enriques or Kummer surfaces, and for many abelian varieties. In particular, for the sub tensor category of $\mathcal{M}_{\mathrm{rat}}(k)$ generated by elliptic curves one gets $\sim_{\mathrm{rat}} = \sim_{\mathrm{hom}}$, and hence the validity of Murre's conjecture.

Theorem 6.5.12. *Under the assumptions of Corollary 6.5.10, the regulator map*

$$H^i_{\mathcal{M}}(X, \mathbb{Q}(j)) \otimes \mathbb{Q}_\ell \xrightarrow{\sim} H^i(\overline{X}, \mathbb{Q}_\ell(j))^\Gamma$$

is an isomorphism for all $i, j \in \mathbb{Z}$, where $\Gamma = \mathrm{Gal}(\overline{k}/k)$ and $\ell \neq$ char k.

Proof This is clear from 6.5.10 and the fact that $H^i(\overline{X}, \mathbb{Q}_\ell(j))^\Gamma = 0$ for $i \neq 2j$ by Deligne's proof of the Weil conjectures. For the definition and properties of these regulator maps we refer to [Ja1, ch. 8], where they are deduced from Chern characters on higher algebraic K-theory constructed by Gillet. They exist for any smooth variety U over k instead of X, and coincide with the cycle maps for $i = 2j$, via the isomorphisms $K_0(U)^{(j)} \cong \mathrm{CH}^j(U)$. □

The following application will be used in the next section.

Corollary 6.5.13. *If C is an elliptic curve or a rational curve, and X is a product of elliptic curves, then, for every open $U \subseteq C$, and all $i, j \in \mathbb{Z}$, the regulator map*

$$H^i_{\mathcal{M}}(X \times U, \mathbb{Q}(j)) \otimes_{\mathbb{Q}} \mathbb{Q}_\ell \to H^i(\overline{X \times U}, \mathbb{Q}_\ell(j))^\Gamma$$

is an isomorphism, and the eigenvalue 1 of F on $H^i(\overline{X \times U}, \mathbb{Q}_\ell(j))$ is semi-simple.

This is a special case of the following conjecture (in which k is still a finite field).

Conjecture 6.5.14. ([Ja1, 12.4]) For any separated scheme of finite type Z over k, and all $a, b \in \mathbb{Z}$, the regulator map

$$H^M_a(Z, \mathbb{Q}(b)) \otimes_{\mathbb{Q}} \mathbb{Q}_\ell \to H^{\text{ét}}_a(\overline{Z}, \mathbb{Q}_\ell(b))^\Gamma$$

is an isomorphism, and the eigenvalue 1 of Frobenius on $H^{\text{ét}}_a(\overline{Z}, \mathbb{Q}_\ell(b))$ is semi-simple.

Again we refer to [Ja1, ch. 8] for the definition and properties of these homological versions of the regulator maps. Corollary 6.5.13 now follows from the following lemma, because the assertion of 6.5.12 holds for $X \times C$ and for $X \times \mathrm{Spec}(k(x))$, for any closed point $x \in C - U$.

Lemma 6.5.15. *a) ([Ja1, 8.4]) If Z is smooth and of pure dimension d, then the regulator map in 6.5.14 coincides with the regulator map*

$$H^{2d-a}_{\mathcal{M}}(Z, \mathbb{Q}(d-b)) \to H^{2d-a}_{\text{ét}}(\overline{Z}, \mathbb{Q}_\ell(d-b))^\Gamma.$$

b) ([Ja1, Th. 12.7 b]) If $Z' \subseteq Z$ is closed, $U = Z - Z'$, and Conjecture 6.5.14 holds for two of the three schemes Z, Z', U, then it also holds for the third one.

Although in this paper, we always used Chow groups and motivic cohomology groups with $\mathbb{Q}$-coefficients, we note that we can also get consequence for groups with $\mathbb{Z}$-coefficients, as in Soulé's paper [So1]. Recall that, for a smooth variety X over a field L one has motivic cohomology with $\mathbb{Z}$-coefficients, which can for example be defined as $H^i_{\mathcal{M}}(X, \mathbb{Z}(j)) =$

$\mathsf{CH}^j(X, 2j-i)$, where the latter groups are the higher Chow groups as defined by Bloch [Bl]. By definition, these groups vanish for $j < 0$ or $i > 2j$ or $i > d+j$, where $d = \dim(X)$, and it is known that $\mathsf{CH}^a(X,b) \otimes_{\mathbb{Z}} \mathbb{Q} \cong K_b(X)^{(a)}$ so that $H^i_{\mathcal{M}}(X, \mathbb{Z}(j)) \otimes_{\mathbb{Z}} \mathbb{Q} = H^i_{\mathcal{M}}(X, \mathbb{Q}(j))$. Moreover $H^{2j}(X, \mathbb{Z}(j)) = \mathsf{CH}^j(X, 0) = \mathsf{CH}^j(X)_{\mathbb{Z}}$, the usual Chow groups with integral coefficients. Finally, for an irreducible smooth projective variety X of dimension d, the group $\mathsf{CH}^d(X \times X, \mathbb{Z})$ of integral correspondences is a ring and acts on the motivic cohomology $H^i_{\mathcal{M}}(X, \mathbb{Z}(j))$. The additive category of integral motives (modulo rational equivalence) is defined by the same formalism as recalled in section 1. We denote the objects as $(X, p, m)_{\mathbb{Z}}$ where p is now an integral idempotent correspondence, and define $h(X)_{\mathbb{Z}} = (X, id, 0)_{\mathbb{Z}}$, the integral motive corresponding to X and $1(j)_{\mathbb{Z}} = (\mathrm{Spec}(k), id, j)_{\mathbb{Z}}$, the j-fold Tate twist of the trivial motive 1.

Corollary 6.5.16. *Under the assumptions of Theorem 6.5.8, the groups $H^i_{\mathcal{M}}(X, \mathbb{Z}(j))$ have finite exponent for $i \neq 2j$. For $H^{2j}_{\mathcal{M}}(X, \mathbb{Z}(j)) = \mathsf{CH}^j(X)_{\mathbb{Z}}$, the subgroup $\mathsf{CH}^j(X)_{\mathbb{Z},\mathrm{num}}$ of numerically trivial cycles has finite exponent, and the quotient group $A^j(X)_{\mathbb{Z},\mathrm{num}} = \mathsf{CH}^j(X)_{\mathbb{Z}}/\mathsf{CH}^j(X)_{\mathbb{Z},\mathrm{num}}$ is isomorphic to $\mathbb{Z}^\rho$, where $\rho = \dim_{\mathbb{Q}_\ell} H^{2j}(X_{\overline{k}}, \mathbb{Q}_\ell(j))^{\mathrm{Gal}(\overline{F}/F)}$.*

Proof In the proof of Theorem 6.5.8 it was shown that there is an isomorphism of $\mathbb{Q}$-linear motives

$$h(X) \cong M_1 \oplus 1(-j)^\rho,$$

with $Q(F)M_1 = 0$ for a polynomial $Q(t) \in \mathbb{Z}[t]$ with $Q(q^j) \neq 0$. Then there are morphisms

$$h(X) \xrightarrow{\alpha} 1(-j)^\rho \xrightarrow{\beta} h(X)$$

with $\alpha\beta = id$, and for the idempotent $\beta\alpha$ one has $Q(F)(id - \beta\alpha) = 0$ ($id - \beta\alpha$ is the idempotent corresponding to M_1). Thus there exist integers $N_1, N_2, N_3 > 0$ such that $N_1\alpha$ and $N_2\beta$ lift to integral correspondences $\tilde{\alpha}$ and $\tilde{\beta}$, and $N_3Q(F)(N_1N_2 - \tilde{\beta}\tilde{\alpha}) = 0$ in $\mathsf{CH}^d(X \times X)_{\mathbb{Z}}$. Now the argument of Soulé for Chow groups ([Sol, Prop. 2 i)]) immediately extends to higher Chow groups to show that F acts on $H^i_{\mathcal{M}}(X, \mathbb{Z}(j))$ as multiplication by q^j. We conclude

$$N_3\tilde{\beta}\tilde{\alpha}H^i_{\mathcal{M}}(X, \mathbb{Z}(j)) = N_1N_2N_3Q(F)H^i_{\mathcal{M}}(X, \mathbb{Z}(j)) = NH^i_{\mathcal{M}}(X, \mathbb{Z}(j))$$

with the non-zero integer $N = N_1N_2N_3Q(q^j)$. But the composition

$$H^i_{\mathcal{M}}(X, \mathbb{Z}(j)) \xrightarrow{\tilde{\alpha}_*} H^i_{\mathcal{M}}(1(-j)^\rho, \mathbb{Z}(j)) \xrightarrow{\tilde{\beta}_*} H^i_{\mathcal{M}}(X, \mathbb{Z}(j))$$

is zero for $i \neq 2j$, because $H^i_{\mathcal{M}}(1(-j), \mathbb{Z}(j)) = H^{i-2j}_{\mathcal{M}}(\mathrm{Spec}(k), \mathbb{Z})$, which is known to be zero for $i \neq 2j$ if k is a finite field. For $i = 2j$ we have $H^{i-2j}_{\mathcal{M}}(\mathrm{Spec}(k), \mathbb{Z}) = \mathbb{Z}$, and thus $NH^{2j}_{\mathcal{M}}(X, \mathbb{Z}(j))$ is isomorphic to $\mathbb{Z}^{\rho}$ (note that $\tilde{\alpha}_*\tilde{\beta}_* = N_1N_2$, which can be checked after tensoring with $\mathbb{Q}$, where it holds by definition). We deduce that the torsion group of $\mathsf{CH}^j(X)_{\mathbb{Z}}$ is killed by N, and coincides with $\mathsf{CH}^j(X)_{\mathbb{Z},\mathrm{num}}$: it is contained in the latter group, and the quotient embeds into the group $\mathsf{CH}^j(X)_{\mathbb{Q}} = A_{\mathrm{num}}(X)_{\mathbb{Q}}$. □

Corollary 6.5.17. *Let X be a smooth projective variety over the finite field k such that the associated motive (with $\mathbb{Q}$-coefficients) is finite-dimensional (or that $J(X)$ is a nil ideal). Then the group $H^i_{\mathcal{M}}(X, \mathbb{Z}(j))$ has finite exponent for $j > d = \dim(X)$, and $H^i_{\mathcal{M}}(X, \mathbb{Z}(d))$ has finite exponent for $i < 2d$.*

Proof The last statement follows from 6.5.16, because the Tate conjecture and the semi-simplicity hold for H^0 and H^{2d}. Formally, the first statement follows as well, since the condition on the Tate conjecture is empty here, but we give a simpler direct proof: Let the integral polynomial $P(t) = \prod_{i=0}^{2d} P_i(t)$ be as in the proof of Theorem 6.5.5. Then the assumption implies that $P(F)^r = 0$ in $\mathsf{CH}^d(X \times X, \mathbb{Q})$, for some integer $r \geq 1$. Therefore $NP(F)^r = 0$ in $\mathsf{CH}^d(X \times X, \mathbb{Z})$ for some integer $N \geq 1$. Because F acts as q^j on $H^i_{\mathcal{M}}(X, \mathbb{Z}(j))$, the integer $NP(q^j)$ annihilates this group, but one has $P(q^j) \neq 0$ for $j \notin \{0, \ldots, 2d\}$ by Deligne's proof of the Weil conjectures. □

6.6 Global function fields

Using the last section, we will deduce some results for global function fields. The most complete results are obtained for certain isotrivial varieties. Let F be a finite field, let C be a smooth projective geometrically irreducible curve over F, and let $k = F(C)$ be its function field.

Theorem 6.6.1. *Let W be a smooth projective variety over k, and assume that, after possibly passing to a finite extension F'/F, W is isomorphic to $Y \times_F k$, where Y/F is a smooth projective variety such that the assumptions of Corollary 6.5.10 hold for $X = Y \times_F C$. Then the strong Tate conjecture holds for W, i.e., the cycle maps induce isomorphisms*

$$A^j_{\mathrm{hom}}(W) \otimes_{\mathbb{Q}} \mathbb{Q}_\ell \xrightarrow{\sim} H^{2j}_{\text{ét}}(W_{\overline{k}}, \mathbb{Q}_\ell(j))^{\mathrm{Gal}(\overline{k}/k)},$$

for all $j \geq 0$, and the Abel-Jacobi map

$$\mathsf{CH}^j(W)_{\mathrm{hom}} \otimes_{\mathbb{Q}} \mathbb{Q}_\ell \xrightarrow{\sim} H^1_{\mathrm{cont}}(G_k, H^{2j-1}_{\text{ét}}(W_{\overline{k}}, \mathbb{Q}_\ell(j)))$$

is an isomorphism for all $j \geq 0$. Furthermore Murre's conjecture holds for W, with the filtration $F^1\mathsf{CH}^j(W) = \mathsf{CH}^j(W)_{\mathrm{hom}}$ and $F^2\mathsf{CH}^j(W) = 0$, and numerical and homological equivalence agree on W (i.e., on all Chow groups of W). Finally one has $H^i_{\mathcal{M}}(W, \mathbb{Q}(j)) = 0$ for $2j - i \neq 0, 1$, i.e., $K_m(W)_{\mathbb{Q}} = 0$ for $m \geq 2$.

Remarks 6.6.2. The assumptions of the theorem hold, if C is a rational or elliptic curve and X is a product of rational or elliptic curves (or the motive of X is contained in the rigid tensor subcategory generated by elliptc curves and Artin motives).

We will prove a somewhat more general result. For any smooth variety W over $k = F(C)$ define the *arithmetic étale cohomology* as

$$H^i_{\mathrm{ar}}(W, \mathbb{Q}_\ell(j)) := \varinjlim H^i(X_{U'} \times_F \overline{F}, \mathbb{Q}_\ell(j))^{\mathrm{Gal}(\overline{F}/F)},$$

where $U \subseteq C$ is some non-empty open, $X \to U$ is a smooth model for W ($W \cong X \times_U k$), and the limit is over all non-empty open subschemes $U' \subseteq U$, with $X_{U'} = X \times_U U'$. By standard limit theorems this cohomology does not depend on the choice of U and X. Moreover, this cohomology is functorial in W, receives a cycle class and allows an action of Chow correspondences if W is smooth and proper. In fact, there are regulator maps

$$H^i_{\mathcal{M}}(W, \mathbb{Q}(j)) \longrightarrow H^i_{\mathrm{ar}}(W, \mathbb{Q}_\ell(j))$$

by taking the limit of the regulator maps

$$H^i_{\mathcal{M}}(X'_U, \mathbb{Q}(j)) \longrightarrow H^i(X'_U \times_F \overline{F}, \mathbb{Q}_\ell(j))^{\mathrm{Gal}(\overline{F}/F)}$$

discussed at the end of the previous section and noting that motivic cohomology commutes with filtered inductive limits, so that the limit on the left hand side is $H^i_{\mathcal{M}}(W, \mathbb{Q}(j))$.

Let V be an ℓ-adic representation of G_k (i.e., a finite-dimensional $\mathbb{Q}_\ell$-vector space with continuous action of G_k). Call V *arithmetic*, if it comes from a representation of the fundamental group $\pi_1(U, \overline{\eta})$, where $U \subseteq C$ is a non-empty open and $\overline{\eta} = \mathrm{Spec}(\overline{k})$ the geometric generic point of U. Then we define the *arithmetic Galois cohomology* of V as

$$H^i_{\mathrm{ar}}(G_k, V) := \varinjlim H^1(\pi_1(U'_{\overline{F}}, \overline{\eta}), V)^{\mathrm{Gal}(\overline{F}/F)},$$

where the limit is again over the non-empty open $U' \subseteq U$. (This is $H^i_{\mathrm{ar}}(\mathrm{Spec}(k), \mathcal{F})$ for the ℓ-adic sheaf $\mathcal{F}$ on $\mathrm{Spec}(k)$ corresponding to V, if one defines arithmetic étale cohomology more generally for an *arithmetic* ℓ-adic sheaf $\mathcal{G}$ on W, i.e., one that extends to some model X over some $U \subseteq C$ as above, cf. [Ja1, 11.7, 12.15].) This definition is functorial in V. With these notations we have the following.

Theorem 6.6.3. *Let W be a smooth projective variety over k. Assume the condition*

(∗) There is a scheme X of finite type over C with generic fiber $W = X \times_C k$ such that for some non-empty open $U \subseteq C$, Conjecture 6.5.14 holds for $X_U = X \times_C U$ and the fibres $X_t = X \times_C t$ for all closed points $t \in U$.

Then the regulator maps induce isomorphisms

$$H^i_{\mathcal{M}}(W, \mathbb{Q}(j)) \otimes_{\mathbb{Q}} \mathbb{Q}_\ell \xrightarrow{\sim} H^i_{\text{ar}}(W, \mathbb{Q}_\ell(j)), \tag{6.1}$$

for all $i, j \in \mathbb{Z}$. Moreover, $H^i_{\mathcal{M}}(W, \mathbb{Q}(j)) = 0$ for $i - 2j \neq 0, 1$, i.e., $K_m(W)_{\mathbb{Q}} = 0$ for $m > 1$. For $i - 2j = 1$ one has isomorphisms $H^i_{\text{ar}}(W, \mathbb{Q}_\ell(j)) \cong H^1_{\text{ar}}(G_k, H^i(W_{\overline{k}}, \mathbb{Q}_\ell(j)))$. For $i - 2j = 0$, the cycle maps induce isomorphisms for all $j \geq 0$

$$A^j_{\text{hom}}(W) \otimes_{\mathbb{Q}} \mathbb{Q}_\ell \xrightarrow{\sim} H^{2j}_{\text{ét}}(W_{\overline{k}}, \mathbb{Q}_\ell(j))^{\text{Gal}(\overline{k}/k)} \tag{6.2}$$

(strong Tate conjecture)

$$\mathsf{CH}^j(W)_{\text{hom}} \otimes_{\mathbb{Q}} \mathbb{Q}_\ell \xrightarrow{\sim} H^1_{\text{cont}}(G_k, H^{2j-1}_{\text{ét}}(W_{\overline{k}}, \mathbb{Q}_\ell(j))) \tag{6.3}$$

(Abel-Jacobi map).

If W has a Chow-Künneth decomposition (e.g., if standard conjecture $C(W)$ holds and condition (∗) also hold for $W \times_k W$), then Murre's conjecture holds for W, with $F^1\mathsf{CH}^j(W) = \mathsf{CH}^j(W)_{\text{hom}}$ and $F^2\mathsf{CH}^j(W) = 0$.

Proof The first isomorphism is clear from the above, since the maps

$$H^i_{\mathcal{M}}(X_{U'}, \mathbb{Q}(j)) \longrightarrow H^i(X_{U'} \times_F \overline{F}, \mathbb{Q}_\ell(j))^{\text{Gal}(\overline{F}/F)}$$

are isomorphisms for all sufficiently small $U' \subseteq U$, by Lemma 6.5.15 (a) and (b). The next three claims follow from [Ja1, Thm. 12.16, diagram (12.16.3) and Rem. 12.17 b)].

Now assume that W is of pure dimension d and has a Chow-Künneth decomposition, i.e., the Künneth projectors π_i are algebraic, and lift to an orthogonal set of idempotents $\tilde{\pi}_i$ in $\mathsf{CH}^d(W \times W)$. Since the cycle maps (6) and (7) are functorial with respect to correspondences, it follows that, for the filtration $F^\nu\mathsf{CH}^j(W)$ defined in the theorem, the action of correspondences on $\text{Gr}^\nu_F \mathsf{CH}^j(W)$ factors through homological equivalence, and that $\pi_i = \delta_{i,2j-\nu}\text{id}$ (Kronecker symbol) on $\text{Gr}^\nu_F \mathsf{CH}^j(W)$. From this the remaining parts of Murre's conjecture follow easily, including the given description of the filtration.

Finally assume that the Künneth components π_i are algebraic and that condition $(*)$ holds for $W \times W$. For any smooth projective variety V over k let F_ℓ^ν be the descending filtration on the continuous étale cohomology $H^i_{\mathrm{cont}}(V, \mathbb{Q}_\ell(j))$ coming from the Hochschild-Serre spectral sequence

$$E_2^{p,q} = H^p_{\mathrm{cont}}(G_k, H^q(V_{\overline{k}}, \mathbb{Q}_\ell(j))) \Rightarrow H^{p+q}_{\mathrm{cont}}(V, \mathbb{Q}_\ell(j)).xs$$

Then, by (7) for $W \times W$, the cycle map $\mathsf{CH}^d(W \times W) \to H^{2d}((W \times W)_{\overline{k}}, \mathbb{Q}_\ell(d))$ (which is compatible with the cycle map (5) for $W \times W$ and $(i,j) = (2d, d)$) induces an injection

$$\begin{aligned}\mathsf{CH}^d(W \times W)_{\mathrm{hom}} &\to \mathrm{Gr}^1_{F_\ell} H^{2d}_{\mathrm{cont}}(W \times W, \mathbb{Q}_\ell(d)) \\ &\cong H^1(G_k, H^{2d-1}((W \times W)_{\overline{k}}, \mathbb{Q}_\ell(d))).\end{aligned}$$

On the other hand, the filtration F_ℓ^ν is respected under the action of correspondences, and $F^\nu.F^\mu \subseteq F^{\nu+\mu}$ under cup product. This shows that $J(W) = \mathsf{CH}^d(W \times W)$ is an ideal of square zero. Hence W has a Chow-Künneth decomposition. □

Proof of Theorem 6.6.1: We may assume that Y is of pure dimension d. Next we observe that a finite constant field extension does not matter, because we have Galois descent for étale cohomology with $\mathbb{Q}_\ell$-coefficients and motivic cohomology with $\mathbb{Q}$-coefficients. Thus we may assume that $W = Y \times_F k$. Then it is clear that Theorem 6.6.1 follows from 5.3, except possibly for the statement on Murre's conjecture. But, in the situation of 6.6.1, the pull-back via the morphism $W = Y_k \to Y$ induces isomorphism $H^i(Y_{\overline{F}}, \mathbb{Q}_\ell) \xrightarrow{\sim} H^i(W_{\overline{k}}, \mathbb{Q}_\ell)$ by proper and smooth base change. This shows that the projectors $\tilde{\pi}_i^Y$ of a Chow-Künneth decomposition for Y (which exist by the assumptions on Y) map to idempotents lifting the Künneth components of W under the pull back $\mathsf{CH}^d(X \times X) \to \mathsf{CH}^d(W \times W)$. Therefore W has a Chow-Künneth decomposition, and we can apply 6.6.3.

While the emphasis of this paper was to investigate conjectures, results and conditions for fixed varieties, we conclude with statements on all varieties over a given field. From Theorem 6.6.3 we get:

Corollary 6.6.4. *If conjecture 6.5.14 holds for all (smooth) varieties over F, then the results of theorem 6.6.3 hold for all smooth projective varieties W over function fields k in one variable over F. In particular, the strong Tate conjecture and the Murre's conjecture hold over such k.*

The reduction to the smooth case is done by lemma 6.5.15 and induction on dimension. Moreover, we note:

Proposition 6.6.5. *Conjecture 6.5.14 holds for all varieties over F if and only if the following holds for all smooth projective varieties X over F:*

i) Tate's conjecture (surjectivity of (5)),
ii) the eigenvalue 1 is semi-simple on $H^{2j}(X_{\overline{F}}, \mathbb{Q}_\ell(j))$ for all j,
iii) the Chow motive $h_{\mathrm{rat}}(X)$ is finite-dimensional.

Proof First we note that the properties (i)–(iii) for all $X \in SP_F$ are equivalent to conjecture 6.5.14 for all $X \in SP_F$. This follows from theorems 6.5.12 and 6.4.9, and the fact that $S(X)$ holds for all $X \in SP_F$. Secondly, conjecture 6.5.14 holds for all varieties if it holds for smooth projective varieties. The proof goes like in [Ja1, 12.7], but instead of assuming resolution of singularities, one may use de Jong's version: Let Z be any reduced separated algebraic F-scheme. By [dJ] there is a smooth projective variety X and a morphism $f : X \to Z$ which is generically étale. Choose a dense smooth open $U \subseteq Z$ such that the restriction $g : V = f^{-1}(U) \to U$ is finite étale. By induction on dimension and lemma 6.5.15 (b) it suffices to prove conjecture 6.5.14 for U, and we may assume that it holds for V. But g induces degree-respecting pull-backs g^* and push-forwards g_* in motivic and étale cohomology making the diagrams

$$\begin{array}{ccc} H^i_{\mathcal{M}}(V, \mathbb{Q}(j)) \otimes \mathbb{Q}_\ell & \longrightarrow & H^i_{\text{ét}}(V_{\overline{F}}, \mathbb{Q}_\ell(j))^{\mathrm{Gal}(\overline{F}/F)} \\ \downarrow g_* & & \downarrow g_* \\ H^i_{\mathcal{M}}(U, \mathbb{Q}(j)) \otimes \mathbb{Q}_\ell & \longrightarrow & H^i_{\text{ét}}(U_{\overline{F}}, \mathbb{Q}_\ell(j))^{\mathrm{Gal}(\overline{F}/F)} \end{array}$$

commutative; similarly with g^*. On the other hand, one has $g_* g^* = m$ on both sides, where m is the degree of g. This implies that the bottom line is a retract of the top line, and hence that conjecture 6.5.14 for V implies conjecture 6.5.14 for U. □

Finally we indicate that the above results can easily be generalized to function fields k of arbitrary transcendence degree over F, by replacing C by any variety over F and using the same definitions of arithmetic étale and Galois cohomology, and a corresponding Hochschild-Serre spectral sequence

$$E_2^{p,q} = H^p_{\mathrm{ar}}(G_k, H^q(W_{\overline{k}}, \mathbb{Q}_\ell(j))) \Rightarrow H^{p+q}_{\mathrm{ar}}(W, \mathbb{Q}_\ell(j)).$$

This gives the following result.

Proposition 6.6.6. *If conjecture 6.5.14 holds for all (smooth projective) varieties over $\mathbb{F}_p$, then the strong Tate conjecture, the equality of numerical and homological equivalence and the conjecture of Bloch-Beilinson-Murre hold over all fields of characteristic p.*

References

[An1] André, Y.: Motifs de dimension finie, d'aprés S.-I. Kimura, P.O'Sullivan…, Sem. Bourbaki **929**, March 2004.

[An2] André, Y.: *Une introduction aux motifs*, Panoramas et synthèses **17**. Société de Mathématique de France, Paris, 2004.

[AK] André, Y., B. Kahn and P. O'Sullivan: Nilpotence, radicaux et structures monoïdales., Rend. Sem. Mat. Univ. Padova **108** (2002), 107–291.

[Bl] Bloch, S.: Algebraic cycles and higher K-theory, Adv. in Math. **61** (1986), no. 3, 267–304.

[dJ] de Jong, A. J.: Smoothness, sem-stability and alterations, Inst. Hautes Études Sci. Publ. Math. **83** (1996), 51–93.

[DM] Deligne, P. and J.S. Milne: Tannakian categories., in *Hodge Cycles, Motives, and Shimura Varieties*, Lecture Notes in Mathematics, **900**. Springer-Varlag, Berlin, 1982, 101–228.

[DP] Del Padrone, A and C. Mazza: Schur finiteness and nilpotency. C.R.Acad.Sci. Paris, Ser. I **341** (2005), 283–286.

[Gei] Geisser, Th.:Tate's conjecture, algebraic cycles, and rational K-theory in characteristic p, K-Theory **13** (1998), 109–122.

[GM] Gordon, B. B. and J.P. Murre: Chow motives of elliptic modular threefolds, J. Reine Angew. Math. **514** (1999), 145–164.

[GP] Guletskii, V. and Pedrini, C.: Finite-dimensional motives and the Conjectures of Beilinson and Murre, K-Theory **30**, No.3 (2003), 243–263.

[Ja1] Jannsen, U.:*Mixed Motives and Algebraic K-Theory*, Lecture Notes in Mathematics, **1400**. Springer-Verlag, Berlin, 1990. xiv + 246pp.

[Ja2] Jannsen, U.: Motivic sheaves and filtrations on Chow groups, in *Motives (Seattle, WA, 1991)*, 245–302, Proc. Sympos. Pure Math., **55**, Part 1, Amer. Math. Soc., Providence, RI, 1994.

[Ja3] Jannsen, U.: Equivalence relations on algebraic cycles, in *The arithmetic and geometry of algebraic cycles (Banff, AB, 1998)*, 225–260, NATO Sci. Ser. C Math. Phys. Sci., **548**, Kluwer Acad. Publ., Dordrecht, 2000.

[Ka] Kahn, B.: Équivalences rationnelle et numérique sur certaines variétés de type abélien sur un corps fini, Ann. Sci. École Norm. Sup. (4) **36** (2003), no. 6, 977–1002 (2004).

[KM] Katz, N. and W. Messing: Some consequences of the Riemann hypothesis for varieties over finite fields, Invent. math. **23**, (1974), 73–77.

[Ki] Kimura, S.-I.: Motives are finite-dimensional, in some sense, Math. Ann. **331** (2005), 173–201.

[Kl] Kleiman, S.: Algebraic cycles and the Weil conjectures in *Dix esposés sur la cohomologie des schémas*, 359–386. North-Holland, Amsterdam; Masson, Paris, 1968.

[Mi] Milne, J. S.: Motives over finite fields, in *Motives (Seattle, WA, 1991)*, 401–459, Proc. Sympos. Pure Math., **55**, Part 1, Amer. Math. Soc., Providence, RI, 1994.

[Mu1] Murre, J. P. On a conjectural filtration on the Chow groups of an algebraic variety I, The general conjecture and some examples. Indag. Math. (N.S.) **4** (1993) 2, 177–188.

[Mu2] Murre, J. P.: On a conjectural filtration on the Chow groups of an algebraic variety II, Verification of the conjectures for threefolds which are the product on a surface and a curve, Indag. Math. (N.S.) **4** (1993), 189–201.

[OSu] O'Sullivan, P.: Letters to Y. André and B. Kahn, 29/4/02 and 12/5/02.

[So1] Soulé, C.: Groupes de Chow et K-théorie de variétés sur un corps fini, Math. Ann. **268** (1984), 317–345.

[So2] Soulé, C.: Opérations en K-théorie algébrique,Canad. J. Math. **37** (1985), no. 3, 488–550.

[Sp] Spiess, M.: Proof of the Tate conjecture for products of elliptic curves over finite fields, Math. Ann. **314** (1999), 285–290.

[Ta] Tate, J.: Conjectures on algebraic cycles in ℓ-adic cohomology, in *Motives (Seattle, WA, 1991)*, 71–83, Proc. Sympos. Pure Math., **55**, Part 1, Amer. Math. Soc., Providence, RI, 1994.

[Voe] Voevodsky, V.: A nilpotence theorem for cycles algebraically equivalent to zero, Internat. Math. Res. Notices 1995,**4**, 187–198.

7

On the Transcendental Part of the Motive of a Surface

Bruno Kahn†

Institut de Mathématiques de Jussieu, 175–179 rue du Chevaleret, 75013 Paris, France,
kahn@math.jussieu.fr

Jacob P. Murre

Universiteit Leiden, Mathematical Institute, P.O. Box 9512, 2300 RA Leiden, the Netherlands,
murre@math.leidenuniv.nl

Claudio Pedrini‡

Università di Genova, Dipartimento di Matematica, Via Dodecaneso 35, 16146 Genova, Italy,
pedrini@dima.unige.it

Introduction

Bloch's conjecture on surfaces [B1], which predicts the converse to Mumford's famous necessary condition for finite-dimensionality of the Chow group of 0-cycles [Mum1], has been a source of inspiration in the theory of algebraic cycles ever since its formulation in 1975. It is known fosr surfaces not of general type by [B-K-L] (see also [G-P1]), for certain generalised Godeaux surfaces [Voi] and in a few other scattered cases. Thirty years later, it remains open.

As was seen at least implicitely by Bloch himself early on, his conjecture is of motivic nature (see [B2, 1.11]). This was made explicit independently by Beilinson and the second author [Mu1]: we refer to Jannsen's article [J2] for an excellent overview. In particular, the Chow-Künneth decomposition for a surface S constructed in [Mu1] easily shows that the information necessary to study Bloch's conjecture is concentrated in the summand $h_2(S)$§ of the Chow motive of S.

† supported by RTN Network HPRN-CT-2002-00287
‡ supported by RTN Network HPRN-CT-2002-00287 and partially by the Italian MIUR
§ In this article we adopt a covariant convention for motives, hence write $h_i(S)$ rather than $h^i(S)$.

The main purpose of this article is to introduce and study a finer invariant of S: the *transcendental part* $t_2(S)$ *of* $h_2(S)$. Let us immediately clarify to the reader that we will not give a proof of Bloch's conjecture! Instead, we study the endomorphism ring of $t_2(S)$ for a general S and prove the following two formulas (Theorems 7.4.3 and 7.4.8):

$$\mathrm{End}_{\mathcal{M}_{\mathrm{rat}}}(t_2(S)) \simeq \frac{A_2(S \times S)}{\mathcal{J}(S,S)} \simeq \frac{T(S_{k(S)})}{T(S_{k(S)}) \cap H_{\leq 0}}. \tag{7.1}$$

Here k is the base field, $\mathcal{M}_{\mathrm{rat}}$ is the category of Chow motives over k with rational coefficients, A_* denotes Chow groups tensored with $\mathbb{Q}$, and

- $\mathcal{J}(S,S)$ is the subgroup of $A_2(S \times S)$ generated by those correspondences which are not dominant over S via either the first or the second projection;
- $T(S)$ is the Albanese kernel;
- $H_{\leq 1}$ is the subgroup of $A^2(S_{k(S)})$ generated by the images of the $A^2(S_L)$, where L runs through the subextensions of $k(S)/k$ of transcendence degree ≤ 1.

The first formula is a higher-dimensional analogue of a classical result of Weil concerning divisorial correspondences. The second formula provides a variant of Bloch's proof of Mumford's theorem in [B2, App. to Lecture 1] (with actually a slightly more precise result), cf. Corollary 7.4.9.

A conjecture generalising Bloch's conjecture:

$$\mathrm{End}_{\mathcal{M}_{\mathrm{rat}}}(t_2(S)) \overset{?}{\simeq} \mathrm{End}_{\mathcal{M}_{\mathrm{hom}}}(t_2^{\mathrm{hom}}(S))$$

(here hom stands for homological equivalence) may therefore be reformulated as saying that *the cycles homologically equivalent to* 0 *should be contained in* $\mathcal{J}(S,S)$. This conjecture, in turn, appears in a wider generality in Beilinson's article [Bei]; a link with this point of view is outlined in the last section, via the theory of *birational motives* ([K-S], see §7.5 here; in the context of Bloch's conjecture this point of view goes right back to Bloch, Colliot-Thélène and Sansuc, see [B2, App. to Lect. 1]. In particular, it is proven that $t_2(S)$ does not depend on the choice of the refined Chow-Künneth decomposition of Propositions 7.2.1 and 7.2.3, and is functorial in S for the action of correspondences (Corollary 7.8.10).

We now describe the contents of this paper in more detail. Section 7.1 fixes notation and reviews motives. Section 7.2 reviews the Chow-Künneth decomposition of a surface S (Proposition 7.2.1) and introduces the hero of our story, $t_2(S)$ (Proposition 7.2.3). Section 7.3 reviews the conjectures of [Mu2] on the Chow-Künneth decomposition as well as some of their consequences established by Jannsen in [J2], and proves a part of these consequences for the case of a product of two surfaces (Theorem 7.3.10). In Section 7.4 the isomorphisms (7.1) are established; they are reinterpreted in the next section in terms of birational motives. Section 7.6 studies the relationship of the previous results with Kimura's notion of finite-dimensional motives [Ki, G-P1, A-K]. The next section discusses some (conditional) higher-dimensional generalizations. Finally, Section 7.8 reproves and generalizes some of the previous results from a categorical viewpoint.

This article may be seen as a convergence point of the ways its 3 authors understand the theory of motives. The styles of the various sections largely reflect the styles of the various authors: we didn't attempt (too much) to homogenize them.

Acknowledgements

This collaboration developed over a series of conferences: one on algebraic cycles in Morelia in 2003 and three on motives and homotopy theory of schemes in Oberwolfach, on K-theory and algebraic cycles in Sestri-Levante and on cycles and algebraic geometry in Leiden in 2004. We would like to thank their organisers for these opportunities and also for providing excellent mathematical environments. To add more acknowledgements, the first author also gratefully acknowledges the hospitality of TIFR, Mumbai, where he stayed in 2005 during the completion of this work. The second author thanks Uwe Jannsen for a helpful discussion.

7.1 Definitions and Notation

7.1.1 Categories

Throughout this paper we shall use interchangeably the notations $\mathrm{Hom}_{\mathcal{C}}(X,Y)$ and $\mathcal{C}(X,Y)$ for Hom sets between objects of a category $\mathcal{C}$. In particular, the notation $\mathcal{C}(-,-)$ is more convenient when the symbol designating $\mathcal{C}$ is long, but the notation $\mathrm{End}_{\mathcal{C}}(X)$ may be more evocative than $\mathcal{C}(X,X)$.

7.1.2 Pure motives

Let k be a field and let $\mathcal{V} = \mathcal{V}_k$ be the category of smooth projective varieties over k. We shall sometimes write $X = X_d$ to say that $X \in \mathcal{V}$ is irreducible (or equidimensional) of dimension d. We denote by $A^i(X) = A_{d-i}(X)$ the group $CH^i(X) \otimes \mathbb{Q}$ of cycles of codimension i (or dimension $d-i$) on X, modulo rational equivalence, with $\mathbb{Q}$ coefficients.

We shall assume that the reader is familiar with the definition of pure motives and will only give minimal recollections on it, except for one thing. In [Mu1, Mu2, J1] and [Sch], pure motives are defined in the Grothendieck tradition so that the natural functor sending a variety to its motive is contravariant. On the contrary, here we are going to consider *covariant* motives, in order to be compatible with Voevodsky's convention that his triangulated motives are covariant. Moreover, we change the sign of the weight. Since the translation thoroughly confused all three authors, we prefer to give a dictionary of how to pass from one convention to the other, for clarity and the benefit of the reader:

To start with, for $X, Y \in \mathcal{V}$ we introduce as in [Sch, p. 165] the groups of *contravariant Chow correspondences*

$$\mathrm{Corr}^i(X,Y) = \bigoplus_\alpha A^{d_\alpha+i}(X_\alpha \times Y)$$

if $X = \coprod_\alpha X_\alpha$ with X_α equidimensional of dimension d_α; composition of correspondences is given by the usual formula (ibid.):

$$g \circ_{\mathrm{contr}} f = (p_{13})_*(p_{12}^* f \cdot p_{23}^* g).$$

Let us denote by $\mathsf{CH}_{\mathcal{M}}(k) = \mathsf{CH}_{\mathcal{M}}$ the category of Chow motives considered in [Mu1, Mu2, J1] or [Sch]. Thus an object of $\mathsf{CH}_{\mathcal{M}}$ is a triple $M = (X, p, m)$ where $X \in \mathcal{V}$, p is an idempotent in $\mathrm{Corr}^0(X,X)$ and $m \in \mathbb{Z}$, while morphisms are given by

$$\mathsf{CH}_{\mathcal{M}}((X,p,m),(Y,q,n)) = q\,\mathrm{Corr}^{n-m}(X,Y)p.$$

To $X \in \mathcal{V}$ we associate $\mathrm{ch}(X) = (X, 1_X, 0) \in \mathsf{CH}_{\mathcal{M}}$ and to a morphism $f : X \to Y$ we associate $\mathrm{ch}(f) = [\Gamma_f]^t$, where Γ_f is the graph of f and γ^t denotes the transpose of a correspondence γ†: this defines a contravariant functor $\mathrm{ch} : \mathcal{V} \to \mathsf{CH}_{\mathcal{M}}$.

We could merely define $\mathcal{M}_{\mathrm{rat}}$ as the opposite category to $\mathsf{CH}_{\mathcal{M}}$. However it is much more comfortable to have an explicit description of it:

Definition 7.1.1. i) The groups of *covariant Chow correspondences* are defined as follows: for $X, Y \in \mathcal{V}$

$$\mathrm{Corr}_i(X,Y) = \mathrm{Corr}^{-i}(Y,X).$$

Composition of covariant correspondences is given by the formula

$$g \circ_{\mathrm{cov}} f = (f^t \circ_{\mathrm{contr}} g^t)^t.$$

(From now on, we drop the index $_{\mathrm{cov}}$ for the composition sign.)

† In [Sch, p. 166], Scholl writes $\mathrm{ch}(f) = [\Gamma_f]$, which is slightly misleading.

ii) The category of *covariant Chow motives* $\mathcal{M}_{\mathrm{rat}}(k) = \mathcal{M}_{\mathrm{rat}}$ has objects triples $M = (X, p, m)$ as above, while morphisms are given by

$$\mathcal{M}_{\mathrm{rat}}((X,p,m),(Y,q,n)) = q\,\mathrm{Corr}_{m-n}(X,Y)p.$$

iii) The "covariant motive" functor $h : \mathcal{V} \to \mathcal{M}_{\mathrm{rat}}$ is given by the formulas

$$h(X) = (X, 1_X, 0)$$
$$h(f) = [\Gamma_f].$$

Note that, by definition

$$\mathrm{Corr}_i(X,Y) = \bigoplus_{\alpha} A_{d_\alpha+i}(X_\alpha \times Y)$$

if $X = \coprod X_\alpha$ with $\dim X_\alpha = d_\alpha$. The reader will easily check the following

Lemma 7.1.2. *There is an anti-isomorphism of categories* $F : \mathsf{CH}_{\mathcal{M}} \to \mathcal{M}_{\mathrm{rat}}$ *defined by*

$$F(X,p,m) = (X, p^t, -m)$$
$$F(\gamma) = \gamma^t.$$

One has the formula $F \circ \mathrm{ch} = h$.

□

Except at the beginning of Section 7.2, we shall never mention the category $\mathsf{CH}_{\mathcal{M}}$ again and will work only with $\mathcal{M}_{\mathrm{rat}}$. Let us review a few features of this category:

7.1.2.1 Effective motives

Let $\mathcal{M}_{\mathrm{rat}}^{\mathrm{eff}}$ be the full subcategory of $\mathcal{M}_{\mathrm{rat}}$ consisting of the $(X, 1_X, 0)$ for $X \in \mathcal{V}$ (see [Sch]): this is the category of *effective Chow motives*.

7.1.2.2 Tensor structure

The product of varieties and of correspondences defines on $\mathcal{M}_{\mathrm{rat}}$ a tensor structure (= a symmetric monoidal structure which is distributive with respect to direct sums); $\mathcal{M}_{\mathrm{rat}}^{\mathrm{eff}}$ is stable under this tensor structure. Then $\mathcal{M}_{\mathrm{rat}}$ is an additive, $\mathbb{Q}$-linear, pseudoabelian tensor category. It is also rigid, in the sense that there exist internal Homs and dual objects $M^\vee$ satisfying suitable axioms. Namely one has

$$(X,p,m)^\vee = (X, p^t, -d-m)$$

if $\dim X = d$, and

$$\gamma^\vee = \gamma^t$$

if $\gamma \in \mathrm{Corr}_n(X,Y) = \mathcal{M}_{\mathrm{rat}}(X, 1_X, n), (Y, 1_Y, 0))$.

7.1.2.3 The unit motive and the Lefschetz motive

The *unit motive* is $\mathbf{1} = (\mathrm{Spec}(k), 1, 0)$: it is a unit for the tensor structure. The *Lefschetz motive* $\mathbb{L}$ is defined via the motive of the projective line over k:

$$h(\mathbb{P}^1_k) = \mathbf{1} \oplus \mathbb{L}.$$

We then have an isomorphism $\mathbb{L} \simeq (\mathrm{Spec}(k), 1, 1)$.

7.1.2.4 Tate twists

For every motive $M = (X, p, m)$ we define the Tate twist $M(r)$ to be the motive $(X, p, m + r)$. Note that, with our conventions, $M(r) \simeq M \otimes \mathbb{L}^{\otimes r}$ for $r \geq 0$.

7.1.2.5 Inverse image morphisms

For a morphism $f : X \to Y$ in $\mathcal{V}$, one often writes f_* instead of $h(f)$. One may also consider the map in $\mathcal{M}_{\mathrm{rat}}$:

$$f^* = [\Gamma_f^t] \in A_d(Y \times X) : h(Y) \to h(X)(e - d)$$

where $d = \dim X$, $e = \dim Y$.

7.1.2.6 Action of correspondences on Chow groups

Observe that, by definition

$$A^i(X) = \mathcal{M}_{\mathrm{rat}}(h(X), \mathbb{L}^i)$$
$$A_i(X) = \mathcal{M}_{\mathrm{rat}}(\mathbb{L}^i, h(X))$$

for any $X \in \mathcal{V}$. This gives us a way to let correspondences act on Chow groups:

- On the left: if $\alpha \in A_i(X)$ and $\gamma \in \mathrm{Corr}_n(X,Y)$, then $\gamma_*\alpha = \gamma \circ \alpha \in A_{i+n}(Y)$. (In terms of cycles: $\gamma_*(\alpha) = (p_2)_*(p_1^*(\alpha) \cdot \gamma)$.)
- On the right: if $\gamma \in \mathrm{Corr}_n(X,Y)$ and $\alpha \in A^i(Y)$, then $\gamma^*\alpha = \alpha \circ \gamma \in A^{i-n}(X)$.

Remark 7.1.3. Suppose that $\dim X = d$. If $\alpha \in A_i(X)$ is interpreted as a morphism from $\mathbb{L}^i$ to $h(X)$, then the dual morphism $\alpha^\vee : h(X)(-d) \to \mathbb{L}^{-i}$ is nothing else than α. The same applies to $\gamma_*\alpha$ and $\gamma^*\alpha$, with notation as above. If we view $\alpha = \alpha^\vee$ in $A^{d-i}(X)$, we thus get a formula comparing left and right actions:

$$\gamma^*\alpha = (\gamma^*\alpha)^\vee = (\alpha \circ \gamma)^\vee = \gamma^\vee \circ \alpha^\vee = \gamma^t \circ \alpha = (\gamma^t)_*\alpha.$$

In the same vein, note the formula

$$A^i(M) = A_{-i}(M^\vee).$$

7.1.2.7 Chow groups of motives

We now extend the functors $A^i : \mathcal{V} \to \mathrm{Vec}_\mathbb{Q}$ (contravariant) and $A_i : \mathcal{V} \to \mathrm{Vec}_\mathbb{Q}$ (covariant) to functors on $\mathcal{M}_{\mathrm{rat}}$:

$$A^i(X,p,m) = \mathcal{M}_{\mathrm{rat}}((X,p,m),\mathbb{L}^i) = p^*A^{i-m}(X)$$
$$A_i(X,p,m) = \mathcal{M}_{\mathrm{rat}}(\mathbb{L}^i,(X,p,m)) = p_*A_{i-m}(X).$$

7.1.3 **Weil cohomology theories**

If $\sim$ is any *adequate equivalence relation* on cycles (see [J3]), then a similar definition yields the category $\mathcal{M}_\sim$. In particular we will consider the cases where $\sim$ equals homological equivalence or numerical equivalence.

We give ourselves a Weil cohomology theory H^* on $\mathcal{V}$, as defined in [Kl] or [An, 3.3]; we shall denote its field of coefficients by K (by convention it is of characteristic 0). We also define

$$H_i(X) = \bigoplus_\alpha H^{2d_\alpha - i}(X_\alpha)$$

if $X = \coprod_\alpha X_\alpha$ with $\dim X_\alpha = d_\alpha$.

For an element $\alpha \in A^i(X)$ we denote by $\mathrm{cl}^i(\alpha)$ its image under the cycle map in $A^i(X) \to H^{2i}(X)$; we write

$$A^i(X)_{\mathrm{hom}} = \mathrm{Ker}(\mathrm{cl}^i)$$
$$A^i_{\mathrm{hom}}(X) = A^i(X)/A^i(X)_{\mathrm{hom}} = \mathrm{Coim}(\mathrm{cl}^i)$$
$$\bar{A}^i(X) = \mathrm{Im}(\mathrm{cl}^i) \simeq A^i_{\mathrm{hom}}(X).$$

Equivalently, we have "homological" cycle maps $\mathrm{cl}_i : A_i(X) \to H_{2i}(X)$ and vector spaces $A_i(X)^{\mathrm{hom}}, A_i^{\mathrm{hom}}(X)$ and $\bar{A}_i(X)$. One easily checks that the Künneth formula and Poincaré duality carry over to homology without any change.

We denote by $\mathcal{M}_{\mathrm{hom}}$ the (covariant) category of homological motives, which is is defined as above by considering correspondences modulo homological equivalence, and by h_{hom} the functor which associates to every $X \in \mathcal{V}_k$ its motive in $\mathcal{M}_{\mathrm{hom}}$.

7.1.3.1 Action of correspondences on cohomology

Let $X, Y \in \mathcal{V}$, equidimensional of dimensions d and e for simplicity. The cycle class map gives us a homomorphism

$$\operatorname{Corr}_i(X,Y) = A_{d+i}(X \times Y) = A^{e-i}(X \times Y) \xrightarrow{\mathrm{cl}^{e-i}} H^{2e-2i}(X \times Y).$$

Since we have

$$\begin{aligned} H^{2e-2i}(X \times Y) \simeq \bigoplus_j H^j(X) \otimes H^{2e-2i-j}(Y) &\simeq \bigoplus_j H^j(X) \otimes H^{2i+j}(Y)^* \\ &\simeq \prod_j \operatorname{Hom}(H^{2i+j}(Y), H^j(X)) \end{aligned}$$

by the Künneth formula and Poincaré duality, we get a contravariant action of correspondences

$$\gamma^* : H^*(Y) \to H^{*-2i}(X)$$

for $\gamma \in \operatorname{Corr}_i(X,Y)$, extending the action of morphisms. Similarly, using the homological cycle map, we get a covariant action

$$\gamma_* : H_*(X) \to H_{*+2i}(Y)$$

by means of the composition

$$\begin{aligned} \operatorname{Corr}_i(X,Y) = A_{d+i}(X \times Y) &\xrightarrow{\mathrm{cl}_{d+i}} H_{2d+2i}(X \times Y) \\ \simeq \bigoplus_j H_{2d+2i-j}(X) \otimes H_j(Y) &\simeq \bigoplus_j H_{j-2i}(X)^* \otimes H_j(Y) \\ &\simeq \prod_j \operatorname{Hom}(H_{j-2i}(X), H_j(Y)). \end{aligned}$$

7.1.3.2 Homology and cohomology of motives

As in 7.1.2.7, we may use 7.1.3.1 to extend H^i, H_i to tensor functors

$$\begin{aligned} H^* &: \mathcal{M}_{\mathrm{hom}} \to \mathrm{Vec}_K^{\mathrm{gr}} \text{ (contravariant)} \\ H_* &: \mathcal{M}_{\mathrm{hom}} \to \mathrm{Vec}_K^{\mathrm{gr}} \text{ (covariant)} \end{aligned}$$

with values in graded vector spaces: the first functor corresponds to [An, 4.2.5.1]. Explicitly, $H^i(X,p,m) = p^* H^{i-2m}(X)$, $H_i(X,p,m) = p_* H_{i-2m}(X)$ and $H^*(\gamma) = \gamma^*$, $H_*(\gamma) = \gamma_*$. We also have

$$H_i(M) = H^{-i}(M^\vee).$$

Definition 7.1.4 (cf. [An, 3.4]). A Weil cohomology theory H is *classical* if:

(i) $\operatorname{char} k = 0$ and H is algebraic de Rham cohomology, l-adic cohomology for some prime number l or Betti cohomology relative to a complex embedding of k, or

(ii) $\operatorname{char} k = p > 0$ and H is crystalline cohomology (if k is perfect) or l-adic cohomology for some prime number $l \neq p$.

(Recall that, if the prime number l is different from $\operatorname{char} k$, l-adic cohomology is defined by $H^i_l(X) = H^i_{et}(X \otimes_k k_s, \mathbb{Q}_l)$, where k_s is some separable closure of k.)

Lemma 7.1.5. *If* $\operatorname{char} k = 0$, *homological equivalence does not depend on the choice of a classical Weil cohomology theory. Moreover, for any smooth projective* X, Y, *the Hom groups* $\mathcal{M}_{\text{hom}}(h_{\text{hom}}(X), h_{\text{hom}}(Y))$ *are finite-dimensional* $\mathbb{Q}$*-vector spaces.*

Proof Suppose first that k admits a complex embedding. Then the first statement follows from the comparison theorems between classical cohomology theories [An, 3.4.2]; the second one follows from taking Betti cohomology, which has rational coefficients.

In general, X and Y are defined over some finitely generated subfield k_0 of k, and k_0 has a complex embedding. If we pick two classical Weil cohomology theories over k, then the base change comparison theorems show that we may compare them with the corresponding ones over k_0; in turn we may compare the latter two with the help of the complex embedding. □

Remark 7.1.6. In arbitrary characteristic, the dimension of $H^i(M)$ for $M \in \mathcal{M}_{\text{rat}}$ is independent of the choice of the (classical) Weil cohomology H [An, 4.2.5.2], and the Euler characteristic $\sum(-1)^i \dim H^i(M)$ is independent of the choice of the (arbitrary) Weil cohomology H because it equals the trace $tr(1_M)$ computed in the rigid category $\mathcal{M}_{\text{rat}}$. For example, for a curve C of genus g one always has $\dim H^1(C) = 2g$.

Unless otherwise specified, all Weil cohomology theories considered in this paper will be classical.

7.1.4 Chow-Künneth decompositions

Let $X \in \mathcal{V}$, $X = X_d$. We say that X has a *Chow-Künneth decomposition in* $\mathcal{M}_{\text{rat}}$ (C-K for short) if there exist orthogonal projectors $\pi_i = \pi_i(X) \in \operatorname{Corr}_0(X, X) = A^d(X \times X)$, for $0 \leq i \leq 2d$, such that $\operatorname{cl}^d(\pi_i)$ is the

$(i, 2d-i)$-component of Δ_X in $H^{2d}(X \times X)$ and

$$[\Delta_X] = \sum_{0 \le i \le 2d} \pi_i.$$

This implies that in $\mathcal{M}_{\mathrm{rat}}$ the motive $h(X)$ decomposes as follows:

$$h(X) = \bigoplus_{0 \le i \le 2d} h_i(X) \tag{7.2}$$

where $h_i(X) = (X, \pi_i, 0)$. Moreover

$$H^*(h_i(X)) = H^i(X), \qquad H_*(h_i(X)) = H_i(X)$$

(see 7.1.3.2).

If we have $\pi_i = \pi_{2d-i}^t$ for all i, we say that the C-K decomposition is *self-dual.*

7.1.5 Triangulated motives

Let $DM_{\mathrm{gm}}^{\mathrm{eff}}(k)$ be the triangulated category of effective geometrical motives constructed by Voevodsky [Voev2]: there is a covariant functor $M : Sm/k \to DM_{\mathrm{gm}}^{\mathrm{eff}}(k)$ where Sm/k is the category of smooth schemes of finite type over k. We shall write $DM_{\mathrm{gm}}^{\mathrm{eff}}(k, \mathbb{Q})$ for the pseudo-abelian hull of the category obtained from $DM_{\mathrm{gm}}^{\mathrm{eff}}(k)$ by tensoring morphisms with $\mathbb{Q}$, and usually abbreviate it into $DM_{\mathrm{gm}}^{\mathrm{eff}}$. By abuse of notation, we shall denote by $\mathbb{Q}(1)$ the image of $\mathbb{Z}(1)$ in $DM_{\mathrm{gm}}^{\mathrm{eff}}(k, \mathbb{Q})$ under the natural functor $DM_{\mathrm{gm}}^{\mathrm{eff}}(k) \to DM_{\mathrm{gm}}^{\mathrm{eff}}(k, \mathbb{Q})$.

By [Voev2, p. 197], M induces a covariant functor $\Phi : \mathcal{M}_{\mathrm{rat}}^{\mathrm{eff}} \to DM_{\mathrm{gm}}^{\mathrm{eff}}$ which is a full embedding by [Voev3] and sends $\mathbb{L}$ to $\mathbb{Q}(1)[2]$ (this functor is already defined and fully faithful on the level of Chow motives with integral coefficients). As in [Voev2], we denote by $DM_{\mathrm{gm}} := DM_{\mathrm{gm}}(k, \mathbb{Q})$ the category obtained from $DM_{\mathrm{gm}}^{\mathrm{eff}}$ by inverting $\mathbb{Q}(1)$.

The category $DM_{\mathrm{gm}}^{\mathrm{eff}}$ admits a natural full embedding, as a tensor triangulated category, into the category $DM_{-}^{\mathrm{eff}} := DM_{-}^{\mathrm{eff}}(k) \otimes \mathbb{Q}$ of (bounded above) motivic complexes [Voev2, 3.2]. If the motive $h(X)$ of a smooth projective variety X has a Chow-Künneth decomposition in $\mathcal{M}_{\mathrm{rat}}$ as in (7.2), we write $M_i(X) = \Phi(h_i(X))$, so that in the category $DM_{\mathrm{gm}}^{\mathrm{eff}}$ we have the following decomposition

$$M(X) = \bigoplus_{0 \le i \le 2d} M_i(X).$$

7.1.6 Abelian varieties

We denote by Ab(k) or Ab the category whose objects are abelian k-varieties and, for $A, B \in \mathrm{Ab}$, $\mathrm{Ab}(A, B) = \mathrm{Hom}(A, B) \otimes \mathbb{Q}$. Recall that these are finite-dimensional $\mathbb{Q}$-vector spaces (see [Mum2, p. 176]).

7.2 Chow-Künneth decomposition for surfaces

In this section we adapt in Proposition 7.2.1 the construction of a suitable Chow-Künneth decomposition (see [Mu1] and [Sch]) for a smooth projective surface S to our covariant setting for $\mathcal{M}_{\mathrm{rat}}$. Then we refine this decomposition in Proposition 7.2.3.

7.2.1 The covariant Chow-Künneth decomposition for surfaces

For a smooth projective variety $X = X_d$, we denote by Alb_X and by Pic^0_X the Albanese variety and the Picard variety of X; $T(X)$ denotes the Albanese kernel of X, i.e the kernel of the map $A^d(X)_0 \to \mathrm{Alb}_X(k)_{\mathbb{Q}}$, where $A^d(X)_0$ is the group of 0-cycles of degree 0.

From [Mu1], [Mu2] and [Sch, §4] it follows that, in $\mathsf{CH}_{\mathcal{M}}$, there exist projectors p_0, p_1, p_{2d-1}, p_{2d} in $\mathrm{End}_{\mathsf{CH}_{\mathcal{M}}}(\mathrm{ch}(X)) = A^d(X \times X)$ with the following properties:

i) $p_0 = (1/n)[P \times X]$ and $p_{2d} = (1/n)[X \times P]$, P a closed point on X of degree n with separable residue field.

ii) p_{2d} operates as 0 on $A^i(X)$ for $i \neq d$; on $A^d(X)$ we have $F^1 A^d(X) = \mathrm{Ker}\, p_{2d} = A^d(X)_{\mathrm{num}}$.

iii) p_1, the "Picard projector", operates as 0 on all $A^i(X)$ with $i \neq 1$; its image on $A^1(X)$ is $A^1(X)_{\mathrm{hom}}$, hence $A^1(\mathrm{ch}^1(X)) = \mathrm{Pic}^0_X(k)_{\mathbb{Q}}$ where $\mathrm{ch}^1(X) = (X, p_1, 0)$. Moreover, on $A^1(X)_{\mathrm{hom}}$ p_1 operates as the identity.

iv) p_{2d-1}, the "Albanese projector", operates as 0 on all $A^i(X)$ with $i \neq d$; on $A^d(X)$ its image lies in $A^d(X)_{\mathrm{hom}}$ and on $A^d(X)_{\mathrm{hom}}$ its kernel is $T(X)$; hence $A^d(\mathrm{ch}^{2d-1}(X)) = \mathrm{Alb}_X(k)_{\mathbb{Q}}$ where $\mathrm{ch}^{2d-1}(X) = (X, p_{2d-1}, 0)$.

v) $F^2 A^d(X) = \mathrm{Ker}(p_{2d-1}) \cap F^1 = T(X)$.

vi) $p_0, p_1, p_{2d-1}, p_{2d}$ are mutually orthogonal.

vii) $p_{2d-1} = p_1^t$.

Note that P exists by [EGA4, 17.15.10 (iii)]. Also, the motive $\mathrm{ch}^0(X)$ is in general not isomorphic to $\mathbf{1}$, but rather to $\mathrm{ch}(\mathrm{Spec}\, k')$ where k' is the *field of constants* of X. If X is equidimensional but reducible, to get the "right"

p_0 and p_{2d} we need to take the sum of the corresponding projectors for all irreducible components X_α of X; then we get

$$\mathrm{ch}^0(X) = \bigoplus \mathrm{ch}^0(X_\alpha) \simeq \bigoplus \mathrm{ch}(\operatorname{Spec} k'_\alpha) = \mathrm{ch}(\coprod \operatorname{Spec} k'_\alpha)$$

where k'_α is the field of constants of X_α: thus $\mathrm{ch}^0(X)$ is an *Artin motive* (cf. [An, 4.1.6.1]). Note finally that the existence of $p_0, p_1, p_{2d-1}, p_{2d}$ with properties (i)–(vii) is part of a conjectural Chow-Künneth decomposition in $\mathsf{CH}_{\mathcal{M}}$ for any variety X (see [Mu2] and §7.3).

In the case of a smooth projective surface S a Chow-Künneth decomposition of the motive $\mathrm{ch}(S)$ always exists by [Mu1] and [Sch, §4].

The following proposition is just a translation of these results in $\mathcal{M}_{\mathrm{rat}}$.

Proposition 7.2.1. *Let S be a smooth projective connected surface over k and let $P \in S$ be a separable closed point. There exists a Chow-Künneth decomposition of $h(S)$ in $\mathcal{M}_{\mathrm{rat}}$: $h(S) = \bigoplus_{0 \le i \le 4} h_i(S)$, with $h_i(S) = (S, \pi_i, 0)$, $\pi_i = \pi_i(S) \in A^2(S \times S)$, with the following properties:*

i) $\pi_i = \pi_{4-i}^t$ *for* $0 \le i \le 4$.

ii) $\pi_0 = (1/\deg(P))[S \times P]$, $\pi_4 = (1/\deg(P))[P \times S]$.

iii) There exists a curve $C \subset S$ of the form $C_1 \cup C_2$, where $C_1 = S \cdot H$ is a general (smooth) hyperplane section of S, such that π_1 is supported on $S \times C$ (and hence π_3 is supported on $C \times S$).

iv) Let $i_1 : C_1 \to S$ be the inclusion and $\xi = (i_1)_* i_1^* : h(S) \to h(S)(1)$. Then the compositions

$$h_3(S) \to h(S) \xrightarrow{\xi} h(S)(1) \to h_1(S)(1)$$

$$h_4(S) \to h(S) \xrightarrow{\xi^2} h(S)(2) \to h_0(S)(2)$$

are isomorphisms.

We set $\pi_2 := \Delta(S) - \pi_0 - \pi_1 - \pi_3 - \pi_4$ and $h_2(S) = (S, \pi_2, 0)$. We have the following tables:

$M =$	$h_0(S)$	$h_1(S)$	$h_2(S)$	$h_3(S)$	$h_4(S)$
$A^0(M) =$	$A^0(S)$	0	0	0	0
$A^1(M) =$	0	$\mathrm{Pic}^0_S(k)_{\mathbb{Q}}$	$\mathrm{NS}(S)_{\mathbb{Q}}$	0	0
$A^2(M) =$	0	0	$T(S)$	$\mathrm{Alb}_S(k)_{\mathbb{Q}}$	$A^2_{\mathrm{num}}(S)$

$M =$	$h_0(S)$	$h_1(S)$	$h_2(S)$	$h_3(S)$	$h_4(S)$
$A_0(M) =$	$A_0^{\mathrm{num}}(S)$	$\mathrm{Alb}_S(k)_{\mathbb{Q}}$	$T(S)$	0	0
$A_1(M) =$	0	0	$\mathrm{NS}(S)_{\mathbb{Q}}$	$\mathrm{Pic}^0_S(k)_{\mathbb{Q}}$	0
$A_2(M) =$	0	0	0	0	$A_2(S)$

and

$$\mathrm{End}_{\mathcal{M}_{\mathrm{rat}}}(h_1(S)) = \mathrm{End}_{\mathrm{Ab}}(\mathrm{Alb}_S)$$
$$\mathrm{End}_{\mathcal{M}_{\mathrm{rat}}}(h_3(S)) = \mathrm{End}_{\mathrm{Ab}}(\mathrm{Pic}_S).$$

Proof In conformity with Lemma 7.1.2, we take $\pi_i = p_i^t$, where p_i are the projectors defined in [Sch, §4]. (The first table is copied from [Sch, p.178]; by Remark 7.1.3, we have the general formula $A_i(h_j(X)) = A^{d-i}(h_{2d-j}(X))$ for a self-dual C-K decomposition.) □

For later use we add some precisions on the construction of the p_i, hence of the π_i, especially concerning rationality issues, making [Sch, 4.2] more specific. Contrary to the case of P, the curve C_1 may always be chosen as defined over k and geometrically connected: this is clear if k is infinite by Bertini's theorem [Ha, p. 179, Th. 8.18 and p. 245, Rem. 7.9.1], and in case k is finite this can also be achieved up to enlarging the projective embedding of S as in [Del-I, 5.7].

In Proposition 7.2.1 (ii) the curve C_2 enters the picture because, in the case of a surface, in order to get mutually orthogonal idempotents π_1 and π_3, one has to introduce a "correction term" (see [Mu1] and [Sch, p. 177]) as follows: in the contravariant setting one first takes projectors $\overset{?}{p_1}, \overset{?}{p_3}$ defined in [Mu1] verifying $\overset{?}{p_3} = (\overset{?}{p_1})^t$ and $\overset{?}{p_3}\overset{?}{p_1} = 0$, and then one corrects them by $p_1 = \overset{?}{p_1} - \frac{1}{2}(\overset{?}{p_1} \circ \overset{?}{p_3})$ and $p_3 = \overset{?}{p_3} - \frac{1}{2}(\overset{?}{p_1} \circ \overset{?}{p_3})$.† The correction term $\frac{1}{2}(\overset{?}{p_1} \circ \overset{?}{p_3})$ is supported on $C_2 \times C_2$ and is of the form $\sum A_\lambda \times A'_\lambda$, where each A_λ, A'_λ is a divisor on S, homologically equivalent to 0.

Supposing k perfect, the projector π_1 may be described more precisely as follows. Let $i : C \to S$ be the closed embedding. Replace C by its normalization $\tilde{C}$, which is smooth. There is a divisor class $D \in A^1(S \times \tilde{C})$ such that $\pi_1 = (1_S \times \tilde{\imath})(D)$, where $\tilde{\imath} : \tilde{C} \to S$ is the proper morphism induced by the projection $\tilde{C} \to C$ and $\tilde{\imath}_*$ is the correspondence given by the graph $\Gamma_{\tilde{\imath}}$. With this description we then have $\pi_3 = D^t \circ \tilde{\imath}^*$, where $\tilde{\imath}^* = \Gamma_{\tilde{\imath}}^t$. Note

† This correction is the one from [Sch] which is different from the one in [Mu1]: its advantage is that $p_3 = p_1^t$.

that $D^t(R)$ is a divisor homologically equivalent to 0 on S for every divisor R on C.

7.2.2 The refined Chow-Künneth decomposition

We now introduce the motive $t_2(S)$, whose construction had been outlined in [An, 11.1.3] in a special case. We start with a well-known lemma (cf. [Bo, §4, no 2, Formula (12) p. 77]):

Lemma 7.2.2. *Let $\mu : V \times W \to \mathbb{Q}$ be a perfect pairing between two finite dimensional $\mathbb{Q}$-vector spaces V, W. Let $(e_i)_{1\le i\le n}$ be a basis of V and let (e_i^*) be the dual basis of W with respect to this pairing ($\mu(e_i, e_j^*) = \delta_{ij}$). Then, in $\mathrm{Hom}(V \otimes W, \mathbb{Q}) \simeq \mathrm{Hom}(V, W^*)$, we have the identity*

$$\mu^{-1} = \sum e_i \otimes e_i^*$$

where we have viewed μ as an isomorphism in $\mathrm{Hom}(W^*, V)$. In particular, the right hand side is independent of the choice of the basis (e_i). □

Proposition 7.2.3. *Let S be a surface provided with a C-K decomposition as in Proposition 7.2.1. Let k_s be a separable closure of k, $G_k = \mathrm{Gal}(k_s/k)$ and*

$$\underline{\mathrm{NS}}_S = \mathrm{NS}(S \otimes_k k_s)_{\mathbb{Q}}$$

be the ($\mathbb{Q}$-linear, geometric) Néron-Severi group of S viewed as a G_k-module. Then there is a unique splitting

$$\pi_2 = \pi_2^{\mathrm{alg}} + \pi_2^{tr}$$

which induces a decomposition

$$h_2(S) \simeq h_2^{\mathrm{alg}}(S) \oplus t_2(S)$$

where $h_2^{\mathrm{alg}}(S) = (S, \pi_2^{\mathrm{alg}}, 0) \simeq h(\underline{\mathrm{NS}}_S)(1)$ and $t_2(S) = (S, \pi_2^{\mathrm{tr}}, 0)$. Here $h(\underline{\mathrm{NS}}_S)$ is the Artin motive associated to $\underline{\mathrm{NS}}_S$. Moreover the tables of Proposition 7.3.6 refine as follows:

$M =$	$h_2^{\mathrm{alg}}(S)$	$t_2(S)$
$A^0(M) =$	0	0
$A^1(M) =$	$\mathrm{NS}(S)_{\mathbb{Q}}$	0
$A^2(M) =$	0	$T(S)$

$M =$	$h_2^{\mathrm{alg}}(S)$	$t_2(S)$
$A_0(M) =$	0	$T(S)$
$A_1(M) =$	$\mathrm{NS}(S)_{\mathbb{Q}}$	0
$A_2(M) =$	0	0

Finally

$$H^2(S) = H^2_{\mathrm{alg}}(S) \oplus H^2_{\mathrm{tr}}(S) = \pi_2^{\mathrm{alg}} H^2(S) \oplus \pi_2^{\mathrm{tr}} H^2(S)$$
$$= (\mathrm{NS}(S) \otimes K) \oplus H^2(t_2(S))$$

where $H^2_{\mathrm{tr}}(S)$ is (by definition) the "transcendental cohomology".

Proof Choose a finite Galois extension E/k such that the action of G_k on $\underline{\mathrm{NS}}_S$ factors through $G = \mathrm{Gal}(E/k)$. Let $[D_i]$ be an orthogonal basis of $\underline{\mathrm{NS}}_S = \mathrm{NS}(S_E)_{\mathbb{Q}}$. It follows from Lemma 7.2.2 that

$$\sum_i \frac{1}{\langle [D_i], [D_i] \rangle} [D_i] \otimes [D_i] \in \mathrm{NS}(S_E)_{\mathbb{Q}} \otimes \mathrm{NS}(S_E)_{\mathbb{Q}}$$

is G-invariant, where $\langle [D_i], [D_i] \rangle$ are the intersection numbers. By Proposition 7.2.1, the k-rational projector π_2 defines a G-equivariant section σ of the projection $A^1(S_E) \to \mathrm{NS}(S_E)_{\mathbb{Q}}$. The composition

$$\lambda : \mathrm{NS}(S_E)_{\mathbb{Q}} \otimes \mathrm{NS}(S_E)_{\mathbb{Q}} \xrightarrow{\sigma \otimes \sigma} A^1(S_E) \otimes A^1(S_E) \xrightarrow{\cap} A^2((S \times S)_E)$$

is also G-equivariant. It follows that

$$\pi_2^{\mathrm{alg}} = \lambda\Big(\sum_i \frac{1}{\langle [D_i], [D_i] \rangle} [D_i] \otimes [D_i]\Big) \in A^2((S \times S)_E)$$

is a G-invariant cycle, hence descends uniquely to a correspondence $\pi_2^{\mathrm{alg}} \in A^2(S \times S)$.

Over E, the correspondences

$$\alpha_i = \frac{1}{\langle [D_i], [D_i] \rangle} [D_i \times D_i] = \lambda\Big(\frac{1}{\langle [D_i], [D_i] \rangle} [D_i] \otimes [D_i]\Big)$$

are mutually orthogonal idempotents, are orthogonal to π_j for $j \neq 2$, and verify

$$\pi_2 \circ \alpha_i = \alpha_i \circ \pi_2 = \alpha_i.$$

It follows that their sum π_2^{alg} is an idempotent orthogonal to π_j for $j \neq 2$, and that $\pi_2 \circ \pi_2^{\text{alg}} = \pi_2^{\text{alg}} \circ \pi_2 = \pi_2^{\text{alg}}$. We define

$$\pi_2^{\text{tr}} := \pi_2 - \pi_2^{\text{alg}}.$$

In order to prove the isomorphism $h_2^{\text{alg}}(S) \simeq h(\underline{\text{NS}}_S)(1)$, it is enough to show that

$$M_i := (S_E, \alpha_i, 0) \simeq \mathbb{L} \text{ for } 1 \leq i \leq \rho,$$

where ρ is the Picard number. Define $f_i : \mathbb{L} \to M_i$ by $f_i = \alpha_i \circ [D_i] \circ 1_{\operatorname{Spec} E} = [D_i]$. The transpose f_i^t is a morphism $M_i \to \mathbb{L}$ and, by taking $g_i = \frac{1}{\langle [D_i, D_i] \rangle} f_i^t$, we get $g_i \circ f_i = 1_{\mathbb{L}}$ and $f_i \circ g_i = \alpha_i$, hence the required isomorphism.

From the construction above we also get $A^*(h_2^{\text{alg}}(S)) = A^1(h_2(S)) = \text{NS}(S)_{\mathbb{Q}}$, $A^*(t_2(S)) = T(S)$. This shows that

$$\mathcal{M}_{\text{rat}}(\mathbb{L}, t_2(S_E)) = \mathcal{M}_{\text{rat}}(t_2(S_E), \mathbb{L}) = 0$$

for any extension E/k. Taking E as above, we get that

$$\mathcal{M}_{\text{rat}}(h_2^{\text{alg}}(S_E), t_2(S_E)) = \mathcal{M}_{\text{rat}}(t_2(S_E), h_2^{\text{alg}}(S_E)) = 0$$

hence by descent that

$$\mathcal{M}_{\text{rat}}(h_2^{\text{alg}}(S), t_2(S)) = \mathcal{M}_{\text{rat}}(t_2(S), h_2^{\text{alg}}(S)) = 0.$$

Therefore

$$\operatorname{End}_{\mathcal{M}_{\text{rat}}}(h_2(S)) = \operatorname{End}_{\mathcal{M}_{\text{rat}}}(h_2^{\text{alg}}(S)) \times \operatorname{End}_{\mathcal{M}_{\text{rat}}}(t_2(S))$$

which implies the uniqueness of the decomposition $\pi_2 = \pi_2^{\text{alg}} + \pi_2^{\text{tr}}$.

The assertions on cohomology immediately follow from the definition of π_2^{alg} and π_2^{tr}. □

Corollary 7.2.4. $t_2(S) = 0 \Rightarrow H^2_{\text{tr}}(S) = 0$. *If* $\text{char}\, k = 0$, *this implies* $p_g = 0$.

Proof The first assertion is obvious from Proposition 7.2.3. The second one is classical [B2]. □

Definition 7.2.5. For a surface S, we call the set of projectors

$$\{\pi_0, \pi_1, \pi_2^{\text{alg}}, \pi_2^{\text{tr}}, \pi_3, \pi_4\}$$

the *refined Chow-Künneth decomposition* associated to the C-K decomposition $\{\pi_0, \pi_1, \pi_2, \pi_3, \pi_4\}$.

7.3 On some conjectures

In this section, we show in Theorem 7.3.10 that part of the results proved by U. Jannsen in [J2, Prop 5.8] hold unconditionally for the Chow motives coming from surfaces. We first recall the conjectures about the existence of a C-K decomposition as formulated in [Mu2] (see also [J2]).

Conjecture 7.3.1 (see Conj. A in [Mu2]). Every smooth projective variety X has a Chow-Künneth decomposition.

The conjecture is true in particular for curves, surfaces, abelian varieties, uniruled 3-folds and Calabi-Yau 3-folds.

If X and Y have a C-K decomposition, with projectors $\pi_i(X)$ and $\pi_j(Y)$ ($0 \leq i \leq 2d, d = \dim X$ and $0 \leq j \leq 2e, e = \dim Y$) then $Z = X \times Y$ also has a C-K decomposition with projectors $\pi_m(Z)$ given by $\pi_m(Z) = \sum_{r+s=m} \pi_r(X) \times \pi_s(Y)$ whith $0 \leq m \leq 2(d+e)$.

If the motive $h(X)$ is finite-dimensional in the sense of Kimura [Ki] and the Künneth components of the diagonal are algebraic (i.e., are classes of algebraic cycles), then $h(X) = \bigoplus_{0 \leq i \leq 2d} h_i(X)$ and the motives $h_i(X)$ are unique, up to isomorphism as follows from the results of [Ki] (see §7.6).

Now let X have a C-K decomposition and consider the action of the correspondence $\pi_i(X)$ on the Chow groups $A^j(X)$. Then Conjecture B in [Mu2] translates as follows in our covariant setting (identical statement):

Conjecture 7.3.2 (Vanishing Conjecture). The correspondences $\pi_i(X)$ act as 0 on $A^j(X)$ for $i < j$ and for $i > 2j$.

Assuming that Conjectures 7.3.1 and 7.3.2 hold, one may define a decreasing filtration $F^\bullet$ on $A^j(X)$ as follows:

$$F^1 A^j(X) = \operatorname{Ker} \pi_{2j}, F^2 A^j(X) = \operatorname{Ker} \pi_{2j} \cap \operatorname{Ker} \pi_{2j-1}, \ldots$$
$$F^\nu A^j(X) = \operatorname{Ker} \pi_{2j} \cap \operatorname{Ker} \pi_{2j-1}, \cap \cdots \cap \operatorname{Ker} \pi_{2j-\nu+1}.$$

Note that, with the above definitions, $F^{j+1} A^j(X) = 0$.

Also it easily follows from the definition of $F^\bullet$ (see [Mu2, 1.4.4]) that:

$$F^1 A^j(X) \subset A^j(X)_{\text{hom}}.$$

Conjecture 7.3.3 (see Conj. D in [Mu2]). $F^1 A^j(X) = A^j(X)_{\text{hom}}$, for all j.

Finally we mention:

Conjecture 7.3.4 (see Conj. C in [Mu2]). The filtration $F^\bullet$ is independent of the choice of the $\pi_i(X)$.

Remark 7.3.5. Jannsen [J2] has shown that if the Conjectures 7.3.1 ... 7.3.4 hold for every smooth projective variety over k, then the filtration $F^\bullet$ satisfies *Beilinson's Conjecture.* The converse also holds [J2, 5.2].

Now let X and Y be two smooth projective varieties: the following result, due to U. Jannsen, relates Conjectures 7.3.2 and 7.3.3 for $Z = X \times Y$, with the groups $\mathcal{M}_{\mathrm{rat}}(h_i(X), h_j(Y))$.

Proposition 7.3.6 ([J2, Prop. 5.8]). *Let X and Y be smooth projective varieties of dimensions respectively d and e, provided with C-K decompositions, and let $Z = X \times Y$ be provided with the product C-K decomposition.*

i) If Z satisfies the vanishing Conjecture 7.3.2, then:

$$\mathcal{M}_{\mathrm{rat}}(h_i(X), h_j(Y)) = 0 \textit{ if } j < i;\ 0 \le i \le 2d;\ 0 \le j \le 2e.$$

ii) If Z satisfies Conjecture 7.3.3 then

$$\mathcal{M}_{\mathrm{rat}}(h_i(X), h_i(Y)) \simeq \mathcal{M}_{\mathrm{hom}}(h_i^{\mathrm{hom}}(X), h_i^{\mathrm{hom}}(Y)).$$

In particular, if $X \times X$ satifies Conjecture 7.3.3, then b) implies that the $\mathbb{Q}$-vector space $\mathrm{End}_{\mathcal{M}_{\mathrm{rat}}}(h_i(X))$ has finite dimension for $0 \le i \le 2d$, at least if char$k = 0$ and the Weil cohomology is classical (cf. Lemma 7.1.5).

Remark 7.3.7. Note that, because of our covariant definition of the functor $h : \mathcal{V} \to \mathcal{M}_{\mathrm{rat}}$, in a) we have $j < i$, while in the contravariant setting (as in [J2, 5.8]) one has $i < j$.

Corollary 7.3.8. *Let S be a smooth projective surface and C a smooth projective curve. Let $\Delta_S = \sum_{0 \le i \le 4} \pi_i(S)$ and $\Delta_C = \sum_{0 \le i \le 2} \pi_j(C)$ be C-K decompositions respectively for S and for C. Then*

$$\textit{(i)}\ \pi_j(C) \cdot \Gamma \cdot \pi_i(S) = 0 \textit{ if } \begin{cases} i > j \textit{ and } \Gamma \in A^1(S \times C) \\ i = j \textit{ and } \Gamma \in A^1(S \times C)_{\mathrm{hom}} \end{cases}$$

$$\textit{(ii)}\ \pi_j(S) \cdot \Gamma \cdot \pi_i(C) = 0 \textit{ if } \begin{cases} i > j \textit{ and } \Gamma \in A^2(C \times S) \\ i = j \textit{ and } \Gamma \in A^2(C \times S)_{\mathrm{hom}} \end{cases}$$

$$\textit{(iii)}\ \pi_r(S) \cdot \Gamma^t \cdot \pi_s(C) = 0 \textit{ if } \begin{cases} r < 2 + s \textit{ and } \Gamma \in A^1(S \times C) \\ r = 2 + s \textit{ and } \Gamma \in A^1(S \times C)_{\mathrm{hom}} \end{cases}$$

$$\textit{(iv)}\ \pi_r(C) \cdot \Gamma^t \cdot \pi_s(S) = 0 \textit{ if } \begin{cases} r + 2 < s \textit{ and } \Gamma \in A^2(C \times S) \\ r + 2 = s \textit{ and } \Gamma \in A^2(C \times S)_{\mathrm{hom}}. \end{cases}$$

Proof By the results in [Mu2, Prop. 4.1], Conjectures 7.3.1, 7.3.2 and 7.3.3 hold for the product $Z = S \times C$. Therefore Proposition 7.3.6 applies to

$S \times C$ (and $C \times S$). Then (1) follows from the fact that

$$\pi_j(C) \cdot \Gamma \cdot \pi_i(S) \in \mathcal{M}_{\text{rat}}(h_i(S), h_j(C)) \subseteq A_2(S \times C)$$

and similarly for (2), (3) and (4). □

Corollary 7.3.9. *Let S and C be as in Corollary 7.3.8. Then:*

i) $\mathcal{M}_{\text{rat}}(h_1(C)(1), h_2(S)) = 0$;
ii) $\mathcal{M}_{\text{rat}}(h_2(S)), h_1(C)) = 0$.

Proof The first assertion follows from the equality:

$$\mathcal{M}_{\text{rat}}(h_1(C)(1), h_2(S)) = \pi_2(S) \circ A^1(C \times S) \circ \pi_1(C)$$

by applying (3) of Cor. 7.3.8 to Γ^t for any $\Gamma \in A^1(S \times C)$.
b) follows from

$$\mathcal{M}_{\text{rat}}(h_2(S), h_1(C)) = \pi_1(C) \circ A^1(S \times C) \circ \pi_2(S)$$

and from (1). □

The next result shows that, in the case of two surfaces S and S', part of Proposition 7.3.6 holds without assuming any conjecture for $S \times S'$.†

Theorem 7.3.10. *Let S and S' be smooth projective surfaces over the field k. Then for any C-K decompositions as in Proposition 7.2.1*

$$h(S) = \bigoplus_{0 \le i \le 4} h_i(S); \qquad h(S') = \bigoplus_{0 \le j \le 4} h_j(S')$$

where $h_i(S) = (S, \pi_i(S), 0)$ and $h_j(S') = (S', \pi'_j(S'), 0)$, we have

i) $\mathcal{M}_{\text{rat}}(h_i(S), h_j(S')) = 0$ *for all* $j < i$ *and* $0 \le i \le 4$
ii) $\mathcal{M}_{\text{rat}}(h_i(S), h_i(S')) \simeq \mathcal{M}_{\text{hom}}(h_i^{\text{hom}}(S), h_i^{\text{hom}}(S'))$ *for* $i \neq 2$.

Proof Let $\pi_i = \pi_i(S)$ and $\pi'_j = \pi_j(S')$. Then $S \times S'$ has a C-K decomposition defined by the projectors $\sum_{r+s=m} \pi_r \times \pi'_s$. For any correspondence $Z \in A^2(S \times S')$ let us define

$$\alpha_{ji}(Z) = \pi'_j \circ Z \circ \pi_i \text{ for } 0 \le i, j \le 4.$$

Then, in order to prove part (i) it is enough to show that $\alpha_{ji}(Z) = 0$ for $j < i$.

† Kenichiro Kimura recently informed the second author that he had also found a proof, for the case of the product of two surfaces, of conjecture 7.3.2 and of part of conjecture 7.3.3.

We will show that $\alpha_{12}(Z) = \alpha_{23}(Z) = 0$: the other cases are easier and follow from the same type of arguments.

Let $\alpha_{12} = \pi'_1 \circ Z \circ \pi_2$: from the construction of the projectors $\{\pi_i\}$ and $\{\pi'_j\}$ in Proposition 1, it follows that $\pi'_1 = j_* \circ D$ where $j : C' \to S'$ is the closed embedding of the curve C' in S' and $D \in A^1(S' \times C')$. By possibly taking a desingularization for each irreducible component of C' we get a morphism $Y' \to S'$ where Y' is a smooth projective curve. Also, by arguing componentwise, we may as well assume that Y' is irreducible and we may replace C' by such a Y'. Then $\alpha_{12} = j_* \circ D \circ Z \circ \pi_2 = j_* \circ D_1 \circ \pi_2$, with $D_1 = D \circ Z \in A^1(S \times Y')$ and $D \circ Z \circ \pi_2 \in \operatorname{Hom}_{\mathcal{M}_{\mathrm{rat}}}(h_2(S), h(Y'))$.

Let us take a C-K decomposition $h(Y') = \sum_{0 \le j \le 2} h_j(Y')$, where $h_j(Y') = (Y', \pi_j(Y'), 0)$. By applying Corollary 7.3.8 to $S \times Y'$, we get $\pi_j(Y') \circ D_1 \circ \pi_2(S) = 0$ for $j = 0, 1$ so that $D \circ Z \circ \pi_2(S) = \pi_2(Y') \circ D \circ Z \circ \pi_2(S)$. If $\pi_2(Y') = [R' \times Y']$, with R' a chosen rational point on Y', then $\pi_2(Y') \circ D_1 = D_1^t(R') \times Y'$ and $D_1^t(R') = Z^t(D^t(R'))$. From the chosen normalization in the construction of the projectors $\{\pi_i(S)\}$ and $\{\pi_j(S')\}$ (see the proof of Proposition 7.2.1) it follows that $D^t(R') \in A^1(S')_{\mathrm{hom}}$ and $D_1^t(R') \in A^1(S)_{\mathrm{hom}}$. Therefore we get:

$$\begin{aligned} D \circ Z \circ \pi_2(S) &= \pi_2(Y') \circ D \circ Z \circ \pi_2(S) = \pi_2(Y') \circ D_1 \circ \pi_2(S) \\ &= (D_1^t(R') \times Y') \circ \pi_2(S) = \pi_2(S)(D_1^t(R')) \times Y' = 0 \end{aligned}$$

since $\pi_2(S)(D_1^t(R')) = 0$ because $\pi_2(S)(A^1(S)_{\mathrm{hom}}) = 0$. Therefore $\alpha_{12}(Z) = 0$.

To show that $\alpha_{23}(Z) = \pi_2(S') \circ Z \circ \pi_3(S) = 0$ it is enough to look at the transpose correspondence α_{23}^t. Then $\alpha_{23}^t(Z) = \pi_1(S) \circ Z^t \circ \pi_2(S')$. By applying the previous case to $S' \times S$ we get $\alpha_{23}^t(Z) = 0$, hence $\alpha_{23}(Z) = 0$.

We now prove part (ii): for Z homologically equivalent to 0, $\alpha_{11}(Z) = \alpha_{33}(Z) = 0$ follows from the definition of $\{\pi_i(S)\}$ and $\{\pi_j(S')\}$ in Proposition 7.2.1 and from the following result in [Sch, 4.5] (by interchanging π_1 and π_3 because of our covariant set-up):

$$\begin{aligned} \mathcal{M}_{\mathrm{rat}}(h_1(S), h_1(S')) &= \mathrm{Ab}(\mathrm{Alb}_S, \mathrm{Alb}_{S'}) \\ \mathcal{M}_{\mathrm{rat}}(h_3(S), h_3(S')) &= \mathrm{Ab}(\mathrm{Pic}^0_S, \mathrm{Pic}^0_{S'}). \end{aligned}$$

Both equalities hold also with $\mathcal{M}_{\mathrm{rat}}$ replaced by $\mathcal{M}_{\mathrm{hom}}$ and therefore we get (ii).

The equalities $\alpha_{00}(Z) = \alpha_{44}(Z) = 0$ are trivial because

$$\mathcal{M}_{\mathrm{rat}}(h_j(S), h_j(S')) = \mathcal{M}_{\mathrm{hom}}(h_j^{\mathrm{hom}}(S), h_j^{\mathrm{hom}}(S')) \simeq \mathbb{Q}$$

for $j = 0, 4$. □

Summarizing what we have done so far with Proposition 7.3.6 in mind, let us display our information on the Hom groups $\mathcal{M}_{\text{rat}}(h_i(S), h_j(S'))$ in matrix form ($r = h$ means "rational equivalence = homological equivalence"):

$$\begin{pmatrix} r=h & 0 & 0 & 0 & 0 \\ * & r=h & 0 & 0 & 0 \\ * & * & ? & 0 & 0 \\ * & * & * & r=h & 0 \\ * & * & * & * & r=h \end{pmatrix}$$

In the next section we study the remaining group on the diagonal: the one marked with a '?'.

7.4 The group $\mathcal{M}_{\text{rat}}(h_2(S), h_2(S'))$

Let S, S' be smooth projective surfaces over the field k: from Proposition 7.2.1 and from Theorem 7.3.10 it follows that for any C-K decompositions $h(S) = \bigoplus_{0 \le i \le 4} h_i(S)$ and $h(S') = \bigoplus_{0 \le i \le 4} h_i(S')$ as in Proposition 7.2.1 in $\mathcal{M}_{\text{rat}}$, the $\mathbb{Q}$-vector spaces $\mathcal{M}_{\text{rat}}(h_i(S), h_i(S'))$ are finite dimensional for $i \neq 2$. In fact we have

$$\mathcal{M}_{\text{rat}}(h_0(S), h_0(S')) \simeq \mathcal{M}_{\text{rat}}(h_4(S), h_4(S')) \simeq \mathbb{Q}$$

(if S and S' are geometrically connected), and

$$\mathcal{M}_{\text{rat}}(h_1(S), h_1(S')) \simeq \text{Ab}(\text{Alb}_S, \text{Alb}_{S'})$$
$$\mathcal{M}_{\text{rat}}(h_3(S), h_3(S')) \simeq \text{Ab}(\text{Pic}_S, \text{Pic}_{S'}).$$

Moreover, from Proposition 7.3.6 (ii) it follows that, if $S \times S'$ satisfies Conjecture 7.3.3, then $\mathcal{M}_{\text{rat}}(h_2(S), h_2(S'))$ is also a finite dimensional $\mathbb{Q}$-vector space, at least in characteristic 0 and for a classical Weil cohomology.

In the case $k = \mathbb{C}$, if the surface S has geometric genus 0 then the isomorphism $\mathcal{M}_{\text{rat}}(h_2(S), h_2(S)) \simeq \mathcal{M}_{\text{hom}}(h_2^{\text{hom}}(S), h_2^{\text{hom}}(S))$ in Proposition 7.3.6 (ii) holds if and only if Bloch's conjecture holds for S i.e. if and only if the Albanese kernel $T(S)$ vanishes (see §7.6).

It is therefore natural to ask how the group $\mathcal{M}_{\text{rat}}(h_2(S), h_2(S'))$ may be computed. We have

Lemma 7.4.1. *There is a canonical isomorphism*

$$\mathcal{M}_{\text{rat}}(h_2(S), h_2(S')) \simeq \mathcal{M}_{\text{rat}}(h_2^{\text{alg}}(S), h_2^{\text{alg}}(S')) \oplus \mathcal{M}_{\text{rat}}(t_2(S), t_2(S'))$$

where $t_2(S)$ and $t_2(S')$ are defined in Proposition 7.2.3.

Proof It suffices to see that

$$\mathcal{M}_{\mathrm{rat}}(t_2(S), h_2^{\mathrm{alg}}(S')) = \mathcal{M}_{\mathrm{rat}}(h_2^{\mathrm{alg}}(S), t_2(S')) = 0$$

which follows immediately from Proposition 7.2.3 (see its proof). □

Since $\mathcal{M}_{\mathrm{rat}}(h_2^{\mathrm{alg}}(S), h_2^{\mathrm{alg}}(S')) \simeq \mathbb{Q}^{\rho\rho'}$, *this lemma reduces the study of* $\mathcal{M}_{\mathrm{rat}}(h_2(S), h_2(S'))$ *to that of* $\mathcal{M}_{\mathrm{rat}}(t_2(S), t_2(S'))$. In this section we give *two* descriptions of this group: one as a quotient of $A_2(S \times S')$ (Theorem 7.4.3) and the other in terms of Albanese kernels (Theorem 7.4.8).

Then, in §7.5, we will relate these results with the *birational motives* of S and S' i.e. with the images of $h(S)$ and $h_2(S')$ in the category $\mathcal{M}_{\mathrm{rat}}^{\mathrm{o}}(k)$ of birational motives of [K-S].

7.4.1 First description of $\mathcal{M}_{\mathrm{rat}}(t_2(S), t_2(S'))$

We start with

Definition 7.4.2. Let $X = X_d$ and $Y = Y_e$ be smooth projective varieties over k: we denote by $\mathcal{J}(X, Y)$ the subgroup of $A_d(X \times Y)$ generated by the classes supported on subvarieties of the form $X \times N$ or $M \times Y$, with M a closed subvariety of X of dimension $< d$ and N a closed subvariety of Y of dimension $< e$.

In other words: $\mathcal{J}(X, Y)$ is generated by the classes of correspondences which are not dominant over X and Y by either the first or the second projection.

Note that $\mathcal{J}(X, Y) = A_d(X \times Y)$ if $d < e$ (project to Y). In the case $X = Y$ $\mathcal{J}(X, X)$ is a two-sided ideal in the ring of correspondences $A_d(X \times X)$ (see [Fu, p. 309]).

Now let S and S' be smooth projective surfaces over k and let $\{\pi_i = \pi_i(S)\}$ and $\{\pi_i' = \pi_i(S')\}$, for $0 \leq i \leq 4$, be projectors giving C-K decompositions respectively for S and for S' as in Proposition 7.2.1. Then, as in Proposition 7.2.3, $\pi_2(S) = \pi_2^{\mathrm{alg}}(S) + \pi_2^{\mathrm{tr}}(S)$, $h_2(S) \simeq \rho\mathbb{L} \oplus t_2(S)$, where $t_2(S) = (S, \pi_2^{\mathrm{tr}}(S), 0)$ and ρ is the Picard number of S. Similarly $\pi_2(S') = \pi_2^{\mathrm{alg}}(S') + \pi_2^{\mathrm{tr}}(S')$, $h_2(S') \simeq \rho'\mathbb{L} \oplus t_2(S')$ where ρ' is the Picard number of S'. Let us define a homomorphism

$$\Phi : A_2(S \times S') \to \mathcal{M}_{\mathrm{rat}}(t_2(S), t_2(S'))$$

as follows: $\Phi(Z) = \pi_2^{\mathrm{tr}}(S') \circ Z \circ \pi_2^{\mathrm{tr}}(S)$. Then we have the following result.

Theorem 7.4.3. *The map* Φ *induces an isomorphism*

$$\bar{\Phi} : \frac{A_2(S \times S')}{\mathcal{J}(S, S')} \simeq \mathcal{M}_{\mathrm{rat}}(t_2(S), t_2(S')).$$

Proof From the definition of the motives $t_2(S)$ and $t_2(S')$ it follows that

$$\mathcal{M}_{\mathrm{rat}}(t_2(S), t_2(S')) = \{\pi_2^{\mathrm{tr}}(S') \circ Z \circ \pi_2^{\mathrm{tr}}(S) \mid Z \in A_2(S \times S')\}.$$

We first show that $\mathcal{J}(S, S') \subset \operatorname{Ker} \Phi$. Let $Z \in \mathcal{J}(S, S')$: we may assume that Z is irreducible and supported either on $S \times Y'$ with $\dim Y' \leq 1$ or on $Y \times S'$ with $\dim Y \leq 1$.

Suppose Z is supported on $S \times Y'$. The case $\dim Y' = 0$ being easy, let us assume that Y' is a curve which, by possibly taking a desingularization (compare proof of Proposition 7.2.1), we may take to be smooth and irreducible. Let $j : Y' \to S'$; then $Z = j_* \circ D$, where j_* is the graph Γ_j and $D \in A^1(S \times Y')$. Using the identity $\pi_2(S) \circ \pi_2^{\mathrm{tr}}(S) = \pi_2^{\mathrm{tr}}(S) \circ \pi_2^{\mathrm{tr}}(S) = \pi_2^{\mathrm{tr}}(S)$ we get

$$\pi_2^{\mathrm{tr}}(S') \circ Z \circ \pi_2^{\mathrm{tr}}(S) = \pi_2^{\mathrm{tr}}(S') \circ j_* \circ D \circ \pi_2(S) \circ \pi_2^{\mathrm{tr}}(S).$$

Let $\Delta_{Y'} = \pi_0(Y') + \pi_1(Y') + \pi_2(Y')$ be a C-K decomposition. By Corollary 7.3.8 (a)

$$\pi_1(Y') \circ D \circ \pi_2(S) = \pi_0(Y') \circ D \circ \pi_2(S) = 0$$

hence $D \circ \pi_2(S) = \pi_2(Y') \circ D \circ \pi_2(S)$. Let R' be a rational point on Y' such that $\pi_2(Y') = [R' \times Y']$; then $\pi_2(Y') \circ D = [D(R')^t \times Y']$ and $D \circ \pi_2^{\mathrm{tr}}(S) = D \circ \pi_2(S) \circ \pi_2^{\mathrm{tr}}(S) = [D(R')^t \times Y'] \circ \pi_2^{\mathrm{tr}}(S) = [\pi_2^{\mathrm{tr}}(S)(D(R')^t) \times Y']$. From $A^2(t_2(S)) = T(S)$ it follows that $\pi_2^{\mathrm{tr}}(S)$ acts as 0 on divisors, hence $[\pi_2^{\mathrm{tr}}(S)(D(R')^t) \times Y'] = 0$ and

$$\pi_2^{\mathrm{tr}}(S') \circ Z \circ \pi_2^{\mathrm{tr}}(S) = 0.$$

This completes the proof in the case Z has support on $S \times Y'$.

Let us now consider the case when Z is supported on $Y \times S'$, Y a curve on S. In order to show that $\pi_2^{\mathrm{tr}}(S') \circ Z \circ \pi_2^{\mathrm{tr}}(S) = 0$ we can just take the transpose. Then we get $\pi_2^{\mathrm{tr}}(S) \circ Z^t \circ \pi_2^{\mathrm{tr}}(S')$ and this brings us back to the previous case.

Therefore Φ induces a map

$$\bar{\Phi} : A_2(S \times S')/\mathcal{J}(S, S') \to \mathcal{M}_{\mathrm{rat}}(t_2(S), t_2(S'))$$

which is clearly surjective, and we are left to show that $\bar{\Phi}$ is injective.

Let $Z \in A_2(S \times S')$ be such that $\pi_2^{\mathrm{tr}}(S') \circ Z \circ \pi_2^{\mathrm{tr}}(S) = 0$: we claim that $Z \in \mathcal{J}(S, S')$.

Let ξ be the generic point of S. To prove our claim we are going to evaluate

$$(\pi_2^{\mathrm{tr}}(S') \circ Z \circ \pi_2^{\mathrm{tr}}(S))(\xi)$$

over $k(\xi)$. By using Chow's moving lemma on $S \times S'$ we may choose a cycle in the class of Z in $A_2(S \times S')$ (which we will still denote by Z) such that $\pi_2^{tr}(S') \circ Z \circ \pi_2^{tr}(S)$ is defined as a cycle and $\pi_2^{tr}(S') \circ Z \circ \pi_2^{tr}(S)(\xi)$ can be evaluated using the formula $(\alpha \circ \beta)(\xi) = \alpha(\beta(\xi))$, for $\alpha, \beta \in A_2(S \times S')$. From the definition of the projector $\pi_2^{tr}(S)$ in Proposition 7.2.3, we have

$$\pi_2^{tr}(S) = \Delta_S - \pi_0(S) - \pi_1(S) - \pi_2^{alg}(S) - \pi_3(S) - \pi_4(S)$$

where $\pi_2^{alg}(S), \pi_3(S)$ and $\pi_4(S)$ act as 0 on 0-cycles, while $\pi_0(S)(\xi) = P$ if $\pi_0(S) = [S \times P]$ and $\pi_1(S)(\xi) = D_\xi$, where D_ξ is a divisor (defined over $k(\xi)$) on the curve $C = C(S)$ used to construct $\pi_1(S)$. Therefore

$$\pi_2^{tr}(S))(\xi) = \xi - P - D_\xi$$

and

$$(\pi_2^{tr}(S') \circ Z \circ \pi_2^{tr}(S))(\xi) = \pi_2^{tr}(S')(Z(\xi) - Z(P) - Z(D_\xi)) = 0.$$

By the same argument as before, applied to the projectors $\{\pi_i(S')\}$, we get

$$\begin{aligned} 0 &= (\pi_2^{tr}(S')(Z(\xi) - Z(P) - Z(D_\xi)) \\ &= (Z(\xi) - Z(P) - Z(D_\xi)) - mP' - \pi_1(S')(Z(\xi) - Z(P) - Z(D_\xi)) \end{aligned}$$

where P' is a rational point defining $\pi_0(S')$ and m is the degree of the 0-cycle $Z(\xi) - Z(P) - Z(D_\xi)$. The cycle $\pi_1(S')(Z(P) + Z(D_\xi)) = D'_\xi$ is a divisor (defined over $k(\xi)$) on the curve $C' = C(S')$ appearing in the construction of $\pi_1(S')$. Therefore we get from $\pi_2^{tr}(S') \circ Z \circ \pi_2^{tr}(S) = 0$:

$$Z(\xi) = Z(P) + Z(D_\xi) + mP' + \pi_1(S')(Z(\xi)) - D'_\xi.$$

The cycle on the right hand side is supported on a curve $Y' \subset S'$, with Y' the union of $Z(C)$ and C'. Therefore, by taking the Zariski closure in $S \times S'$ of both sides of the above formula we get:

$$Z = Z_1 + Z_2$$

where $Z_1, Z_2 \in A_2(S \times S')$, Z_1 is supported on $S \times Y'$ with $\dim Y' \leq 1$ and Z_2 is a cycle supported on $Y \times S$ with $Y \subset S$, $\dim Y \leq 1$. Therefore $Z \in \mathfrak{J}(S, S')$.

□

Remark 7.4.4. Theorem 7.4.3 is an analogue in the case of surfaces of a well-known result for curves, namely the isomorphism:

$$\frac{A_1(C \times C')}{\mathfrak{J}(C, C')} \simeq \mathcal{M}_{rat}(h_1(C), h_1(C'))$$

which immediately follows from the definitions of $\mathfrak{J}(C, C')$ and of the motives $h_1(C)$ and $h_1(C')$. In this case $\mathfrak{J}(C, C')$ is the subgroup generated by the classes which are represented by "horizontal" and "vertical" divisors on $C \times C'$. The equivalence relation defined by $\mathfrak{J}(C, C')$ is denoted in [Weil1, Chap. 6] as "three line equivalence". In the case of curves, since $h_1(C)$ and $h_1(C')$ have a "realization" as the Jacobians $J(C)$ and $J(C')$, the following result in [Weil1, Ch. 6, Thm 22] holds:

$$\frac{A_1(C \times C')}{\mathfrak{J}(C, C')} \simeq \mathrm{Ab}(J(C), J(C'))$$

where $J(C)$, $J(C')$ are the Jacobians.

Corollary 7.4.5. *Keep the same notation and let* $\Pi_r = \sum_{i+j=r} \pi_i(S) \times \pi_j(S')$ *be the Chow-Künneth projectors on* $S \times S'$ *deduced from those of* S *and* S'. *Let* $F^\bullet$ *be the filtration on* $A_j(S \times S')$ *defined by the* Π_r. *Then*

$$\mathfrak{J}(S, S') \cap A_2(S \times S')_{\mathrm{hom}} \simeq F^1 A_2(S \times S') = \mathrm{Ker}\, \Pi_4.$$

Therefore $S \times S'$ *satisfies Conjecture 7.3.3 if and only if* $A_2(S \times S')_{\mathrm{hom}} \subset \mathfrak{J}(S, S')$.

Proof For simplicity let us drop (S) and (S') from the notation for projectors and use π_i etc. for those of S and π'_i etc. for those of S'. Let $\Gamma \in A_2(S \times S')_{\mathrm{hom}}$: from Theorem 7.3.10 $\pi'_j \circ \Gamma \circ \pi_j = 0$ for $j \neq 2$. Therefore $\Gamma \in \mathrm{Ker}\, \Pi_4$ if and only if $\pi'_2 \circ \Gamma \circ \pi_2 = 0$. By Lemma 7.4.1, it suffices to consider separately the algebraic and transcendental parts.

Let $\Gamma \in \mathfrak{J}(S, S') \cap A_2(S \times S')_{\mathrm{hom}}$: then $(\pi_2^{\mathrm{tr}})' \circ \Gamma \circ \pi_2^{\mathrm{tr}} = 0$, because $\Gamma \in \mathfrak{J}(S, S')$. Since $\Gamma \in A_2(S \times S')_{\mathrm{hom}}$ we also have: $(\pi_2^{\mathrm{alg}})' \circ \Gamma \circ \pi_2^{\mathrm{alg}} = 0$. This follows from the isomorphism $h_2(S) = \rho\mathbb{L} \oplus t_2(S)$ where $\rho\mathbb{L} \simeq (S, \pi_2^{\mathrm{alg}}, 0)$, and the same for S'. In fact we have $\mathcal{M}_{\mathrm{rat}}(\rho\mathbb{L}, \rho'\mathbb{L}) \simeq \mathcal{M}_{\mathrm{hom}}(\rho\mathbb{L}, \rho'\mathbb{L})$, so that, if $\Gamma \in A_2(S \times S')_{\mathrm{hom}}$, then $(\pi_2^{\mathrm{alg}})' \circ \Gamma \circ \pi_2^{\mathrm{alg}}$ yields the 0 map in $\mathcal{M}_{\mathrm{hom}}(\rho\mathbb{L}, \rho'\mathbb{L})$, hence it is 0. Therefore $\pi'_2 \circ \Gamma \circ \pi_2 = 0$ which proves that $\Gamma \in \mathrm{Ker}\, \Pi_4$.

Conversely let $\Gamma \in F^1 A_2(S \times S') = \mathrm{Ker}\, \Pi_4$: then $\Gamma \in A_2(S \times S')_{\mathrm{hom}}$ by [Mu2, 1.4.4]. By Theorem 7.4.3, we also have $\Gamma \in \mathfrak{J}(S, S')$ because

$$\pi'_2 \circ \Gamma \circ \pi_2 = (\pi_2^{\mathrm{alg}})' \circ \Gamma \circ \pi_2^{\mathrm{alg}} + (\pi_2^{\mathrm{tr}})' \circ \Gamma \circ \pi_2^{\mathrm{tr}} = 0$$

and $(\pi_2^{\mathrm{alg}})' \circ \Gamma \circ \pi_2^{\mathrm{alg}} = 0$, since $\Gamma \in A_2(S \times S')_{\mathrm{hom}}$. □

7.4.2 Second description of $\mathcal{M}_{\mathrm{rat}}(t_2(S), t_2(S'))$

Let us still keep the same notation.

Definition 7.4.6. We denote by $H_{\leq 1}$ be the subgroup of $A^2(S'_{k(S)})$ generated by the subgroups $A^2(S'_L)$, when L runs through all the subfields of $k(S)$ containing k and which are of transcendence degree ≤ 1 over k.

Theorem 7.4.8 below will give a description of $\mathcal{M}_{\mathrm{rat}}(t_2(S), t_2(S'))$ in terms of $T(S'_{k(S)})$ and $H_{\leq 1}$. We need a preparatory lemma:

Lemma 7.4.7. *Let S, S' and $H_{\leq 1}$ be as above. Let ξ be the generic point of S and let $Z \in A_2(S \times S')$. Then $Z \in \mathfrak{J}(S, S')$ if and only if $Z(\xi) \in H_{\leq 1}$.*

Proof Let us denote – by abuse – with the same letter Z both a cycle class and a suitable cycle in this class. Let $Z \in \mathfrak{J}(S, S') \subset A_2(S \times S') = A^2(S \times S')$. If Z has support on $Y \times S$ with Y closed in S and of dimension ≤ 1, then $Z(\xi) = 0$. Therefore we may assume that Z has support on $S \times Y'$ and by linearity we may take Z to be represented by a k-irreducible subvariety of $S \times Y'$. Furthermore, by taking its desingularization if necessary, we may also assume that Y' is a smooth curve. Let $j : Y' \to S'$ be the corresponding morphism and $j_* : A^1(Y'_{k(\xi)}) \to A^2(S'_{k(\xi)})$ the induced homomorphim on Chow groups. Then $Z(\xi) = j_* D(\xi)$ where D is a k-irreducible divisor on $S \times Y'$. Then $D(\xi)$ has a smallest field of rationality L, in the sense of [Weil2, Cor 4 p. 269] with $k \subset L \subset k(\xi)$. $D(\xi)$ consists of a finite number of points $P_1, \ldots, P_m$ on Y' each one conjugate to the others over L, and with the same multiplicity. Moreover L is contained in the algebraic closure of $k(P_1)$, where $P_1 \in Y'$. Therefore $tr \deg_k L \leq 1$. We have $D(\xi) \in A^1(Y'_L)$ and $Z(\xi) = \tilde{j}_* D(\xi)$ where $\tilde{j}_* : A^1(Y'_L) \to A^2(S'_L)$. Therefore $Z(\xi) \in H_{\leq 1}$.

Conversely suppose that $Z(\xi) \in H_{\leq 1}$; because of the definition of $H_{\leq 1}$ we may assume that $Z(\xi)$ is a cycle defined over a field L with $k \subset L \subset k(\xi)$ and $t = tr \deg_k L \leq 1$. If $t = 0$ then $Z(\xi)$ is defined over an algebraic extension extension of k, hence $Z \in \mathfrak{J}(S, S')$. Assume $t = 1$ and let C be a smooth projective curve with function field L. Since $L \subset k(\xi)$ there is a dominant rational map f from S to C. Let $U \subset S$ be an open subset such that f is a morphism on U. Then $\eta = f(\xi)$ is the generic point of C. Moreover, $Z(\xi) \in A^2(S'_{k(\eta)}) \subset A^2(S'_{k(\xi)})$. Let Z' be the closure of $Z(\xi)$ in $C \times S'$ so that $Z'(\eta) = Z(\xi)$. Let $Y' \subset S'$ be the projection of Z': then $\dim Y' \leq 1$. Consider the morphism

$$(f_{|U} \times id_{S'})^* : A^2(C \times S') \to A^2(U \times S')$$

and let Z_1 be the cycle in $A^2(S \times S')$ obtained by taking the Zariski closure of $(f_{|U} \times id_{S'})^*(Z')$. Then Z_1 has support on $S \times Y'$ and $Z(\xi) = Z'(\eta) = Z_1(\xi)$. Therefore $Z = Z_1 + Z_2$ where Z_2 has support on $Y \times S'$, Y a curve on S, hence $Z \in \mathfrak{J}(S \times S')$. □

Theorem 7.4.8. *Let S and S' be smooth projective surfaces over k and let $T(S'_{k(S)})$ and $H_{\leq 1}$ be as above. Let us define $H = T(S'_{k(S)}) \cap H_{\leq 1}$. Then there is an isomorphism*

$$\mathcal{M}_{\text{rat}}(t_2(S), t_2(S')) \simeq \frac{T(S'_{k(S)})}{H}.$$

Proof Let us define a homomorphism

$$A_2(S \times S') \xrightarrow{\beta} T(S'_{k(S)})$$

by $\beta(Z) = ((\pi_2^{\text{tr}})' \circ Z \circ \pi_2^{\text{tr}})(\xi)$, with ξ the generic point of S. By Lemma 7.4.7, β induces a map

$$\bar{\beta} : \frac{A_2(S \times S')}{\mathcal{J}(S, S')} \to \frac{T(S'_{k(S)})}{H}$$

and, by Theorem 7.4.3

$$\frac{A_2(S \times S')}{\mathcal{J}(S, S')} \simeq \mathcal{M}_{\text{rat}}(t_2(S), t_2(S')).$$

Therefore we are left to show that $\bar{\beta}$ is an isomorphism.

Let $[\sigma] \in \frac{T(S'_{k(S)})}{H}$, σ a representative in $T(S'_{k(S)})$ and Z the Zariski closure of σ in $S \times S'$: then $Z(\xi) = \sigma$. Let $Z_1 = (\pi_2^{\text{tr}})' \circ Z \circ \pi_2^{\text{tr}}$ and $Z_2 = Z - Z_1$: then $(\pi_2^{\text{tr}})' \circ Z_2 \circ \pi_2^{\text{tr}} = 0$. From Theorem 7.4.3 and Lemma 7.4.7 we get $Z_2 \in \mathcal{J}(S, S')$ and $Z_2(\xi) \in H_{\leq 1}$. On the other hand, both $Z(\xi) = \sigma \in T(S'_{k(S)})$ and $Z_1(\xi) \in T(S'_{k(S)})$, hence $Z_2(\xi) \in H = T(S'_{k(S)}) \cap H_{\leq 1}$. Therefore we get $\bar{\beta}(Z) = [Z_1(\xi)] = [Z(\xi) - Z_2(\xi)] = [\sigma - Z_2(\xi)] = [\sigma]$ and this shows that $\bar{\beta}$ is surjective.

Let $Z \in A_2(S \times S')$ be such that $\beta(Z) \in H$. Let $Z_1 = (\pi_2^{\text{tr}})' \circ Z \circ \pi_2^{\text{tr}}$: then $Z_1 \in H_{\leq 1}$, and by Lemma 7.4.7 $Z_1 \in \mathcal{J}(S, S')$. By taking $Z_2 = Z - Z_1$ as before we have $Z_2 \in \mathcal{J}(S, S')$, hence $Z = Z_1 + Z_2 \in \mathcal{J}(S, S')$. Therefore $\bar{\beta}$ is injective. □

Corollary 7.4.9. *a) $t_2(S) = 0 \Leftrightarrow T(S_{k(S)}) \subset H_{\leq 1} \Leftrightarrow T(S_{k(S)}) = 0$.*

b) Suppose that k is algebraically closed and has infinite transcendence degree over its prime subfield. Then $t_2(S) = 0 \Leftrightarrow T(S) = 0$.

c) With the same assumption as in b), $T(S) = 0$ implies $H^2_{\text{tr}}(S) = 0$, and $p_g = 0$ if $\operatorname{char} k = 0$.

Proof From Proposition 7.2.3 we get the second implication in the following:

$$t_2(S) = 0 \Rightarrow t_2(S_{k(S)}) = 0 \Rightarrow T(S_{k(S)}) = 0 \Rightarrow T(S_{k(S)}) \subset H_{\leq 1}$$

the other two being obvious. The implication $T(S_{k(S)}) \subset H_{\leq 1} \Rightarrow t_2(S) = 0$ follows from Theorem 7.4.8, hence a). To see b), in view of Proposition 7.2.3 we need only show that $T(S) = 0 \Rightarrow t_2(S) = 0$. Note that there exists a finitely generated subfield $k_0 \subset k$ and a smooth projective k_0-surface S_0 such that $S \simeq S_0 \times_{k_0} k$. The assumption on k implies that the inclusion $k_0 \subset k$ extends to an inclusion $k_0(S_0) \subset k$. A standard transfer argument shows that $T((S_0)_{k_0(S_0)}) \to T(S)$ is injective. So $t_2(S_0) = 0$ by a) and therefore $t_2(S) = 0$. Finally, c) follows from b) and Corollary 7.2.4. □

7.5 The birational motive of a surface

In this section we first recall some definitions and results from [K-S] on the category of birational Chow motives (with rational coefficients) over k: this category is denoted there (see 6.1) by $\mathrm{CH}^{\mathrm{o}}(k, \mathbb{Q})$ or $\mathrm{Mot}^{\mathrm{o}}_{\mathrm{rat}}(k, \mathbb{Q})$ while we shall denote it here by $\mathcal{M}^{\mathrm{o}}_{\mathrm{rat}}(k)$ or even $\mathcal{M}^{\mathrm{o}}_{\mathrm{rat}}$. Then we compute the group $\mathcal{M}^{\mathrm{o}}_{\mathrm{rat}}(\bar{h}(S), \bar{h}(S))$ of a surface S, where $\bar{h}(S)$ is the image of $h(S)$ in $\mathcal{M}^{\mathrm{o}}_{\mathrm{rat}}$.

Lemma 7.5.1. *For all smooth and projective varieties X and Y, with $\dim X = d$, let $\mathcal{I}$ be the subgroup of $\mathcal{M}^{\mathrm{eff}}_{\mathrm{rat}}(h(X), h(Y)) = A_d(X \times Y)$ defined as follows:*

$$\mathcal{I}(X, Y) = \{f \in A_d(X \times Y) \mid f \text{ vanishes on } U \times Y, U \text{ open in } X\}.$$

Then $\mathcal{I}$ is a two-sided tensor ideal in $\mathcal{M}^{\mathrm{eff}}_{\mathrm{rat}}$. In particular for any smooth projective variety X there is an exact sequence of rings:

$$0 \to \mathcal{I}(X, X) \to A^d(X \times X) \xrightarrow{\phi} A_0(X_{k(X)}) \to 0 \tag{7.3}$$

where $d = \dim X$ and $k(X)$ is the function field of X. If we denote by $\bullet$ the multiplication in $A_0(X_{k(X)})$ defined via (7.3) then, if P and Q are two rational points of X, we have:

$$[P] \bullet [Q] = [P]$$

in $A_0(X_{k(X)})$.

Proof The fact that $\mathcal{I}$ is a tensor ideal in $\mathcal{M}^{\mathrm{eff}}_{\mathrm{rat}}$ is proven in [K-S, 5.3]. We review the proof:

If X, Y, Z are smooth projective varieties and $U \subset X$ is open, then the usual formula defines a composition of correspondences:

$$A_{\dim X}(U \times Y) \times A_{\dim Y}(Y \times Z) \to A_{\dim X}(U \times Z)$$

and this composition is compatible with the restriction to any open subset $V \subset U$. Passing to the limit, since:

$$A_0(Y_{k(X)}) = A^{\dim Y}(Y_{k(X)}) = \lim_{U \subset X} A^{\dim Y}(U \times Y) = \lim_{U \subset X} A_{\dim X}(U \times Y)$$

we get a composition

$$A_0(Y_{k(X)}) \times A_{\dim Y}(Y \times Z) \to A_0(Z_{k(X)}).$$

If $\alpha \in A_0(Y_{k(X)})$ and $\beta \in \mathcal{J}(Y, Z)$, i.e. if β has support on a closed subset $M \times Z$ of $Y \times Z$ then $\beta \circ \alpha = 0 \in A_0(Z_{k(X)})$ as one sees by moving α away from M. Therefore we get a a pairing

$$A_0(Y_{k(X)}) \times A_0(Z_{k(Y)}) \to A_0(Z_{k(X)}) \tag{7.4}$$

which, in the case $X = Y = Z$ yields a multiplication $\bullet$ in $A_0(X_{k(X)})$ defined by

$$\bar{\beta} \bullet \bar{\alpha} = \overline{\beta \circ \alpha}$$

where for a correspondence Γ in $A^d(X \times X)$, $\bar{\Gamma}$ denotes its class in $A_0(X_{k(X)})$. Let η be the class of the generic point of X in $A_0(X_{k(X)})$, which is the image of the cycle $[\Delta_X]$ of $A^d(X \times X)$ under the map ϕ in (7.3): then η is the identity for $\bullet$. Let P and Q be closed points in X, and let $[P]$ and $[Q]$ be the corresponding elements in $A_0(X_{k(X)})$. By choosing representatives $[X \times P]$ and $[X \times Q]$ in $A^d(X \times X)$ we get $[X \times P] \circ [X \times Q] = [X \times P]$ in $A^d(X \times X)$. This shows that

$$[P] \bullet [Q] = [P]$$

in $A_0(X_{k(X)})$. □

Definition 7.5.2. We denote by $\mathcal{M}^{o}_{rat}$ the category of *birational Chow motives*, i.e the pseudo-abelian envelope of the factor category $\mathcal{M}^{eff}_{rat}/\mathcal{J}$ and, if $M \in \mathcal{M}^{eff}_{rat}$, by $\bar{M}$ its image in $\mathcal{M}^{o}_{rat}$. We also denote by $\bar{h}$ the (covariant) composite functor $\mathcal{V} \xrightarrow{h} \mathcal{M}^{eff}_{rat} \to \mathcal{M}^{o}_{rat}$.

Note that under the functor $\mathcal{M}^{eff}_{rat} \to \mathcal{M}^{o}_{rat}$ the Lefschetz motive $\mathbb{L}$ goes to 0. By Lemma 7.5.1, one has the following isomorphism in $\mathcal{M}^{o}_{rat}$:

$$\mathcal{M}^{o}_{rat}(\bar{h}(X), \bar{h}(Y)) \simeq A_0(Y_{k(X)}) \tag{7.5}$$

for $X, Y \in \mathcal{V}$. We also have:

Proposition 7.5.3 ([K-S, 5.3 and 5.4]). *A morphism f in $\mathcal{M}^{eff}_{rat}$ belongs to the ideal $\mathcal{J}$ if and only if it factors through an object of the form $M(1)$.*

Remark 7.5.4. The proof in [K-S, 5.4] is not correct because Chow's moving lemma is applied on a singular variety. However, N. Fakhruddin pointed out that it is sufficient to take the subvariety Z appearing in this proof minimal to repair it, and moreover Chow's moving lemma is then avoided. This correction will appear in the final version.

Definition 7.5.5. For all $n \geq 0$, we let

a) $d_{\leq n}\mathcal{M}_{\text{rat}}^{\text{eff}}$ denote the thick subcategory of $\mathcal{M}_{\text{rat}}^{\text{eff}}$ generated by motives of varieties of dimension $\leq n$ (thick means full and stable under direct summands).
b) $d_{\leq n}\mathcal{M}_{\text{rat}}^{\text{o}}$ denote the thick image of $d_{\leq n}\mathcal{M}_{\text{rat}}^{\text{eff}}$ in $\mathcal{M}_{\text{rat}}^{\text{o}}$.
c) $\mathcal{K}_{\leq n}$ denote the ideal of $\mathcal{M}_{\text{rat}}^{\text{eff}}$ consisting of those morphisms that factor through an object of $d_{\leq n}\mathcal{M}_{\text{rat}}^{\text{eff}}$.
d) $\mathcal{K}_{\leq n}^{\text{o}}$ denote the thick image of $\mathcal{K}_{\leq n}$ in $\mathcal{M}_{\text{rat}}^{\text{o}}$.

For simplicity, we write $\mathcal{K}_{\leq n}(X,Y)$ and $\mathcal{K}_{\leq n}^{\text{o}}(X,Y)$ for two varieties X, Y instead of $\mathcal{K}_{\leq n}(h(X),h(Y))$ and $\mathcal{K}_{\leq n}^{\text{o}}(h(X),h(Y))$.

Lemma 7.5.6. a) *The functor*

$$D^{(n)} : \mathcal{M}_{\text{rat}} \to \mathcal{M}_{\text{rat}}$$
$$M \mapsto \underline{\text{Hom}}(M, \mathbb{L}^n),$$

where $\underline{\text{Hom}}(M,\mathbb{L}^n) = M^\vee \otimes \mathbb{L}^n$ *is the internal* Hom *in* $\mathcal{M}_{\text{rat}}$*, sends* $d_{\leq n}\mathcal{M}_{\text{rat}}^{\text{eff}}$ *to itself and defines a self-duality of this category such that* $D^{(n)}(h(X)) = h(X)$ *for any n-dimensional X. Moreover, for X, Y purely of dimension n,*

b) *The map* $D^{(n)} : A_n(X \times Y) \to A_n(Y \times X)$ *is the transposition of cycles and in particular*

$$D^{(n)}(\mathcal{J}(X,Y)) = \mathcal{J}(Y,X)$$

where $\mathcal{J}(X,Y)$ is the subgroup of Definition 7.4.2.

c) $D^{(n)}(\mathcal{I}(X,Y)) = \mathcal{K}_{\leq n-1}(Y,X)$ *and* $D^{(n)}(\mathcal{K}_{\leq n-1}(X,Y)) = \mathcal{I}(Y,X)$*, where $\mathcal{I}$ is as in Lemma 7.5.1.* d) *For X, Y purely of dimension n we have*

$$\mathcal{J}(X,Y) = \mathcal{I}(X,Y) + \mathcal{K}_{\leq n-1}(X,Y). \tag{7.6}$$

Proof a) and b) are obvious. For c), the argument in [K-S, proof of 5.4] (see Remark 7.5.4) implies that $\mathcal{I}(X,Y)$ consists of those morphisms that factor through some $h(Z)(1)$, where $\dim Z = n-1$. A similar argument shows that $\mathcal{K}_{\leq n-1}(X,Y)$ consists of those morphisms that factor through some $h(Z)$ with $\dim Z = n-1$. The claim is now obvious. Finally, d) follows immediately from c) and the definition of $\mathcal{J}$. $\square$

Lemma 7.5.7. *Let S, S' be smooth projective surfaces over a field k. For all C-K decompositions as in Proposition 7.2.1*

$$h(S) = \bigoplus_{0 \le i \le 4} h_i(S), \qquad h(S') = \bigoplus_{0 \le i \le 4} h_i(S'),$$

we have

$$\mathcal{M}^o_{\mathrm{rat}}(\bar{h}_1(S), \bar{h}_2(S')) \oplus \mathcal{M}^o_{\mathrm{rat}}(\bar{h}_2(S), \bar{h}_2(S')) \simeq T(S'_{k(S)})/T(S')$$

and

$$\mathcal{M}^o_{\mathrm{rat}}(\bar{h}_2(S), \bar{h}_2(S')) = \mathcal{M}^o_{\mathrm{rat}}(\overline{t_2(S)}, \overline{t_2(S')}).$$

Proof Let $\bar{\pi}_i = \bar{\pi}_i(S)$, $\bar{\pi}'_j = \bar{\pi}_j(S')$, $\bar{h}_i(S)$, $\bar{h}_j(S')$, be the images in $\mathcal{M}^o_{\mathrm{rat}}$ of the projectors π_i, π'_j and of the corresponding motives $h_i(S)$, $h_j(S')$) for the surfaces S and S' (as defined in Proposition 7.2.1).

It follows from Proposition 7.5.3 and Proposition 7.2.1 (iii) that $\bar{\pi}_3 = \bar{\pi}'_3 = \bar{\pi}_4 = \bar{\pi}'_4 = 0$.

From Proposition 7.2.3 we get isomorphisms: $h_2(S) \simeq \rho\mathbb{L} \oplus t_2(S)$ and $h_2(S') \simeq \rho'\mathbb{L} \oplus t_2(S')$. It follows that $\bar{h}_2(S) \simeq \overline{t_2(S)}$, in $\mathcal{M}^o_{\mathrm{rat}}$ and similarly for S': $\bar{h}_2(S') \simeq \overline{t_2(S')}$.

Therefore in $\mathcal{M}^o_{\mathrm{rat}}$ we have

$$\bar{h}(S) = 1 \oplus \bar{h}_1(S) \oplus \bar{h}_2(S) = 1 \oplus \bar{h}_1(S) \oplus \overline{t_2(S)}$$

and

$$\bar{h}(S') = 1 \oplus \bar{h}_1(S') \oplus \bar{h}_2(S') = 1 \oplus \bar{h}_1(S') \oplus \overline{t_2(S')}.$$

According to (7.5) we have

$$\mathcal{M}^o_{\mathrm{rat}}(\bar{h}(S'), \bar{h}(S')) = A_0(S'_{k(S)})$$

and

$$\mathcal{M}^o_{\mathrm{rat}}(1, \bar{h}(S')) = \mathcal{M}^o_{\mathrm{rat}}(\bar{h}(\operatorname{Spec} k), \bar{h}(S)) \simeq A_0(S').$$

From Proposition 7.2.1 it follows:

$$\mathcal{M}^{\mathrm{eff}}_{\mathrm{rat}}(h_1(S), 1) = A^0(S)\pi_1 = 0; \mathcal{M}^{\mathrm{eff}}_{\mathrm{rat}}(h_2(S), 1) = A^0(S)\pi_2 = 0.$$

Therefore we get:

$$A_0(S'_{k(S)})/A_0(S') \simeq \mathcal{M}^o_{\mathrm{rat}}(\bar{h}_1(S) \oplus \bar{h}_2(S), \bar{h}_1(S') \oplus h_2(S')).$$

Theorem 7.3.10 (i) yields : $\mathcal{M}_{\text{rat}}(h_2(S), h_1(S')) = 0$ while from [Sch, prop. 4.5] it follows:

$$\mathcal{M}_{\text{rat}}(h_1(S), h_1(S')) \simeq \operatorname{Ab}(\operatorname{Alb}_S, \operatorname{Alb}_{S'}).$$

Therefore we have:

$$A_0(S'_{k(S)})/A_0(S') \simeq \operatorname{Ab}(\operatorname{Alb}_S, \operatorname{Alb}_{S'}) \oplus \mathcal{M}^{\text{o}}_{\text{rat}}(\bar{h}_1(S), \bar{h}_2(S')) \oplus \mathcal{M}^{\text{o}}_{\text{rat}}(\bar{h}_2(S), \bar{h}_2(S')). \quad (7.7)$$

There is a canonical map $\alpha : A_0(S'_{k(S)}) \to \operatorname{Ab}(\operatorname{Alb}_S, \operatorname{Alb}_{S'})$ which is 0 on $A_0(S')$ (see [K-S, (9.5))]) as well as an isomorphism:

$$\operatorname{Ab}(\operatorname{Alb}_S, \operatorname{Alb}_{S'}) \simeq \frac{\operatorname{Alb}_{S'}(k(S))_{\mathbb{Q}}}{\operatorname{Alb}_{S'}(k)_{\mathbb{Q}}}.$$

Therefore we get the following exact sequence:

$$0 \to T(S'_{k(S)})/T(S') \to A_0(S'_{k(S)})/A_0(S') \to \frac{\operatorname{Alb}_{S'}(k(S))_{\mathbb{Q}}}{\operatorname{Alb}_{S'}(k)_{\mathbb{Q}}} \to 0.$$

Hence:

$$A_0(S'_{k(S)})/A_0(S') \simeq (\operatorname{Alb}_{S'}(k(S))/\operatorname{Alb}_{S'}(k))_{\mathbb{Q}} \oplus T(S'_{k(S)})/T(S'). \quad (7.8)$$

From (7.7) and (7.8) we get;

$$\mathcal{M}^{\text{o}}_{\text{rat}}(\bar{h}_1(S), \bar{h}_2(S')) \oplus \mathcal{M}^{\text{o}}_{\text{rat}}(\bar{h}_2(S), \bar{h}_2(S')) \simeq T(S'_{k(S)})/T(S').$$

□

Proposition 7.5.8. *With the same notation as in Lemma 7.5.7, the projection map*

$$\Psi : \mathcal{M}^{\text{eff}}_{\text{rat}}(t_2(S), t_2(S')) \to \mathcal{M}^{o}_{\text{rat}}(\overline{t_2(S)}, \overline{t_2(S')})$$

is an isomorphism.

Proof The map Ψ of the proposition is clearly surjective, and we have to show that it is injective.

Let $f \in \mathcal{M}^{\text{eff}}_{\text{rat}}(t_2(S), t_2(S'))$ be such that $\Psi(f) = 0$. Then f, as a correspondence in $A^2(S \times S')$, belongs to the subgroup $\mathcal{I}(S, S')$: from the definition of $\mathcal{I}(S, S')$ and $\mathcal{J}(S, S')$ (see Definition 7.4.2) it follows that $\mathcal{I}(S, S') \subset \mathcal{J}(S, S')$. Thus $f \in \mathcal{J}(S, S')$ and from Theorem 7.4.3 we get that $(\pi_2^{\text{tr}})' \circ f \circ \pi_2^{\text{tr}} = 0$. Since $f \in \mathcal{M}^{\text{eff}}_{\text{rat}}(t_2(S), t_2(S'))$, we also have $(\pi_2^{\text{tr}})' \circ f \circ \pi_2^{\text{tr}} = f$, hence $f = 0$.

□

Lemma 7.5.9. *Let S be a smooth projective surface and C a smooth projective curve. Then for all C-K decompositions $h(S) = \bigoplus_{0\le i\le 4} h_i(S)$ and $h(C) = \bigoplus_{0\le j\le 2} h_j(C)$ as in Proposition 7.2.1, we have*

$$A_0(C_{k(S)})/A_0(C) \simeq \mathcal{M}^o_{\mathrm{rat}}(\bar h_1(S), \bar h_1(C))$$

where $\bar h_i(X)$ and $\bar h_j(C)$ are the images in $\mathcal{M}^o_{\mathrm{rat}}$.

Proof We have in $\mathcal{M}^o_{\mathrm{rat}}$:

$$\bar h(S) = 1 \oplus \bar h_1(S) \oplus \bar h_2(S); \qquad \bar h(C) = 1 \oplus \bar h_1(C)$$

and, by Proposition 7.5.3,

$$A_0(C_{k(S)}) \simeq \mathcal{M}^o_{\mathrm{rat}}(\bar h(S), \bar h(C)); \qquad A_0(C) \simeq \mathcal{M}^o_{\mathrm{rat}}(1, \bar h(C))$$

with $A_0(C) \simeq \mathbb{Q} \oplus J_C(k)_{\mathbb{Q}}$, where J_C is the Jacobian of C. Therefore

$$A_0(C_{k(S)})/A_0(C) \simeq \mathcal{M}^o_{\mathrm{rat}}(\bar h_1(S), \bar h(C)) \oplus \mathcal{M}^o_{\mathrm{rat}}(\bar h_2(S), \bar h(C))$$

and

$$\begin{aligned}\mathcal{M}^o_{\mathrm{rat}}(\bar h_i(S), \bar h(C)) &= \mathcal{M}^o_{\mathrm{rat}}(\bar h_i(S), 1) \oplus \mathcal{M}^o_{\mathrm{rat}}(\bar h_i(S), \bar h_1(C)) \\ &= \mathcal{M}^o_{\mathrm{rat}}(\bar h_i(S), \bar h_1(C))\end{aligned}$$

because $\mathcal{M}^o_{\mathrm{rat}}(\bar h_i(S), 1) = 0$ for $i = 1, 2$.

From Corollary 7.3.9 (ii) we get $\mathcal{M}_{\mathrm{rat}}(h_2(S), h_1(C)) = 0$ hence

$$\mathcal{M}^o_{\mathrm{rat}}(\bar h_2(S), \bar h(C)) = 0.$$

Therefore

$$A_0(C_{k(S)})/A_0(C) \simeq \mathcal{M}^o_{\mathrm{rat}}(\bar h_1(S), \bar h_1(C)).$$

□

The following Theorem 7.5.10 is a reintepretation of Theorem 7.4.8 in terms of the birational motives $\bar h_2(S)$ and $\bar h_2(S')$.

Let $d_{\le 1}\mathcal{M}^o_{\mathrm{rat}}$ be the thick subcategory of $\mathcal{M}^o_{\mathrm{rat}}$ generated by motives of curves: by a result in [K-S, 9.5], $d_{\le 1}\mathcal{M}^o_{\mathrm{rat}}$ is equivalent to the category $\mathrm{AbS}(k)$ of abelian k-schemes (extensions of a lattice by an abelian variety) with rational coefficients.

Theorem 7.5.10. *Let S, S' be smooth projective surfaces over k. Given any two refined C-K decompositions as in Propositions 7.2.1 and 7.2.3, there are two isomorphisms*

$$\mathcal{M}_{\mathrm{rat}}(t_2(S), t_2(S')) \simeq \mathcal{M}^o_{\mathrm{rat}}(\overline{t_2(S)}, \overline{t_2(S')}) \simeq \frac{T(S'_{k(S)})}{\mathcal{K}^o_{\le 1}(S, S') \cap T(S'_{k(S)})}$$

and $\mathcal{K}^o_{\leq 1}(S, S') = H_{\leq 1}$. *(See Definition 7.5.5 (iv) for the definition of* $\mathcal{K}^o_{\leq 1}$ *and Definition 7.4.6 for the definition of* $H_{\leq 1}$.*)*

Proof Let $\{\pi_i\}$ and $\{\pi'_i\}$, for $0 \leq i \leq 4$, be the projectors giving a refined C-K decomposition respectively for S and S'. From Proposition 7.5.8 it follows that

$$\mathcal{M}_{\mathrm{rat}}(t_2(S), t_2(S')) \simeq \mathcal{M}^o_{\mathrm{rat}}(\overline{t_2(S)}, \overline{t_2(S')}).$$

From Lemma 7.5.1 we get:

$$A_0(S'_{k(S)}) \simeq \frac{A_2(S \times S')}{\mathcal{I}(S, S')}$$

and from Lemma 7.5.6 (d):

$$\mathcal{J}(S, S') = \mathcal{I}(S, S') + \mathcal{K}_{\leq 1}(S, S').$$

From Theorems 7.4.3 and 7.4.8, the map

$$\beta : A_2(S \times S') \to T(S'_{k(S)})$$

defined by

$$\beta(Z) = ((\pi_2^{tr})' \circ Z \circ \pi_2^{tr})(\xi),$$

where ξ is the generic point of S, induces isomorphisms:

$$\mathcal{M}_{\mathrm{rat}}(t_2(S), t_2(S')) \simeq \frac{A_2(S \times S')}{\mathcal{J}(S, S')} \simeq \frac{T(S'_{k(S)})}{H_{\leq 1} \cap T(S'_{k(S)})}.$$

Moreover, it follows from Lemma 7.4.7 that, if $T \in A_2(S \times S')$ then $T \in \mathcal{J}(S, S')$ if and only if $T(\xi) \in H_{\leq 1}$. Hence

$$\begin{aligned} \beta(Z) \in H_{\leq 1} \cap T(S'_{k(S)}) &\iff (\pi_2^{tr})' \circ Z \circ \pi_2^{tr} \in \mathcal{J}(S, S') \\ &\iff (\pi_2^{tr})' \circ Z \circ \pi_2^{tr} = \Gamma_1 + \Gamma_2 \end{aligned}$$

where $\Gamma_1 \in \mathcal{I}(S.S')$ and $\Gamma_2 \in \mathcal{K}_{\leq 1}(S, S')$. Since $\Gamma_1(\xi) = 0$ we get, for any $Z \in A_2(S \times S')$

$$\beta(Z) \in H_{\leq 1} \cap T(S'_{k(S)}) \iff (\pi_2^{tr})' \circ Z \circ \pi_2^{tr}(\xi) = \Gamma_2(\xi)$$

with $\Gamma_2 \in \mathcal{K}_{\leq 1}(S, S')$. This proves that the image of $\mathcal{K}_{\leq 1}(S, S')$ under the map β is $\mathcal{K}^o_{\leq 1}(S, S') \cap T(S'_{k(S)})$ and coincides with $H_{\leq 1} \cap T(S'_{k(S)})$.

Therefore we get:

$$\mathcal{M}_{\mathrm{rat}}(t_2(S), t_2(S')) \simeq \frac{T(S'_{k(S)})}{\mathcal{K}^o_{\leq 1}(S, S') \cap T(S'_{k(S)})}$$

and $\mathcal{K}^o_{\leq 1}(S, S') \cap T(S'_{k(S)}) = H_{\leq 1} \cap T(S'_{k(S)})$.

So we are left to show that

$$\mathcal{K}^{o}_{\leq 1}(S,S') = H_{\leq 1}.$$

From the definitions of $\mathcal{K}^{o}_{\leq 1}(S,S')$ and $H_{\leq 1}$ it follows that $H_{\leq 1} \subset \mathcal{K}^{o}_{\leq 1}(S,S')$.

We have $\mathcal{M}^{o}_{\text{rat}}(\bar{h}(S),\bar{h}(S')) \simeq A_0(S'_{k(S)})$, $\mathcal{M}^{o}_{\text{rat}}(1,\bar{h}(S')) \simeq A_0(S')$ and

$$\mathcal{M}^{o}_{\text{rat}}(\bar{h}_1(S),\bar{h}_1(S')) = \{\bar{\pi}'_1 \circ \bar{\Gamma} \circ \bar{\pi}_1 | \Gamma \in A_0(S'_{k(S)})\}.$$

From the construction of the projector $\pi_1(S')$ as in Proposition 7.2.3, it follows that there exists a curve $C' \subset S'$ such that $\pi_1(S')$, as a map in $\mathcal{M}_{\text{rat}}(h(S'),h(S'))$, factors through the motive $h_1(C')$. Therefore, every map α in $\mathcal{M}^{o}_{\text{rat}}(\bar{h}_1(S),\bar{h}_1(S'))$ factors trough the birational motive of a curve C', i.e. it is in the image $\mathcal{K}^{o}_{\leq 1}(S,S')$ of $\mathcal{K}_{\leq 1}(S,S')$. Moreover, the same argument, as in the proof of lemma 7.4.7 shows that $\alpha \in H_{\leq 1}$.

From Corollary 7.3.9 (ii) it follows that the only map in the group $\mathcal{M}^{o}_{\text{rat}}(\bar{h}_2(S),\bar{h}_2(S'))$ that factors through $\bar{h}(C)$ for some curve C is 0. Therefore we get:

$$\mathcal{K}^{o}_{\leq 1}(S,S') = \mathcal{M}^{o}_{\text{rat}}(1,\bar{h}(S')) + \mathcal{M}^{o}_{rat}(\bar{h}_1(S),\bar{h}_1(S')) + \mathcal{M}^{o}_{\text{rat}}(\bar{h}_1(S),\bar{h}_2(S'))$$

because $\mathcal{M}^{o}_{\text{rat}}(\bar{h}_2(S),\bar{h}_1(S')) = 0$. Furthermore

$$\mathcal{M}^{o}_{\text{rat}}(1,\bar{h}(S')) + \mathcal{M}^{o}_{\text{rat}}(\bar{h}_1(S),\bar{h}_1(S') \subset H_{\leq 1}.$$

From Lemma 7.5.7

$$\mathcal{M}^{o}_{\text{rat}}(\bar{h}_1(S),\bar{h}_2(S')) \oplus \mathcal{M}^{o}_{\text{rat}}(\bar{h}_2(S),\bar{h}_2(S')) \simeq T(S'_{k(S)})/T(S')$$

hence:

$$\mathcal{M}^{o}_{\text{rat}}(\bar{h}_1(S),\bar{h}_2(S')) = \frac{\mathcal{K}^{o}_{\leq 1}(S,S') \cap T(S'_{k(S)})}{T(S')} = \frac{H_{\leq 1} \cap T(S'_{k(S)})}{T(S')}.$$

This proves that $\mathcal{K}^{o}_{\leq 1}(S,S') \subset H_{\leq 1}$. □

Remarks 7.5.11. 1) According to Proposition 7.3.6 (ii), if S and S' are surfaces such that $S \times S'$ satisfies Conjecture 7.3.3 then the group $\mathcal{M}_{\text{rat}}(t_2(S), t_2(S'))$ has finite rank. From Theorem 7.4.3 and Theorem 7.5.10 it follows that this group is isomorphic to a quotient of the group $T(S'_{k(S)})/T(S')$. The following example, suggested to us by Schoen and Srinivas, shows that, if S is a surface, the group $T(S_{k(S)})/T(S)$ may have infinite rank.

Let $E \subset \mathbb{P}^2_{\bar{\mathbb{Q}}}$ denote the elliptic curve defined by $X^3 + Y^3 + Z^3 = 0$, $L = \bar{\mathbb{Q}}(E)$ and $S = E \times E$. Then from the results in [Schoen] it follows that the group $A^2(S_L)_{\deg 0}/A^2(S)_{\deg 0}$ has infinite rank. Now, applying the exact sequence (7.8) with $X = E, Y = S = E \times E$ and $k = \bar{\mathbb{Q}}$, we get an exact sequence

$$0 \to T(S_L)/T(S) \to A_0(S_L)/A_0(S) \to \mathrm{Ab}(E, E \times E) \to 0.$$

Since $\mathrm{Ab}(E, E \times E)$ has finite rank, $T(S_L)/T(S)$ has infinite rank; since $L \subset k(S)$, so does $T(S_{k(S)})/T(S)$.

2) In [B2, 1.8] (see also [J2, 1.12]) Bloch conjectured that, if S is a smooth projective surface and $\Gamma \in A_2(S \times S)_{\mathrm{hom}}$, then Γ acts trivially on $T(S_\Omega)$, where Ω is a universal domain containing k. This conjecture implies that, if $H^2_{\mathrm{tr}}(S) = 0$, then the Albanese kernel $T(S)$ vanishes. We claim that, from the results in §§7.4 and 7.5, it follows that the above conjecture also implies $A_2(S \times S)_{\mathrm{hom}} \subset \mathcal{J}(S, S)$, hence that $\mathrm{End}_{\mathcal{M}_{\mathrm{rat}}}(t_2(S)) \simeq A_2(S \times S)/\mathcal{J}(S, S)$ (Theorem 7.4.3) is finite-dimensional as a quotient of $A^2_{\mathrm{hom}}(S \times S)$ (at least in characteristic 0 for a "classical" Weil cohomology in Bloch's conjecture).

To show the claim, observe that if $\alpha \in A_0(S_{k(S)})$, then $\alpha(\beta) = \beta \circ \alpha$ for every $\alpha \in A_0(S_{k(S)})$ (see (7.4)). Therefore, if $\Gamma \in A_2(S \times S)_{\mathrm{hom}}$, then $\bar{\Gamma}(\pi_2^{\mathrm{tr}}) = 0$ because $\pi_2^{\mathrm{tr}}(\xi) \in T(S_{k(S)})$ and $k(S) \subset \Omega$. This implies that $\bar{\pi}_2^{\mathrm{tr}} \circ \bar{\Gamma} \circ \pi_2^{\mathrm{tr}} = 0$ in $\mathrm{End}_{\mathcal{M}^{\circ}_{\mathrm{rat}}}(\bar{t}_2(S))$. From Theorem 7.4.3 and Proposition 7.5.8 it follows that $\Gamma \in \mathcal{J}(S, S)$.

7.6 Finite-dimensional motives

In this section we first recall from [Ki] and [G-P2] some definitions and results on finite dimensional motives. Then we relate the finite dimensionality of the motive of a surface S with Bloch's Conjecture on the vanishing of the Albanese kernel and with the results in §§7.4 and 7.5.

Let $\mathcal{C}$ be a pseudoabelian, $\mathbb{Q}$-linear, rigid tensor category and let X be an object in $\mathcal{C}$. Let Σ_n be the symmetric group of order n: any $\sigma \in \Sigma_n$ defines a map $\sigma : (x_1, \dots, x_n) \to (x_{\sigma(1)}, \dots, x_{\sigma(n)})$ on the n-fold tensor product $X^{\otimes n}$ of X by itself. There is a one-to-one correspondence between all irreducible representations of the group Σ_n (over $\mathbb{Q}$) and all partitions of the integer n. Let V_λ be the irreducible representation corresponding to a partition λ of n and let χ_λ be the character of the representation V_λ. Let

$$d_\lambda = \frac{\dim(V_\lambda)}{n!} \sum_{\sigma \in \Sigma_n} \chi_\lambda(\sigma) \cdot \Gamma_\sigma$$

where Γ_σ is the correspondence associated to σ. Then $\{d_\lambda\}$ is a set of pairwise orthogonal idempotents in $\mathrm{Hom}_{\mathcal{C}}(X^{\otimes n}, X^{\otimes n}))$ such that $\sum d_\lambda = \Delta_{X^{\otimes n}}$. The category $\mathcal{C}$ being pseudoabelian, they give a decomposition of $X^{\otimes n}$. The n-th symmetric product $S^n X$ of X is then defined to be the image

$\mathrm{Im}(d_\lambda)$ when λ corresponds to the partition (n), and the n-th exterior power $\wedge^n X$ is $\mathrm{Im}(d_\lambda)$ when λ corresponds to the partition $(1,\dots,1)$.

If $\mathcal{C} = \mathcal{M}_{\mathrm{rat}}$ and $M = h(X) \in \mathcal{M}_{\mathrm{rat}}$ for a smooth projective variety X, then $\wedge^n M$ is the image of $M(X^n)$ under the projector $(1/n!)(\sum_{\sigma\in\Sigma_n} sgn(\sigma)\Gamma_\sigma)$, while $S^n M$ is its image under the projector $(1/n!)(\sum_{\sigma\in\Sigma_n}\Gamma_\sigma)$.

Definition 7.6.1 (see [Ki] and [G-P1]). The object X in $\mathcal{C}$ is said to be *evenly (oddly) finite-dimensional* if $\wedge^n X = 0$ ($S^n X = 0$) for some n. An object X is finite-dimensional if it can be decomposed into a direct sum $X_+ \oplus X_-$ where X_+ is evenly finite-dimensional and X_- is oddly finite-dimensional.

Kimura's nilpotence theorem [Ki, 7.2] says that if M is finite-dimensional, any numerically trivial endomorphism of M is nilpotent. We shall need the following more precise version in the proof of Theorem 7.6.9:

Theorem 7.6.2. *Let $M \in \mathcal{M}_{\mathrm{rat}}$ be a finite-dimensional motive. Then the ideal of numerically trivial correspondences in* $\mathrm{End}_{\mathcal{M}_{\mathrm{rat}}}(M, M)$ *is nilpotent.*

Recall [A-K, 9.1.4] that the proof is simply this: Kimura's argument shows that the nilpotence level is uniformly bounded. On the other hand, a theorem of Nagata and Higman says that if I in a non unital and not necessarily commutative ring such that there exists $n > 0$ for which $f^n = 0$ for all $f \in I$, then I is nilpotent.

Examples 7.6.3. 1) If two motives are finite-dimensional so is their direct sum and their tensor product.

2) A direct summand of an evenly (oddly) finite-dimensional motive is evenly (oddly) finite-dimensional. If a motive M is evenly and oddly finite-dimensional then $M = 0$ [Ki, 6.2]. A direct summand of a finite-dimensional motive is finite-dimensional [Ki, 6.9].
3) The dual motive M^* is finite-dimensional if and only if M is finite-dimensional.
4) The motive of a smooth projective curve is finite-dimensional: hence the motive of an abelian variety X is finite-dimensional. Also if X is the quotient of a product $C_1 \times \cdots \times C_n$ of curves under the action of a finite group G acting freely on $C_1 \times \cdots \times C_n$ then $h(X)$ is finite-dimensional.

More generally:

Proposition 7.6.4 (Kimura's lemma, [Ki, 6.6 and 6.8]). *If $f : X \to Y$ is a surjective morphism of smooth projective varieties, then $h(Y)$ is a* *direct summand of $h(X)$. Hence, by Example 7.6.3 2), if $h(X)$ is finite-dimensional then $h(Y)$ is also finite-dimensional.*

Here is a simple proof, in the spirit of Kimura's: let $g = 1_Y \times f : Y \times X \to Y \times Y$ and $T = g^{-1}(\Delta)$, where Δ is the diagonal of $Y \times Y$. Pick a closed point p of the generic fibre of $g_{|T} : T \to \Delta$: the closure of p in T defines a closed subvariety Z in $Y \times X$ which is finite surjective over Δ. Then Z defines a correspondence $[Z]$ from Y to X, and one checks immediately that the composition

$$h(Y) \xrightarrow{[Z]} h(X) \xrightarrow{f_*} h(Y)$$

is multiplication by the generic degree of Z over Δ. □

We also have Kimura's conjecture:

Conjecture 7.6.5 ([Ki]). Any motive in $\mathcal{M}_{\mathrm{rat}}$ is finite-dimensional.

Lemma 7.6.6. *For any smooth projective surface S, the motives $h_0(S)$, $h_1(S)$, $h_2^{\mathrm{alg}}(S)$, $h_3(S)$ and $h_4(S)$ appearing in Propositions 7.2.1 and 7.2.3 are finite-dimensional. Hence all direct summands of $h(S)$ appearing in these propositions are finite-dimensional, except perhaps $t_2(S)$.*

Proof The lemma is clear for $h_0(S)$, $h_2^{\mathrm{alg}}(S)$ and $h_4(S)$ since these motives are tensor products of Artin motives and Tate motives. Since $h_3(S) \simeq h_1(S)(1)$, it remains to deal with $h_1(S)$. But the construction of the projector defining $h_1(S)$ in [Mu1, Sch] shows that it is a direct summand of the motive of a curve; the claim therefore follows from Examples 7.6.3 2) and 4). □

Lemma 7.6.7. *Let U be a group acting transitively on a set E. Suppose that the following condition is verified:*

(*) If $e \in E$, $u \in U$ and $n \geq 1$ are such that $u^n e = e$, then $ue = e$.

On the other hand, let G be a group "acting on this action": there is an action of G on U and an action of G on E such that

$${}^g(ue) = {}^g u \, {}^g e$$

for any $(g, u, e) \in G \times U \times E$. Suppose moreover that G is finite, U has a finite G-invariant composition series $\{1\} = Z_r \subset \cdots \subset Z_1 = U$ such that, for all i,

a) $Z_i' \lhd U$;
b) Z_i/Z_{i+1} is central in U/Z_{i+1} and uniquely divisible.

Then G has a fixed point e on E. If f is another fixed point, there exists $u \in U$, invariant by G, such that $f = ue$.

Proof We argue by induction on r, the case $r = 1$ being trivial. Suppose $r > 1$ and let $Z = Z_{r-1}$. Since G preserves Z, it acts on U/Z and E/Z, preserving the induced action. Moreover, the fact that Z is central in U and divisible implies that Condition (*) is preserved.

By induction, there is $e \in E$ such that

$$^g e = z_g e \quad \forall g \in G$$

with $z_g \in Z$.

Let U_e be the stabilizer of e in U. Since the z_g are central, U_e is stable under the action of G; in particular, G acts on $Z/Z \cap U_e$. An easy computation shows that, for all $g, h \in G$:

$$z_{gh}^{-1} {}^g z_h z_g \in Z \cap U_e.$$

Now $Z/Z \cap U_e$ is divisible as a quotient of Z, and moreover Condition (*) implies that it is torsion-free. Therefore, $H^1(G, Z/Z \cap U_e) = 0$ and there is some $z \in Z$ such that $z_g \equiv {}^g z^{-1} z \pmod{Z \cap U_e}$ for all $g \in G$. Then ze is G-invariant.

For uniqueness, we argue in the same way. By induction, there exists $u_0 \in U$ such that $f = u_0 e$ and ${}^g u_0 = z_g u_0$ for all $g \in G$, with $z_g \in Z$. Applying $g \in G$ to the equation $f = u_0 e$ shows that $z_g \in U_f$. Thus, $g \mapsto z_g$ defines a 1-cocycle with values in $Z \cap U_f$. Since Z and $Z/Z \cap U_f$ are uniquely divisible, so is $Z \cap U_f$, hence this 1-cocycle is a 1-coboundary and we are done. □

Lemma 7.6.8. *Let A be a $\mathbb{Q}$-algebra, π a subset of A and ν a nilpotent element of A. Suppose that there exists a polynomial $P = \sum a_i t^i$, with $a_1 \neq 0$, such that $P(\nu)$ commutes with all the elements of π. Then ν commutes with all the elements of π.*

Proof Let us denote by C the centralizer of π and let r be such that $\nu^r = 0$. We prove that $\nu^i \in C$ for all i by descending induction on i. The case $i \geq r$ is clear. Note that we may (and do) assume that $P(0) = 0$. Then $P(\nu)$ is nilpotent. Let $i < r$: then

$$C \ni P(\nu)^i = a_1^i \nu^i + \dots$$

where the next terms are higher powers of ν. By induction, $a_1^i \nu^i \in C$, hence $\nu^i \in C$. □

The following is a slight improvement of [G-P2, Th. 3]:

Theorem 7.6.9. *Let X be a smooth projective variety over k of dimension d, such that the the Künneth components of the diagonal are algebraic. Assume that the motive $h(X) \in \mathcal{M}_{\text{rat}}$ is finite-dimensional. Then*

a) *$h(X)$ has a Chow-Künneth decomposition*

$$h(X) = \bigoplus_{0 \le i \le 2d} h_i(X)$$

with $h_i(X) = (X, \pi_i, 0)$.

If $\{\tilde{\pi}_i\}$ is another set of such orthogonal idempotents, then there exists a nilpotent correspondence n on X such that

$$\tilde{\pi}_i = (1+n)\pi_i(1+n)^{-1} \tag{7.9}$$

for all i. In particular,

$$h_i(X) \simeq \tilde{h}_i(X)$$

in $\mathcal{M}_{\text{rat}}$, where $\tilde{h}_i(X) = (X, \tilde{\pi}_i, 0)$.

b) *Moreover, the π_i may be chosen so that $\pi_i^t = \pi_{2d-i}$. If $\{\tilde{\pi}_i\}$ is another such choice, there exists a nilpotent correspondence n on X such that (7.9) holds and moreover, $(1+n)^t = (1+n)^{-1}$.*

Proof a) The existence and "uniqueness" of the π_i follow immediately from Kimura's nilpotence theorem (Theorem 7.6.2) and from [J2, 5.4]. For b), let $\mathcal{N} = A_d(X \times X)_{\text{hom}}$: this is a nilpotent ideal of $\text{End}_{\mathcal{M}_{\text{rat}}}(h(X))$ by Theorem 7.6.2. We apply Lemma 7.6.7 with

$$\begin{aligned} U &= 1 + \mathcal{N} \\ Z_i &= 1 + \mathcal{N}^i \\ E &= \{\{\pi_i\} \mid \pi_i \mapsto \pi_i^{\text{hom}}\} \\ G &\simeq \mathbb{Z}/2. \end{aligned}$$

We let U act on E by conjugation: this action is transitive by a). We let G act on this action as follows: if g is the nontrivial element of G, then

$${}^g u = (u^{-1})^t; \qquad {}^g\{\pi_i\} = \{\pi_i{}^t\}.$$

Note that the action on E exists because the π_i^{hom} are stable under transposition (Poincaré duality). We now check the hypotheses of Lemma 7.6.7: clearly the Z_i are normal in U, G-invariant and verify the centrality assumption. Moreover, $Z_i/Z_{i+1} \simeq \mathcal{N}^i/\mathcal{N}^{i+1}$ is uniquely divisible. It remains to verify Condition (*): but this follows from Lemma 7.6.8 applied to $P(\nu) = (1+\nu)^n$. The proof is complete. □

Theorem 7.6.10 ([G-P2, Th. 7]). *Let S be a smooth projective surface over an algebraically closed field k of characteristic 0 with $p_g(S) = 0$, and suppose that k has infinite transcendence degree over $\mathbb{Q}$: then the motive $h(S)$ is finite-dimensional if and only if the Albanese kernel $T(S)$ vanishes.*

Proof "If" follows from Corollary 7.4.9 b) (see proof of (1) $\Leftrightarrow$ (2) in Theorem 7.6.12 below). For "only if", note that the hypothesis $p_g = 0$ implies $H^2_{\mathrm{tr}}(S) = 0$ and therefore $(\pi_2^{\mathrm{tr}}(S))^{\mathrm{hom}} = 0$. By Kimura's nilpotence theorem (Theorem 7.6.2), the finite-dimensionality hypothesis now implies that $\pi_2^{\mathrm{tr}}(S) = 0$, and we conclude by Proposition 7.2.3. □

The following corollary may be viewed as a "birational" version of a result by S. Bloch in [B2, Lect. 1, Prop. 2].

Corollary 7.6.11. *Let S be a smooth projective surface over an algebraically closed field k of characteristic 0 and infinite transcendence degree over $\mathbb{Q}$. Then the following conditions are equivalent:*

i) $p_g(S) = 0$ *and the motive* $h(S)$ *is finite-dimensional;*
ii) the Albanese Kernel $T(S)$ *vanishes;*
iii) $t_2(S) = 0$*;*
iv) $\bar{t}_2(S) = 0$ *in* $\mathcal{M}^o_{\mathrm{rat}}$*;*
v) the motive $\bar{h}(S)$ *in* $\mathcal{M}^o_{\mathrm{rat}}$ *is a direct summand of the birational motive of a curve.*

Proof By Theorem 7.6.10, (i) $\Rightarrow$ (ii). The equivalence of (ii) and (iii) has been seen in Corollary 7.4.9 b) and the equivalence of (iii) and (iv) follows from Proposition 7.5.8. If $t_2(S) = 0$, then $p_g = 0$ by Corollary 7.4.9 c) and $h(S)$ is finite-dimensional by Lemma 7.6.6. Thus we have (i) $\Leftrightarrow$ (ii) $\Leftrightarrow$ (iii) $\Leftrightarrow$ (iv).

Note that in general, $\bar{h}_2^{\mathrm{alg}}(S) = \bar{h}_3(S) = \bar{h}_4(S) = 0$ by Proposition 7.2.1 (iv). Therefore (iv) $\Rightarrow$ (v) (see proof of Lemma 7.6.6). Conversely, if $\bar{h}(S)$ is a direct summand of the birational motive of a (not necessarily connected) curve D, then so is $\bar{t}_2(S)$. But Corollary 7.3.9 (ii) implies that $\mathcal{M}_{\mathrm{rat}}(t_2(S), h(D))) = 0$, hence $\mathcal{M}^o_{\mathrm{rat}}(\bar{t}_2(S), \bar{h}(D)) = 0$, which implies that $\mathrm{End}_{\mathcal{M}^o_{\mathrm{rat}}}(\bar{t}_2(S)) = 0$ and therefore that $\bar{t}_2(S) = 0$. So (v) $\Rightarrow$ (iv) and the proof is complete. □

Theorem 7.6.12. *Let S be a smooth projective surface and let $h(S) = \bigoplus_{0 \le i \le 4} h_i(S) = \bigoplus_{0 \le i \le 4}(S, \pi_i, 0)$ be a refined Chow-Künneth decomposition as in Prop. 7.2.1 and 7.2.3. Let us consider the following conditions:*

(i) the motive $h(S)$ *is finite-dimensional;*
(ii) the motive $t_2(S)$ *is evenly finite-dimensional;*

(iii) *every endomorphism* $f \in \mathrm{End}_{\mathcal{M}_{\mathrm{rat}}}(h(S))$ *which is homologically trival is nilpotent;*

(iv) *for every correspondence* $\Gamma \in A_2(S \times S)_{\mathrm{hom}}$, $\alpha_{i,i} = \pi_i \circ \Gamma \circ \pi_i = 0$, *for* $0 \leq i \leq 4$;

(v) *for all* i, *the map* $\mathrm{End}_{\mathcal{M}_{\mathrm{rat}}}(h_i(S)) \to \mathrm{End}_{\mathcal{M}_{\mathrm{hom}}}(h_i^{\mathrm{hom}}(S))$ *is an isomorphism (hence* $\mathrm{End}_{\mathcal{M}_{\mathrm{rat}}}(h_i(S))$ *has finite rank in characteristic* 0*);*

(vi) *the map* $\mathrm{End}_{\mathcal{M}_{\mathrm{rat}}}(t_2(S)) \to \mathrm{End}_{\mathcal{M}_{\mathrm{hom}}}(t_2^{\mathrm{hom}}(S))$ *is an isomorphism;*

(vii) *let* $\mathfrak{J}(S)$ *be the 2-sided ideal of* $A_2(S \times S)$ *defined in Definition 7.4.2: then* $A_2(S \times S)_{\mathrm{hom}} \subset \mathfrak{J}(S)$.

Then (i) $\Leftrightarrow$ *(ii)* $\Rightarrow$ *(iii)* $\Leftarrow$ *(iv)* $\Leftrightarrow$ *(v)* $\Leftrightarrow$ *(vi)* $\Leftrightarrow$ *(vii).*

Proof (i) $\Leftrightarrow$ (ii) by Lemma 7.6.6.($t_2(S)$ is evenly finite dimensional because it is a direct summand of $h_2(S)$.)

(i) $\Rightarrow$ (iii) follows from [Ki, 7.2] (see Theorem 7.6.2).

(iv) $\Rightarrow$ (iii): (iv) implies that $h_i(S)$, for $0 \leq i \leq 4$, satisfy (i) and (ii) in Theorem 7.3.10. Therefore, by [G-P2, Cor. 3], every endomorphism $f \in \mathrm{End}_{\mathcal{M}_{\mathrm{rat}}}(h(S))$ which is homologically trivial is nilpotent.

(iv) $\Rightarrow$ (v). We have

$$\mathrm{End}_{\mathcal{M}_{\mathrm{rat}}}(h_1(S)) \simeq \mathrm{End}_{\mathrm{Ab}}(\mathrm{Alb}_S) \simeq \mathrm{End}_{\mathcal{M}_{\mathrm{hom}}}(h_1^{\mathrm{hom}}(S)).$$

By duality the same result holds for $\mathrm{End}_{\mathcal{M}_{\mathrm{rat}}}(h_3(S))$. From (4) it follows that also the map

$$\mathrm{End}_{\mathcal{M}_{\mathrm{rat}}}(h_2(S)) \to \mathrm{End}_{\mathcal{M}_{\mathrm{hom}}}(h_2^{\mathrm{hom}}(S))$$

is an isomorphism.

(v) $\Rightarrow$ (vi) is obvious.

(vi) $\Rightarrow$ (vii). If $\Gamma \in A_2(S \times S)_{\mathrm{hom}}$ then $\pi_2^{\mathrm{tr}} \circ \Gamma \circ \pi_2^{\mathrm{tr}}$ yields the 0 map in $\mathrm{End}_{\mathcal{M}_{\mathrm{hom}}}(t_2^{\mathrm{hom}}(S))$, therefore it is 0. Since $\Gamma \in A_2(S \times S)_{\mathrm{hom}}$ we also have $\pi_2^{\mathrm{alg}} \circ \Gamma \circ \pi_2^{\mathrm{alg}} = 0$, hence $\Gamma \in \mathfrak{J}(S)$ (see Lemma 7.4.1).

(vii) $\Rightarrow$ (iv). Let $\Gamma \in A_2(S \times S)_{\mathrm{hom}}$: then $\Gamma \in \mathfrak{J}(S)$ which proves that $\pi_2^{\mathrm{tr}} \circ \Gamma \circ \pi_2^{\mathrm{tr}} = 0$. □

Remark 7.6.13 (Abelian varieties and Kummer surfaces). Let A be an abelian variety of dimension d. Then $h(A)$ has a Chow-Künneth decomposition (see [Sch]) $h(A) = \bigoplus_{0 \leq i \leq 2d} h_i(A)$ where $h_i(A) = (A, \pi_i^A, 0)$ and $n^* \circ \pi_i^A = n^i \pi_i^A$, for every $n \in \mathbb{Z}$. Here $n^* = (id \times n)^*$ is the correspondence induced by multiplication by n on A. The motive $h(A)$ is finite-dimensional, hence the above decomposition is unique (up to isomorphism).

Now suppose that $d = 2$, and let S be the Kummer surface associated to the involution $a \to -a$ on A (with singularities resolved). The rational map $f : A \to S$ induces an isomorphism between the Albanese kernels : $T(A) \simeq T(S)$ (see [B-K-L, A.11]). Let $h(S) = \bigoplus_{0 \le i \le 4} h_i(S)$, with $h_i(S) = (S, \pi_i^S, 0)$. Reasoning as in [A-J, Th 3.2], we get that the formula

$$\pi_i^S = (1/2)(f \times f)_* \pi_i^A$$

defines a C-K decomposition on S. From the exact sequence in (7.3) and from Proposition 7.5.3 it follows that the map f induces homomorphisms $f^* : A_0(S_{k(S)}) \to A_0(A_{k(A)})$ and $f_* : A_0(A_{k(A)}) \to A_0(S_{k(S)})$. Then

$$f^*(f_*(\alpha)) = \alpha + [-1] \cdot \alpha$$

for all $\alpha \in A_0(A_{k(A)})$. From the equality $n^* \circ \pi_2^A = n^2 \pi_2^A$ it follows that $\bar{\pi}_2^A \in A_0(S_{k(S)})$, where $\bar{\pi}_2^A$ is the image of π_2^A under the map in (7.3). From the isomorphism

$$A_0(S_{k(S)}) \simeq \mathrm{End}_{\mathcal{M}^{\circ}_{\mathrm{rat}}}(\bar{h}(S))$$

we get $\bar{h}_2(S) \simeq \bar{h}_2(A)$ in $\mathcal{M}^{\circ}_{\mathrm{rat}}$. The Kummer surface S has $q = \dim H^1(S, \mathcal{O}_S)$ $= 0$, hence $h_1(S) = h_3(S) = 0$. Therefore we get:

$$\bar{h}(S) \simeq 1 \oplus \bar{h}_2(A)$$

and:

$$\bar{h}_2(A) \simeq 1 \oplus \bar{h}_1(A) \oplus \bar{h}_2(A).$$

In particular, f induces an isomorphism $f_* : t_2(A) \to t_2(S)$.

7.7 Higher-dimensional refinements

The next results extend, under the assumption of certain conjectures, some of the properties proven in Propositions 7.2.1 and 7.2.3 for the refined Chow-Künneth decomposition of the motive of a surface to varieties of higher dimension. In particular these results apply to abelian varieties. To avoid questions of rationality we shall assume that k is separably closed; the reader will have no difficulties to extend these results to the general case along the lines of the proof of Proposition 7.2.3.

In the following we will denote by

$$\bar{A}^i(X) \subset H^{2i}(X)$$

the image of the cycle class map $\mathrm{cl}^i : A^i(X) \to H^{2i}(X)$ for a smooth projective variety X (see §7.1.3).

Definition 7.7.1. We say that the Hard Lefschetz theorem holds for the Weil cohomology H if, for any smooth projective variety X of dimension

d, any smooth hyperplane section $W \subset X$ and any $i \leq d$, the Lefschetz operator

$$L^{d-i} : H^i(X) \to H^{2d-i}(X)$$

given by cup product by $\mathrm{cl}(W)^{d-i}$, is an isomorphism.

It is known that every classical Weil cohomology satisfies the Hard Lefschetz theorem.

Let us choose a classical Weil cohomology theory H. Following [Kl], let $B(X)$ and $Hdg(X)$ denote respectively the Lefschetz standard conjecture and the Hodge standard conjecture for a smooth projective variety X. The conjecture $B(X)$ is equivalent to the following for any L as in Definition 7.7.1 (see [Kl, 4.1]):

$\theta(X)$ For each $i \leq d$, there exists an algebraic correspondence θ^i inducing the isomorphism $H^{2d-i}(X) \to H^i(X)$ inverse to L^{d-i}.

Recall also the conjectures:

$A(X, L)$ The restriction $L^{d-2i} : \bar{A}^i(X) \to \bar{A}^{d-i}(X)$ is an isomorphism for all i.
$C(X)$ The Künneth projectors are algebraic.
$D(X)$ Numerical equivalence equals homological equivalence.

Under $D(X)$, $A^*_{\mathrm{hom}}(X)$ is a finite-dimensional $\mathbb{Q}$-vector space. By [Kl, 4.1 and 5.1], we have the following implications (for any L):

$$\begin{gathered} A(X \times X, L \otimes 1 + 1 \otimes L) \Rightarrow B(X) \Rightarrow A(X, L) \\ B(X) \Rightarrow C(X) \\ A(X, L) + Hdg(X) \Rightarrow D(X) \Rightarrow A(X, L). \end{gathered} \tag{7.10}$$

Finally, $B(X)$ is satisfied by curves, surfaces†, abelian varieties and it is stable under products and hyperplane sections [Kl, 4.1 and 4.3]. Also $Hdg(X)$ is true in characteristic 0 and holds in arbitrary characteristic if X is a surface [Kl, §5].

We shall also need the following easy lemma:

Lemma 7.7.2. *Let H be a classical Weil cohomology theory. Let $M = (X_d, p, m) \in \mathcal{M}^{\mathrm{eff}}_{\mathrm{hom}}$. Then*

a) $m \geq -d$.
b) If $p^* H^i(X) \neq 0$ then we have the sharper inequality $m \geq -[i/2]$.

† In [Kl, 4.3], Kleiman requires that $\dim H^1(X) = 2 \dim \mathrm{Pic}^0_X$, but this assumption is verified by all classical Weil cohomologies.

Proof **a)** Let $\alpha : M \to h(Y)$ and $\beta : h(Y) \to M$ be two morphisms such that $\beta \circ \alpha = 1_M$. In particular, $0 \neq \alpha \in \mathrm{Corr}_m(X, Y) = A_{d+m}(X \times Y)$, hence $d + m \geq 0$.

b) We have $H^{i+2m}(M) = p^* H^i(X) \neq 0$. On the other hand, the correspondence α of a) realises $H^{i+2m}(M)$ as a direct summand of $H^{i+2m}(Y)$. The inequality follows. □

Theorem 7.7.3. *Let X be a smooth projective variety of dimension d such that Conjecture $B(X)$ holds and that the ideal $\mathrm{Ker}(\mathrm{End}_{\mathcal{M}_{\mathrm{rat}}}(h(X)) \to \mathrm{End}_{\mathcal{M}_{\mathrm{hom}}}(h_{\mathrm{hom}}(X)))$ is nilpotent (by Theorem 7.6.2, this is true if the motive $h(X)$ is finite-dimensional). Let $X \hookrightarrow \mathbb{P}^N$ be a fixed projective embedding. Then*

i) There exists a self-dual C-K decomposition $h(X) = \bigoplus h_i(X)$ ($\pi_i^t = \pi_{2d-i}$).

ii) Let $i : W \hookrightarrow X$ be a smooth hyperplane section of X and $L = i_* i^* : h(X) \to h(X)(1)$ be the corresponding "Lefschetz operator". Then, for each $i \in [0, d]$, the composition

$$\ell_i : h_{2d-i}(X) \to h(X) \xrightarrow{L^{d-i}} h(X)(d-i) \to h_i(X)(d-i) \tag{7.11}$$

is an isomorphism.

If moreover Conjecture $D(X \times X)$ holds, then:

iii) For each $i \in [0, 2d]$ there exists a further decomposition

$$h_i(X) \simeq \bigoplus_{j=0}^{[i/2]} h_{i,j}(X)(j) \tag{7.12}$$

such that, for each j, $h_{i,j}^{\mathrm{hom}}(X)$ is effective but $h_{i,j}^{\mathrm{hom}}(X)(-1)$ is not effective. Moreover, the isomorphism from (ii) *induces isomorphisms*

$$h_{2d-i,d-i+j}(X) h_{i,j}(X).$$

iv) Let $(\pi_{i,j})$ be the orthogonal set of projectors defining this decomposition. If $(\pi'_{i,j})$ is another such set of projectors, then there exists a correspondence n, homologically equivalent to 0, such that

$$\pi'_{i,j} = (1+n)\pi_{i,j}(1+n)^{-1} \text{ for all } (i,j).$$

In particular, the $h_{i,j}(X)$ are unique up to isomorphism.

Proof We first prove (i), (ii), (iii) and (iv) modulo homological equivalence. (i) is immediate since $B(X) \Rightarrow C(X)$ (see (7.10)). The homological version of (ii) follows immediately from the form $\theta(X)$ of Conjecture $B(X)$.

We now come to the homological versions of (iii) and (iv). By $D(X \times X)$ and Jannsen's theorem [J1], the algebra $\mathrm{End}_{\mathcal{M}_{\mathrm{hom}}}(h_{\mathrm{hom}}(X))$ is semi-simple. Given $i \in [0, 2d]$, write $M = h_i^{\mathrm{hom}}(X)$ as the direct sum of its isotypical components M_α: for each α, we have $M_\alpha \simeq S_\alpha^{n_\alpha}$ where S_α is a simple motive and $n_\alpha > 0$. By Lemma 7.7.2 b), the largest integer j_α such that $S_\alpha(-j_\alpha)$ is effective exists and verifies $j_\alpha \leq i/2$. We set

$$h_{i,j}^{\mathrm{hom}}(X) = \bigoplus_{j_\alpha = j} M_\alpha(-j).$$

This proves the first claim of (iii). Moreover, this construction shows that the homological version of (7.12) is unique; in particular, the corresponding projectors $\pi_{i,j}^{\mathrm{hom}}$ are central in $\mathrm{End}_{\mathcal{M}_{\mathrm{hom}}}(h_i^{\mathrm{hom}}(X))$. This proves [the homological version of] (iv).

To see the second claim in (iii) (still in its homological version), let S_α be a simple summand of $h_{2d-i}^{\mathrm{hom}}(X)$; then clearly $\ell_i(S_\alpha)(-j_\alpha)$ is effective but $\ell_i(S_\alpha)(-j_\alpha - 1)$ is not effective. This proves that $\ell_i(h_{2d-i,d-i+j}^{\mathrm{hom}}(S)(d-i+j)) = h_{i,j}^{\mathrm{hom}}(S)(j)$, hence an isomorphism

$$h_{2d-i,d-i+j}^{\mathrm{hom}}(S) h_{i,j}^{\mathrm{hom}}(S).$$

Lifting these results from homological equivalence to rational equivalence follows from the nilpotency hypothesis, as in the proof of Theorem 7.6.9 a). The fact that ℓ_i still induces an isomorphism $h_{2d-i,d-i+j}(S) h_{i,j}(S)$ is also a standard consequence of nilpotence (cf. [An, 5.1.3.3]): we first note that numerically trivial endomorphisms of $h_{2d-i,d-i+j}(S)$ and $h_{i,j}(S)$ are nilpotent as both motives are direct summands of $h(X)$, up to Tate twists. Let θ be a cycle giving the inverse isomorphism to ℓ_i in $\mathcal{M}_{\mathrm{hom}}$. Then, in $\mathcal{M}_{\mathrm{rat}}$, $m = \ell_i \circ \theta - 1$ and $n = \theta \circ \ell_i - 1$ are numerically equivalent to 0, hence nilpotent. But then, $1 + m$ and $1 + n$ are isomorphisms. It follows that ℓ_i is left and right invertible, hence is an isomorphism. □

Moreover, we have:

Theorem 7.7.4. *Let X verify the hypotheses of Theorem 7.7.3. Then, for a C-K decomposition $h(X) = \bigoplus_{0 \leq i \leq 2d} h_i(X)$ as in this theorem, for every $i < d$, the projector π_i factors through $h(Y_i)$, where $Y_i = X \cdot H_1 \cdot H_2 \cdot \cdots \cdot H_{d-i}$ is a smooth hyperplane section of X of dimension i (with H_i hyperplanes). Hence $h_i(X)$ is a direct summand of $h(Y_i)$ for all $i < d$. Similarly, h_{2d-i} is a direct summand of $h(Y_i)$.*

Proof By $B(X)$, for each $i \leq d$ there exists an algebraic correspondence θ^i inducing the isomorphism $H^{2d-i}(X) \to H^i(X)$ inverse to the isomorphism

$L^{d-i}: H^i(X) \to H^{2d-i}(X)$. Let $j: Y_i \to X$ be the closed embedding and let $\Gamma_i = j^* \circ \theta^i \in A^{d-i}(X \times Y_i)$ and $q_i = j_* \circ \Gamma_i \in A^d(X \times X)$. Then Γ_i and hence also q_i factor trough Y_i: furthermore $q_i^{\hom}$ operates as the identity on $H^{2d-i}(X)$ because $j_* \cdot j* = L^{d-i}$. Let

$$f_i = \pi_i \circ q_i \circ \pi_i \in \mathcal{M}_{rat}(h_i(X), h_i(X)).$$

Then $f_i^{\hom}$ is a projector on $H^*(X)$ and in fact is the $(i, 2d-i)$-Künneth projector. Therefore the map $a_i = \pi_i - f_i$ is homologically trivial, hence nilpotent by hypothesis, i.e. $a_i^n = 0$ for some $n > 0$. Let $b_i = (1 + a_i + a_i^2 + \cdots + a_i^{n-1}) = (1 - a_i)^{-1}$. We have: $a_i \circ \pi_i = \pi_i \circ a_i = a_i$. Therefore $(1 - a_i) \circ \pi_i = \pi_i - a_i = f_i$ and we get

$$\pi_i = (1 - a_i)^{-1} \circ f_i = (1 + a_i + a_i^2 + \cdots + a_i^{n-1}) \circ f_i = b_i \circ f_i;$$
$$\pi_i = f_i \circ (1 - a_i)^{-1} = f_i \circ b_i.$$

Since q_i and therefore also f_i factor trough $h(Y_i)$ it follows that π_i factors trough $h(Y_i)$. Let $g_i = \Gamma_i \circ \pi_i : h(X) \to h(Y_i)$ and $g_i' = b_i \circ \pi_i \circ j_* : h(Y_i) \to h_i(X)$: then we have $g_i' \circ g_i = \pi_i$, hence g_i has a left inverse. Therefore $h_i(X)$ is a direct summand of $h(Y_i)$ for all $i < d$.

The case of π_{2d-i} follows from the above since the C-K decomposition of Theorem 7.7.3 is self-dual. □

Remark 7.7.5. Theorem 7.7.3 notably applies to abelian varieties in characteristic 0. Also note that Theorem 7.7.4 answers – in a slightly weaker form – a question raised in [Mu2, p. 187].

We shall complement Theorem 7.7.3 with a somewhat more explicit result. For this we need Lemma 7.7.6 which is just a reformulation of a result in [Ki, Prop. 2.11]:

Lemma 7.7.6. *Let X be a smooth projective variety of dimension d and let $a \in A^i(X)$, $b \in A^{d-i}(X)$, with $i \leq d$, be such that $\langle b, a\rangle = \deg(a \cdot b) = 1$. Let $\alpha = p_1^* a \cdot p_2^* b \in A^d(X \times X)$, where $p_i : X \times X \to X$ are the projections. Then α is a projector and the motive $M = (X, \alpha, 0)$ is isomorphic to $\mathbb{L}^i$.*

Proof We have $\alpha \circ \alpha = \langle a, b\rangle \alpha = \alpha$, hence α is a projector. On the other hand

$$\mathcal{M}_{\mathrm{rat}}(M, \mathbb{L}^i) = A_{d-i}(X) \circ \alpha = A^i(X) \circ \alpha$$

and

$$\mathcal{M}_{\mathrm{rat}}(\mathbb{L}^i, M) = \alpha \circ A_i(X) = \alpha \circ A^{d-i}(X).$$

Therefore $\alpha \circ b \in \mathcal{M}_{\mathrm{rat}}(\mathbb{L}^i, M)$ and $a \circ \alpha \in \mathcal{M}_{\mathrm{rat}}(M, \mathbb{L}^i)$.

Considering a as an element of $A_{d-i}(X \times \operatorname{Spec} k)$ and b as an element of $A_i(\operatorname{Spec} k \times X)$, we have $a \circ \alpha = \langle a, b \rangle a = a$ and $\alpha \circ b = \langle a, b \rangle b = b$. Moreover $a \circ b = 1_{\operatorname{Spec} k}$ and $b \circ a = p_1^* a \cdot p_2^* b = \alpha = 1_M$. Hence a and b yield an isomorphism between M and $\mathbb{L}^i$. □

Theorem 7.7.7. *Keep the notation and hypotheses of Theorem 7.7.3. For all $i \in [0, d]$, the motive $h_{2i,i}(X)$ contains $h(\underline{\bar{A}}_{i,X})$ as a direct summand, where $h(\underline{\bar{A}}_{i,X})$ is the Artin motive associated to the finite-dimensional vector space $\underline{\bar{A}}_{i,X}$.*

Proof We prove this for $i \leq d/2$; the result then follows for $i \geq d/2$ by Poincaré duality.

We proceed as in the proof of Proposition 7.2.3. By $B(X)$, the homomorphism

$$L^{d-2i} : H^{2i}(X) \to H^{2d-2i}(X)$$

restricts to an isomorphism

$$L^{d-2i} : \bar{A}^i(X) \bar{A}^{d-i}(X)$$

where we use the same notation for the restriction of L^{d-2i}.

From $D(X \times X)$ it follows that the restriction of the Poincaré duality pairing on $H^{2i}(X) \times H^{2d-2i}(X)$ is still nondegenerate on $\bar{A}^i(X) \times \bar{A}^{d-i}(X)$. Therefore there exist elements $a_{i,\alpha} \in A^i(X)$ and $b_{i,\alpha} \in A^{d-i}(X)$ ($\alpha = 0, \ldots, \rho_i$) such that

1) $\tilde{e}_{i,\alpha} = \mathrm{cl}^i(a_{i,\alpha})$ and $\tilde{e}_{i,\alpha} = \mathrm{cl}^i(b_{i,\alpha})$ form dual bases for Poincaré duality.

Now we claim that we can choose the $a_{i,\alpha}$ and $b_{i,\alpha}$ so that they also satisfy

2) $\pi_k^t(a_{i,\alpha}) = \pi_k(b_{i,\alpha}) = 0$ for $k \neq 2i$; $\pi_{2i,j}^t(a_{i,\alpha}) = \pi_{2i,j}(b_{i,\alpha}) = 0$ for $j < i$.

To prove the claim, let $\pi' = \sum_{k \neq 2i} \pi_k$ and $\pi'' = \sum_{j<i} \pi_{2i,j}$: then π'_{hom} acts as 0 on $H^{2d-2i}(X)$ and π''_{hom} acts as 0 on $\bar{A}^{d-i}(X)$, so replacing $b_{i,\alpha}$ by $b_{i,\alpha} - (\pi' + \pi'')(b_{i,\alpha})$ we do not change the $\tilde{e}_{i,\alpha}$. Similarly for the $a_{i,\alpha}$.

Let us take $a_{i,\alpha}$ and $b_{i,\alpha}$ satisfying 1) and 2) and define

$$p_{i,\alpha} = a_{i,\alpha} \times b_{i,\alpha} \in A^d(X \times X).$$

We have $\langle a_{i,\alpha}, b_{i\alpha} \rangle = 1$ and, by Lemma 7.7.6, the $p_{i,\alpha}$ are projectors and each motive $M_{i,\alpha} = (X, p_{i,\alpha}, 0)$ is isomorphic to $\mathbb{L}^i$. Moreover the $p_{i,\alpha}$ are pairwise orthogonal. Let $\pi_{2i}^{\mathrm{alg}} = \sum_{1\le\alpha\le\rho_i} p_{i,\alpha}$: by Condition 2) we have

$$\pi_k \circ \pi_{2i}^{\mathrm{alg}} = \pi_{2i}^{\mathrm{alg}} \circ \pi_k = 0 \text{ for } k \neq 2i;$$
$$\pi_{2i,j} \circ \pi_{2i}^{\mathrm{alg}} = \pi_{2i}^{\mathrm{alg}} \circ \pi_{2i,j} = 0 \text{ for } j < i.$$

Therefore

$$\pi_{2i,i} \circ \pi_{2i}^{\mathrm{alg}} = \pi_{2i}^{\mathrm{alg}} \circ \pi_{2i,i} = \pi_{2i}^{\mathrm{alg}}$$

and we can split $\pi_{2i,i}$ as a sum $\pi_{2i}^{\mathrm{alg}} + p$ of two orthogonal projectors. The theorem is proven. □

7.8 Return to birational motives

Throughout this section, we assume that k is perfect. We recover and strengthen some of the previous results in two steps:

(i) In §7.8.2 we show that, for a surface S provided with a refined C-K decomposition as in Propositions 7.2.1 and 7.2.3, the image of $t_2(S)$ under the full embedding of [Voev2, p. 197 and 3.2]

$$\Phi : \mathcal{M}_{\mathrm{rat}}^{\mathrm{eff}} \to DM_{\mathrm{gm}}^{\mathrm{eff}} \to DM_{-}^{\mathrm{eff}} := DM_{\mathrm{gm}}^{\mathrm{eff}}(k, \mathbb{Q}) \qquad (7.13)$$

is a birational motive. See Theorem 7.8.4. This gives back some of the previous computations.

(ii) In §7.8.3, we interpret the computation of the endomorphism ring of $t_2(S)$ as the existence of adjoints between certain categories of motives: see Theorem 7.8.8.

Finally, in §7.8.4 we show that "nothing more happens" for surfaces when we pass from the category of pure motives to Voevodsky's triangulated category of motives: see Corollary 7.8.13.

7.8.1 Categorical trivialities

Let $\mathcal{A}$ be a pseudo-abelian additive category and $\mathcal{B}$ be a thick subcategory of $\mathcal{A}$ (thick means full and closed under direct summands). To $\mathcal{B}$ one may associate the following ideal $\mathcal{I}$ of $\mathcal{A}$ (cf. [A-K, 1.3.1]):

$$\mathcal{I}(A, A') = \{f : A \to A' \mid f \text{ factors through an object of } \mathcal{B}\}.$$

Let $\mathcal{C} = (\mathcal{A}/\mathcal{I})^\natural$ be the pseudo-abelian envelope of the corresponding factor category, and $P : \mathcal{A} \to \mathcal{C}$ the corresponding projection functor. Recall that, for two objects $A, A' \in \mathcal{A}$,

$$\mathcal{C}(PA, PA') = \mathcal{A}(A, A')/\mathcal{I}(A, A').$$

Let us now define

$$\begin{aligned} {}^{\perp}\mathcal{I} &= \{A \in \mathcal{A} \mid \mathcal{I}(\mathcal{A}, A) = 0\} \\ &= \{A \in \mathcal{A} \mid \mathcal{A}(\mathcal{B}, A) = 0\} \\ &= \{A \in \mathcal{A} \mid \forall A' \in \mathcal{A}, P : \mathcal{A}(A', A)\mathcal{C}(PA', PA)\}. \end{aligned} \tag{7.14}$$

Note that ${}^{\perp}\mathcal{I}$ is stable under direct sums and direct summands: we view it as a thick subcategory of $\mathcal{A}$.

As usual, we say that "the" right adjoint of P is *defined* at an object $C \in \mathcal{C}$ if the functor

$$\mathcal{A} \ni A \mapsto \mathcal{C}(PA, C) \tag{7.15}$$

is representable. Let $\mathcal{C}'$ be the full subcategory of $\mathcal{C}$ consisting of such objects: it is a thick subcategory of $\mathcal{C}$.

The following is an abstraction of the arguments in [K-S, proof of 9.5]:

Proposition 7.8.1. a) *If $P^{\#}$ is "the" partial right adjoint of P (defined on $\mathcal{C}'$), then $P^{\#}(\mathcal{C}') \subseteq {}^{\perp}\mathcal{I}$.*

b) *For any $C \in \mathcal{C}'$, the counit map of the adjunction*

$$\varepsilon : PP^{\#}C \to C$$

is an isomorphism.

c) *$\mathcal{C}'$ coincides with the essential image of $P' = P_{|{}^{\perp}\mathcal{I}}$.*

d) *For any $B \in {}^{\perp}\mathcal{I}$, the unit morphism*

$$\eta : B \to P^{\#}PB$$

is an isomorphism. In particular, $P^{\#}(\mathcal{C}') = {}^{\perp}\mathcal{I}$ and the functors

$$P' : {}^{\perp}\mathcal{I} \to \mathcal{C}'$$
$$P^{\#} : \mathcal{C}' \to {}^{\perp}\mathcal{I}$$

form a pair of quasi-inverse equivalences of categories.

Proof **a)** is obvious.

b) By definition of adjunction, for any $A \in \mathcal{A}$ the composition

$$\mathcal{A}(A, P^{\#}C) \xrightarrow{P} \mathcal{C}(PA, PP^{\#}C) \xrightarrow{\varepsilon_*} \mathcal{C}(PA, C)$$

is an isomorphism. Since $P^{\#}C \in {}^{\perp}\mathcal{I}$ by a), the left map is an isomorphism and hence so is the right one. It follows that, for any $C' \in \mathcal{C}$ (which may be written as a direct summand of PA for some A), the map

$$\mathcal{C}(C', PP^{\#}C) \xrightarrow{\varepsilon_*} \mathcal{C}(C', C)$$

is an isomorphism. By Yoneda's lemma, this implies that ε is an isomorphism.

c) Let for a moment $\mathcal{C}''$ denote the essential image of P'. If $C = PB \in \mathcal{C}''$, with $B \in {}^{\perp}\mathcal{I}$, then clearly the functor (7.15) is represented by B, so $\mathcal{C}'' \subseteq \mathcal{C}'$. Conversely, if $C \in \mathcal{C}'$, then $C \in \mathcal{C}''$ by a) and b).

d) Note that, by c), $P^{\#}PB$ is defined. Let $A \in \mathcal{A}$. As in any adjunction the composition

$$\mathcal{A}(A, B) \xrightarrow{\eta_*} \mathcal{A}(A, P^{\#}PB)\mathcal{C}(PA, PB)$$

is equal to P. Since it is an isomorphism, so is η_* and hence η is an isomorphism by Yoneda. The other conclusions follow immediately. □

Corollary 7.8.2. *P has an everywhere defined right adjoint if and only if $\mathcal{A} = \mathcal{B} \oplus {}^{\perp}\mathcal{I}$.* □

For future reference, we state the dual results (same proofs):

Proposition 7.8.3. *Let $\mathcal{I}^{\perp} = \{A \in \mathcal{A} \mid \mathcal{I}(A, \mathcal{A}) = 0\}$, viewed as a thick subcategory of $\mathcal{A}$. Then the thick subcategory $\mathcal{C}'$ of $\mathcal{C}$ of those objects where a left adjoint ${}^{\#}P$ of P is defined coincides with the essential image of $P' = P_{|\mathcal{I}^{\perp}}$; P' is an equivalence of categories, ${}^{\#}P(\mathcal{C}') = \mathcal{I}^{\perp}$ and ${}^{\#}P$ is a quasi-inverse of P'. Finally, ${}^{\#}P$ is everywhere defined if and only if $\mathcal{A} = \mathcal{B} \oplus \mathcal{I}^{\perp}$.* □

7.8.2 $t_2(S)$ as a birational motive

Recall from [Voev2] that the category $DM_{-}^{\text{eff}}(k)$ admits a partially defined internal Hom

$$\underline{\text{Hom}} : DM_{\text{gm}}^{\text{eff}}(k) \times DM_{-}^{\text{eff}}(k) \to DM_{-}^{\text{eff}}(k)$$

which extends by $\mathbb{Q}$-linearity to an internal Hom

$$\underline{\text{Hom}} : DM_{\text{gm}}^{\text{eff}}(k, \mathbb{Q}) \times DM_{-}^{\text{eff}}(k)_{\mathbb{Q}} \to DM_{-}^{\text{eff}}(k)_{\mathbb{Q}}.$$

This gives a meaning to the following

Theorem 7.8.4. a) *For any smooth projective variety X, one has*

$$\underline{\text{Hom}}(\mathbb{Q}(1), \Phi(h_i(X))) = 0 \text{ for } i = 0, 1$$

in $DM_{-}^{\text{eff}} = DM_{-}^{\text{eff}}(k, \mathbb{Q})$, where Φ is the full embedding of (7.13).

b) *Let S be a surface provided with a refined C-K decomposition as in Propositions 7.2.1 and 7.2.3. Then*

$$\underline{\text{Hom}}(\mathbb{Q}(1), \Phi(t_2(S))) = 0.$$

Therefore, $\Phi(h_0(X)), \Phi(h_1(X))$ and $\Phi(t_2(S))$ belong to the image of the inclusion functor $i : DM_-^o \to DM_-^{\mathrm{eff}}$, where $DM_-^o := DM_-^o(k)_{\mathbb{Q}}$ is the category of [K-S, 6.1].

Proof We do the proof for $t_2(S)$: the other cases are similar and easier (reduce to X a curve).

By definition, $\underline{\mathrm{Hom}}(\mathbb{Q}(1), \Phi(t_2(S)))$ is a complex of sheaves on the category of smooth k-schemes provided with the Nisnevich topology. The fact that it is 0 may be checked locally; moreover, using [Voev1, Prop. 4.20], it suffices to check that for any function field extension K/k we have

$$\mathbb{H}^*(K, \underline{\mathrm{Hom}}(\mathbb{Q}(1), \Phi(t_2(S)))) = 0.$$

($\mathbb{H}^*$ denotes Nisnevich hypercohomology.)

Since $t_2(S)$ is a direct summand of $h(S)$, $\underline{\mathrm{Hom}}(\mathbb{Q}(1), \Phi(t_2(S)))$ is a direct summand of $\underline{\mathrm{Hom}}(\mathbb{Q}(1), M(S))$. By [H-K, Lemma B.1], we have an isomorphism

$$\underline{\mathrm{Hom}}(\mathbb{Q}(1), M(S)) \simeq \underline{\mathrm{Hom}}(M(S), \mathbb{Q}(1)[4]).$$

This isomorphism is induced by the duality isomorphism $M(S) \simeq M(S)^*(2)[4]$. The latter is the image under Φ of the duality isomorphism $\theta : h(S) \simeq h(S)^\vee(2)$ in $\mathcal{M}_{\mathrm{rat}}^{\mathrm{eff}}$; since θ carries $t_2(S)$ to $t_2(S)^\vee(2)$, $\Phi(\theta)$ carries $\Phi(t_2(S))$ to $\Phi(t_2(S))^*(2)[4]$ and thus $\underline{\mathrm{Hom}}(\mathbb{Q}(1), \Phi(t_2(S)))$ is isomorphic to the direct summand $\underline{\mathrm{Hom}}(t_2(S), \mathbb{Q}(1)[4])$ of the complex $\underline{\mathrm{Hom}}(M(S), \mathbb{Q}(1)[4])$.

Let U be a smooth k-scheme. We have

$$\begin{aligned}\mathbb{H}^q_{Nis}(U, \underline{\mathrm{Hom}}(M(S), \mathbb{Q}(1)[4])) &\simeq H^{q+4}_{Nis}(U \times S, \mathbb{Q}(1)) \\ &= H^{q+4}_{Zar}(U \times S, \mathbb{Q}(1)) = H^{q+3}_{Zar}(U \times S, \mathbb{G}_m)_{\mathbb{Q}}.\end{aligned}$$

Passing to the function field K of U we get that, for $q \in \mathbb{Z}$, the group $\mathbb{H}^q(K, \underline{\mathrm{Hom}}(\mathbb{Q}(1), t_2(S)))$ is a direct summand of $H^{q+3}_{Zar}(S_K, \mathbb{G}_m)_{\mathbb{Q}}$.

It is therefore 0 except perhaps for $q = -3, -2$. But for $q = -3$, $H^0(S_K, \mathbb{G}_m) = K^*$ is "caught" by the direct summand $\mathbf{1}$ of $h(S)$. For $q = -2$, $H^1(S_K, \mathbb{G}_m)_{\mathbb{Q}} = \mathrm{Pic}(S_K)_{\mathbb{Q}}$ decomposes into $\mathrm{Pic}^0(S_K)_{\mathbb{Q}} \oplus \mathrm{NS}(S_K)_{\mathbb{Q}}$. The first summand is obtained from $h_1(S)$ and the second from $h(\underline{\mathrm{NS}}_S)(1)$ as a direct summand of $h_2(S)$. Hence the vanishing. The last claim now follows from [K-S, 6.2]. □

Corollary 7.8.5. *Keep the notation of Theorem 7.8.4. Then*

(i) $h_0(X), h_1(X), t_2(S) \in {}^\perp\mathcal{I}$, where $\mathcal{I}$ is the ideal of Lemma 7.5.1 and ${}^\perp\mathcal{I}$ is defined in (7.14). *(Here, $\mathcal{A} = \mathcal{M}_{\mathrm{rat}}^{\mathrm{eff}}$, $\mathcal{B} = \mathcal{A} \otimes \mathbb{L}$.)*

(ii) $h_i(X) \in \mathcal{K}^{\perp}_{\leq i-1}$ $(i = 0, 1)$ *and* $t_2(S) \in \mathcal{K}^{\perp}_{\leq 1}$*, where* $\mathcal{K}_{\leq n}$ *is the ideal of Definition 7.5.5 (iii) and* $\mathcal{K}^{\perp}_{\leq n}$ *is defined in Proposition 7.8.3.*

Moreover, for any smooth projective variety Y *of dimension* d*, one has*

$$\mathcal{M}^{\mathrm{eff}}_{\mathrm{rat}}(h(Y), h_0(X)) \simeq A^{\mathrm{num}}_0(X_{k(Y)})$$
$$\mathcal{M}^{\mathrm{eff}}_{\mathrm{rat}}(h(Y), h_1(X)) \simeq \mathrm{Alb}_X(k(Y))$$
$$\mathcal{M}^{\mathrm{eff}}_{\mathrm{rat}}(h(Y), t_2(S)) \simeq T(S_{k(Y)}).$$

Proof (1) is just a special case of Theorem 7.8.4 by the full faithfulness of (7.13), via Proposition 7.5.3. (2) follows from (1) by duality. For the isomorphisms, let us treat the case of $t_2(S)$. We first observe that

$$\mathcal{M}^{\mathrm{eff}}_{\mathrm{rat}}(\mathbf{1}, t_2(S)) = T(S) \tag{7.16}$$

(see Proposition 7.2.3). Then the isomorphism follows from (1), Proposition 7.5.3 and (7.16) applied over the function field of Y. The cases of $h_0(X)$ and $h_1(X)$ are similar. □

Corollary 7.8.6. *a) Let* $P : \mathcal{M}^{\mathrm{eff}}_{\mathrm{rat}} \to \mathcal{M}^{o}_{\mathrm{rat}}$ *denote the projection functor. Then the right adjoint* $P^{\#}$ *of* P *is defined on* $d_{\leq 2}\mathcal{M}^{o}_{\mathrm{rat}}$*.*

b) Let $S : \mathcal{M}^{o}_{\mathrm{rat}} \to (\mathcal{M}^{o}_{\mathrm{rat}}/\mathcal{K}^{o}_{\leq 1})^{\natural}$ *be the projection functor. The the left adjoint* ${}^{\#}S$ *of* S *is defined on the thick image of* $d_{\leq 2}\mathcal{M}^{o}_{\mathrm{rat}}$ *by* S*.*

Proof **a)** Consider the thick subcategory $d^{o}_{\leq 2}\mathcal{M}^{\mathrm{eff}}_{\mathrm{rat}}$ of $\mathcal{M}^{\mathrm{eff}}_{\mathrm{rat}}$ generated by those motives of the form $h_0(X), h_1(X)$ and $t_2(S)$ as in Theorem 7.8.4. Clearly $P(d^{o}_{\leq 2}\mathcal{M}^{\mathrm{eff}}_{\mathrm{rat}}) = d_{\leq 2}\mathcal{M}^{o}_{\mathrm{rat}}$, and Corollary 7.8.5 (1) gives the inclusion $d^{o}_{\leq 2}\mathcal{M}^{\mathrm{eff}}_{\mathrm{rat}} \subset {}^{\perp}\mathcal{I}$. The conclusion now follows from Proposition 7.8.1.

b) The proof is the same, using this time Corollary 7.8.5 (2) (or rather its projection into $\mathcal{M}^{o}_{\mathrm{rat}}$) and Proposition 7.8.3. □

Remark 7.8.7. It is natural to ask what is the largest full subcategory of $\mathcal{M}^{o}_{\mathrm{rat}}$ on which $P^{\#}$ is defined. We don't know the answer to this question but at least, $P^{\#}$ is *not defined on* $\bar{h}(X)$ for any 3-fold X such that the group $A^2_{\mathrm{alg}}(X)$ of codimension 2 cycles modulo algebraic equivalence is not finitely generated (cf. Griffiths' examples). This will be proven in the final version of [K-S], the core of the argument being due to Joseph Ayoub.

On the other hand we expect that the functor S_n of (7.18) below always has a left adjoint: this will be the object of a further work.

7.8.3 Birational motives and motives at the generic point

Let

$$d_n\mathcal{M}^{\mathrm{eff}}_{\mathrm{rat}} = \left(\frac{d_{\leq n}\mathcal{M}^{\mathrm{eff}}_{\mathrm{rat}}}{\mathcal{K}_{\leq n-1} \cap d_{\leq n}\mathcal{M}^{\mathrm{eff}}_{\mathrm{rat}}}\right)^\natural \tag{7.17}$$

$$d_n\mathcal{M}^{\mathrm{o}}_{\mathrm{rat}} = \left(\frac{d_{\leq n}\mathcal{M}^{\mathrm{o}}_{\mathrm{rat}}}{\mathcal{K}^{\mathrm{o}}_{\leq n-1} \cap d_{\leq n}\mathcal{M}^{\mathrm{o}}_{\mathrm{rat}}}\right)^\natural$$

where $^\natural$ means taking the pseudo-abelian envelope (cf. Definition 7.5.5 for the definitions of the objects appearing in (7.17).) We thus have a diagram of categories and functors

$$\begin{array}{ccc} d_{\leq n}\mathcal{M}^{\mathrm{eff}}_{\mathrm{rat}} & \xrightarrow{P_n} & d_{\leq n}\mathcal{M}^{\mathrm{o}}_{\mathrm{rat}} \\ {\scriptstyle R_n}\downarrow & & {\scriptstyle S_n}\downarrow \\ d_n\mathcal{M}^{\mathrm{eff}}_{\mathrm{rat}} & \xrightarrow{Q_n} & d_n\mathcal{M}^{\mathrm{o}}_{\mathrm{rat}}. \end{array} \tag{7.18}$$

(With a previous notation, $P_n(M) = \bar{M}$ for $M \in d_{\leq n}\mathcal{M}^{\mathrm{eff}}_{\mathrm{rat}}$.)

Note that the duality $D^{(n)}$ of Lemma 7.5.6 acts on (7.18) by exchanging the categories $d_{\leq n}\mathcal{M}^{\mathrm{o}}_{\mathrm{rat}}$ and $d_n\mathcal{M}^{\mathrm{eff}}_{\mathrm{rat}}$ which are therefore anti-equivalent, and also induces a duality on $d_n\mathcal{M}^{\mathrm{o}}_{\mathrm{rat}}$.

If $\dim X \leq n$, then we write $h_{\mathrm{gen}}(X)$ for $S_n(\bar{h}(X)) = S_nP_n(h(X))$: this is the *motive of X at the generic point* (relative to dimension n). We have $h_{\mathrm{gen}}(X) = 0$ if $\dim X < n$. Lemma 7.5.6 c) shows that, for two n-dimensional smooth projective k-varieties X, Y,

$$d_n\mathcal{M}^{\mathrm{o}}_{\mathrm{rat}}(h_{\mathrm{gen}}(X), h_{\mathrm{gen}}(Y)) = A_n(X \times Y)/\mathcal{J}(X, Y) \tag{7.19}$$

where $\mathcal{J}$ is the ideal of Definition 7.4.2.

The name "motive at the generic point" is in reference to Beilinson's paper [Bei], where he calls the right hand side of (7.19) *correspondences at the generic point.* His prediction (conjecture) (*) on p. 35†, deduced from some "standard" conjectures on mixed motives, implies that, for $X = Y$, *this $\mathbb{Q}$-algebra is semi-simple finite-dimensional and that $A_n(X \times X)_{\mathrm{hom}} \subset \mathcal{J}(X, X)$.* (Compare with Theorem 7.6.12 (7) in the case of a surface.)

Theorem 7.8.8. a) *Suppose that $n \leq 2$. In* (7.18), *P_n and Q_n have a right adjoint while R_n and S_n have a left adjoint. All these adjoints are right inverse to the corresponding functors. In particular, the functor*

$$S_nP_n = Q_nR_n : d_{\leq n}\mathcal{M}^{\mathrm{eff}}_{\mathrm{rat}} \to d_n\mathcal{M}^{o}_{\mathrm{rat}}$$

† Also due to Rovinski and Bloch as he points out.

has a canonical section

$$\Sigma_n = P_n^\# \circ {}^\# S_n = {}^\# R_n \circ Q_n^\#.$$

b) *Suppose $n = 1$: if C is a smooth projective curve, then for any C-K decomposition of C we have*

$$\begin{aligned} P_1^\# \bar{h}(C) &\simeq h_0(C) \oplus h_1(C) \\ Q_1^\# h_{\mathrm{gen}}(C) &\simeq R_2(h_1(C)) \\ {}^\# R_1 R_1(h(C)) &\simeq h_1(C) \oplus h_2(C) \\ {}^\# S_1 h_{\mathrm{gen}}(C) &\simeq \bar{h}_1(C) \\ \Sigma_1(h_{\mathrm{gen}}(C)) &\simeq h_1(C). \end{aligned}$$

c) *Suppose $n = 2$: if S is a smooth projective surface, then for any refined C-K decomposition of S as in Propositions 7.2.1 and 7.2.3, we have*

$$\begin{aligned} P_2^\# \bar{h}(S) &\simeq h_0(S) \oplus h_1(S) \oplus t_2(S) \\ Q_2^\# h_{\mathrm{gen}}(S) &\simeq R_2(t_2(S)) \\ {}^\# R_2 R_2(h(S)) &\simeq t_2(S) \oplus h_3(S) \oplus h_4(S) \\ {}^\# S_2 h_{\mathrm{gen}}(S) &\simeq \bar{t}_2(S) \\ \Sigma_2(h_{\mathrm{gen}}(S)) &\simeq t_2(S). \end{aligned}$$

Note that, as a composite of a left and a right adjoint, Σ_n has no special adjunction property.

Proof a) For P_n and S_n this follows immediately from Corollary 7.8.6; the cases of Q_n and R_n follow by using the duality $D^{(n)}$.

b) and c) Let us prove the first formula in c): the other cases are similar. Writing $h(S) = \bigoplus_{i=0}^4 h_i(S)$, we have $\bar{h}(S) \simeq \bar{h}_0(S) \oplus \bar{h}_1(S) \oplus \bar{t}_2(S)$ (see proof of Lemma 7.5.7). By Corollary 7.8.5 (1), $h_0(S) \oplus h_1(S) \oplus t_2(S) \in {}^\perp\mathcal{J}$; the conclusion then follows from Proposition 7.8.1 d). □

Corollary 7.8.9. *The functors P_n, Q_n, R_n, S_n are essentially surjective for $n \le 2$.* □

(This fact is not obvious a priori since we added projectors when defining the quotient categories: it amounts to saying that this operation was not necessary.)

From Theorem 7.8.8 a), we have for $n \leq 2$ a commutative diagram of natural transformations in $d_{\leq n}\mathcal{M}_{\mathrm{rat}}^{\mathrm{eff}}$:

$$\begin{array}{ccc} \mathrm{id} & \to & P_n^{\#}P_n \\ \uparrow & & \uparrow \\ R_n^{\#}R_n & \to R_n^{\#}Q_n^{\#}Q_nR_n = & P_n^{\#}S_n^{\#}S_nP_n \end{array}$$

given by the units and counits of the respective adjunctions. Applying this diagram to $h(C)$ (resp. $h(S)$) for C a curve (resp. S a surface) and using Theorem 7.8.8 b) (resp. c)), we get the following corollary, which gives a partial positive answer to Conjecture 7.3.4:

Corollary 7.8.10. a) *Given a curve C, in the diagram*

$$\begin{array}{ccc} h(C) & \to & h_0(C) \oplus h_1(C) \\ \uparrow & & \uparrow \\ h_1(C) \oplus h_2(C) & \to & h_1(C) \end{array}$$

all maps and objects are independent of the choice of a C-K decomposition, and are natural in C for the action of correspondences.
b) *Given a surface S, in the diagram*

$$\begin{array}{ccc} h(S) & \to & h_0(S) \oplus h_1(S) \oplus t_2(S) \\ \uparrow & & \uparrow \\ t_2(S) \oplus h_3(S) \oplus h_4(S) & \to & t_2(S) \end{array}$$

all maps and objects are independent of the choice of a refined C-K decomposition as in Propositions 7.2.1 and 7.2.3, and are natural in S for the action of correspondences. □

In the case of a surface S, we may think of $h_{bir}(S) := P_2^{\#}\bar{h}(S)$ as the *largest birational quotient* of $h(S)$ and think of $\Sigma_2(h_{\mathrm{gen}}(S))$ as the *largest submotive at the generic point* of $h_{bir}(S)$. Similarly, $R_2^{\#}R_2(h(S))$ may be thought of as the largest subobject of $h(S)$ "purely of dimension 2". Note that both maps $h(S) \to t_2(S)$ and $t_2(S) \to h(S)$ given by a projector π_2^{tr} from a refined C-K decomposition do depend on the choice of this decomposition. Nevertheless, it is unambiguous to write $t_2(S)$ for $\Sigma_2(h_{\mathrm{gen}}(S))$, viewed as a functor in S.

Corollary 7.8.11. *Let $d_{\leq 2}Sm$ be the category of smooth (open) varieties of dimension ≤ 2 over k. Assume that k is of characteristic* 0. *There are functors*

$$h_{bir}, t_2 : d_{\leq 2}Sm \to \mathcal{M}_{\mathrm{rat}}^{\mathrm{eff}}$$

extending the above functors. These functors are (stably) birationally invariant. There are similar contravariant functors starting from the category $d_{\leq 2}$place *of function fields of transcendence degree* ≤ 2 *over* k*, with morphisms the* k*-places.*

Proof By [K-S, 5.6] there are canonical functors

$$d_{\leq 2}T^{-1}\text{place}^{op} \to d_{\leq 2}S_r^{-1}Sm \to d_{\leq 2}\mathcal{M}^{o}_{\text{rat}}$$

the latter extending the natural functor from smooth projective varieties. Here T^{-1}place denotes the category of finitely generated extensions of k with morphisms the k-places, localized by inverting morphisms of the form $K \hookrightarrow K(t)$, $S_r^{-1}Sm$ denotes the category of smooth k-varieties localized by inverting the dominant morphisms which induce a purely transcendental extension of function fields, and $d_{\leq 2}$ denotes the full subcategories respectively of function fields of transcendence degree ≤ 2 and of varieties of dimension ≤ 2. □

7.8.4 The triangulated birational motive of a surface

Suppose k perfect. Recall from [K-S] that the projection functor $P : \mathcal{M}^{\text{eff}}_{\text{rat}} \to \mathcal{M}^{o}_{\text{rat}}$ inserts into a naturally commutative diagram of categories and functors

$$\begin{array}{ccc} \mathcal{M}^{\text{eff}}_{\text{rat}} & \xrightarrow{\Phi} & DM^{\text{eff}}_{-} \\ {\scriptstyle P}\downarrow & & \downarrow{\scriptstyle \nu_{\leq 0}} \\ \mathcal{M}^{o}_{\text{rat}} & \xrightarrow{\bar{\Phi}} & DM^{o}_{-} \end{array}$$

where DM^{o}_{-} is a birational analogue of DM^{eff}_{-}, and that $\nu_{\leq 0}$ has an everywhere-defined right adjoint/right inverse i (in fact DM^{o}_{-} is a priori defined as the full subcategory of DM^{eff}_{-} consisting of those objects C such that $\underline{\text{Hom}}(\mathbb{Q}(1), C) = 0$, and it is then proven that the inclusion functor i has a left adjoint $\nu_{\leq 0}$ which inserts itself in the above commutative diagram). If X is a smooth projective variety, we set

$$\bar{M}(X) = \bar{\Phi}\bar{h}(X) = \nu_{\leq 0}M(X) \in DM^{o}_{-}.$$

Suppose that $P^{\#}$ is defined at some Chow birational motive $\bar{M} \in \mathcal{M}^{o}_{\text{rat}}$. Starting from the natural isomorphism $\nu_{\leq 0}\Phi = \bar{\Phi}P$, the two adjunctions give a "base change" morphism

$$\Phi P^{\#}\bar{M} \to i\nu_{\leq 0}\Phi P^{\#}\bar{M} = i\bar{\Phi}PP^{\#}\bar{M} \to i\bar{\Phi}\bar{M}. \tag{7.20}$$

Proposition 7.8.12. *Let $\bar{M} \in \mathcal{M}^o_{\text{rat}}$ be such that $P^\# \bar{M}$ is defined. Then the following conditions are equivalent:*

i) (7.20) *is an isomorphism.*
ii) $\Phi P^\# \bar{M}$ *is in the essential image of* i *(i.e.* $\underline{\text{Hom}}(\mathbb{Q}(1), \Phi P^\# \bar{M}) = 0$*).*

Proof (i) $\Rightarrow$ (ii) is obvious. Conversely, suppose that $\Phi P^\# \bar{M} \simeq iN$ for some $N \in DM^{\text{eff}}_-$. Since i is right inverse to $\nu_{\leq 0}$, we find first

$$N \simeq \nu_{\leq 0} iN \simeq \nu_{\leq 0} \Phi P^\# \bar{M} \simeq \bar{\Phi} P P^\# \bar{M}.$$

Since i is fully faihtful, the morphism $iN \to i\bar{\Phi}\bar{M}$ comes from a unique morphism $\bar{\Phi} P P^\# \bar{M} \simeq N \to \bar{\Phi}\bar{M}$, which clearly is the morphism induced by the counit $PP^\# \bar{M} \to \bar{M}$. This counit is an isomorphism by Proposition 7.8.1 b), hence (7.20) is an isomorphism. □

By Theorem 7.8.4 and Corollary 7.8.6 a), $P^\# \bar{M}$ is defined and Condition (ii) of Proposition 7.8.12 is verified for all $\bar{M} \in d_{\leq 2}\mathcal{M}^o_{\text{rat}}$. Hence (i) holds for them. Taking $\bar{M} = \bar{h}(C)$ ($\bar{h}(S)$) for a curve C (a surface S), we get:

Corollary 7.8.13. a) *For any curve C, $i\bar{M}(C) \simeq \Phi(P^\# \bar{h}(C))$. Given a C-K decomposition, this motive is isomorphic to $M_0(C) \oplus M_1(C)$.*
b) *For any surface S, $i\bar{M}(S) \simeq \Phi(P^\# \bar{h}(S))$. Given a C-K decomposition as in Propositions 7.2.1 and 7.2.3, this motive is isomorphic to $M_0(S) \oplus M_1(S) \oplus \Phi(t_2(S))$.*

□

In particular, $i\bar{M}(S)$ is a direct summand of $M(S)$ for any surface S, which answers positively a question of Ayoub.

References

[A-J] R. Akhtar and R. Joshua *Künneth decomposition for quotient varieties*, preprint, 2003, 18 pages.

[An] Y. André, Une introduction aux motifs, Panoramas et synthèses, SMF, 2004.

[A-K] Y. André and B. Kahn, *Nilpotence, radicaux et structures monoïdales* (with an appendix by P. O'Sullivan), Rend. Sem. Mat. Univ. Padova **108** (2003), 106–291.

[Bei] A. Beilinson *Remarks on n-motives and correspondences at the generic point, in* Motives, Polylogarithms and Hodge Theory, F. Bogomolov and L. Katzarkov, eds., International Press, 2002, 33–44.

[B1] S. Bloch K_2 of Artinian $\mathbb{Q}$-algebras, with application to algebraic cycles. Comm. Algebra **3** (1975), 405–428.

[B2] S. Bloch Lectures on Algebraic Cycles, Duke University Mathematics Series IV, Duke University Press, Durham U.S.A., 1980.

[B-K-L] S. Bloch, A. Kas and D. Lieberman *Zero cycles on surfaces with $p_g = 0$*, Compositio Math. **33** (1976), 135–145.

[Bo] N. Bourbaki Éléments de mathématique: Algèbre, Ch. II, Hermann, 1970.

[Del-I] P. Deligne *La conjecture de Weil, I*, Publ. Math. IHÉS **43** (1974), 273–307.

[Fu] W. Fulton Intersection Theory, Ergeb. Math. Grenzgeb., Springer-Verlag, Berlin, 1984.

[G-P1] V. Guletskii and C. Pedrini *The Chow motive of the Godeaux Surface*, *in* Algebraic Geometry: A volume in memory of Paolo Francia, 179–195 W. de De Gruyter, 2002.

[G-P2] V. Guletskii and C. Pedrini *Finite-dimensional Motives and the Conjectures of Beilinson and Murre*, K-Theory **30** (2003), 243–263.

[Ha] R. Hartshorne Algebraic geometry, Springer, 1977.

[H-K] A. Huber, B. Kahn *The slice filtration and mixed Tate motives*, preprint, 2005, `http://www.math.uiuc.edu/K-theory/0719`.

[J1] U. Jannsen *Motives, numerical equivalence and semi-simplicity*, Invent. Math. **107** (1992), 447–452.

[J2] U. Jannsen *Motivic Sheaves and Filtrations on Chow groups*, Proceedings of Symposia in Pure mathematics **55** (I) (1994), 245–302.

[J3] U. Jannsen *Equivalence realtions on algebraic cycles*, *in* The Arithmetic and Geometry of Agebraic Cycles, NATO Sci. Ser. C Math. Phys. Sc. **548** Kluwer Ac. Publ. Co., 2000, 225–260.

[K-S] B. Kahn and R. Sujatha *Birational Motives, I*, preliminary version, preprint, 2002.

[Ki] S.I. Kimura *Chow groups can be finite-dimensional, in some sense*, Math. Ann. **331** (2005), 173–201.

[Kl] S. Kleiman *The standard conjectures*, Proceedings of Symposia in Pure mathematics **55** (I) (1994), 3–20.

[Mum1] D. Mumford *Rational equivalence of* 0*-cycles on surfaces*, J. Math. Kyoto Univ. **9** 1968, 195–204.

[Mum2] D. Mumford Abelian Varieties, Oxford University Press, Ely house, London, 1974.

[Mu1] J. Murre *On the motive of an algebraic surface*, J. Reine angew. Math. **409** (1990), 190–204.

[Mu2] J. Murre *On a conjectural filtration on the Chow groups of an algebraic variety*, Part I and II, Indagationes Mathem., n.S. **4(2)** (1993), 177–201.

[Schoen] C. Schoen *Zero cycles modulo rational equivalence for some varieties over fields of transcendence degree* 1, Proc. Symposia in Pure Math. **46** (1987), 463–473.

[Sch] A.J. Scholl *Classical motives*, Proc. Symposia in Pure Math. **55** (I) (1994), 163–187.

[Voev1] V. Voevodsky *Cohomological theory of presheaves with transfers*, *in* Cycles, transfers and motivic homology theories, Ann. Math. Studies **143**, Princeton University Press, 2000.

[Voev2] V. Voevodsky *Triangulated categories of motives over a field*, *in* E. Friedlander, A. Suslin, V. Voevodsky Cycles, transfers and motivic cohomology theories, Ann. Math. Studies **143**, Princeton University Press, 2000, 188–238.

[Voev3] V. Voevodsky *Motivic cohomology groups are isomorphic to higher Chow groups in any characteristic*, Int. Math. Res. Not. **2002**, 351–355.

[Voi] C. Voisin *Sur les zéro-cycles de certaines hypersurfaces munies d'un automorphisme*, Ann. Scuola Norm. Sup. Pisa **19** (1992), 473–492.

[Weil1] A. Weil Variétés abéliennes et courbes algébriques, Hermann, Paris, 1948.

[Weil2] A. Weil Foundations of Algebraic Geometry, American Math. Soc. Providence, Rhode Island U.S.A., 1962.

[EGA4] A. Grothendieck, A. Dieudonné *Éléments de géométrie algébrique, IV: étude locale ds schémas et des morphismes de schémas (quatrième partie)*, Publ. Math. IHÉS **32** (1967).

8

A note on finite dimensional motives

Shun-ichi Kimura

Department of Mathematics Graduate School of Science, Hiroshima University
Higashi-Hiroshima 739-8526, Japan
shunkimura@mac.com

To Jacob for his 75th birthday

Abstract

Let $\mathcal{C}$ be a pseudo-abelian tensor $\mathbb{Q}$-linear category. We consider the following problems.

(i) Characterize 1-dimensional objects in $\mathcal{C}$.
(ii) When $A \in \mathcal{C}$ is Schur finite, study the set $\{\lambda | S_\lambda A = 0\}$.

For (i), we prove that if the unit object **1** has no non-trivial direct summand, then invertible objects are 1-dimensional. If moreover $\mathcal{C}$ is rigid, then the converse holds. For (ii), under some technical condition, we prove the existence of a minimal Young diagram that kills A.

8.1 Introduction

Let $\mathcal{C}$ be a pseudo-abelian tensor category with $\mathbb{Q}$-coefficients, namely each idempotent endo-morphism has Kernel and Cokernel, each $\mathrm{Hom}(A, B)$ is a $\mathbb{Q}$-vector space with $\mathbb{Q}$-bilinear compositions ($A, B \in \mathcal{C}$), and has a tensor product functor $\otimes : \mathcal{C} \times \mathcal{C} \to \mathcal{C} : (A, B) \to A \otimes B$ so that (1) for $A_1, \ldots, A_N \in \mathcal{C}$, the tensor product $A_1 \otimes \cdots \otimes A_N$ is well-defined, independent of the order of taking the tensor product, and (2) there is a natural isomorphism $A \otimes B \simeq B \otimes A$ such that for each $\sigma \in \mathfrak{S}_N$, the morphism $A_1 \otimes \cdots \otimes A_N \to A_{\sigma(1)} \otimes \cdots \otimes A_{\sigma(N)}$ is well-defined, independent of the order of composing the transpositions.

In [5], we have introduced the notion of Kimura finiteness. For $A \in \mathcal{C}$ and a permutation $\sigma \in \mathfrak{S}_N$, we define $\sigma_A : A^{\otimes N} \to A^{\otimes N}$ by sending $a_1 \otimes \cdots \otimes a_N$ to $a_{\sigma(1)} \otimes \cdots \otimes a_{\sigma(N)}$. Then $\frac{1}{N!} \sum_{\sigma \in \mathfrak{S}_N} \mathrm{sgn}(\sigma) \sigma_A : A^{\otimes N} \to A^{\otimes N}$ (resp. $\frac{1}{N!} \sum_{\sigma \in \mathfrak{S}_N} \sigma_A : A^{\otimes N} \to A^{\otimes N}$) is an idempotent, and its image, which exists by the assumption that $\mathcal{C}$ is pseudo-abelian, is denoted by $\bigwedge^N A$ (resp.

$\mathrm{Sym}^N A$). An object A is called evenly finite dimensional (resp. oddly finite dimensional) when $\bigwedge^N A = 0$ (resp. $\mathrm{Sym}^N A = 0$) for some $N > 0$. An object $A \in \mathcal{C}$ is called Kimura finite if it is a direct sum of an evenly finite dimensional object and an oddly finite dimensional object. For example, when $\mathcal{C}$ is the category of $\mathbb{Z}/2$-graded vector spaces, $\bigwedge^{d+1} V = 0$ for an even degree d-dimensional vector space V, and $\mathrm{Sym}^{e+1} W = 0$ for an odd degree e-dimensional vector space W. Hence Kimura finite objects in $\mathcal{C}$ are exactly finite dimensional objects in $\mathcal{C}$ as vector spaces. The notion of Kimura finiteness is applied to Chow motives in [5], also see [4].

In [7], C. Mazza introduced a weaker finiteness condition based on Schur idempotents. When $\lambda = (\lambda_1, \ldots, \lambda_k)$ with $\lambda_1 \geq \cdots \geq \lambda_k$ is a partition of N (or equivalently, a Young diagram with N boxes), then λ determines a Young symmetrizer $c_\lambda := \sum_{\sigma \in \mathfrak{S}_N} c_{\lambda,\sigma}\sigma \in \mathbb{Q}[\mathfrak{S}_N]$ so that $\mathbb{Q}[\mathfrak{S}_n] \cdot c_\lambda$ is isomorphic to the irreducible representation of $\mathfrak{S}_N$ corresponding to λ. The Young symmetrizer c_λ is an idempotent up to scalar, hence

$$\mathrm{Im} \sum_{\sigma \in \mathfrak{S}_N} c_{\lambda,\sigma}\sigma_A : A^{\otimes N} \to A^{\otimes N}$$

is well-defined, which image is denoted by $S_\lambda A$. When $\lambda = \underbrace{(1, 1, \ldots, 1)}_{N\text{-times}}$, then $S_\lambda A = \bigwedge^N A$, and when $\lambda = (N)$, then $S_\lambda A = \mathrm{Sym}^N A$. We define $A \in \mathcal{C}$ to be Schur finite if $S_\lambda A = 0$ for some λ. Kimura finite objects are Schur finite, but not vice versa (see [7, Example 3.3] for such an example due to P. O'Sullivan).

In this paper, we consider the following two problems.

1) Characterize 1-dimensional objects.
2) For a Schur finite object A, describe $\{\lambda | S_\lambda A = 0\}$.

For (1), we prove that if the unit object $\mathbf{1}$ has no non-trivial direct summand, then invertible objects are 1-dimensional (Proposition 8.2.6). If moreover $\mathcal{C}$ is rigid, then the converse holds (Proposition 8.2.9). In particular, in the category of Chow motives, a motive is 1-dimensional if and only if it is invertible.

For (2), assume that in $\mathcal{C}$, $A^{\otimes N} = 0$ implies $A = 0$ (for example when $\mathcal{C}$ is rigid, see Lemma 8.3.13). Then we prove that for a Schur finite object $A \in \mathcal{C}$, there exists a minimal Young diagram λ that kills A. Namely, for a Young diagram μ, we have $S_\mu A = 0$ if and only if $\mu \supset \lambda$ (Corollary 8.3.10). We call this λ the Schur dimension of A (Definition 8.3.11).

Finally in the last section, we apply these results to study the Chow motives of smooth hypersurfaces.

Acknowledgements The author is very much grateful for the invaluable discussion with Prof. Yves André. He is also grateful for the discussion with Prof. Uwe Jannsen and Prof. Carlo Mazza. The referee's comments, which greatly clarified the arguments and generalized the results, were so much helpful that the author feels this paper is virtually coauthored by the referee. Special thanks go to the referee.

8.2 1-dimensional finite objects

Definition 8.2.1. Let $\mathcal{C}$ be a tensor category. An object $\mathbf{1} \in \mathcal{C}$ is called *a unit object* if there exists a natural isomorphism $g_X : X \simeq \mathbf{1} \otimes X$, which is compatible with the associativity law and the commutativity law (see [8, 1.3, 2.2, 2.3, 2.4]).

Remark 8.2.2. An identity object in a tensor category is unique up to a canonical isomorphism which preserves the isomorphism $\mathbf{1} \to \mathbf{1} \otimes \mathbf{1}$, if it exists (see [2, Prop. 1.3], and [8]).

In this section, we work in the category of pseudo-abelian $\mathbb{Q}$-linear tensor category $\mathcal{C}$ with a fixed unit object $\mathbf{1}$.

Definition 8.2.3. A unit $A \in \mathcal{C}$ is invertible when for some $B \in \mathcal{C}$, we have $A \otimes B \simeq \mathbf{1}$.

One can easily see that $A \in \mathcal{C}$ is invertible, if and only if the functor of tensoring A, namely $\mathcal{C} \to \mathcal{C}$ defined by $X \to X \otimes A$, is an equivalence of categories.

Definition 8.2.4. A non-zero object $A \in \mathcal{C}$ is called *evenly* 1*-dimensional* if $\bigwedge^2 A = 0$, and *oddly* 1*-dimensional* if $\mathrm{Sym}^2 A = 0$. A is called 1*-dimensional* when A is either evenly or oddly 1-dimensional.

Example 8.2.5. The unit object $\mathbf{1}$ is 1-dimensional. In fact, when $\sigma : \mathbf{1} \otimes \mathbf{1} \to \mathbf{1} \otimes \mathbf{1}$ is the commutativity law morphism, then the compatibility of g_X and the commutativity law says that the following diagram commutes (see [8, 2.3]):

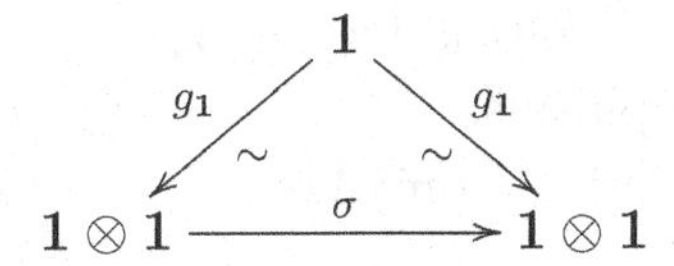

The commutativity of the diagram implies that $\sigma = \mathrm{id}_{\mathbf{1}\otimes\mathbf{1}} : \mathbf{1} \otimes \mathbf{1} \to \mathbf{1} \otimes \mathbf{1}$, and hence $\bigwedge^2 \mathbf{1} = 0$.

Proposition 8.2.6. *Suppose that* End($\mathbf{1}$) *has no non-trivial idempotent. Then every invertible object of* $\mathcal{C}$ *is* 1*-dimensional.*

Proof Let A be invertible in $\mathcal{C}$. Then $A\otimes A$ is again invertible. In particular, $\mathrm{End}(A \otimes A) \simeq \mathrm{End}(\mathbf{1})$. So $A \otimes A$ has no non-trivial direct factors. As $A \otimes A \simeq \mathrm{Sym}^2 A \oplus \bigwedge^2 A$, one of $\mathrm{Sym}^2 A$ or $\bigwedge^2 A$ is zero. □

Remark 8.2.7. In an earlier version of this paper, the author did not notice that $\mathbf{1}$ is always 1-dimensional. Moreover, he assumed only this condition for Proposition 8.2.6. The referee pointed out that the condition always holds, and gave the counterexample $(\mathbb{Q}, \mathbb{Q}[+1])$ in the category of a pair of graded $\mathbb{Q}$-vector spaces, which is invertible but not 1-dimensional. Finally the referee gave the above elegant proof.

Remark 8.2.8. The converse of Proposition 8.2.6 does not hold in general. In the category of R-modules, for a non-zero ideal $I \subset R$, the R-module R/I is 1-dimensional, but not invertible. Also in the construction of Chow motives, if one forgets to invert the Lefschetz motive $\mathbb{L} := h^1(\mathbb{P}^1)$, then $\mathbb{L}$ is 1-dimensional, but not invertible.

Once we invert the Lefschetz motive, the category becomes rigid, and under the assumption of the rigidness, the converse of Proposition 8.2.6 holds, as Proposition 8.2.9 below. (In an earlier version, Proposition 8.2.9 was proved only for Chow motives, and the referee generalized it to rigid tensor categories.)

Proposition 8.2.9. *Let* A *be a* 1*-dimensional object of a rigid tensor category* $\mathcal{C}$*. Then the evaluation morphism* $\check{A} \otimes A \to \mathbf{1}$ *identifies* $\check{A} \otimes A$ *with a non-zero direct summand of* $\mathbf{1}$*. In particular, if* End($\mathbf{1}$) *has no non-trivial idempotents, then* A *is invertible.*

Proof Let $\sigma : \check{A} \otimes A \to A \otimes \check{A}$ be the permutation of the factors, sending (x, y) to (y, x). This σ is an isomorphism, so it is enough to show that the composition

$$\check{A} \otimes A \xrightarrow{\mathrm{ev}} \mathbf{1} \xrightarrow{\delta} A \otimes \check{A}$$

is equal to the permutation of the factors σ, up to signature. Let c be the composition of these morphisms.

Sending c by the adjunction $\mathrm{Hom}(\check{A} \otimes A, \_) \to \mathrm{Hom}(\check{A}, \_ \otimes \check{A})$, c is identified with the composition

$$\check{A} = \check{A} \xrightarrow{\delta \otimes \mathrm{id}_{\check{A}}} A \otimes \check{A}$$

On the other hand, sending σ by the adjunction, one gets

$$\check{A} \xrightarrow{\mathrm{id}_{\check{A}} \otimes \delta} \check{A} \otimes A \otimes \check{A} \xrightarrow{\sigma \otimes \mathrm{id}_{\check{A}}} A \otimes \check{A} \otimes \check{A}$$

These two adjunction morphisms are equal, after composing $\mathrm{id}_A \otimes \check{\sigma}$, where $\check{\sigma} : \check{A} \otimes \check{A} \to \check{A} \otimes \check{A}$ is the permutation of the factors. But because A is 1-dimensional, $\check{A}$ is also 1-dimensional, and hence $\check{\sigma}$ is either identity (when $\check{A}$ is evenly 1-dimensional) or the multiplication by -1 (when $\check{A}$ is oddly 1-dimensional). Hence we obtain $c = \pm\sigma$. We are done. □

8.3 Schur dimension of a Schur finite object

Definition 8.3.1. For a partition λ, we write $|\lambda|$ for $\sum \lambda_i$. If μ is another partition, we define $\lambda \cap \mu$ to be the partition corresponding to the intersection of two Young diagrams, as in the diagram below.

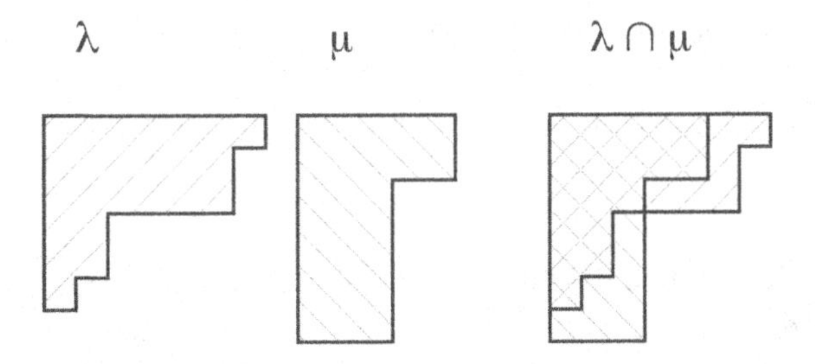

In other words, for $\lambda = (\lambda_1, \ldots, \lambda_k)$ and $\mu = (\mu_1, \ldots, \mu_\ell)$, let $m = \min(k, \ell)$, then $\lambda \cap \mu := (\min(\lambda_1, \mu_1), \min(\lambda_2, \mu_2), \ldots, \min(\lambda_m, \mu_m))$. When $\lambda = \lambda \cap \mu$, we write $\lambda \subset \mu$.

Proposition 8.3.2. *If $A \in \mathcal{C}$ is an objet of a tensor category $\mathcal{C}$ with $S_\lambda A = 0$, then for a partition μ with $\lambda \subset \mu$, we have $S_\mu A = 0$.*

Proof Set $N = |\mu| - |\lambda|$, then by Littlewood-Richardson rule below (or simply by Pieri's formula), $S_\mu A$ is a direct summand of $S_\lambda A \otimes A^{\otimes N}$, hence it is 0. □

Let us recall Littlewood-Richardson rule, the key tool in this section.

Definition 8.3.3. (Littlewood-Richardson rule) Let λ, μ, ν be Young diagrams, and $\mu = (\mu_1, \mu_2, \ldots, \mu_k)$. Then the Littlewood-Richardson number

$N_{\lambda\mu\nu}$ is the number of ways to add $|\nu| - |\lambda|$ boxes to the Young diagram λ to make the Young diagram ν, to each of which boxes some integer is attached, subject to the following conditions:

1) The attached numbers are from 1 to k, and the number i is attached to μ_i boxes.
2) If two numbered boxes are in the same column, then their numbers are different.
3) In each row, the attached numbers are weakly increasing from the left to the right.
4) If the numbered boxes are listed from right to left, starting with the top row and working down, and one looks at the first t entries in this list (for any t between 1 and ν), each integer p between 1 and $k-1$ occurs at least as many times as the next integer $p+1$.

Theorem 8.3.4. *When $A \in \mathcal{C}$ is an object of a tensor category C, and λ, μ are Young diagrams, then we have $S_\lambda A \otimes S_\mu A = \oplus S_\nu^{\oplus N_{\lambda\mu\nu}}$.*

See [6, I-9] for a proof, where Theorem 8.3.4 is proved in terms of Schur polynomials.

Definition 8.3.5. For a Young diagram $\lambda = (\lambda_1, \ldots, \lambda_k)$ and a pair of non-zero integers $(a, b) \in \mathbb{Z}^2_{\geq 0}$, we define $n_{(a,b)}(\lambda) := \sum_{i=a}^{k} \max(\lambda_i - b, 0)$. This is the number of boxes in the Young diagram λ, below the $a-1$-st row and beyond the $b-1$-st column.

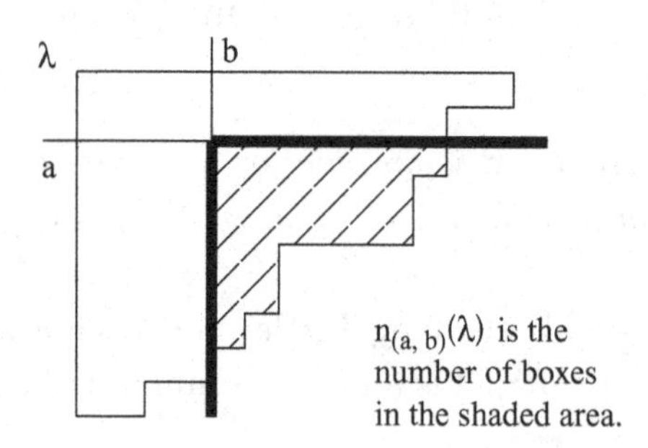

Lemma 8.3.6. *Let λ and μ be two Young diagrams with $n_{(a,b)}(\mu) > 0$. Let τ be a Young diagram such that the coefficient $N_{\lambda\mu\tau}$ in the Littlewood-Richardson rule is non-zero. Then $n_{(a,b)}(\tau) \geq n_{(a,b)}(\lambda) + 1$.*

Proof By the assumption $n_{(a,b)}(\mu) > 0$, we have $\mu_a \geq b$. Because the Littlewood-Richardson coefficient $N_{\lambda\mu\tau}$ is positive, one can add μ_k boxes, numbered k, according to the Littlewood-Richardson rule ($k = 1, 2, \ldots,$), to the Young diagram λ to make the Young diagram τ. In particular, we add at least b boxes numbered a to λ by the rule (1). By the rules (3) and (4), these boxes are below the $a-1$-st row. And they are in different columns by the rule (2), so at least one of these boxes should be beyond $b-1$-st column. Therefore, τ has at least one more box than λ below the $a-1$-st row and beyond $b-1$-st row, which implies $n_{(a,b)}(\tau) > n_{(a,b)}(\lambda)$. □

Corollary 8.3.7. *Let λ be a Young diagram with $n_{(a,b)}(\lambda) > 0$. Decompose $S_\lambda^{\otimes N} = \sum_\nu c_\nu S_\nu$. If $c_\nu > 0$, then $n_{(a,b)}(\nu) \geq N$.*

Proof We proceed by induction on N. When $N = 1$, there is nothing to prove. Assume that this Corollary holds for $N-1$. When $S_\lambda^{\otimes N} = \sum_\nu c_\nu S_\nu$ with $c_\nu > 0$, there is some μ such that $S_\lambda^{\otimes N-1} = \sum_\mu d_\mu S_\mu$ with $d_\mu > 0$ and that S_ν is a direct summand of $S_\mu \otimes S_\lambda$, namely $N_{\nu:\lambda,\mu} > 0$. By inductive assumption, $n_{(a,b)}(\mu) \geq N-1$, hence by Lemma 8.3.6, it follows that $n_{(a,b)}(\nu) \geq N$. □

Lemma 8.3.8. *Let λ and μ be Young diagrams. For integer N, write $(S_{\lambda\cap\mu})^{\otimes N} = \sum c_\nu S_\nu$. If N is large enough and $c_\nu > 0$, then ν contains either λ or μ.*

Proof Assume not. Then for each integer N, there exists a Young diagram ν_N such that ν_N is a direct summand of $(S_{\lambda\cap\mu})^{\otimes N}$, and ν_N does not contain λ nor μ, namely for some i_N and j_N, we have $\lambda_{i_N} > (\nu_N)_{i_N}$ and $\mu_{j_N} > (\nu_N)_{j_N}$. Because there are finitely many possible i_N's and j_N's, some pair (i_N, j_N) appears infinitely many times. Each $\nu_{N'}$ with $N' > N$ contains some direct summand of $(S_{\lambda\cap\mu})^{\otimes N}$, hence by replacing ν_N if necessary, we may assume that the same pair $(i,j) = (i_N, j_N)$ works for all N. Because $\nu_1 = \lambda\cap\mu$, we have $\lambda_i > \min(\lambda_i, \mu_i)$ (hence $\lambda_i > \mu_i$) and $\mu_j > \min(\lambda_j, \mu_j)$ (hence $\mu_j > \lambda_j$). In particular, we have $i \neq j$. By symmetry, we may assume that $i < j$, and hence $\mu_i < \lambda_i$. We have $n_{(i,\mu_i)}(\lambda\cap\mu) > 0$, hence by Corollary 8.3.7, $n_{(i,\mu_i)}(\nu_N) \geq N$. On the other hand, because $(\nu_N)_i < \lambda_i$ and $(\nu_N)_j < \mu_j \leq \mu_i$, there are at most $(\lambda_i - \mu_i)\cdot(j-i)$ boxes in ν_N below the $i-1$-th row and beyond the $\mu_i - 1$-th column (see the picture below), hence $n_{(i,\mu_i)}(\nu_N) \leq (\lambda_i - \mu_i)\cdot(j-i)$, a contradiction. □

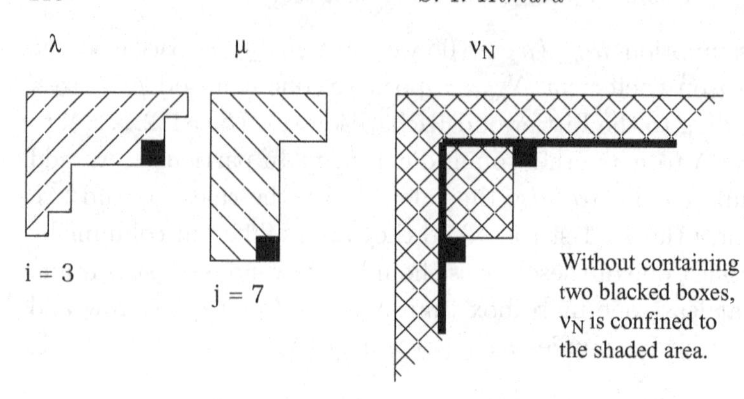

Proposition 8.3.9. *Let $A \in \mathcal{C}$ be a Schur finite object in a tensor category $\mathcal{C}$. Suppose that two Young diagrams λ and μ satisfy $S_\lambda A = 0$ and $S_\mu A = 0$. Then there exists some integer $N > 0$ such that $S_{\lambda\cap\mu}^{\otimes N} A = 0$.*

Proof By Lemma 8.3.8, there exists some N such that each direct summand $S_\nu A$ of $S_{\lambda\cap\mu}^{\otimes N} A$, satisfies either $\nu \supset \lambda$ or $\nu \supset \mu$. By Proposition 8.3.2, we have $S_\nu A = 0$ for each such ν, hence $S_{\lambda\cap\mu}^{\otimes N} A = 0$. □

Corollary 8.3.10. *Assume that the tensor category $\mathcal{C}$ satisfies the condition that $A^{\otimes N} = 0$ implies $A = 0$ for $A \in \mathcal{C}$ and $N > 0$. Then for each Schur finite objet $A \in \mathcal{C}$, there exists a smallest Young diagram λ which kills A, namely, a Young diagram μ satisfies $S_\mu A = 0$ if and only if $\mu \supset \lambda$.*

Proof Let λ_A be the intersection of all the Young diagrams μ with $S_\mu A = 0$. Then by Proposition 8.3.9, we have $S_{\lambda_A} A = 0$. By the construction, if $S_\mu A = 0$, then $\mu \supset \lambda$. Conversely, if $\mu \supset \lambda_A$, then by Proposition 8.3.2, we have $S_\mu A = 0$. □

The following definition is suggested by the referee.

Definition 8.3.11. Let $\mathcal{C}$ be a tensor $\mathbb{Q}$-linear category verifying the condition of Corollary 8.3.10. For a Schur finite object $A \in \mathcal{C}$, we define *the Schur dimension of A* to be the smallest Young diagram λ_A which satisfies $S_{\lambda_A} A = 0$; its existence is guaranteed by Corollary 8.3.10.

Remark 8.3.12. When V is a (d, e)-dimensional supervector space, then its Schur dimension is $\overbrace{(e+1, e+1, \ldots, e+1)}^{d+1}$, one more each, for row and column, than the classical dimension.

For the condition that "$A^{\otimes N} = 0 \Rightarrow A = 0$", we have the following criterion, whose proof is due to the referee.

Lemma 8.3.13. *Let $\mathcal{C}$ be a rigid $\mathbb{Q}$-linear tensor category. If A is an object of $\mathcal{C}$ such that $A^{\otimes N} = 0$ for some $N > 0$, then $A = 0$.*

Proof By adjunction, $\mathrm{id}_A : A \to A$ corresponds to $\delta : \mathbf{1} \to A \otimes \check{A}$, and the composition $\mathrm{id}_A \circ \mathrm{id}_A \circ \cdots \circ \mathrm{id}_A$ corresponds to the morphism $\delta \otimes \cdots \otimes \delta : \mathbf{1} \to (A \otimes \check{A})^{\otimes N}$, composed with the morphism $\mathrm{id}_A \otimes \mathrm{ev} \otimes \cdots \otimes \mathrm{ev} \otimes \mathrm{id}_{\check{A}}$, where $\mathrm{ev} : A \otimes \check{A} \to \mathbf{1}$ is the evaluation morphism . But by the assumption, $(A \otimes \check{A})^{\otimes N}$ is already zero, hence $\mathrm{id}_A = \mathrm{id}_A \circ \mathrm{id}_A \circ \cdots \circ \mathrm{id}_A$ is zero, which implies $A = 0$. □

Remark 8.3.14. Just assuming that $\mathcal{C}$ is pseudo-abelian closed tensor category is not good enough for Lemma 8.3.13. For example, in the category of $\mathbb{Z}$-modules, we have $(\mathbb{Q}/\mathbb{Z})^{\otimes 2} = 0$.

Remark 8.3.15. Over $\mathbb{C}$, if one assumes Hodge conjecture, one can even say that a Chow motive is 1-dimensional if and only if $M \simeq \mathbb{L}^{\otimes n}$ for some n (see [5]). Over $\mathbb{Q}$, let $X = \operatorname{Spec} \mathbb{Q}[\sqrt{-1}]$, and $\iota : X \to X$ be the complex conjugate, and define $M = (X, \frac{1}{2}([\Delta_X] - [\Gamma_\iota]), 0)$. Then M is 1-dimensional, but not isomorphic to $\mathbb{L}^{\otimes n}$ for any n, because $H^*(M) = 0$.

By inverting the Lefschetz motive, the category of Chow motives becomes rigid, and it satisfies the assumption of Lemma 8.3.13 (and hence Corollary 8.3.10).

8.4 Chow Motive of a hypersurface

Let $X \subset \mathbb{P}^{n+1}$ be a hypersurface of degree d, with $h \in CH^1X$ the hyperplane section.

Definition 8.4.1. Define the primitive part of the motive $h(X)$ to be

$$h(X)_{\mathrm{pr}} := (X, [\Delta_X] - \frac{1}{d} \sum_{i=0}^{n} h^i \times h^{n-i}, 0)$$

Remark 8.4.2. The cohomology of $h(X)_{\mathrm{pr}}$ is concentrated on the mid-dimensional part H^n.

Definition 8.4.3. Let us denote $\dim H^*(h(X)_{\mathrm{pr}})$ by N. We define the determinant of $h(X)_{\mathrm{pr}}$ by

$$\det h(X)_{\mathrm{pr}} := \begin{cases} \bigwedge^N h(X)_{\mathrm{pr}} & \text{(if } n \text{ is even)} \\ \mathrm{Sym}^N h(X)_{\mathrm{pr}} & \text{(if } n \text{ is odd)} \end{cases}.$$

Remark 8.4.4. If we assume the finite dimensionality of the motive $h(X)$, then $\det h(X)_{\mathrm{pr}}$ is a 1-dimensional motive, because its cohomology group is 1-dimensional, and the Kimura dimension of a Kimura finite motive is same as the dimension of its cohomology group ([5, Cor. 7.4]). In particular, $\det h(X)_{\mathrm{pr}}$ is invertible by Proposition 8.2.9. If moreover we work over $\mathbb{C}$ and assume the Hodge conjecture, then $H^*(\det h(X)_{\mathrm{pr}}) \subset H^*(X^N)$ is spanned by an algebraic cycle α. When we choose its lift to the Chow group $\tilde{\alpha} \in CH_*(X^N)$, we can write $\det h(X)_{\mathrm{pr}} = (X^N, c(\tilde{\alpha} \times \tilde{\alpha}), 0)$ for some $c \in \mathbb{Q}$ (see [5, Prop. 10. 3]).

Let us consider the following question: If conversely we know that $\det h(X)_{\mathrm{pr}}$ can be written as $(X^N, c(\alpha \times \alpha), 0)$ for some $c \in \mathbb{Q}$ and $\alpha \in CH_*(X^N)$. Can we conclude that $h(X)$ is finite dimensional?

Proposition 8.4.5. *When n is even, and we have $\det h(X)_{\mathrm{pr}} = (X^N, c\alpha \times \alpha, 0)$, then the motive $h(X)$ is finite dimensional. When n is odd and, if moreover, $\det h(X)_{\mathrm{pr}} = (X^N, c(\alpha \times \alpha), 0)$, then $h(X)$ is Schur finite. More precisely, we have $S_{(N+1,1)} h(X)_{\mathrm{pr}} = 0$.*

Proof Because $h(X) = (X, \frac{1}{d}\sum_{i=0}^n h^i \times h^{n-i}, 0) \oplus h(X)_{\mathrm{pr}}$, the finite dimensionality (resp. Schur finiteness) of $h(X)$ is equivalent to the finite dimensionality (resp. Schur finiteness) of $h(X)_{\mathrm{pr}}$. When n is even, the assumption says that $\bigwedge^N h(X)_{\mathrm{pr}} = (X^N, c(\alpha \times \alpha), 0)$, which is evenly 1-dimensional, hence we have $0 = \bigwedge^2(\bigwedge^N h(X)_{\mathrm{pr}}) \twoheadrightarrow \bigwedge^{2N} h(X)_{\mathrm{pr}}$, therefore $h(X)_{\mathrm{pr}}$ is evenly finite dimensional (actually, N dimensional).

When n is odd, the assumption implies that

$$0 = \bigwedge^2 (\mathrm{Sym}^N h(X)_{\mathrm{pr}}) = \bigoplus_{\substack{0<i\leq N \\ i \text{ is odd}}} S_{(2N-i,i)} h(X)_{\mathrm{pr}}$$

which implies the Schur finiteness of $h(X)_{\mathrm{pr}}$. By Corollary 8.3.10, we can conclude that $S_{(N+1,1)} h(X)_{\mathrm{pr}} = 0$. □

Conjecture 8.4.6. When M is a Schur finite Chow motive, then Schur dimension of M is a rectangle.

Assuming the finite dimensionality of M, using the nilpotency theorem, Schur dimension of M is the same as the $\mathbb{Z}/2$-graded vector space $H^*(M)$, whose Schur dimension is always rectangle (Remark 8.3.12).

Proposition 8.4.7. *If we assume Conjecture 8.4.6, then assuming* $\det h(X)_{\text{pr}} = (X^N, c(\alpha \times \alpha), 0)$, *we obtain the finite dimensionality of* $h(X)$.

Proof By Proposition 8.4.5 and its proof, we may assume that n is odd, and prove the finite dimensionality of $h(X)_{\text{pr}}$, which satisfies $S_{(N+1,1)}h(X)_{\text{pr}} = 0$. By Conjecture 8.4.6, the smallest Young diagram to kill $h(X)_{\text{pr}}$ must be a rectangle, which is contained in either $(N+1)$ or $(1,1)$, anyway $h(X)_{\text{pr}}$ is oddly of evenly finite dimensional, hence finite dimensional. In fact, then we have nilpotency theorem, so $h(X)_{\text{pr}}$ is oddly finite dimensional. □

References

[1] Deligne, P.: Catégories tensorielles, Mosc. Math. J. **2** (2002) 227–248.

[2] Deligne, P. and J. S. Milne Tannakian Categories, in *Hodge Cycles, Motives and Shimura Varieties* 101–228, Springer Lecture Notes **900**.

[3] Fulton, W. and J. Harris. *Representation Theory*, Springer GTM **129**.

[4] Guletskii, V. andC. Pedrini: The Chow motive of the Godeaux surface, in *Algebraic geometry*, 179–195, de Gruyter Berlin, 2002.

[5] S. Kimura: Chow groups are finite dimensional, in some sense, Math. Ann. (2005).

[6] Macdonald, I. G. : *Symmetric Functions and Hall Polynomials*, Oxford University Press (1995).

[7] Mazza, C.: Schur functors and motives, K-theory **33** (2004) pp 89–106.

[8] Saavedra, R.: *Catégories Tannakiennes*, Springer Lecture NoteS **265**.

9
Real Regulators on Milnor Complexes, II

James D. Lewis,[†]

632 Central Academic Building, University of Alberta
Edmonton, Alberta T6G 2G1, CANADA
lewisjd@ualberta.ca

To Jacob Murre on the occasion of his 75th birthday.

With admiration!

Abstract

Let X be a projective algebraic manifold, and let $\mathcal{K}_{k,X}^M$ be the k-th Milnor K-theory sheaf on X. In some earlier work, we constructed a real regulator r from the Zariski cohomology of $\mathcal{K}_{k,X}^M$ (in all degrees) to a certain quotient of real Deligne cohomology. For certain classes of X, and from the work of others, this real regulator is known to satisfy a Noether-Lefschetz property, viz., r is 'trivial'. In this paper we construct a twisted variant of Milnor K-theory, and corresponding twisted regulator $\underline{r}$ and arrive at a corresponding complex whose regulator images are nontrivial, even in the cases where r has trivial image.

9.1 Introduction

Let $X/_{\mathbb{C}}$ be a projective algebraic manifold of dimension n, $\mathcal{O}_X$ the sheaf of regular functions on X, with sheaf of units $\mathcal{O}_X^\times$. We put

$$\mathcal{K}_{k,X}^M := \underbrace{\mathcal{O}_X^\times \otimes \cdots \otimes \mathcal{O}_X^\times}_{k\text{-times}} / \mathcal{J}, \qquad \text{(Milnor sheaf)}$$

where $\mathcal{J}$ is the subsheaf of the tensor product generated by sections of the form:

$$\big\{\tau_1 \otimes \cdots \otimes \tau_k \;\big|\; \tau_i + \tau_j = 1, \quad \text{for some } i \neq j\big\}.$$

† Partially supported by a grant from the Natural Sciences and Engineering Research Council of Canada

[For example, $\mathcal{K}_{1,X}^{M} = \mathcal{O}_X^{\times}$.] For $0 \leq m \leq 2$, it is known [MS2] that $\mathsf{CH}^k(X,m) \simeq H_{\text{Zar}}^{k-m}(X, \overline{\mathcal{K}}_{k,X}^{M})$, where $\mathsf{CH}^k(X,m)$ is Bloch's higher Chow group [Blo1], and where $\overline{\mathcal{K}}_{k,X}^{M} = \text{Image}\ (\mathcal{K}_{k,X}^{M} \to K_k^M(\mathbb{C}(X)))$. There is a real regulator [Blo2]

$$r_{k,m} : \mathsf{CH}^k(X,m) \to H_{\mathcal{D}}^{2k-m}(X, \mathbb{R}(k)), \tag{9.1}$$

which is known to satisfy a Noether-Lefschetz theorem for X a sufficiently general complete intersection of high enough multidegree [MS1]. In the case where $k = m = 2$, the vanishing of the real regulator in (9.1) for a sufficiently general curve X of genus $g(X) > 1$ is due to Collino [Co]. (One can also consult [Ke1, Ke2] for similar vanishing results for the Milnor regulator.) Apart from one's natural inclination to further extend the results of [Lew1] on the real Milnor regulator on $H_{\text{Zar}}^{k-m}(X, \mathcal{K}_{k,X}^{M})$ to other geometrically interesting Chow groups, the purpose of this paper is to construct a twisted variant $H^{k-m}(\underline{K}_{k-\bullet,X}^{M})$ of $H_{\text{Zar}}^{k-m}(X, \mathcal{K}_{k,X}^{M})$, for all k and m, and which has nontrivial regulator images, even in those cases where the standard regulator in (9.1) fails to be "nontrivial". This however, comes at a price of giving up some functorial properties. There is a natural map $H_{\text{Zar}}^{k-m}(X, \mathcal{K}_{k,X}^{M}) \to H^{k-m}(\underline{K}_{k-\bullet,X}^{M})$ which turns out to be neither injective (modulo torsion) nor surjective (see §9.8). Towards this goal of constructing a twisted variant $H^{k-m}(\underline{K}_{k-\bullet,X}^{M})$ of $H_{\text{Zar}}^{k-m}(X, \mathcal{K}_{k,X}^{M})$, we introduce the notion of a flat holomorphic line bundle over a complex projective variety Z, being essentially a pair $(L, \| \ \|_L)$, where $\| \ \|_L$ is a flat metric, and a meromorphic section $\sigma \in \text{Rat}^*(L)$, where $\text{Rat}^*(L)$ denotes the non-zero rational sections of L over Z, and where we view holomorphic line bundles in terms of their algebraic counterparts. The role of flatness of lines bundles is crucial to our construction of a regulator current that descends to the level of cohomology (see §9.3). We formulate a variant of Milnor K-theory based on contructing "symbols" out of rational sections of flat line bundles, and construct a twisted "Gersten-Milnor" complex

$$\underline{K}_{k-\bullet,X}^{M} : \quad \underline{K}_{k,X}^{M} \overset{T^{(k)}}{\to} \bigoplus_{\text{codim}_X Z=1} \underline{K}_{k-1,Z}^{M} \overset{T^{(k-1)}}{\to} \cdots$$

$$\cdots \overset{T^{(j+2)}}{\to} \bigoplus_{\text{codim}_X Z=k-j-1} \underline{K}_{j+1,Z}^{M} \overset{T^{(j+1)}}{\to} \bigoplus_{\text{codim}_X Z=k-j} \underline{K}_{j,Z}^{M} \overset{T^{(j)}}{\to} \cdots$$

$$\cdots \overset{T^{(3)}}{\to} \bigoplus_{\text{codim}_X Z=k-2} \underline{K}_{2,Z}^{M} \overset{T^{(2)}}{\to} \bigoplus_{\text{codim}_X Z=k-1} \underline{K}_{1,Z}^{M} \overset{T^{(1)}}{\to} \bigoplus_{\text{codim}_X Z=k} \underline{K}_{0,Z}^{M} \to 0,$$

and accordingly define

$$H^{k-m}(\underline{K}^M_{k-\bullet,X}) := \frac{\ker T^{(m)} : \bigoplus_{\mathrm{codim}_X Z = k-m} \underline{K}^M_{m,Z} \to \bigoplus_{\mathrm{codim}_X Z = k-m+1} \underline{K}^M_{m-1,Z}}{T^{(m+1)}\left(\bigoplus_{\mathrm{codim}_X Z = k-2} \underline{K}^M_{m+1,Z}\right)}.$$

The main result of this paper, which describes the twisted regulator map (with this group as its source), can be found in 9.6 (Theorem 9.6.6).

In §9.7, we provide some examples where the twisted regulator is non-trivial, in the situation where the real regulator in (9.1) vanishes (modulo $D(k,m)$).

We are indebted to the referee for suggesting the need for improvements to this paper, and to the organizers Jan Nagel and Chris Peters for their excellent job in organizing this conference.

This paper is the end result of some fruitful collaboration we had with Brent Gordon some years ago [GL]. Since then, Brent Gordon has moved into the area of artificial intelligence. We are grateful for his contributions.

Notation. $X/_{\mathbb{C}}$ is a projective algebraic manifold, of dimension n. For a subring $\mathbb{A} \subset \mathbb{R}$, put $\mathbb{A}(k) = (2\pi\sqrt{-1})^k\mathbb{A}$.

$$\pi_{k-1} : \mathbb{C} = \mathbb{R}(k) \oplus \mathbb{R}(k-1) \to \mathbb{R}(k-1),$$

is the projection. Denote by E^k_X (resp. $E^{p,q}_X$), the (global) complex-valued C^∞-forms of degree k (resp. Hodge type (p,q)) on X. The Picard variety of X is denoted by $\mathrm{Pic}^0(X)$. The subspace of algebraic cocycles in $H^{2k-2}(X,\mathbb{Q})$ is denoted by $H^{2k-2}_{\mathrm{alg}}(X,\mathbb{Q})$. Finally, $N^{k-m}H^{2k-m-1}(X,\mathbb{R})$ denotes the $(k-m)$-th coniveau filtration subspace of $H^{2k-m-1}(X,\mathbb{R})$.

9.2 Real Deligne cohomology

For a subring $\mathbb{A} \subset \mathbb{R}$, we introduce the Deligne complex

$$\mathbb{A}_{\mathcal{D}}(k) : \quad \mathbb{A}(k) \to \underbrace{\mathcal{O}_X \to \Omega^1_X \to \cdots \to \Omega^{k-1}_X}_{\text{call this } \Omega^{\bullet<k}_X}.$$

Definition 9.2.1. Deligne cohomology is given by the hypercohomology:

$$H^i_{\mathcal{D}}(X, \mathbb{A}(k)) = \mathbb{H}^i(\mathbb{A}_{\mathcal{D}}(k)).$$

Note that there is a short exact sequence:

$$0 \to \Omega^{\bullet<k}_X[-1] \to \mathbb{A}_{\mathcal{D}}(k) \to \mathbb{A}(k) \to 0.$$

For $m \geq 1$ and from Hodge theory, we have the isomorphisms:

$$\begin{aligned} H_{\mathcal{D}}^{2k-m}(X, \mathbb{R}(k)) &\simeq \frac{H^{2k-m-1}(X, \mathbb{C})}{F^k H^{2k-m-1}(X, \mathbb{C}) + H^{2k-m-1}(X, \mathbb{R}(k))} \\ &\underset{\simeq}{\xrightarrow{\pi_{k-1}}} \frac{H^{2k-m-1}(X, \mathbb{R}(k-1))}{\pi_{k-1}(F^k H^{2k-m-1}(X, \mathbb{C}))} \end{aligned}$$

For example:

$$\begin{aligned} H_{\mathcal{D}}^{2k-1}(X, \mathbb{R}(k)) &\simeq \frac{H^{2k-2}(X, \mathbb{C})}{F^k H^{2k-2}(X, \mathbb{C}) + H^{2k-2}(X, \mathbb{R}(k))} \\ &\underset{\simeq}{\xrightarrow{\pi_{k-1}}} H^{k-1,k-1}(X, \mathbb{R}) \otimes \mathbb{R}(k-1) \\ =: H^{k-1,k-1}(X, \mathbb{R}(k-1)) &\simeq \left\{ H^{n-k+1,n-k+1}(X, \mathbb{R}(n-k+1)) \right\}^{\vee}. \end{aligned}$$

9.3 A regulator

Using the isomorphism:

$$H_{\mathcal{D}}^{2k-1}(X, \mathbb{R}(k)) \simeq H^{k-1,k-1}(X, \mathbb{R}(k-1)),$$

in this section we give a description of the real regulator,

$$\underline{r} : \underline{z}^k(X, 1) \to H^{k-1,k-1}(X, \mathbb{R}(k-1)) \simeq H^{n-k+1,n-k+1}(X, \mathbb{R}(n-k+1))^{\vee},$$

where $\underline{z}^k(X, 1)$ is a certain twisted group of K_1 classes. Towards this goal, we introduce the notion of a flat holomorphic line bundle, being essentially a pair $(L, \| \ \|_L)$, where $\| \ \|_L$ is a flat metric on L, and the group

$$\underline{z}^k(X, 1) = \left\{ \sum_i (\sigma_i, \| \ \|_{L_i}) \otimes E_i \,\middle|\, \operatorname{codim}_X E_i = k-1, \ L_i/E_i \text{ flat}, \right. \\ \left. \sigma_i \in \operatorname{Rat}^*(L_i), \sum_i \operatorname{div}(\sigma_i) = 0 \right\},$$

where $\operatorname{div}(\sigma)$ is the divisor associated to the meromorphic section $\sigma \in \operatorname{Rat}^*(L)$, where $\operatorname{Rat}^*(L)$ denotes the nonzero rational sections of L over E, and where we view holomorphic line bundles in terms of their algebraic counterparts, in the Zariski topology. Recall that [GH, Sou]:

$$H^{n-k+1,n-k+1}(X) \simeq \frac{E_{X,d\text{-closed}}^{n-k+1,n-k+1}}{\partial\bar{\partial} E_X^{n-k,n-k}}.$$

Now assume given a subvariety $Z \subset X$ of codimension $k-1$, an algebraic line bundle L over Z with a hermitian metric, and $\sigma \in \mathrm{Rat}^*(L_i)$. For $\omega \in E_X^{n-k+1,n-k+1} \otimes \mathbb{R}(n-k+1)$, one has the L^1-integral:

$$\frac{1}{(2\pi\sqrt{-1})^{n-k+1}} \int_Z \omega \log \|\sigma\|_L \in \mathbb{R}, \tag{9.2}$$

where we use the fact that Z_{sing} has *real codimension* ≥ 2 in Z. This integral naturally extends to elements of the form

$$\xi = \sum_{\substack{\mathrm{codim}_X Z = k-1 \\ \{L, \| \ \|_L\}/Z}} (\sigma, \| \ \|_L) \otimes Z.$$

Thus ξ defines a current on $E_X^{n-k+1,n-k+1} \otimes \mathbb{R}(n-k+1)$. That this current descends on the cohomology level, viz., it is $\partial\bar{\partial}$-closed, requires some assumptions on L and the metric. For this, we introduce the notion of a flat line bundle. Let Z be a projective variety and L a(n algebraic) line bundle on Z. Recall that if Z is smooth, a metric on L is equivalent to a collection of C^∞ functions $\{\rho_\alpha : U_\alpha \to (0, \infty)\}$ satisfying $\rho_\beta = \rho_\alpha |g_{\alpha\beta}|^2$ on $U_\alpha \cap U_\beta$, where $\{U_\alpha\}$ is taken to be a Zariski open cover of Z where $L|_{U_\alpha} \simeq U_\alpha \times \mathbb{C}$ trivializes, and where the transition functions of L with respect to this cover are $\{g_{\alpha\beta}\}$. The curvature form is the global closed real $(1,1)$-form ν_L given locally on U_α by

$$\nu_L = \frac{1}{2\pi\sqrt{-1}} \partial\bar{\partial} \log \rho_\alpha,$$

and it's image $c_1(L) = [\nu_L]$ in $H^2(X, \mathbb{Z})$ is (up to torsion) the first Chern class of L.

Lemma 9.3.1. *Let us assume given a line bundle L as above on a smooth projective Z, with first Chern class $c_1(L) = 0$. Then one can find a corresponding metric for which the curvature form $\nu_L = 0$.*

Proof This can be deduced from [GH, Proposition, p. 148]. Since the proof is easy, and for convenience to the reader, we supply it here. From Hodge theory one has $\nu_L \in \partial\bar{\partial} E_Z^0$. Thus $\nu_L = \frac{1}{2\pi\sqrt{-1}} \partial\bar{\partial}\psi$ for some real-valued global C^∞ function ψ on Z. Now if we set $\rho = \exp(\psi)$, then $\partial\bar{\partial} \log \rho_\alpha = \partial\bar{\partial} \log \rho$ over each U_α. Now replace each ρ_α by $\tilde{\rho}_\alpha := \rho_\alpha/\rho$. Then $\tilde{\rho}_\alpha$ transforms accordingly over $U_\alpha \cap U_\beta$ and hence defines a metric on L; moreover $\{\partial\bar{\partial} \log \tilde{\rho}_\alpha\} = 0$, which is what we needed to show. □

Now assume that Z is a projective variety, not necessarily smooth, with smooth part $Z_{\mathrm{reg}} \subset Z$. Also assume given a line bundle L on Z. A metric

$\|\ \|_L$ is defined in the same way as given by a collection of functions $\{\rho_\alpha : U_\alpha \to (0,\infty)\}$, $\rho_\beta = \rho_\alpha |g_{\alpha\beta}|^2$, where the ρ_α pullback to C^∞ functions with respect to any desingularization $\tilde{Z} \to Z$. The curvature form ν_L is defined in the same way on Z_{reg}.

Definition 9.3.2. The metric $\|\ \|_L$ is flat if $\nu_L = 0$. A flat line bundle L is a pair $(L, \|\ \|_L)$ where $\|\ \|_L$ is flat.

Note that if the metric is flat, then for any desingularization $\tilde{Z} \to Z$, L pulls back to a line bundle with trivial curvature form, hence trivial first Chern class.

The presence of the Tate twist $\mathbb{R}(j)$ for the remaining part of this paper does not form any useful role. Thus we will ignore twists by supressing the factor $1/(2\pi\sqrt{-1})^{n-k+1}$ in (9.2). The reason for working with flat line bundles comes from our next result, viz.,

Proposition 9.3.3. *Consider* $\xi = \sum_i (\sigma_i, \|\ \|_{L_i}) \otimes Z_i \in \underline{z}^k(X,1)$ *and* $\omega \in E^{n-k+1,n-k+1}_{X,d\text{-closed}}$. *Then the current defined by*

$$\omega \longmapsto \underline{r}(\xi)(\omega) := \sum_i \int_{Z_i} \omega \log \|\sigma_i\|_{L_i},$$

is $\partial\overline{\partial}$*-closed.*

Proof Let $\omega \in \partial\overline{\partial} E^{n-k,n-k}_X$. Then we can write $\omega = d\overline{\partial}\eta$ for some $\eta \in E^{n-k,n-k}_X$. Also let

$$\xi = \sum_i (\sigma_i, \|\ \|_{L_i}) \otimes Z_i \in \underline{z}^k(X,1)$$

be given as above, and consider the corresponding integral

$$\sum_i \int_{Z_i} \omega \log \|\sigma_i\|_{L_i}.$$

By Stokes' theorem and a standard calculation (below):

$$\begin{aligned}\int_{Z_i} (d\overline{\partial}\eta) \log \|\sigma_i\|_{L_i} &= \int_{Z_i} \overline{\partial}\eta \wedge d \log \|\sigma_i\|_{L_i} \\ &= \int_{Z_i} \overline{\partial}\eta \wedge \partial \log \|\sigma_i\|_{L_i} = \int_{Z_i} d\eta \wedge \partial \log \|\sigma_i\|_{L_i},\end{aligned}$$

where the latter two equalities follow by Hodge type, and the former uses

$$d\big((\overline{\partial}\eta) \log \|\sigma\|_L\big) = (d\overline{\partial}\eta) \log \|\sigma\|_L - \overline{\partial}\eta \wedge d \log \|\sigma\|_L.$$

More specifically, we make use of the following facts. By taking ϵ-tubes about the components $D \subset \operatorname{div}(\sigma)$, and using that $\dim D = n - k$ and $\overline{\partial}\eta \in E_X^{n-k,n-k+1}$, and that $\log \|\sigma\|$ is locally L^1, in the limit as $\epsilon \to 0$

$$\lim_{\epsilon \to 0} \int_{\operatorname{Tube}_\epsilon((\sigma))} (\overline{\partial}\eta) \log \|\sigma\|_L = 0,$$

(using estimates involving $\lim_{\epsilon \to 0^+} \epsilon \log \epsilon = 0$).

Note that $\overline{\partial}\eta \wedge \overline{\partial} \log \|\sigma\|_L \in E_{X,L^1}^{n-k,n-k+2}$ (locally L^1 forms) and that $\dim Z_i = n - k + 1$. Thus we are left with an integral of the form $\int_Z d\eta \wedge \partial \log \|\sigma\|_L$ as indicated above. Next,

$$d(\eta \wedge \partial \log \|\sigma\|_L) = d\eta \wedge \partial \log \|\sigma\|_L,$$

since $\overline{\partial}\partial \log \|\sigma\|_L = 0$, by the key ingredient of flatness of L. Thus

$$\begin{aligned} \int_{Z_i} (d\overline{\partial}\eta) \log \|\sigma_i\|_{L_i} &= \int_{Z_i} d(\eta \wedge \partial \log \|\sigma_i\|_L) \\ &= \lim_{\epsilon \to 0} \int_{\operatorname{Tube}_\epsilon((\sigma_i))} \eta \wedge \partial \log \|\sigma_i\|_L. \end{aligned}$$

If we put $Z = Z_i$ for a given i, with z a local coordinate on Z_{reg}, then we have the residue integral:

$$\lim_{\epsilon \to 0} \int_{|z|=\epsilon} \eta \wedge \partial \log |z|^2 = \lim_{\epsilon \to 0} \int_{|z|=\epsilon} \eta \wedge \frac{dz}{z}$$

$$= 2\pi\sqrt{-1} \int_{\{z=0\} \cap Z} \eta|_{\{z=0\}\cap Z}, \ \eta|_{\{z=0\}\cap Z} = \operatorname{Residue}_{\{z=0\}\cap Z} \left(\eta \wedge \frac{dz}{z}\right),$$

(i.e. taking "tubes" is dual to taking "residues"). Then by a residue calculation, linearity, and Stokes' theorem, we arrive at the formula:

$$\underline{r}(\xi)(\omega) = \frac{-2\pi\sqrt{-1}}{2} \sum_D \left[\left(\sum_i \nu_D(\sigma_i) \right) \int_D \eta \right].$$

(We note that there remains the possibility that $Z = Z_i$ is singular along D. To remedy this, one may pass to a normalization of Z with the same calculations above.) Therefore $\sum_i \operatorname{div}(\sigma_i) = 0$, hence $\underline{r}(\xi)(\omega) = 0$ and we are done. □

9.3.4. Independence of the metric. Now consider $Z \subset X$ an irreducible subvariety of codimension $k - 1$ in X, and L/Z a flat (algebraic) line bundle given by $\{g_{\alpha\beta}, U_\alpha\}$, and $\{\rho_\alpha\}$ a flat metric, i.e., $\partial\overline{\partial} \log \rho_\alpha = 0$. Let $\{\tilde{\rho}_\alpha\}$ be another flat metric. Then

$$\frac{\tilde{\rho}_\beta}{\rho_\beta} = \frac{\tilde{\rho}_\alpha |g_{\alpha\beta}|^2}{\rho_\alpha |g_{\alpha\beta}|^2},$$

hence $\rho := (\tilde{\rho}_\alpha/\rho_\alpha) > 0$ extends globally over Z; moreover, $\partial\overline{\partial}\log\rho = 0$. It follows that $\lambda := \frac{1}{2}\log\rho \in \mathbb{R}$, being harmonic and global, is necessarily constant. Therefore

$$\int_Z \omega \log \|\sigma\|_{\tilde{L}} = \int_Z \omega(\lambda + \log\|\sigma\|_L).$$

But $\lambda \int_Z(\cdots)$ defines a class in $H^{k-1,k-1}_{\text{alg}}(X,\mathbb{R})$. Thus if

$$\xi = \sum_j (\sigma_j, \|\ \|_{L_j}) \otimes Z_j,$$

then in

$$\frac{H^{k-1,k-1}(X,\mathbb{R})}{H^{k-1,k-1}_{\text{alg}}(X,\mathbb{R})},$$

the regulator $\underline{r}(\xi)$ is independent of the choice of flat metric $\|\ \|_{L_j}$ on L_j/Z_j.

9.4 Some Hodge theory

The goal of this section is to describe the $\partial\overline{\partial}$-closed regulator current $\underline{r}(\xi)$ given in Proposition 9.3.3, from the point of view of de Rham cohomology. A good reference for this section is [GH, Chapter 0] and [Sou, Chapter II]. Let $\mathcal{D}_\ell(X)$, $\mathcal{D}_{r,s}(X)$ be the spaces of currents acting on E^ℓ_X and $E^{r,s}_X$ rspectively, and write $\mathcal{D}^{2n-\ell}(X) = \mathcal{D}_\ell(X)$, $\mathcal{D}^{n-r,n-s}(X) = \mathcal{D}_{r,s}(X)$. One has a corresponding decomposition

$$\mathcal{D}^k(X) = \bigoplus_{p+q=k} \mathcal{D}^{p,q}(X).$$

Lemma 9.4.1 ($\partial\overline{\partial}$-Lemma). *If $T \in \mathcal{D}^{p,q}_X$ is a coboundary, then $T = \partial\overline{\partial}T_0$ for some $T_0 \in \mathcal{D}^{p-1,q-1}(X)$.*

Corollary 9.4.2.

$$H^{p,q}(X) \simeq \frac{E^{p,q}_{X,d\text{-closed}}}{\partial\overline{\partial}E^{p-1,q-1}_X} \simeq \frac{\mathcal{D}^{p,q}_{d\text{-closed}}(X)}{\partial\overline{\partial}\mathcal{D}^{p-1,q-1}(X)}.$$

Lemma 9.4.3. *The natural inclusion*

$$\mathcal{D}^{p,q}_{d\text{-closed}}(X) \to \mathcal{D}^{p,q}_{\partial\overline{\partial}\text{-closed}}(X),$$

induces an isomorphism

$$\frac{\mathcal{D}^{p,q}_{d\text{-closed}}(X)}{\partial\overline{\partial}\mathcal{D}^{p-1,q-1}(X)} \simeq \frac{\mathcal{D}^{p,q}_{\partial\overline{\partial}\text{-closed}}(X)}{\partial\mathcal{D}^{p-1,q}(X) + \overline{\partial}\mathcal{D}^{p,q-1}(X)}.$$

Proof Let $T \in \mathcal{D}^{p,q}_{\partial\overline{\partial}\text{–closed}}(X)$. Then $\partial T \in \mathcal{D}^{p+1,q}(X)$ and $\overline{\partial}T \in \mathcal{D}^{p,q+1}(X)$ are both d-closed. Therefore, from Hodge theory, $\partial T = dS_1$ and $\overline{\partial}T = dS_2$ for some $S_1 \in F^{p+1}\mathcal{D}^{p+q}(X)$, and $S_2 \in \overline{F^{q+1}\mathcal{D}^{p+q}(X)}$. Thus $d(T-S_1-S_2) = 0$ and moreover by the Hodge (p,q) decomposition theorem, we can modify S_j within it's Hodge type, such that the cohomology class $[T - S_1 - S_2]$ is of type (p,q). (More explicitly, we can write

$$[T - S_1 - S_2] = [A_1] \oplus [B] \oplus [A_2],$$

where

$$[A_1] \in F^{p+1}H^{p+q}(X,\mathbb{C}), \quad [A_2] \in \overline{F^{q+1}H^{p+q}(X,\mathbb{C})}, \quad [B] \in H^{p,q}(X),$$

are represented by d-closed currents (or forms) A_1, A_2, B of the corresponding Hodge types. Now replace S_j by $S_j - A_j$, and relabel it S_j.) Hence there exists T_0 such that $T - S_1 - S_2 + dT_0 \in \mathcal{D}^{p,q}_{d\text{–closed}}(X)$. This implies that $T + \partial T_0^{p-1,q} + \overline{\partial}T_0^{p,q-1}$ is d-closed. Next, suppose that $T \in \mathcal{D}^{p,q}_{d\text{–closed}}(X)$ is given such that $T \in \operatorname{Im}\partial + \operatorname{Im}\overline{\partial}$. By the Hodge theorem, T has no harmonic part, and being d-closed implies that it is a coboundary. The lemma easily follows from this. □

To arrive at the same sort of de Rham description of $\underline{r}(\xi)$ that appears in the twisted case in [Ja, p. 349], we for the moment include the twist factor $1/(2\pi\sqrt{-1})^{n-k+1}$ appearing in (9.2). It is obvious that $\underline{r}(\xi)$ given in Proposition 9.3.3 determines an element of $\mathcal{D}_{n-k+1,n-k+1,\partial\overline{\partial}-\text{ closed}}(X, \mathbb{R}(n-k+1))$. It follows easily from the proof of Lemma 9.4.3, that there exists $\psi \in \mathcal{D}_{2n-2k+2,0}(X) \oplus \cdots \oplus \mathcal{D}_{n-k+2,n-k}(X)$ such that $\underline{r}(\xi) + \pi_{n-k+1}(\psi)$ is d-closed. Its action on

$$E^{n-k+1,n-k+1}_{X,d\text{–closed}}/\partial\overline{\partial}E^{n-k,n-k}_X$$

is the same as $\underline{r}(\xi)$. By duality, viz.,

$$H^{k-1,k-1}(X,\mathbb{R}(k-1)) \simeq H^{n-k+1,n-k+1}(X,\mathbf{R}(n-k+1))^{\vee},$$

we end up with a class $\underline{r}(\xi) \in H^{k-1,k-1}(X,\mathbb{R}(k-1))$. Note that likewise

$$[\underline{r}(\xi)] \in \frac{\mathcal{D}^{k-1,k-1}_{\partial\overline{\partial}\text{–closed}}(X)}{\partial\mathcal{D}^{k-2,k-1}(X) + \overline{\partial}\mathcal{D}^{k-1,k-2}(X)} \simeq H^{k-1,k-1}(X).$$

Let

$$Q^{k-1,k-1} = \frac{\mathcal{D}^{k-1,k-1}_{\partial\overline{\partial}\text{–closed}}(X)}{\mathcal{D}^{k-1,k-1}_{X,d\text{–closed}}}.$$

There is a commutative diagram of short exact sequences:

$$\begin{array}{ccccccccc}
 & & 0 & & 0 & & 0 & & \\
 & & \downarrow & & \downarrow & & \downarrow & & \\
0 & \to & \mathrm{Im}\partial\overline{\partial} & \to & \mathrm{Im}\partial + \mathrm{Im}\overline{\partial} & \to & \frac{\mathrm{Im}\partial + \mathrm{Im}\overline{\partial}}{\mathrm{Im}\partial\overline{\partial}} & \to & 0 \\
 & & \downarrow & & \downarrow & & \downarrow & & \\
0 & \to & \mathcal{D}^{k-1,k-1}_{X,d\text{-closed}} & \to & \mathcal{D}^{k-1,k-1}_{\partial\overline{\partial}\text{-closed}} & \to & Q^{k-1,k-1} & \to & 0 \\
 & & \downarrow & & \downarrow & & \downarrow & & \\
0 & \to & H^{k-1,k-1}(X) & = & H^{k-1,k-1}(X) & \to & 0 & & \\
 & & \downarrow & & \downarrow & & & & \\
 & & 0 & & 0. & & & &
\end{array}$$

Thus

$$Q^{k-1,k-1} \simeq \frac{\mathrm{Im}\partial + \mathrm{Im}\overline{\partial}}{\mathrm{Im}\partial\overline{\partial}}.$$

Note that $\underline{r}(\xi)$ determines a class

$$\{\underline{r}(\xi)\} \in Q^{k-1,k-1},$$

which is a measure of how far $\underline{r}(\xi)$ is from being d-closed.

Proposition 9.4.4. *Let X be a projective algebraic manifold of dimension n, and D an algebraic cycle of dimension $k-1$ on X. Next, let $\xi \in \underline{z}^k(X,1) \otimes \mathbb{R}$ be given and consider the corresponding $\underline{r}(\xi)$. Let us write $\underline{r}(\xi) = \sum_{i,\alpha} r_i \int_{Z_{i,\alpha}} \log\|\sigma_{i,\alpha}\| \wedge (?)$, with $r_i \in \mathbb{R}$, and assume that D meets each $Z_{i,\alpha}$ properly (i.e. in a 0-dimensional set), and that $|D| \cap |(\sigma_{i,\alpha})_{Z_{i,\alpha}}| = \varnothing$. Then if we put $[D]$ to be the Poincaré dual of D, the cup product is given by the formula:*

$$\langle \underline{r}(\xi), [D] \rangle = \sum_{i,\alpha} r_i \int_{Z_{i,\alpha}\cap D} \log\|\sigma_{i,\alpha}\|.$$

Proof By desingularization and linearity, we reduce to the case where $j : D \hookrightarrow X$ is a smooth subvariety of X. Let $[\gamma]$ the Poincaré dual of any given cycle γ on X. Then

$$j_*\circ j^*[\gamma] = [\gamma \cap D].$$

This follows from

$$\begin{aligned} j_*\circ j^*[\gamma] &= \langle j_*\circ j^*[\gamma], [X]\rangle = j_*\langle j^*[\gamma], j^*[X]\rangle \\ &= j_*\langle j^*[\gamma], [D]\rangle = \langle [\gamma], j_*[D]\rangle_X = [\gamma \cap D]. \end{aligned}$$

In this case $\underline{r}(\xi) \in H^{k-1,k-1}(X)$ has a well-defined pullback $j^*\underline{r}(\xi) \in H^{k-1,k-1}(D)$, where $\dim_X D = k-1$, and where in this case, $j^*\underline{r}(\xi) = \sum_\alpha \int_{Z_\alpha \cap D} \log \|\sigma_\alpha\|$. Note that j_* is just the trace. The proposition follows from this. □

Remarks 9.4.5. i) It is easy to show that $\underline{r}(\xi)$ is d-closed $\Leftrightarrow$ it is an $\mathbb{R}$ – linear combination of algebraic cycles. This is generalized in Theorem 9.6.6 below.

ii) The formula in Proposition 9.4.4 can be interpreted in terms of height pairings. Let us further assume that $|D| \cap (\bigcup_{i,\alpha} Z_{i,\alpha,\text{ Sing}}) = \varnothing$. Then for a suitable choice of flat metrics, we have:

$$\langle \underline{r}(\xi), [D]\rangle = \sum_{i,\alpha} r_i \langle (\sigma_{i,\alpha})_{\tilde{Z}_{i,\alpha}}, D \cap \tilde{Z}_{i,\alpha}\rangle_{\text{ ht}},$$

where $\langle\ ,\ \rangle_{\text{ ht}}$ is the height pairing on a desingularization $\tilde{Z}$. The version of the definition of height pairing we employ is given in [MS3, Def. 1)]. (One need only show that for a suitable flat metric, $\mathbf{H}(\log \|\sigma\|) = 0$, where $\mathbf{H}(-)$ is the harmonic projection. However if we write $c = \mathbf{H}(\log \|\sigma\|)$, then we know $c \in \mathbb{R}$ is constant. Put $\lambda = \mathrm{e}^{-c} > 0$, and multiply the metric ρ by $\lambda \cdot \rho$.)

9.5 A Tame symbol

Now let $Z \subset X$ be of codimension $k-2$ with given flat bundles $(L_j, \|\ \|_{L_j})$, $j = 1, 2$ and $\sigma_f \in \text{Rat}^*(L_1)$, $\sigma_g \in \text{Rat}^*(L_2)$. As a first step in the direction of constructing a twisted Milnor complex, we define a generalization of the Tame symbol as follows.

$$\begin{aligned} &\underline{T}(\{(\sigma_f, \|\ \|_{L_1}), (\sigma_g, \|\ \|_{L_2})\} \otimes Z) \\ &= \sum_{\operatorname{codim}_Z D = 1} \left((-1)^{\nu_D(\sigma_f)\nu_D(\sigma_g)} \left(\frac{\sigma_f^{\nu_D(\sigma_g)}}{\sigma_g^{\nu_D(\sigma_f)}} \right)_D, \|\ \|_{L_1^{\otimes \nu_D(\sigma_g)} \otimes L_2^{-\nu_D(\sigma_f)}} \right) \otimes D. \end{aligned}$$

Proposition 9.5.1. $\underline{T}(\{(\sigma_f, \|\ \|_{L_1}), (\sigma_g, \|\ \|_{L_2})\} \otimes Z) \subset \underline{z}^k(X, 1)$.

Proof One first shows that $\underline{T}$ takes flat line bundles over Z to flat line bundles over each such $D \subset Z$ of codimension one. If $\{f_{\alpha\beta}\}$, resp. $\{g_{\alpha\beta}\}$ are the transition functions for L_1, resp. L_2, over Z with common open trivializing cover $\{U_\alpha\}_{\alpha\in I}$, and with corresponding $\{\rho_\alpha^{(j)} : U_\alpha \to (0,\infty)\}$, $j = 1,2$ i.e. $\partial\overline{\partial}\rho_\alpha^{(j)} = 0$, we consider the following calculation for codimension one irreducible $D \subset Z$. First, for $\sigma_f = \{f_\alpha\}$ and $\sigma_g = \{g_\alpha\}$ local representations of nonzero meromorphic sections of L_1 and L_2 we have $f_\alpha = f_{\alpha\beta}f_\beta$ and $g_\alpha = g_{\alpha\beta}g_\beta$ over $U_\alpha \cap U_\beta$. Hence over $D \cap U_\alpha \cap U_\beta$,

$$\frac{f_\alpha^{\nu_D(g)}}{g_\alpha^{\nu_D(f)}} = h_{\alpha\beta}\frac{f_\beta^{\nu_D(g)}}{g_\beta^{\nu_D(f)}},$$

where $h_{\alpha\beta} = f_{\alpha\beta}^{\nu_D(\sigma_g)}g_{\alpha\beta}^{-\nu_D(\sigma_f)}$ on $D \cap U_\alpha \cap U_\beta$. Further, we have the metric $\{(\rho_\alpha^{(1)})^{\nu_D(\sigma_g)}(\rho_\alpha^{(2)})^{-\nu_D(\sigma_f)}\}$ associated to the line bundle $\{h_{\alpha\beta}\}$ over D, and with respect to the open cover $\{U_\alpha \cap D\}_{\alpha\in I}$. But

$$\partial\overline{\partial}\log((\rho_\alpha^{(1)})^{\nu_D(\sigma_g)}(\rho_\alpha^{(2)})^{-\nu_D(\sigma_f)}) = \nu_D(\sigma_g)\partial\overline{\partial}\log(\rho_\alpha^{(1)}) - \nu_D(\sigma_f)\partial\overline{\partial}\log(\rho_\alpha^{(2)}) = 0,$$

hence this metric is flat as well. Next, we show that the divisor associated to $\underline{T}(\{(\sigma_f, \|\ \|_{L_1}), (\sigma_g, \|\ \|_{L_2})\}\otimes Z)$ is zero. Choose a Zariski open set $U \subset Z$ for which L_1, L_2 both trivialize over U. Then over U we have (the restriction of) the divisor associated to the usual tame symbol, $T(\{f,g\}\otimes Z)$, which is zero, as required. Here we have identified $f = \sigma_f$ and $g = \sigma_g$ for $f, g \in \mathbb{C}(Z)^\times$. Since Z is covered by such U, we are done. □

Proposition 9.5.2. $\underline{r}(\ Im(\underline{T})) = 0$.

We prove this by first establishing two Lemmas.

Lemma 9.5.3. *Let Z be a smooth subvariety of codimension $k-2$ in X, and let $f, g \in \mathbb{C}(Z)^\times$ be given. Then $\underline{r}(T\{f,g\}) = 0$.*

Proof [Lev]. It is instructive to sketch the proof. By a pushforward of the relevant currents involved, and by a proper modification, it suffices to assume that Z is smooth and that $f, g : Z \to \mathbb{P}^1$ are morphisms. Put $F = (f,g) : Z \to \mathbb{P}^1\times\mathbb{P}^1$, and let $(t,s) = (z_1/z_0, w_1/w_0)$ be affine coordinates for $\mathbb{P}^1 \times \mathbb{P}^1$. Then $F^*T\{t,s\} = T\{f,g\}$. One can explicitly compute:

$$T\{t,s\} = (\infty \times \mathbb{P}^1, s) - (0 \times \mathbb{P}^1, s) + (\mathbb{P}^1 \times 0, t) - (\mathbb{P}^1 \times \infty, t).$$

Next, put

$$\eta = \int_{\infty\times\mathbb{P}^1} \log|s|\wedge ? - \int_{0\times\mathbb{P}^1} \log|s|\wedge ? + \int_{\mathbb{P}^1\times 0} \log|t|\wedge ? - \int_{\mathbb{P}^1\times\infty} \log|t|\wedge ?.$$

It is easy to see that $F^*\eta = \underline{r}(T\{f,g\})$, and that η defines the zero cohomology class. This proves the lemma. □

Now let σ be a section of a flat line bundle over a given subvariety $Z \subset X$. Then $\|\sigma\|$ has at worst pole like growth along the pole set of σ; moreover $\partial\overline{\partial}\log\|\sigma\| = 0$.

Lemma 9.5.4. *In terms of local analytic coordinates, $\|\sigma\|$ is locally a product of the form $\rho = h\overline{h}$, where h is meromorphic.*

Proof Since $\overline{\partial}\partial\log\|\sigma\| = 0$, it follows that $\partial\log\|\sigma\|$ is a meromorphic 1-form. Therefore by the holomorphic Poincaré lemma, and away from divisor set $|(\sigma)|$, locally (in the strong topology) we have

$$\partial\log\|\sigma\| = \partial H,$$

for some holomorphic function H. Therefore locally

$$d\log\|\sigma\| = \partial H + \overline{\partial}\,\overline{H} = (\partial+\overline{\partial})H + (\partial+\overline{\partial})\overline{H} = d(H+\overline{H}).$$

Thus $\log\|\sigma\| = H + \overline{H} + K$ for some $K \in \mathbb{R}$, and hence $\rho := h\overline{h}$ where $h = e^{H+\frac{K}{2}}$. Next, locally over the divisor set $|(\sigma)|$, we can replace ρ by $\tilde{\rho} = |f|^2\rho$, for some meromorphic function f, such that $\log\tilde{\rho}$ is defined and $\partial\overline{\partial}\log\tilde{\rho} = 0$. Since $\tilde{\rho}$ is a product (local) of a holomorphic function times its conjugate, it follows that ρ is a product of a meromorphic function times its conjugate. □

9.5.5. Proof of Proposition 9.5.2 Observe that if $h\overline{h} = k\overline{k}$ on an open set in $\mathbb{C}^n$, with h, k meromorphic, then

$$\frac{h}{k} = \overline{\left(\frac{h}{k}\right)}^{-1},$$

is both ∂ and $\overline{\partial}$-closed. Thus $h = ck$ for some $c \in \mathbb{C}^\times$ with $|c| = 1$. Thus over an analytic cover $\{\Delta_\alpha\}$ of Z, we can write $\|\sigma\||_{\Delta_\alpha} = h_\alpha\overline{h}_\alpha$, where h_α is a meromorphic function on Δ_α. Suppose that there is a finite cover $\tilde{Z} \to Z$ such that by analytic continuation, h becomes a rational function on $\tilde{Z}$. Let's write this as $\tilde{h}$. Then $\|\tilde{\sigma}\| = \tilde{h}\overline{\tilde{h}}$, where $\tilde{\sigma}$ is the pullback of σ to $\tilde{Z}$. This for example would be the case if the c's above are m-th roots of unity for some $m \in \mathbb{N}$. Arriving at this situation would imply Proposition 9.5.2, by putting us in the setting of Lemma 9.5.3. However by a limit argument, we can reduce to this situation. Put $S^1 = \{z \in \mathbb{C} \mid |z| = 1\}$. Then $\{h_\alpha\}$ naturally defines an element in $H^0(Z, \mathcal{M}_Z^\times/S^1)$. We can assume that Z is smooth (and projective). The map $H^1(Z,S^1) \to H^1(Z,\mathcal{M}_Z^\times) = 0$ factors through $H^1(Z,\mathcal{O}_Z^\times) \to H^1(Z,\mathcal{M}_Z^\times)$, which is well-known to be zero [GH]. From the short exact sequence

$$0 \to S^1 \to \mathcal{M}_Z^\times \to \mathcal{M}_Z^\times/S^1 \to 0,$$

we deduce that

$$H^0(Z,\mathcal{M}_Z^\times/S^1) \to H^1(Z,S^1) = H^1(Z,\mathbb{R}/\mathbb{Z}),$$

is surjective, where $e^{\sqrt{-1}t} : \mathbb{R}/\mathbb{Z} \xrightarrow{\sim} S^1$. Since the kernel of the map $H^2(Z,\mathbb{Z}) \to H^2(X,\mathbb{Q})$ is a finite group, there is no loss of generality in identifying $H^1(Z,\mathbb{R}/\mathbb{Z})$ with $H^1(Z,\mathbb{R})/H^1(Z,\mathbb{Z})$, and $H^1(Z,\mathbb{Q}/\mathbb{Z})$ with $H^1(Z,\mathbb{Q})/H^1(Z,\mathbb{Z})$. Next, since $\|\sigma\|\big|_{\Delta_\alpha} = h_\alpha\overline{h}_\alpha$, it follows that $\mathrm{div}(\{h_\alpha\}) = \mathrm{div}(\sigma)$, and hence the class of $\{h_\alpha\}$ in $H^1(Z,\mathbb{R}/\mathbb{Z})$, which we identify with $H^1(Z,\mathbb{R})/H^1(Z,\mathbb{Z}) \simeq H^{0,1}(Z)/H^1(Z,\mathbb{Z}) \simeq \mathrm{Pic}^0(Z)$, is the class corresponding to the flat line bundle associated to σ. Let $\mathcal{L}/\mathrm{Pic}^0(Z) \times Z$ be the Poincaré bundle, and let $\tilde{\sigma}$ be a rational section of $\mathcal{L}/\mathrm{Pic}^0(Z) \times Z$ which doesn't vanish on $\{0\} \times Z$.† Then for t in some polydisk neighbourhood of $0 \in \mathrm{Pic}^0(Z)$, one has a family of flat metrics $\|\ \|_t$ on $L_t := \mathcal{L}\big|_{\{t\}\times Z}$, and if we put $\sigma_t = \tilde{\sigma}\big|_{\{t\}\times Z}$, then locally $\|\sigma_t\|_t = h_{t,\alpha}\overline{h}_{t,\alpha}$, and we arrive at a deformation $\{h_{t,\alpha}\}$ of $\{h_{0,\alpha}\}$. Next, by our identifications above, $H^1(Z,\mathbb{Q}/\mathbb{Z})$ is a dense subset of the torus $H^1(Z,\mathbb{R}/\mathbb{Z})$, and a point in $H^1(Z,\mathbb{Q}/\mathbb{Z})$ corresponds to a class $\{\underline{h}_\alpha\}$ with corresponding c's being m-th roots of unity. Thus with regard to $\|\sigma\|$, we can write $\{h_\alpha\}$ as a limit of classes corresponding to points in $H^1(Z,\mathbb{Q}/\mathbb{Z})$. Namely,

$$\{\underline{\underline{h}}_{t,\alpha} := \big(h_{t,\alpha} \cdot h_{0,\alpha}^{-1}\big) \cdot h_\alpha\} \quad \overset{t\to 0}{\mapsto} \quad \{h_\alpha\},$$

where $\{\underline{\underline{h}}_{t,\alpha}\}$ corresponds to a class in $H^1(Z,\mathbb{Q}/\mathbb{Z})$ for t in a countably dense subset of some polydisk neighbourhood of $0 \in \mathrm{Pic}^0(Z)$. This proves the proposition. □

9.6 A Milnor complex

We first recall the definition of Milnor K-theory [BT]. Let $\mathbb{F}$ be a field with multiplicative group $\mathbb{F}^\times$, and set

$$T(\mathbb{F}^\times) = \bigoplus_{n\geq 0} T^n(\mathbb{F}^\times),$$

† This is easy to arrange. With respect to a projective embedding of $\mathrm{Pic}^0(Z) \times Z$, the twisted bundle $\mathcal{L}(m)$ is very ample for $m \gg 1$. One simply chooses a general section of $\Gamma(\mathcal{L}(m))$ and of $\Gamma(\mathcal{O}(m))$ and assigns σ to be the quotient.

the tensor algebra of the $\mathbb{Z}$-module $\mathbb{F}^\times$. (Here $T^0(\mathbb{F}^\times) := \mathbb{Z}$.) Then $\mathbb{F}^\times \xrightarrow{\sim} T^1(\mathbb{F}^\times)$ by $a \mapsto [a]$. If $a \neq 0,\ 1$, set $r_a = [a] \otimes [1-a]$ in $T^2(\mathbb{F}^\times)$. Then the two-sided ideal R generated by the r_a's is graded, and we put

$$K_*^M\mathbb{F} = T(\mathbb{F}^\times)/R = \bigoplus_{n\geq 0} K_n^M\mathbb{F}.$$

Then $K_*^M\mathbb{F}$ may be presented as a ring with generators $\ell(a)$, for $a \in \mathbb{F}^\times$, subject to the relations

$$\ell(ab) = \ell(a) + \ell(b),$$
$$\ell(a)\ell(1-a) = 0, \quad a \neq 0,\ 1.$$

Observe that $K_j^M(\mathbb{F}) = K_j(\mathbb{F})$ for $j = 0, 1, 2$, where the latter is Quillen K-theory.

9.6.1. A twisted Milnor complex. As before let L be a (flat) line bundle over $Z \subset X$. We view L in terms of a corresponding Cartier divisor, viz., work on the Cartier divisor level. Then we first observe that although the set of nonzero rational sections $\mathrm{Rat}^*(L/Z)$ of L over Z is not a group, by fixing Z, the union $\coprod_{L/Z} \mathrm{Rat}^*(L/Z)$ can be endowed with the structure of a group. More specifically, an element of $\coprod_{L/Z} \mathrm{Rat}^*(L/Z)$ is given by a pair (σ, L), $\sigma \in \mathrm{Rat}^*(L/Z)$, L/Z flat, with product structure $(\sigma_1, L_1) \star (\sigma_2, L_2) = (\sigma_1\sigma_2, L_1 \otimes L_2)$. Thus we assign

$$\underline{K}_{1,L/Z}^M(\mathbb{C}(Z)) := \mathrm{Rat}^*(L/Z).$$

In terms of local trivializations,

$$\sigma \in \mathrm{Rat}^*(L/Z) \Leftrightarrow \{\sigma = \{\sigma_\alpha\} \mid \sigma_\alpha \in K_1(\mathbb{C}(Z)) \text{ and } g_{\alpha\beta}\sigma_\beta = \sigma_\alpha\}, \tag{9.3}$$

where $\{g_{\alpha\beta}\}$ defines L, and where $\{\sigma_\alpha\}$ lies in the direct product $\prod_\alpha K_1(\mathbb{C}(Z))$. By passing to a direct limit over refining open covers of Z, the latter term in (9.3), which will still be denoted by $\underline{K}_{1,L/Z}^M(\mathbb{C}(Z))$, is given the structure we require, and we will assume this to be the case in the discussion below. Next, we consider the group

$$\underline{K}_{1,Z}^M = \coprod_{L/Z \text{ flat}} \underline{K}_{1,L/Z}^M(\mathbb{C}(Z)) = \coprod_{L/Z \text{ flat}} \mathrm{Rat}^*(L/Z).$$

Put

$$\underline{K}_{\bullet,Z}^M = \left(\sum_{j=0}^{\infty} (\underline{K}_{1,Z}^M)^{\otimes_{\mathbf{Z}} j}\right) \Big/ R^\bullet,$$

where $R^\bullet$ is the two-sided ideal generated by

(i) $\{\sigma \otimes (-\sigma) \mid \sigma \in \underline{K}^M_{1,Z}\}$,
(ii) $\{(1-f) \otimes f \mid f \in \mathbb{C}(Z)^\times - \{1\}\}$.

The first relation leads to a desired anticommutative property of products of "symbols", and the second relation incorporates the "usual" Steinberg relation in the case of function fields. Since $R^\bullet = \oplus_{j\geq 2} R^j$ is graded, we can write

$$\underline{K}^M_{\bullet,Z} = \bigoplus_{j=0}^{\infty} \underline{K}^M_{j,Z},$$

where for example $\underline{K}^M_{0,Z} = \mathbb{Z}$, $\underline{K}^M_{1,Z}$ is given above, and $\underline{K}^M_{2,Z} =$

$$\big\{ \text{ group of symbols } \{\sigma_1, \sigma_2\} \ / \ \text{[generalized] Steinberg relations}\big\}.$$

Thus for example, $\{\sigma, -\sigma\} = 1$, hence one can easily check that

$$\{\sigma, \sigma\} = \{\sigma, -1\} = \{-1, \sigma\} = \{\sigma, \sigma^{-1}\}.$$

In general, given a symbol $\{\sigma_1, \sigma_2\}$, and up to rewriting this as a product of other symbols, one can always "factor out" common divisors in the divisor sets $|(\sigma_1)|$, $|(\sigma_2)|$.

9.6.2. We want to build a twisted Milnor-Gersten complex out of this, with the j-th term given by

$$\bigoplus_{\mathrm{codim}_X Z = k-j} \underline{K}^M_{j,Z}.$$

Thus

$$\bigoplus_{\mathrm{codim}_X Z = k} \underline{K}^M_{0,Z} = \bigoplus_{\mathrm{codim}_X Z = k} K_0(\mathbb{C}(Z)),$$

and the first two (generalized) Tame symbols

$$\mathrm{div} := T^{(1)} : \bigoplus_{\mathrm{codim}_X Z = k-1} \underline{K}^M_{1,Z} \to \bigoplus_{\mathrm{codim}_X Z = k} \underline{K}^M_{0,Z},$$

$$T := T^{(2)} : \bigoplus_{\mathrm{codim}_X Z = k-2} \underline{K}^M_{2,Z} \to \bigoplus_{\mathrm{codim}_X Z = k-1} \underline{K}^M_{1,Z},$$

have already been defined. In order to define higher Tame symbols

$$T^{(j+1)} : \bigoplus_{\mathrm{codim}_X Z = k-j-1} \underline{K}^M_{j+1,Z} \longrightarrow \bigoplus_{\mathrm{codim}_X Z = k-j} \underline{K}^M_{j,Z},$$

we digress again [BT]. This time we will assume given a field $\mathbb{F}$ with a discrete valuation $\nu : \mathbb{F}^\times \to \mathbb{Z}$, and the corresponding discrete valuation ring $\mathcal{O} := \{a \in \mathbb{F} \mid \nu(a) \geq 0\}$, where we assign $\nu(0) = \infty$. Let $\pi \in \mathcal{O}$ generate the unique maximal ideal (π), i.e., $\nu(\pi) = 1$, and recall that all other nonzero ideals are of the form (π^m), for $m \geq 0$. Note that $\mathbb{F}^\times = \mathcal{O}^\times \cdot \pi^{\mathbb{Z}}$ (direct product). Next let $\mathbf{k} = \mathbf{k}(\nu)$ be the residue field, and $K_\bullet^M$ Milnor K-theory. Then there is a map

$$d_\pi : \mathbb{F}^\times \to (K_\bullet^M \mathbf{k}(\nu))(\Pi) : u\pi^i \mapsto \ell(\overline{u}) + i\Pi,$$

where $\Pi = \ell(\pi)$, and satisfies $\Pi^2 = \ell(-1)\Pi$. This map induces

$$\partial_\pi : K_\bullet^M \mathbb{F} \to (K_\bullet^M \mathbf{k}(\nu))(\Pi).$$

Next, one defines maps

$$\partial_\pi^0,\ \partial_\nu : K_\bullet^M \mathbb{F} \to K_\bullet^M \mathbf{k}(\nu),$$

by

$$\partial_\pi(x) = \partial_\pi^0(x) + \partial_\nu(x)\Pi, \tag{9.4}$$

which can be shown to be independent of the choice of π such that $\nu(\pi) = 1$. This hinges on an explicit description of ∂_ν [BT, Proposition 4.5]. The (generalized) Tame symbol is the map $\partial_\nu : K_m^M \mathbb{F} \to K_{m-1}^M \mathbf{k}(\nu)$. We also have that multiplication is graded and skew commutative, and that $\deg \ell(a) = 1 = \deg \Pi$ for $a \in \mathbb{F}^\times$. For example, if we let $a = a_0\pi^i$, $b = b_0\pi^j$, so that $\nu(a) = i$, $\nu(b) = j$, and let $\overline{a}_0$, $\overline{b}_0$ be the corresponding values in $\mathbf{k}(\nu)$, then

$$\partial_\pi(a) = d_\pi(a) = \ell(\overline{a}_0) + i\Pi, \quad \partial_\pi(b) = d_\pi(b) = \ell(\overline{b}_0) + j\Pi,$$

so that $\partial_\nu(a) = i$ and $\partial_\nu(b) = j$. Next, as shown in [BT],

$$\begin{aligned}
\partial_\pi(\{a, b\}) &= (\ell(\overline{a}_0) + i\Pi)(\ell(\overline{b}_0) + j\Pi) \\
&= (\ell(\overline{a}_0)\ell(\overline{b}_0)) \ + \ (\ell(\overline{a}_0^j) - \ell(\overline{b}_0^i) + ij\ell(-1))\Pi \\
\partial_\nu(\{a, b\}) &= \ell\left((-1)^{ij}\frac{\overline{a}_0^j}{\overline{b}_0^i}\right) = \ell\left(\overline{(-1)^{ij}\frac{a^j}{b^i}}\right) = \ell(T\{a, b\}),
\end{aligned}$$

where T is the usual Tame symbol. Similarly, for a general product, we have

$$\begin{aligned}\partial_\pi(\{a^{(1)},\dots,a^{(N)}\}) &= (\ell(\overline{a}_0^{(1)})+k_1\Pi)\cdots(\ell(\overline{a}_0^{(N)})+k_j\Pi)\\ &= (\ell(\overline{a}_0^{(1)})+k_1\Pi)\,(\partial_\pi^0(\{a^{(2)},\dots,a^{(N)}\})+\partial_\nu(\{a^{(2)},\dots,a^{(N)}\})\Pi)\\ &= \left(\ell(\overline{a}_0^{(1)})\cdots\ell(\overline{a}_0^{(N)})\right)+\left((-1)^{N-1}k_1\,(\ell(\overline{a}_0^{(2)})\cdots\ell(\overline{a}_0^{(N)}))\right.\\ &\qquad\left.+(\ell(\overline{a}_0^{(1)})+k_1\ell(-1))\partial_\nu(\{a^{(2)},\dots,a^{(N)}\})\right)\Pi,\end{aligned}$$

whence

$$\begin{aligned}&\ell(T^{(N)}\{a^{(1)},\dots,a^{(N)}\}) := \partial_\nu(\{a^{(1)},\dots,a^{(N)}\})\\ &= \ell\left(\left\{\overline{a}_0^{(2)},\dots,\overline{a}_0^{(N)}\right\}^{(-1)^{N-1}k_1}\left\{(-1)^{k_1}\overline{a}_0^{(1)},T^{(N-1)}(\{a^{(2)},\dots,a^{(N)}\})\right\}\right),\end{aligned}$$

where again $T^{(2)} = T$ is the usual Tame symbol above.

For our purposes we define the Tame symbol on $\underline{K}^M_{N,Z}$, by defining its value at the generic point of an irreducible codimension one $D \subset Z$, viz.,

$$\left.\begin{aligned}T_D^{(N)}(\{\sigma^{(1)},\dots,\sigma^{(N)}\}) :=\ & T^{(N)}(\{\sigma^{(1)},\dots,\sigma^{(N)}\})\Big|_D\\ = \left\{\overline{\sigma}_0^{(2)},\dots,\overline{\sigma}_0^{(N)}\right\}^{(-1)^{N-1}\nu_D(\sigma^{(1)})}&\times\\ &\left\{(-1)^{\nu_D(\sigma^{(1)})}\overline{\sigma}_0^{(1)},T^{(N-1)}(\{\sigma^{(2)},\dots,\sigma^{(N)}\})\right\}\end{aligned}\right\}\tag{9.5}$$

where $\sigma^{(j)} = \pi_D^{\nu_D(\sigma^{(j)})}\sigma_0^{(j)}$, π_D a local equation of D in Z, and $\overline{\sigma}_0^{(j)}$ the value of $\sigma_0^{(j)}$ in $\mathbb{C}(D)^\times$. Here it is important to understand that if L_j/Z is the flat line bundle associated to $\sigma^{(j)}$, then $\overline{\sigma}_0^{(j)}$ is a section of $L_j\big|_D$. In other words, this calculation occurs over the generic point of D, where one fixes the choice of local equation of D (but see Proposition 9.6.3(i) below). Now note that $T^{(2)}(\{\sigma_g,\sigma_h\}) = T(\{\sigma_g,\sigma_h\})$ involves a section of a flat line bundle. From this it follows that $T^{(3)}$ can be defined with symbols of sections of flat bundles. By induction, it follows that if $T^{(j+1)}$ involves symbols of sections of flat bundles, since this is the case for $T^{(j)}$. If we work modulo 2-torsion, the formula for T_D in (9.5) simplifies somewhat, viz.,

$$\begin{aligned}&T_D^{(N)}(\{\sigma^{(1)},\dots,\sigma^{(N)}\}) \equiv\\ &\quad\equiv \left\{\overline{\sigma}_0^{(2)},\dots,\overline{\sigma}_0^{(N)}\right\}^{(-1)^{N-1}\nu_D(\sigma^{(1)})}\left\{\overline{\sigma}_0^{(1)},T^{(N-1)}(\{\sigma^{(2)},\dots,\sigma^{(N)}\})\right\}\\ &\quad\equiv \prod_{j=1}^N\left\{\overline{\sigma}_0^{(1)},\dots,\widehat{\overline{\sigma}_0^{(j)}},\dots,\overline{\sigma}_0^{(N)}\right\}^{(-1)^{N-j}\nu_D(\sigma^{(j)})},\end{aligned}$$

where the latter equivalence modulo 2-torsion follows by induction. Since the real regulator is blind to torsion, it makes sense to redefine the Tame symbol by the formula:

$$\left.\begin{array}{l} T_D^{(N)}(\{\sigma^{(1)},\dots,\sigma^{(N)}\}) \\ \quad := \prod_{j=1}^N \left\{\overline{\sigma}_0^{(1)},\dots,\widehat{\overline{\sigma}_0^{(j)}},\dots,\overline{\sigma}_0^{(N)}\right\}^{(-1)^{N-j}\nu_D(\sigma^{(j)})}, \end{array}\right\} \tag{9.6}$$

and accordingly redefine

$$\underline{K}^M_{\bullet,Z} = \left\{\bigoplus_{j=0}^{\infty} \underline{K}^M_{j,Z}\right\} \Big/ \left\{\begin{array}{c}\text{2–torsion} \\ \text{subgroup}\end{array}\right\}. \tag{9.7}$$

Thus $\underline{K}^M_{j,Z}$ will now be interpreted as the corresponding group modulo 2-torsion. Note that $T^{(1)}$ is still the divisor map, $T^{(1)}$ and $\underline{T}$ in Proposition 9.5.1 both agree on $\underline{K}^M_{2,Z}$ (as we are working modulo 2-torsion), and that $T^{(1)}\circ T^{(2)} = 0$. Quite generally, we prove the following:

Proposition 9.6.3. *Assume given our modified definition of $T^{(N)}$ in (9.6) above.*

i) The definition of $T^{(N)}$ does not depend on the local equations defining the codimension one D's in Z.
ii) Up to 2-torsion, $T^{(N)}(R^N) \subset R^{N-1}$ for all N.
iii) $T^{(N)}\circ T^{(N+1)} = 0 \in \bigoplus \underline{K}^M_{N-1,Z}$ for all N.

Proof [Proof of Proposition 9.6.3.] The proof of Proposition 9.6.3 is a straightforward series of calculations. First of all, (i) is true for the same reasons as in the standard case in (9.4) above. If we let $\equiv$ have the meaning "modulo R^{N-1} and 2-torsion", then the proof of (ii) follows from the calculations:

$$T_D\{\sigma, -\sigma, \sigma^{(3)},\dots,\sigma^{(N)}\}$$
$$\equiv \prod_{j=3}^{N} \{\overline{\sigma}_0, -\overline{\sigma}_0, \dots, \widehat{\overline{\sigma}_0^{(j)}},\dots,\overline{\sigma}_0^{(N)}\}^{\nu_D(\sigma^{(j)}(-1)^{N-j}} \equiv 0.$$

Likewise,

$$T_D\{f, 1-f, \sigma^{(3)},\dots,\sigma^{(N)}\} \equiv 0.$$

Here we use the fact that for $D \subset |(f)|$, f itself is a local equation. Hence

$$(\overline{f}_0, \overline{[1-f]_0}) = (1, [\pm]1) := \begin{cases} (1,1) & \text{if } \nu_D(f) \geq 0 \\ (1,-1) & \text{if } \nu_D(f) < 0 \end{cases}.$$

To prove (iii), first observe that if $\sigma^{(1)}$, $\sigma^{(2)}$ are rational sections of a bundle L/Z, then

$$\operatorname{div}\circ T\{\sigma^{(1)},\sigma^{(2)}\} = 0.$$

This we proved earlier. This translates to saying that

$$\sum_{E\subset D\ (\subset Z)} \big[\nu_D(\sigma^{(2)})\nu_E(\overline{\sigma}_0^{(1)}) - \nu_D(\sigma^{(1)})\nu_E(\overline{\sigma}_0^{(2)})\big]\{E\} = 0. \tag{9.8}$$

We now consider $E\subset D\subset Z$ and compute:

$$T_E\circ T_D\{\sigma^{(1)},\dots,\sigma^{(N)}\} = \prod_{j=1}^N T_E\Big\{\overline{\sigma}_0^{(1)},\dots,\widehat{\overline{\sigma}_0^{(j)}},\dots,\overline{\sigma}_0^{(N)}\Big\}^{(-1)^{N-j}\nu_D(\sigma^{(j)})}$$

$$= \prod_{j=1}^N\Bigg(\prod_{i<j}\Big\{\overline{\overline{\sigma}}_{00}^{(1)},\dots,\widehat{\overline{\overline{\sigma}}_{00}^{(i)}},\dots,\widehat{\overline{\overline{\sigma}}_{00}^{(j)}},\dots,\overline{\overline{\sigma}}_{00}^{(N)}\Big\}^{(-1)^{i+j}\nu_D(\sigma^{(j)})\nu_E(\overline{\sigma}_0^{(i)})}$$

$$\times\prod_{i>j}\Big\{\overline{\overline{\sigma}}_{00}^{(1)},\dots,\widehat{\overline{\overline{\sigma}}_{00}^{(j)}},\dots,\widehat{\overline{\overline{\sigma}}_{00}^{(i)}},\dots,\overline{\overline{\sigma}}_{00}^{(N)}\Big\}^{(-1)^{i+j+1}\nu_D(\sigma^{(j)})\nu_E(\overline{\sigma}_0^{(i)})}\Bigg)$$

$$= \prod_{j=1}^N\Bigg(\prod_{i<j}\Big\{\overline{\overline{\sigma}}_{00}^{(1)},\dots,\widehat{\overline{\overline{\sigma}}_{00}^{(i)}},$$

$$\dots,\widehat{\overline{\overline{\sigma}}_{00}^{(j)}},\dots,\overline{\overline{\sigma}}_{00}^{(N)}\Big\}^{(-1)^{i+j}[\nu_D(\sigma^{(j)})\nu_E(\overline{\sigma}_0^{(i)})-\nu_D(\sigma^{(i)})\nu_E(\overline{\sigma}_0^{(j)})]}\Bigg).$$

By summing over $E\subset D\subset Z$ and using (9.8), this proves the proposition. □

It follows then that one has a twisted Milnor complex of the form

$$\underline{K}^M_{k-\bullet,X}:\quad \underline{K}^M_{k,X}\to\bigoplus_{\operatorname{codim}_X Z=1}\underline{K}^M_{k-1,Z}\to$$

$$\cdots\to\bigoplus_{\operatorname{codim}_X Z=k-j-1}\underline{K}^M_{j+1,Z}\to\bigoplus_{\operatorname{codim}_X Z=k-j}\underline{K}^M_{j,Z}\to\cdots$$

$$\cdots\to\bigoplus_{\operatorname{codim}_X Z=k-2}\underline{K}^M_{2,Z}\to\bigoplus_{\operatorname{codim}_X Z=k-1}\underline{K}^M_{1,Z}\to\bigoplus_{\operatorname{codim}_X Z=k}\underline{K}^M_{0,Z}\to 0. \tag{9.9}$$

Let

$$H^{k-m}(\underline{K}^M_{k-\bullet,X}) := \frac{\ker T^{(m)}:\bigoplus_{\operatorname{codim}_X Z=k-m}\underline{K}^M_{m,Z}\to\bigoplus_{\operatorname{codim}_X Z=k-m+1}\underline{K}^M_{m-1,Z}}{T^{(m+1)}\big(\bigoplus_{\operatorname{codim}_X Z=k-2}\underline{K}^M_{m+1,Z}\big)}.$$

Remarks 9.6.4.

i) One should expect the same story, viz., working with the unmodified definition of $T^{(N)}$ in (9.5) and not working modulo 2-torsion, for the existence of such a complex above. Again, since the details would appear to be more involved, and that our regulator is blind to torsion, we opted for the simpler definition of $H^{k-m}(\underline{K}^M_{k-\bullet,X})$.

ii) Warning: It would be pointless to attempt to sheafify the complex in (9.9), as it would lead to the Gersten-Milnor resolution of an untwisted Milnor sheaf (see 9.8).

iii) It is reasonably clear that there is a surjective map $H^k(\underline{K}^M_{k-\bullet,X}) \to \mathrm{CH}^k(X)/\mathrm{CH}^k_{\mathrm{alg}}(X)$, i.e. when $m=0$. [This is because $H^k(\underline{K}^M_{k-\bullet,X})$ is the free abelian group generated by cycles of codimension k in X, modulo the divisors of sections σ of flat line bundles L over codimension $k-1$ subvarieties Z in X. Since a flat line bundle L corresponds to a line bundle with trivial first Chern class on any desingularization $\tilde{Z} \overset{\approx}{\to} Z$, the divisors of such σ are *algebraically* equivalent to zero.] In particular, one has a surjective map $H^k(\underline{K}^M_{k-\bullet,X}) \to L_{n-k}H_{2(n-k)}(X)$, where $L_m H_\ell(X) := \pi_{\ell-2m}\mathcal{Z}_m(X)$, $\ell \geq 2m$, is Lawson homology. (Here: $\mathcal{Z}_m(X) = C_m(X) \times C_m(X)/\sim$, with $(x,y) \sim (x',y') \Leftrightarrow x+y' = y+x'$, and where $C_m(X)$ is the disjoint union of the Chow varieties of effective m-cycles of degree $d = 0,1,2,\ldots$.) With respect to an unmodified definition of $H^{k-m}(\underline{K}^M_{k-\bullet,X})$ as suggested in (i), we pose the following.

Question 9.6.5. Does there exist a map

$$H^{k-m}(\underline{K}^M_{k-\bullet,X}) \to L_{n-k}H_{2n-2k+m}(X) =: \pi_m(\mathcal{Z}_{n-k}(X)) = \pi_m(\mathcal{Z}^k(X))?$$

We now state the main result:

Theorem 9.6.6 (Main Theorem). *Assume $m \geq 1$. The current defined by*

$$\begin{array}{c}\prod_1^m \sigma_j \in \prod_1^m \mathrm{Rat}^*(L_j/Z) \\ \mathrm{codim}_X Z = k-m\end{array} \longmapsto \Bigg[\omega \mapsto \sum_{\ell=1}^m \int_Z (-1)^{\ell-1} \log\|\sigma_\ell\| (d\log\|\sigma_1\| \wedge \cdots \wedge \widehat{d\log\|\sigma_\ell\|} \wedge \cdots \wedge d\log\|\sigma_m\|) \wedge \omega\Bigg]$$

descends to a cohomological map

$$\underline{r}_{k,m} : H^{k-m}(\underline{K}^M_{k-\bullet,X}) \to \frac{\{H^{k-1,k-m}(X) \oplus H^{k-m,k-1}(X)\} \cap H^{2k-m-1}(X,\mathbb{R})}{D(k,m)},$$

where

$$D(k,m) = \begin{cases} H^{2k-2}_{\rm alg}(X,\mathbb{Q})\otimes\mathbb{R} & \textit{if } m=1\\ 0 & \textit{if } m>1.\end{cases}$$

That is, this map does not depend on the choice of flat metrics on the respective flat bundles. Moreover, assume given D*, smooth of dimension* $k-1$*, and a morphism* $f : D \to X$ *such that* $f(D)$ *is in "general position". If* $\omega \in H^{m-1,0}(D) \oplus H^{0,m-1}(D)$ *and* $\xi \in H^{k-m}(\underline{K}^M_{k-\bullet,X})$ *are given, then* $\underline{r}_{k,m}(\xi)(f_*\omega)$ *is induced by*

$$\begin{matrix}\prod_1^m \sigma_j \in \prod_1^m \mathrm{Rat}^*(L_j/Z) \\ \mathrm{codim}_X Z = k-m\end{matrix} \longmapsto$$

$$\Bigg[\omega \mapsto \sum_{\ell=1}^m \int_{Z\cap D:=f^{-1}(Z)} (-1)^{\ell-1}\log\|\sigma_\ell\|(d\log\|\sigma_1\|\wedge\cdots\wedge\widehat{d\log\|\sigma_\ell\|}\wedge$$

$$\cdots\wedge d\log\|\sigma_m\|)\wedge\omega\Bigg].$$

Furthermore, as a current acting on $E_X^{2n-2k+m+1}$*, if* $\underline{r}_{k,m}(\xi)$ *is* d*-closed, then it restricts to the zero class in* $H^{2k-m-1}(X-V)$ *where* $V=\bigcup_\alpha Z_\alpha$ *is the support of* $\xi = \sum_\alpha(\prod_1^m \sigma_{j,\alpha}, Z_\alpha) \in H^{k-m}(\underline{K}^M_{k-\bullet,X})$*. In this case,* $\underline{r}_{k,m}(\xi)$ *lies in the Hodge projected image* $N^{k-m}H^{2k-m-1}(X,\mathbb{R}) \to H^{k-1,k-m}(X)\oplus H^{k-m,k-1}(X)$*.*

Proof While it would be redundant to give a complete proof of this theorem, it is important to explain the new ingredients required to make the proof in the nontwisted case [Lew1] adaptable to the twisted situation. Firstly, the theorem is already proven in the case $m=1$, this being the import of §9.3–9.5. Next, if $\|\sigma_\ell\| \in \mathbb{R}^\times$ is constant for some ℓ, then some standard estimates together with a Stokes' theorem argument implies that the corresponding regulator value on closed forms ω is zero for $m\geq 2$. This leads to independence of the flat metric, after quotienting out by $D(k,m)$, for $m \geq 1$. To show for example that $\{(\sigma,-\sigma,\sigma_3,\dots,\sigma_m),Z\}$ or say $\{(f,1-f,\sigma_3,\dots,\sigma_m),Z\}$ goes to zero under the real regulator, amounts to reducing to the case where the σ_j's are rational functions via some finite covering $Z'\to Z$, similar to what we did earlier in 9.5.5, and then applying the same arguments given in [Lew1]. That the current given in Thm 9.6.6 is $\partial\bar\partial$-closed now follows from the same proof as given in [Lew1]. Finally, the latter statement of the theorem is rather easy to prove. Being d-closed implies that we have a cohomology class on X, which restricts to a cohomology class on $X-V$; moreover the current clearly vanishes on those forms compactly supported on $X-V$. □

9.7 Some examples

In contrast to the various vanishing results in the literature for the regulator in the nontwisted case [MS1, Ke1, Ke2, Co], etc., we exhibit some nonvanishing regulator results in the twisted case.

9.7.1. Regulator on $H^0(\underline{K}^M_{2-\bullet,X})$ for a curve X.
Let X be a compact Riemann surface of genus $g \geq 1$, and let $f, g \in \mathbb{C}(X)^\times$ be given. Write $T\{f,g\} = \sum_{j=1}^M (c_j, p_j)$, where $p_j \in X$ and $c_j \in \mathbb{C}^\times$. Then $\prod_{j=1}^M c_j = 1$ by Weil reciprocity. Now fix $p \in X$, and let L_j be a choice of line bundle corresponding to the zero cycle $p_j - p$. Since $\deg(p_j - p) = 0$, L_j is a flat line bundle. There exists rational sections $\{\sigma_j\}$ of the flat bundles $\{L_j\}$ over X such that $\mathrm{div}(\sigma_j) = p_j - p$. Thus one can easily check by a Tame symbol calculation that $\xi := \{f,g\}\prod_j\{\sigma_j, c_j\} \in H^0(\underline{K}^M_{2-\bullet,X})$. Note that

$$d\big(\log|f|\log|g|\big) = \log|f| d\log|g| + \log|g| d\log|f|,$$

and by a Stokes' theorem argument together with standard estimates,

$$\int_X \log|f| d\log|g| \wedge \omega = -\int_X \log|g| d\log|f| \wedge \omega, \tag{9.10}$$

for any d-closed form $\omega \in E^1_{X,\mathbb{R}}$. Thus if either f or g were constant, then both integrals in (9.10) would vanish. Applying the same reasoning to the terms $\prod_j\{\sigma_j, c_j\}$, it easily follows that

$$\underline{r}_{2,2}(\xi)(\omega) = \int_X \bigg[\log|f| d\log|g| - \log|g| d\log|f|\bigg] \wedge \omega, \tag{9.11}$$

namely the contribution of the terms $\prod_j\{\sigma_j, c_j\}$ to the regulator current is zero. One can always find f and g [Lew1] such that the computation in (9.11) is nonzero for general X. For a simple example, consider this. Let E be a general elliptic curve, D another general curve, and $X \subset E \times D$ a general hyperplane section. Since X dominates E, and since the real regulator for E is nontrivial [Blo3], such an f and g can be found for X via pullback of corresponding rational functions on E, and then by continuity, the same story will hold as X varies with general moduli. Thus in summary, one can find general curves of genus $g \gg 1$ for which the regulator

$$\underline{r}_{2,2} : H^0(\underline{K}^M_{2-\bullet,X}) \to H^1(X, \mathbb{R}),$$

is nontrivial. This is in complete constrast to the situation of the regulator in (9.1), viz., $r_{2,2} : \mathrm{CH}^2(X,2) \simeq H^0_{\mathrm{Zar}}(X, \mathcal{K}_{2,X}) \to H^1(X,\mathbb{R})$, where it is known ([Co]) that $r_{2,2}$ is trivial for sufficiently general X of genus $g > 1$.

Indeed, one can naively carry out the same construction above to arrive at a class $\xi \in H^0_{\text{Zar}}(X, \mathcal{K}_{2,X})$ arising from rational functions f, g on X with $T\{f,g\} = \sum_{j=1}^{M}(c_j, p_j)$, where $p_j \in X$ and $c_j \in \mathbb{C}^\times$ and

$$\int_X \Big[\log|f| d\log|g| - \log|g| d\log|f|\Big] \wedge \omega \neq 0.$$

The issue boils down to finding a *rational* functions h_j on X for which $\text{div}(h_j) = N(p_j - p)$, for some integer $N \neq 0$, which is not in general possible. In fact the difficulty of finding such h_j amounts to finding torsion points on X, and is related to the known affirmative answer to the Mumford-Manin conjecture. For a discussion of the relation of the Mumford-Manin conjecture to this regulator calculation, the reader can consult [Lew1], as well as the references cited there.

9.7.2. Regulator on $H^1(\underline{K}^M_{1-\bullet,X})$ for a surface X.

For a simple example of a nontrivial twisted regulator calculation, where the usual regulator vanishes, consider the case $X = M \times N$, where M and N are smooth curves. If M and N are sufficiently general with $g(M)g(N) \geq 2$, then the image of the regulator in (9.1) vanishes (modulo the group of algebraic cocycles) [C-L1]. Now suppose we are given a curve $C \subset M \times N$, $f \in \mathbb{C}(C)^\times$, and $\omega \in H^1(M,\mathbb{C}) \otimes H^1(N,\mathbb{C}) \bigcap H^{1,1}(X,\mathbb{R}(1))$ for which

$$\int_C \omega \log|f| \neq 0. \tag{9.12}$$

Such a situation in (9.12) is fairly easy to arrive at for general M and N, by a deformation from a special case situation. (For example, one can use a construction in [Lew2, §7], or $M \times N$ can be a general deformation of a product of 2 curves dominating a general Abelian surface, together with the main results of [C-L2].) Write

$$\text{div}(f)_C = \sum_{j=1}^{m}[(p_j, q_j) - (s_j, t_j)].$$

We can write

$$(p_j, q_j) - (s_j, t_j) = [(p_j, q_j) - (p_j, t_j)] + [(p_j, t_j) - (s_j, t_j)].$$

Now put $D_j = \{p_j\} \times N$ and $K_j = M \times \{t_j\}$, and observe that the degree zero divisor $(p_j, t_j) - (p_j, q_j)$ on D_j corresponds to a flat line bundle on D_j, and likewise the degree zero divisor $(s_j, t_j) - (p_j, t_j)$ on K_j corresponds to a flat line bundle on K_j. Consider Cartier divisors σ_j on D_j, η_j on K_j, i.e.

rational sections of the respective flat line bundles, with

$$\operatorname{div}(\sigma_j) = (p_j, t_j) - (p_j, q_j),$$
$$\operatorname{div}(\eta_j) = (s_j, t_j) - (p_j, t_j).$$

Then

$$\xi := (f, C) + \sum_{j=1}^{m} (\sigma_j, D_j) + \sum_{j=1}^{m} (\eta_j, K_j) \in H^1(\underline{K}^M_{1-\bullet,X}).$$

[Note: As in the previous example, observe that one cannot replace the σ_j's (resp. η_j's) by rational functions on the D_j's (resp. K_j's), even if one replaces $(p_j, t_j) - (p_j, q_j)$ and $(s_j, t_j) - (p_j, t_j)$ by nonzero integral multiples.] Moreover the pullback of ω to D_j and K_j is zero. Thus:

$$\underline{r}(\xi)(\omega) = \int_C \omega \log |f| \neq 0.$$

Finally, observe that for general M and N, $\{H^1(M, \mathbb{Q}) \otimes H^1(N, \mathbb{Q})\} \cap H^2_{\text{alg}}(M \times N, \mathbb{Q}) = 0$. Thus while the image of the regulator in (9.1) for $M \times N$ vanishes (modulo the group of algebraic cocycles), the twisted regulator does not vanish.

9.8 Comparison to usual Milnor K-cohomology

Recall that $\overline{\mathcal{K}}^M_{k,X} = \text{Image}\left(\mathcal{K}^M_{k,X} \to K^M_k(\mathbb{C}(X))\right)$. From the works of (Elbaz-Vincent/Müller-Stach, 1998) and (Gabber, 1992) [E-M], there is a flasque Gersten resolution for Milnor K-theory, viz.,

$$0 \to \overline{\mathcal{K}}^M_{k,X} \to K^M_k(X) \to \bigoplus_{\operatorname{codim}_X Z=1} K^M_{k-1}(Z) \to \cdots$$
$$\to \bigoplus_{\operatorname{codim}_X Z=k-1} K^M_1(Z) \to \bigoplus_{\operatorname{codim}_X Z=k} K^M_0(Z) \to 0.$$

The maps in this complex are the [higher] Tame symbols. Taking Zariski cohomologies, we arrive at natural maps

$$H^{k-m}_{\text{Zar}}(X, \overline{\mathcal{K}}^M_{k,X}) \to H^{k-m}(\underline{K}^M_{k-\bullet,X}),$$
$$H^{k-m}_{\text{Zar}}(X, \mathcal{K}^M_{k,X}) \to H^{k-m}_{\text{Zar}}(X, \overline{\mathcal{K}}^M_{k,X}),$$

and hence a map

$$H^{k-m}_{\text{Zar}}(X, \mathcal{K}^M_{k,X}) \to H^{k-m}(\underline{K}^M_{k-\bullet,X}). \tag{9.13}$$

The regulator $\underline{r}_{k,m}$ on $H^{k-m}(\underline{K}^M_{k-\bullet,X})$ pulls back to a corresponding regulator on $H^{k-m}_{\mathrm{Zar}}(X, \mathcal{K}^M_{k,X})$ via the map in (9.13). This corresponding regulator on $H^{k-m}_{\mathrm{Zar}}(X, \mathcal{K}^M_{k,X})$ is induced by the regulator in [Lew1]. The regulator in [Lew1] coincides, up to a real isomorphism on cohomology, (and up to a normalizing constant), with the regulator in (9.1) for $m = 1,\ 2$. But from the examples in §9.7, it is clear that the map in (9.13) cannot be surjective, for otherwise one would have the vanishing of the twisted regulator in the cases where the regulator in (9.1) vanishes. Secondly, for $m = 0$, we have $H^k_{\mathrm{Zar}}(X, \overline{\mathcal{K}}^M_{k,X}) \simeq \mathsf{CH}^k(X)$, whereas from Remarks 9.6.4(iii), $H^k(\underline{K}^M_{k-\bullet,X})$ looks more like $\mathsf{CH}^k(X)/\mathsf{CH}^k_{\mathrm{alg}}(X)$. Thus the map in (9.13) cannot be injective either. It would be nice to have a more precise description of the image and kernel of the map in (9.13), and a possible connection between $H^{k-m}(\underline{K}^M_{k-\bullet,X})$ and Lawson homology as raised in Question 9.6.5. However that line of enquiry will not be pursued here.

References

[BT] Bass, H. and J. Tate: The Milnor ring of a global field, in *Algebraic K-theory II* Lecture Notes in Math. **342** Springer-Verlag (1972) 349–446

[Bei] Beilinson, A.: Higher regulators and values of L-functions J. Soviet Math. **30** (1985) 2036–2070.

[Blo1] Bloch, S.: Algebraic cycles and higher K-theory Adv. Math. **61** (1986) 267–304

[Blo2] ———: Algebraic cycles and the Beilinson conjectures, Contemporary Mathematics Vol. **58**, Part I (1986) 65–79

[Blo3] ———, *Lectures on Algebraic Cycles* Duke University Mathematics Series **IV** (1980)

[C-L1] Chen, X. and J. D. Lewis: Noether-Lefschetz for K_1 of a certain class of surfaces, Bol. Soc. Mexicana (3) **10** (2004)

[C-L2] ————: The Hodge-$\mathfrak{D}$-conjecture for K_3 and Abelian surfaces J. Alg. Geometry **14**, (2005) 213–240.

[Co] Collino, A.: Griffiths' infinitesimal invariant and higher K-theory on hyperelliptic jacobians J. Algebraic Geometry **6** (1997) 393–415

[E-M] Elbaz-Vincent, P. and S. Müller-Stach: Milnor K-theory, higher Chow groups and applications Invent. math. **148** (2002) 177–206

[EV] Esnault, H. and E. Viehweg: Deligne-Beilinson cohomology, in *Beilinson's Conjectures on Special Values of L-Functions* (Rapoport, Schappacher, Schneider, eds.) Perspectives in Math. **4** Academic Press, San Diego (1988) 43–91

[GL] Gordon B. and J. Lewis: Collaboration.

[GH] Griffiths, P. and J. Harris: *Principles of Algebraic Geometry* John Wiley & Sons, New York (1978)

[Ja] Jannsen, U.: Deligne cohomology, Hodge-$\mathfrak{D}$-conjecture, and motives, in *Beilinson's Conjectures on Special Values of L-Functions* (Rapoport, Schappacher,

Schneider, eds.) Perspectives in Math. **4** Academic Press, San Diego (1988) 305–372.

[Ka] Kato, K.: Milnor K-theory and the Chow group of zero cycles, in *Applications of K-theory to Algebraic Geometry and Number Theory, Part I* Contemp. Math. **55** (1986) 241–253

[Ke1] Kerr, M.: Geometric construction of regulator currents with applications to algebraic cycles, Princeton University Thesis, (2003)

[Ke2] ———: A regulator formula for Milnor K-groups. K-Theory **29** (2003) 175–210

[Lev] Levine, M.: Localization on singular varieties Invent. Math. **31** (1988) 423–464

[Lew1] Lewis, J.: Real regulators on Milnor complexes K-Theory **25** (2002) 277–298

[Lew2] ———, Regulators of Chow cycles on Calabi-Yau varieties, in *Calabi-Yau Varieties and Mirror Symmetry* (N. Yui, J. D. Lewis, eds.), Fields Institute Communications **38**, (2003), 87–117

[MS1] Müller-Stach, S.: Constructing indecomposable motivic cohomology classes on algebraic surfaces J. Algebraic Geometry **6** (1997) 513–543

[MS2] ———, Algebraic cycle complexes, in *Arithmetic and Geometry of Algebraic Cycles*, (Gordon, Lewis, Müller-Stach, S. Saito, Yui, eds.), Kluwer Academic Publishers, Dordrecht, The Netherlands, 2000, 285–305

[MS3] ———: A remark on height pairings, in *Algebraic Cycles and Hodge Theory, Torino, 1993* (A. Albano, F. Bardelli, eds.), Lecture Notes in Mathematics **1594** Springer-Verlag (1994), 253–259

[Sou] Soulé, C.: *Lectures on Arakelov Geometry* Cambridge Studies in Advanced Mathematics **33** Cambridge University Press, Cambridge, England (1992)

10

Chow-Künneth decomposition for universal families over Picard modular surfaces

A. Miller

Math. Inst. Univ. Heidelberg,
Im Neuenheimer Feld 288, 69120 Heidelberg
miller@mathi.uni-heidelberg.de

S. Müller-Stach

Math. Inst., Johannes Gutenberg Univ. Mainz,
Staudingerweg 9, 55099 Mainz
mueller-stach@mathematik.uni-mainz.de

S. Wortmann

Math. Inst. Univ. Heidelberg,
Im Neuenheimer Feld 288, 69120 Heidelberg
wortmann@mathi.uni-heidelberg.de

Y.-H.Yang

Max Planck Inst. für Mathematik,
Inselstrasse 22, 04103 Leipzig
yhyang@mail.tongji.edu.cn, yi-hu.yang@mis.mpg.de

K. Zuo

Math. Inst., Johannes Gutenberg Univ. Mainz,†
Staudingerweg 9, 55099 Mainz
!kzuo@mathematik.uni-mainz.de

Dedicated to Jaap Murre

Abstract

We prove existence results for Chow–Künneth projectors on compactified universal families of Abelian threefolds with complex multiplication over a particular Picard modular surface studied by Holzapfel. Our method builds up on the approach of Gordon, Hanamura and Murre in the case of Hilbert modular varieties. In addition we use relatively complete models in the sense of Mumford, Faltings and Chai and prove vanishing results for L^2–Higgs cohomology groups of certain arithmetic subgroups in $SU(2,1)$ which are not cocompact.

† Supported by: DFG Schwerpunkt–Programm, DFG China Exchange program, NSF of China (grant no. 10471105), Max–Planck Gesellschaft.
‡ 1991 AMS Subject Classification 14C25
Keywords: Chow motive, Higgs bundle, Picard modular surface

10.1 Introduction

In this paper we discuss conditions for the existence of absolute Chow-Künneth decompositions for families over Picard modular surfaces and prove some partial existence results. In this way we show how the methods of Gordon, Hanamura and Murre [12] can be slightly extended to some cases but fail in some other interesting cases. Let us first introduce the circle of ideas which are behind Chow–Künneth decompositions. For a general reference we would like to encourage the reader to look into [26] which gives a beautiful introduction to the subject and explains all notions we are using.

Let Y be a smooth, projective k–variety of dimension d and H^* a Weil cohomology theory. In this paper we will mainly be concerned with the case $k = \mathbb{C}$, where we choose singular cohomology with rational coefficients as Weil cohomology. Grothendieck's Standard Conjecture C asserts that the Künneth components of the diagonal $\Delta \subset Y \times Y$ in the cohomology $H^{2d}(Y \times Y, \mathbb{Q})$ are algebraic, i.e., cohomology classes of algebraic cycles. In the case $k = \mathbb{C}$ this follows from the Hodge conjecture. Since Δ is an element in the ring of correspondences, it is natural to ask whether these algebraic classes come from algebraic cycles π_j which form a complete set of orthogonal idempotents

$$\Delta = \pi_0 + \pi_1 + \ldots + \pi_{2d} \in CH^d(Y \times Y)_{\mathbb{Q}}$$

summing up to Δ. Such a decomposition is called a *Chow–Künneth decomposition* and it is conjectured to exist for every smooth, projective variety. One may view π_j as a Chow motive representing the projection onto the j-th cohomology group in a universal way. There is also a corresponding notion for k–varieties which are relatively smooth over a base scheme S. See section 10.3, where also Murre's refinement of this conjecture with regard to the Bloch–Beilinson filtration is discussed. Chow–Künneth decompositions for abelian varieties were first constructed by Shermenev in 1974. Fourier–Mukai transforms may be effectively used to write down the projectors, see [18, 26]. The cases of surfaces was treated by Murre [27], in particular he gave a general method to construct the projectors π_1 and π_{2d-1}, the so–called Picard and Albanese Motives. Aside from other special classes of 3–folds [1] not much evidence is known except for some classes of modular varieties. A fairly general method was introduced and exploited recently by Gordon, Hanamura and Murre, see [12], building up on previous work by Scholl and their own. It can be applied in the case where one has a modular parameter space X together with a universal family $f : A \to X$ of abelian varieties with possibly some additional structure. Examples are given by elliptic and Hilbert modular varieties. The goal of this paper was to extend the range of examples to the case of Picard modular surfaces, which are uniformized by a ball, instead of a product of upper half planes. Let us now describe the general strategy of Gordon, Hanamura and Murre so that we can understand to what extent this approach differs and eventually fails for a general Picard modular surface with sufficiently high level structure.

Let us assume that we have a family $f : a \to X$ of abelian varieties over X. Since all fibers are abelian, we obtain a relative Chow–Künneth decomposition over X in the sense of Deninger/Murre [6], i.e., algebraic cycles Π_j in $a \times_X a$ which sum up to $\Delta \times_X \Delta$. One may view Π_j as a projector related to $R^j f_* \mathbb{C}$. Now let $\overline{f} : \overline{a} \to \overline{X}$ be a compactification of the family. We will use the language of perverse sheaves from [3] in particular also the notion of a stratified map. In [11] Gordon, Hanamura and Murre have introduced the *Motivic Decomposition Conjecture*:

Conjecture 10.1.1. *Let $\overline{a}$ and $\overline{X}$ be quasi-projective varieties over $\mathbb{C}$, $\overline{a}$ smooth, and $\overline{f} : \overline{a} \to \overline{X}$ a projective map. Let $\overline{X} = X_0 \supset X_1 \supset \ldots \supset X_{\dim(X)}$ be a stratification of $\overline{X}$ so that $\overline{f}$ is a stratified map. Then there are local systems $\mathcal{V}^j_\alpha$ on $X^0_\alpha = X_\alpha - X_{\alpha-1}$, a complete set Π^j_α of orthogonal projectors and isomorphisms*

$$\sum_{j,\alpha} \Psi^j_\alpha : \mathbb{R}\overline{f}_* \mathbb{Q}_{\overline{a}} \xrightarrow{\sim} \bigoplus_{j,\alpha} IC_{X_\alpha}(\mathcal{V}^j_\alpha)[-j - \dim(X_\alpha)]$$

in the derived category.

This conjecture asserts of course more than a relative Chow–Künneth decomposition for the smooth part f of the morphism $\overline{f}$. Due to the complicated structure of the strata in general its proof in general needs some more information about the geometry of the stratified morphism $\overline{f}$. In the course of their proof of the Chow–Künneth decomposition for Hilbert modular varieties, see [12], Gordon, Hanamura and Murre have proved the motivic decomposition conjecture in the case of toroidal compactifications for the corresponding universal families. However to complete their argument they need the vanishing theorem of Matsushima–Shimura [21]. This theorem together with the decomposition theorem [3] implies that each relative projector Π^j on the generic stratum X_0 only contributes to one cohomology group of A and therefore, using further reasoning on boundary strata X_α, relative projectors for the family f already induce absolute projectors.

The plan of this paper is to extend this method to the situation of Picard modular surfaces. These were invented by Picard in his study of the family of curves (called Picard curves) with the affine equation

$$y^3 = x(x-1)(x-s)(x-t).$$

The Jacobians of such curves of genus 3 have some additional CM–structure arising from the $\mathbb{Z}/3\mathbb{Z}$ deck transformation group. Picard modular surfaces are certain two dimensional ball quotients $X = \mathbb{B}_2/\Gamma$ and form a particular beautiful set of Shimura surfaces in the moduli space of abelian varieties of dimension 3. A nice class of Picard Modular Surfaces are the above mentioned Jacobians of Picard curves. Many examples are known through the work of Holzapfel [15, 16]. Unfortunately the generalization of the vanishing theorem of Matsushima and Shimura does not hold for Picard modular surfaces and their compactifications. The reason is that $\mathbb{B}_2$ is a homogenous space for the Lie group $G = SU(2,1)$ and general vanishing theorems like Ragunathan's theorem [4, pg. 225] do not hold. If $\mathbb{V}$ is an irreducible, non–trivial representation of any arithmetic subgroup Γ of G, then the intersection cohomology group $H^1(X, \mathbb{V})$ is frequently non–zero, whereas in order to make the method of Gordon, Hanamura and Murre work, we would need its vanishing. This happens frequently for small Γ, i.e., high level. However if Γ is sufficiently big, i.e., the level is small, we can sometimes expect some vanishing theorems to hold. This is the main reason why we concentrate our investigations on one particular example of a Picard modular surface $\overline{X}$ in section 10.4. The necessary vanishing theorems are proved by using Higgs bundles and their L^2–cohomology in section 10.6. Such techniques provide a new method to compute intersection cohomology in cases where the geometry is known. This methods uses a recent proof of the Simpson correspondence in the non–compact case by Jost, Yang and Zuo [17, Thm. A/B]. But even in the case of our chosen surface $\overline{X}$ we are not able to show the complete vanishing result which would be necessary to proceed with the argument of Gordon, Hanamura and Murre. We are however able to prove the existence of a partial set $\pi_0, \pi_1, \pi_2, \pi_3, \pi_7, \pi_8, \pi_9, \pi_{10}$ of orthogonal idempotents under the assumption of the motivic decomposition conjecture 10.1.1 on the universal family $\overline{A}$ over $\overline{X}$:

Theorem 10.1.2. *Assume the motivic decomposition conjecture 10.1.1 for* $\overline{f} : \overline{a} \to \overline{X}$. *Then* $\overline{a}$ *supports a partial set of Chow–Künneth projectors* π_i *for* $i \neq 4, 5, 6$.

Unfortunately we cannot prove the existence of the projectors π_4, π_5, π_6 due to the non–vanishing of a certain L^2–cohomology group, in our case $H^1(X, S^2\mathbb{V}_1)$, where $\mathbb{V}_1$ is (half of) the standard representation. This is special to $SU(2,1)$ and therefore the proposed method has no chance to go through for other examples involving ball quotients. If $H^1(X, S^2\mathbb{V}_1)$ would vanish or consist out of algebraic Hodge $(2,2)$–classes only, then we would obtain a complete Chow–Künneth decomposition. This is an interesting open question and follows from the Hodge conjecture, if all classes in $H^1(X, S^2\mathbb{V}_1)$ would have Hodge type $(2,2)$. We also sketch how to prove the motivic decomposition conjecture in this particular case, see section 10.7.2, however details will be published elsewhere. This idea generalizes the method from [12], since the fibers over boundary points are not anymore toric varieties, but toric bundles over elliptic curves. We plan to publish the full details in a forthcoming publication and prefer to assume the motivic decomposition conjecture 10.1.1 in this paper.

The logical structure of this paper is as follows:

In section 10.2 we present notations, definitions and known results concerning Picard Modular surfaces and the universal Abelian schemes above them. Section 10.3 first gives a short introduction to Chow Motives and the Murre Conjectures and then proceeds to our case in §10.3.2. The remainder of the paper will then be devoted to the proof of Theorem 10.1.2:

In section 10.4 we give a description of toroidal degenerations of families of Abelian threefolds with complex multiplication. In section 10.5 we describe the geometry of a class of Picard modular surfaces which have been studied by Holzapfel. In section 10.6 we prove vanishing results for intersection cohomology using the non–compact Simpson type correspondence between the L^2–Higgs cohomology of the underlying VHS and the L^2–de Rham cohomology resp. intersection cohomology of local systems. In section 10.7 everything is put together to prove the main theorem 10.1.2. The appendix (section 10.8) gives an explicit description of the L^2–Higgs complexes needed for the vanishing results of section 10.6.

10.2 The Picard modular surface

In this section we are going to introduce the (non–compact) Picard modular surfaces $X = X_\Gamma$ and the universal abelian scheme $\mathcal{A}$ of fibre dimension 3 over X. For proofs and further references we refer to [9].

Let E be an imaginary quadratic field with ring of integers $\mathcal{O}_E$. The Picard modular group is defined as follows. Let V be a 3-dimensional E-vector space and $L \subset V$ be an $\mathcal{O}_E$-lattice. Let $J : V \times V \to E$ be a nondegenerate Hermitian form of signature $(2,1)$ which takes values in $\mathcal{O}_E$ if it is restricted to $L \times L$. Now let $G' = \mathrm{SU}(J,V)/\mathbb{Q}$ be the special unitary group of (V,ϕ). This is a semisimple algebraic group over $\mathbb{Q}$ and for any $\mathbb{Q}$-algebra A its group of A-rational points is

$$G'(A) = \{g \in \mathrm{SL}(V \otimes_{\mathbb{Q}} A) \mid J(gu, gv) = J(u, v), \text{ for all } u, v \in V \otimes_{\mathbb{Q}} A\}.$$

In particular one has $G'(\mathbb{R}) \simeq \mathrm{SU}(2,1)$. The symmetric domain $\mathcal{H}$ associated to $G'(\mathbb{R})$ can be identified with the complex 2-ball as follows. Let us fix once and for all an embedding $E \hookrightarrow \mathbb{C}$ and identify $E \otimes_{\mathbb{Q}} \mathbb{R}$ with $\mathbb{C}$. This gives $V(\mathbb{R})$ the structure of a 3-dimensional $\mathbb{C}$-vector space and one may choose a basis of $V(\mathbb{R})$ such that the form J is represented by the diagonal matrix $[1, 1, -1]$. As $\mathcal{H}$ can be identified with the (open) subset of the Grassmannian $\mathrm{Gr}_1(\mathrm{V}(\mathbb{R}))$ of complex lines on which J is negative definite, one has

$$\mathcal{H} \simeq \{(Z_1, Z_2, Z_3) \in \mathbb{C}^3 \mid |Z_1|^2 + |Z_2|^2 - |Z_3|^2 < 0\}/\mathbb{C}^*.$$

This is contained in the subspace, where $Z_3 \neq 0$ and, switching to affine coordinates, can be identified with the complex 2-ball

$$\mathbb{B} = \{(z_1, z_2) \in \mathbb{C}^2 \mid |z_1|^2 + |z_2|^2 < 1\}.$$

Using this description one sees that $G'(\mathbb{R})$ acts transitively on $\mathbb{B}$.

The Picard modular group of E is defined to be $G'(\mathbb{Z}) = \mathrm{SU}(J, L)$, i.e. the elements $g \in G'(\mathbb{Q})$ with $gL = L$. It is an arithmetic subgroup of $G(\mathbb{R})$ and acts properly discontinuously on $\mathbb{B}$. The same holds for any commensurable subgroup $\Gamma \subset G'(\mathbb{Q})$, in particular if $\Gamma \subset G'(\mathbb{Z})$ is of finite index the quotient $X_\Gamma(\mathbb{C}) = \mathbb{B}/\Gamma$ is a non-compact complex surface, the Picard modular surface. Moreover, for torsionfree Γ it is smooth.

We want to describe $X_\Gamma(\mathbb{C})$ as moduli space for polarized abelian 3-folds with additional structure. For this we will give a description of $X_\Gamma(\mathbb{C})$ as the identity component of the Shimura variety $S_K(G, \mathcal{H})$.

Let $G = \mathrm{GU}(J, V)/\mathbb{Q}$ be the reductive algebraic group of unitary similitudes of J, i.e. for any $\mathbb{Q}$-algebra A

$$G'(A) = \{g \in \mathrm{GL}(V \otimes_{\mathbb{Q}} A) \mid \text{there exists } \mu(g) \in A^* \text{ such that}$$
$$J(gu, gv) = \mu(g) J(u, v), \text{ for all } u, v \in V \otimes_{\mathbb{Q}} A\}.$$

As usual $\mathbb{A}$ denotes the $\mathbb{Q}$-adeles and $\mathbb{A}_f$ denotes the finite adeles. Let K be a compact open subgroup of $G(\mathbb{A}_f)$, which is compatible with the integral structure defined by the lattice L. I.e., K should be in addition a subgroup of finite index in $G(\hat{\mathbb{Z}}) := \{g \in G(\mathbb{A}_f) \mid g(L \otimes_{\mathbb{Z}} \hat{\mathbb{Z}}) = L \otimes_{\mathbb{Z}} \hat{\mathbb{Z}}\}$. Then one can define

$$S_K(G, \mathcal{H})(\mathbb{C}) := G(\mathbb{Q}) \backslash \mathcal{H} \times G(\mathbb{A}_f)/K.$$

This can be decomposed as $S_K(G, \mathcal{H})(\mathbb{C}) = \coprod_{j=1}^{n(K)} X_{\Gamma_j}(\mathbb{C})$.

The variety $S_K(G, \mathcal{H})(\mathbb{C})$ has an interpretation as a moduli space for certain 3-dimensional abelian varieties. Recall that over $\mathbb{C}$ an abelian variety A is determined by the following datum: a real vector space $W(\mathbb{R})$, a lattice $W(\mathbb{Z}) \subset W(\mathbb{R})$, and a complex strucuture $j : \mathbb{C}^\times \to \mathrm{Aut}_{\mathbb{R}}(W(\mathbb{R})))$, for which there exists a nondegenerate $\mathbb{R}-$bilinear skew-symmetric form ψ : $W(\mathbb{R}) \times W(\mathbb{R}) \to \mathbb{R}$ taking values in $\mathbb{Z}$ on $W(\mathbb{Z})$ such that the form given by $(w, w') \mapsto \psi(j(i)w, w')$ is symmetric and positive definite. The form ψ is called a Riemann form and two forms ψ_1, ψ_2 are called equivalent if there exist $n_1, n_2 \in \mathbb{N}_{>0}$ such that $n_1\psi_1 = n_2\psi_2$. An equivalence class of Riemann forms is called a homogeneous polarization of A.

An endomorphism of a complex abelian variety is an element of $\mathrm{End}_{\mathbb{R}}(W(\mathbb{R}))$ preserving $W(\mathbb{Z})$ and commuting with $j(z)$ for all $z \in \mathbb{C}^\times$. A homogenously polarized abelian variety $(W(\mathbb{R}), W(\mathbb{Z}), j, \psi)$ is said to have complex multiplication by an order $\mathcal{O}$ of E if and only if there is a homomorphism $m : \mathcal{O} \to \mathrm{End}(A)$ such that $m(1) = 1$, and which is compatible with ψ, i.e. $\psi(m(\alpha^\rho)w, w') = \psi(w, m(\alpha)w')$ where ρ is the Galois automorphism of E induced by complex conjugation (via our fixed embedding $E \hookrightarrow \mathbb{C}$.) We shall only consider the case $\mathcal{O} = \mathcal{O}_E$ in the following.

One can define the signature of the complex multiplication m, resp. the abelian variety $(W(\mathbb{R}), W(\mathbb{Z}), j, \psi, m)$ as the signature of the hermitian form $(w, w') \mapsto \psi(w, iw') + i\psi(w, w')$ on $W(\mathbb{R})$ with respect to the complex structure imposed by m via $\mathcal{O} \otimes_{\mathbb{Z}} \mathbb{R} \simeq \mathbb{C}$. We write $m_{(s,t)}$ if m has signature (s, t).

Finally for any compact open subgroup $K \subset G(\hat{\mathbb{Z}})$ as before one has the notion of a level-K structure on A. For a positive integer n we denote by $A_n(\mathbb{C})$ the group of points of order n in $A(\mathbb{C})$. This group can be identified with $W(\mathbb{Z}) \otimes \mathbb{Z}/n\mathbb{Z}$ and taking the projective limit over the system $(A_n(\mathbb{C}))_{n \in \mathbb{N}_{>0}}$ defines the Tate module of A:

$$T(A) := \varprojlim A_n(\mathbb{C}) \simeq W(\mathbb{Z}) \otimes \hat{\mathbb{Z}}.$$

Now two isomorphisms $\varphi_1, \varphi_2 : W(\mathbb{Z}) \otimes \hat{Z} \simeq L \otimes \hat{\mathbb{Z}}$ are called K-equivalent if there is a $k \in K$ such hat $\varphi_1 = k\varphi_2$ and a K-level structure on A is just a K-equivalence class of these isomorphisms.

Proposition 10.2.1. *For any compact open subgroup $K \subset G(\hat{\mathbb{Z}})$ there is a one-to-one correspondence between*

(i) the set of points of $S_K(G, \mathcal{H})(\mathbb{C})$ and
(ii) the set of isomorphism classes of $(W(\mathbb{R}), W(\mathbb{Z}), j, \psi, m_{(2,1),\varphi})$ as above.

Proof [9, Prop.3.2] □

Remark 10.2.2. If we take

$$K_N := \{g \in G(\mathbb{A}_f) \mid (g-1)(L \otimes_{\mathbb{Z}} \hat{\mathbb{Z}}) \subset N \cdot (L \otimes_{\mathbb{Z}} \hat{\mathbb{Z}})\},$$

then a level-K structure is just the usual level-N structure, namely an isomorphism $A_N(\mathbb{C}) \to L \otimes \mathbb{Z}/N\mathbb{Z}$. Moreover $K_N \subset G(\mathbb{Q}) = \Gamma_N$, where Γ_N is the principal congruence subgroup of level N, i.e. the kernel of the canonical map $G'(\mathbb{Z}) \to G'(\mathbb{Z}/N\mathbb{Z})$. In this case the connected component of the identity of $S_{K_N}(G, \mathcal{H})$ is exactly $X_{\Gamma_N}(\mathbb{C})$.

We denote with $\mathcal{A}_\Gamma$ the universal abelian scheme over $X_\Gamma(\mathbb{C})$. In section 10.4 the compactifications of these varieties will be explained in detail. For the time being we denote them with $\overline{X}_\Gamma$ and $\overline{\mathcal{A}}_\Gamma$.

As the group Γ will be fixed throughout the paper we will drop the index Γ if no confusion is possible.

10.3 Chow motives and the conjectures of Murre

Let us briefly recall some definitions and results from the theory of Chow motives. We refer to [26] for details.

10.3.1

For a smooth projective variety Y over a field k let $\mathsf{CH}^j(Y)$ denote the Chow group of algebraic cycles of codimension j on Y modulo rational equivalence, and let $\mathsf{CH}^j(Y)_{\mathbb{Q}} := \mathsf{CH}^j(Y) \otimes \mathbb{Q}$. For a cycle Z on Y we write $[Z]$ for its class in $\mathsf{CH}^j(Y)$. We will be working with relative Chow motives as well, so let us fix a smooth connected, quasi-projective base scheme $S \to \operatorname{Spec} k$. If $S = \operatorname{Spec} k$ we will usually omit S in the notation. Let Y, Y' be smooth projective varieties over S, i.e., all fibers are smooth. For ease of notation (and as we will not consider more general cases) we may assume that Y is irreducible and of relative dimension g over S. The group of relative correspondences from Y to Y' of degree r is defined as

$$\operatorname{Corr}^r(Y \times_S Y') := \mathsf{CH}^{r+g}(Y \times_S Y')_{\mathbb{Q}}.$$

Every S-morphism $Y' \to Y$ defines an element in $\operatorname{Corr}^0(Y \times_S Y')$ via the class of the transpose of its graph. In particular one has the class $[\Delta_{Y/S}] \in \operatorname{Corr}^0(Y \times_S Y)$ of the relative diagonal. The self correspondences of degree 0 form a ring, see [26, pg. 127]. Using the relative correspondences one proceeds as usual to define the category $\mathcal{M}_S$ of (pure) Chow motives over S. The objects of this pseudoabelian $\mathbb{Q}$-linear tensor category are triples

(Y, p, n) where Y is as above, p is a projector, i.e. an idempotent element in $\mathrm{Corr}^0(Y \times_S Y)$, and $n \in \mathbb{Z}$. The morphisms are

$$\mathrm{Hom}_{\mathcal{M}_S}((Y, p, n), (Y', p', n')) := p' \circ \mathrm{Corr}^{n'-n}(Y \times_S Y') \circ p.$$

When $n = 0$ we write (Y, p) instead of $(Y, p, 0)$, and $h(Y) := (Y, [\Delta_Y])$.

Definition 10.3.1. For a smooth projective variety Y/k of dimension d a *Chow-Künneth-decomposition* of Y consists of a collection of pairwise orthogonal projectors $\pi_0, \ldots, \pi_{2d}$ in $\mathrm{Corr}^0(Y \times Y)$ satisfying

(i) $\pi_0 + \ldots + \pi_{2d} = [\Delta_Y]$ and
(ii) for some Weil cohomology theory H^* one has $\pi_i(H^*(Y)) = H^i(Y)$.

If one has a Chow-Künneth decomposition for Y one writes $h^i(Y) = (Y, \pi_i)$. A similar notion of a *relative Chow-Künneth-decomposition* over S can be defined in a straightforward manner, see also introduction.

Towards the existence of such decomposition one has the following conjecture of Murre:

Conjecture 10.3.2. *Let Y be a smooth projective variety of dimension d over some field k.*

(i) There exists a Chow-Künneth decomposition for Y.
(ii) For all $i < j$ and $i > 2j$ the action of π_i on $CH^j(Y)_{\mathbb{Q}}$ is trivial, i.e. $\pi_i \cdot CH^j(Y)_{\mathbb{Q}} = 0$.
(iii) The induced j step filtration on

$$F^\nu CH^j(Y)_{\mathbb{Q}} := \mathrm{Ker}\pi_{2j} \cap \cdots \cap \mathrm{Ker}\pi_{2j-\nu+1}$$

is independent of the choice of the Chow–Künneth projectors, which are in general not canonical.
(iv) The first step of this filtration should give exactly the subgroup of homological trivial cycles $CH^j(Y)_{\mathbb{Q}}$ in $CH^j(Y)_{\mathbb{Q}}$.

There are not many examples for which these conjectures have been proved, but they are known to be true for surfaces [26], in particular we know that we have a Chow-Künneth decomposition for $\overline{X}$.

In the following theorem we are assuming the motivic decomposition conjecture which was explained in the introduction. The main result we are going to prove in section 10.7 is:

Theorem 10.3.3. *Under the assumption of the motivic decomposition conjecture 10.1.1 $\overline{\mathcal{A}}$ has a partial Chow–Künneth decomposition, including the projectors π_i for $i \neq 4, 5, 6$ as in Part (1) of Murre's conjecture.*

Over the open smooth part $X \subset \overline{X}$ one has the relative projectors constructed by Deninger and Murre in [6], see also [18]: Let S be a fixed base scheme as in section 10.3. We will now state some results on relative Chow motives in the case that A is an abelian scheme of fibre dimension g over S. Firstly we have a functorial decomposition of the relative diagonal $\Delta_{A/S}$.

Theorem 10.3.4. *There is a unique decomposition*

$$\Delta_{A/S} = \sum_{s=0}^{2g} \Pi_i \qquad in \qquad \mathrm{CH}^g(A \times_S A)_{\mathbb{Q}}$$

such that $(id_A \times [n])^* \Pi_i = n^i \Pi_i$ *for all* $n \in \mathbb{Z}$. *Moreover the* Π_i *are mutually orthogonal idempotents, and* $[{}^t\Gamma_{[n]}] \circ \Pi_i = n^i \Pi_i = \Pi_i \circ [{}^t\Gamma_{[n]}]$, *where* $[n]$ *denotes the multiplication by* n *on* A.

Proof [6, Thm. 3.1] □

Putting $h^i(A/S) = (A/S, \Pi_i)$ one has a Poincaré-duality for these motives.

Theorem 10.3.5. (Poincaré-duality)

$$h^{2g-i}(A/S)^\vee \simeq h^i(A/S)(g)$$

Proof [18, 3.1.2] □

10.3.2

We now turn back to our specific situation. From Theorem 10.3.4 we have the decomposition $\Delta_{\mathcal{A}/X} = \Pi_0 + \ldots + \Pi_6$.

We will have to extend these relative projectors to absolute projectors. In order to show the readers which of the methods of [11], where Hilbert modular varieties are considered, go through and which of them fail in our case, we recall the main theorem (Theorem 1.3) from [11]:

Theorem 10.3.6. *Let* $p : \mathcal{A} \to X$ *as above satisfy the following conditions:*

1) *The irreducible components of* $\overline{X} - X$ *are smooth toric projective varieties.*
2) *The irreducible components of* $\overline{\mathcal{A}} - \mathcal{A}$ *are smooth projective toric varieties.*
3) *The variety* $\mathcal{A}/X$ *has a relative Chow-Künneth decomposition.*
4) $\overline{X}$ *has a Chow-Künneth decomposition over* k.

a) *If x is a point of X the natural map.*

$$CH^r(\mathcal{A}) \to H_B^{2r}(\mathcal{A}_x(\mathbb{C}), \mathbb{Q})^{\pi_1^{top}(X,x)}$$

is surjective for $0 \leq r \leq d = \dim \mathcal{A} - \dim X$.

5) *For i odd, $H_B^i(\mathcal{A}_x(\mathbb{C}), \mathbb{Q})^{\pi_1^{top}(X,x)} = 0$.*

6) *Let ρ be an irreducible, non-constant representation of $\pi_1^{top}(X, x)$ and $\mathbb{V}$ the corresponding local system on X. Assume that $\mathbb{V}$ is contained in the i–th exterior power $R^i p_* \mathbb{Q} = \Lambda^i R^1 p_* \mathbb{Q}$ of the monodromy representation for some $0 \leq i \leq 2d$. Then the intersection cohomology $H^q(X, \mathbb{V})$ vanishes if $q \neq \dim X$.*

Under these assumptions $\mathcal{A}$ has a Chow-Künneth decomposition over k.

As it stands we can only use conditions (3),(4) and (5) of this theorem, all the other conditions fail in our case. As for conditions (1) and (2) we will have to weaken them to torus fibrations over an elliptic curve. This will be done in section 4.

Condition (3) holds because of the work of Deninger and Murre ([6]) on Chow-Künneth decompositions of Abelian schemes.

Condition (4) holds in our case because of the existence of Chow-Künneth projectors for surfaces (see [26]).

In order to prove condition (5) and to replace conditions (6) and (7) we will from section 5 on use a non-compact Simpson type correspondence between the L^2-Higgs cohomology of the underlying variation of Hodge structures and the L^2-de Rham cohomology (respectively intersection cohomology) of local systems. This will show the vanishing of some of the cohomology groups mentioned in (6) of Theorem 10.3.6 and enable us to weaken condition (7).

10.4 The universal abelian scheme and its compactification

In this section we show that the two conditions (1) and (2) of Theorem 10.3.6 fail in our case. Instead of tori we get toric fibrations over an elliptic curve as fibers over boundary components. The main reference for this section is [23].

10.4.1 Toroidal compactifications of locally symmetric varieties

In this paragraph an introduction to the theory of toroidal compactifications of locally symmetric varieties as developed by Ash, Mumford, Rapoport and

Tai in [2] is given. The main goal is to fix notation. All details can be found in [2], see the page references in this paragraph.

Let $D = G(\mathbb{R})/K$ be a bounded symmetric domain (or a finite number of bounded symmetric domains, for the following discussion we will assume D to be just one bounded symmetric domain), where $G(\mathbb{R})$ denotes the $\mathbb{R}$-valued points of a semisimple group G and $K \subset G(\mathbb{R})$ is a maximal compact subgroup. Let $\check{D}$ be its compact dual. Then there is an embedding

$$D \hookrightarrow \check{D}. \tag{10.1}$$

Note that $G^0(\mathbb{C})$ acts on $\check{D}$.

We pick a parabolic P corresponding to a rational boundary component F, by Z^0 we denote the connected component of the centralizer $Z(F)$ of F, by P^0 the connected component of P and by Γ a (torsion free, see below for this restriction) congruence subgroup of G. We will explicitly be interested only in connected groups, so from now on we can assume that $G^0 = G$. Set

$N \subset P^0$	the unipotent radical
$U \subset N$	the center of the unipotent radical
$U_{\mathbb{C}}$	its complexification
$V = N/U$	
$\Gamma_0 = \Gamma \cap U$	
$\Gamma_1 = \Gamma \cap P^0$	
$T = \Gamma_0 \backslash U_{\mathbb{C}}.$	

Note that U is a real vector space and by construction, T is an algebraic torus over $\mathbb{C}$. Set

$$D(F) := U_{\mathbb{C}} \cdot D,$$

where the dot $\cdot$ denotes the action of $G^0(\mathbb{C})$ on $\check{D}$. This is an open set in $\check{D}$ and we have the inclusions

$$D \subset D(F) = U_{\mathbb{C}} \cdot D \subset \check{D} \tag{10.2}$$

and furthermore a complex analytic isomorphism

$$U_{\mathbb{C}} \cdot D \simeq U_{\mathbb{C}} \times \mathcal{E}_P \tag{10.3}$$

where $\mathcal{E}_P$ is some complex vector bundle over the boundary component corresponding to P. We will not describe $\mathcal{E}_P$ any further, the interested reader is referred to [2], chapter 3. The isomorphism in (10.3) is complex analytic and takes the $U_{\mathbb{C}}$-action on $\check{D}$ from the left to the translation on $U_{\mathbb{C}}$ on the right.

We once for all choose a boundary component F and denote its stabilizer by P.

From (10.2) we get (see [2], chapter 3 for details, e.g. on the last isomorphism)

$$\Gamma_0 \backslash D \subset \Gamma_0 \backslash (U_{\mathbb{C}} \cdot D) \simeq \Gamma_0 \backslash (U_{\mathbb{C}} \times \mathcal{E}_P) \simeq T \times \mathcal{E},$$

where $\mathcal{E} = \Gamma_0 \backslash \mathcal{E}_P$.

The torus T is the one we use for a toroidal embedding. Furthermore D can be realized as a Siegel domain of the third kind:

$$D \simeq \{(z, e) \in U_{\mathbb{C}} \cdot D \simeq U_{\mathbb{C}} \times \mathcal{E} \mid \operatorname{Im}(z) \in C + h(e)\},$$

where

$$h : \mathcal{E} \to U$$

is a real analytic map and $C \subset U$ is an open cone in U.

A finer description of C which is needed for the most general case can be found in [2].

We pick a cone decomposition $\{\sigma_\alpha\}$ of C such that

$$\begin{aligned} &(\Gamma_1/\Gamma_0) \cdot \{\sigma_\alpha\} = \{\sigma_\alpha\} \text{ with finitely many orbits and} \\ &C \subset \bigcup_\alpha \sigma_\alpha \subset \overline{C}. \end{aligned} \tag{10.4}$$

This yields a torus embedding

$$T \subset X_{\{\sigma_\alpha\}}. \tag{10.5}$$

We can thus partially compactify the open set $\Gamma_0 \backslash (U_{\mathbb{C}} \cdot D)$:

$$\Gamma_0 \backslash (U_{\mathbb{C}} \cdot D) \simeq \Gamma_0 \backslash U_{\mathbb{C}} \times \mathcal{E} \hookrightarrow X_{\{\sigma_\alpha\}} \times \mathcal{E}.$$

The situation is now the following:

$$\begin{array}{c} \Gamma_0 \backslash (U_{\mathbb{C}} \cdot D) \simeq T \times \mathcal{E} \hookrightarrow X_{\{\sigma_\alpha\}} \times \mathcal{E} \\ \cup \\ \Gamma_0 \backslash D. \end{array} \tag{10.6}$$

We proceed to give a description of the vector bundle $\mathcal{E}_P$ in order to describe the toroidal compactification geometrically.

Again from [2] (pp.233) we know that

$$\begin{aligned} D &\cong F \times C \times N \qquad \text{as real manifolds and} \\ D(F) &\cong F \times V \times U_{\mathbb{C}}, \end{aligned}$$

where $V = N/U$ is the abelian part of N.

Now set

$$D(F)' := D(F) \bmod U_{\mathbb{C}}.$$

This yields the following fibration:

$$\begin{array}{c} D(F) \\ \pi_1 \big\downarrow \text{fibres } U_{\mathbb{C}} \\ D(F)' \\ \pi_2 \big\downarrow \text{fibres } V \\ F. \end{array} \qquad (\pi = \pi_2 \circ \pi_1 : D(F) \to F)$$

Taking the quotient by Γ_0 yields a quotient bundle

$$\begin{array}{c} \Gamma_0 \backslash D(F) \\ \overline{\pi_1} \big\downarrow \text{fibres } T := \Gamma_0 \backslash U_{\mathbb{C}} \\ D(F)'. \end{array}$$

So, T is an algebraic torus group with maximal compact subtorus

$$T_{cp} := \Gamma_0 \backslash U.$$

Take the closure of $\Gamma_0 \backslash D$ in $X_{\{\sigma_\alpha\}} \times \mathcal{E}$ and denote by $(\Gamma_0 \backslash D)_{\{\sigma_\alpha\}}$ its interior.

Factor $D \to \Gamma \backslash D$ by

$$D \to \Gamma_0 \backslash D \to \Gamma_1 \backslash D \to \Gamma \backslash D.$$

It is the following situation we aim at obtaining:

$$\begin{array}{ccccccc} (\Gamma_0 \backslash D)_{\{\sigma_\alpha\}} & \hookleftarrow & \Gamma_0 \backslash D & \to & \Gamma_1 \backslash D & \to & \Gamma \backslash D \\ \cup & & \cup & & \cup & & \\ (\Gamma_0 \backslash D(c))_{\{\sigma_\alpha\}} & \hookleftarrow & \Gamma_0 \backslash D(c) & \to & \Gamma_1 \backslash D(c) & \hookrightarrow & \Gamma \backslash D. \end{array} \tag{10.7}$$

Here $D(c)$ is a neighborhood of our boundary component. More precisely for any compact subset K of the boundary component and any $c \in C$ define

$$D(c, K) = \Gamma_1 \cdot \left\{ (z, e) \in U_{\mathbb{C}} \times \mathcal{E} \;\middle|\; \begin{array}{l} \operatorname{Im} z \in C + h(e) + c \\ \text{and } e \text{ lies above } K \end{array} \right\}.$$

Then by reduction theory for c large enough, Γ-equivalence on $D(c, K)$ reduces to Γ_1-equivalence.

This means that we have an inclusion

$$\begin{array}{ccc} \Gamma_1\backslash D(c) & \hookrightarrow & \Gamma\backslash D \\ \downarrow \simeq & & \\ (\Gamma_1/\Gamma_0)\backslash(\Gamma_0\backslash D(c)) & & \end{array}$$

where the quotient by Γ_1/Γ_0 is defined in the obvious way.

Furthermore $\Gamma_0\backslash D \hookrightarrow (\Gamma_0\backslash D)_{\{\sigma_\alpha\}}$ directly induces

$$\Gamma_0\backslash D(c) \hookrightarrow (\Gamma_0\backslash D(c))_{\{\sigma_\alpha\}}.$$

Having chosen $\{\sigma_\alpha\}$ such that $(\Gamma_1/\Gamma_0)\cdot\{\sigma_\alpha\} = \{\sigma_\alpha\}$, we get

$$\Gamma_1\backslash D(c) \hookrightarrow (\Gamma_1/\Gamma_0)\backslash(\Gamma_0\backslash D(c))_{\{\sigma_\alpha\}}$$

which yields the partial compactification and establishes the diagram ((10.7)).

The following theorem is derived from the above.

Theorem 10.4.1. *With the above notation and for a cone decomposition* $\{\sigma_\alpha\}$ *of* C *satisfying the condition* (10.4), *the diagram*

$$\begin{array}{ccc} \Gamma_1\backslash D(c) & \hookrightarrow & \Gamma\backslash D \\ & \searrow & \\ & & (\Gamma_1/\Gamma_0)\backslash(\Gamma_0\backslash D(c))_{\{\sigma_\alpha\}} \end{array} \tag{10.8}$$

yields a (smooth if $\{\sigma_\alpha\}$ *is chosen appropriately) partial compactification of* $\Gamma\backslash D$ *at* F. □

10.4.2 Toroidal compactification of Picard modular surfaces

We will now apply the results of the last paragraph to the case of Picard modular surfaces and give a finer description of the fibres at the boundary.

Theorem 10.4.2. *For each boundary component of a Picard modular surface the following holds. With the standard notations from [2] (see also the last paragraph and [23] for the specific choices of* Γ_0, Γ_1 *etc.)*

$$(\Gamma_1/\Gamma_0)\backslash(\Gamma_0\backslash D)$$

is isomorphic to a punctured disc bundle over a CM elliptic curve A.

A toroidal compactification

$$(\Gamma_1/\Gamma_0)\backslash(\Gamma_0\backslash D)_{\{\sigma_\alpha\}}$$

is obtained by closing the disc with a copy of A (e.g. adding the zero section of the corresponding line bundle).

We now turn to the modification of condition (2). The notation we use is as introduced in chapter 3 of [8]. Let $\tilde{P}$ be a relatively complete model of an ample degeneration datum associated to our moduli problem. As a general reference for degenerations see [25], see [8] for the notion of relatively complete model and [23] for the ample degeneration datum we need here. In [23] the following theorem is proved.

Theorem 10.4.3. *(i) The generic fibre of $\tilde{P}$ is given by a fibre-bundle over a CM elliptic curve E, whose fibres are countably many irreducible components of the form $\mathbb{P}$, where $\mathbb{P}$ is a $\mathbb{P}^1$-bundle over $\mathbb{P}^1$.*
(ii) The special fibre of $\tilde{P}$ is given by a fibre-bundle over the CM elliptic curve E, whose fibres consist of countably many irreducible components of the form $\mathbb{P}^1 \times \mathbb{P}^1$.

Remark 10.4.4. In this paper we work with some very specific Picard modular surfaces and thus the generality of Theorem 10.4.3 is not needed. It will be needed though to extend our results to larger families of Picard modular surfaces, see section 10.7.2.

10.5 Higgs bundles on Picard modular surfaces

In this section we describe in detail the Picard modular surface of Holzapfel which is our main object. We follow Holzapfel [15, 16] very closely. In the remaining part of this section we explain the formalism of Higgs bundles which we will need later.

10.5.1 Holzapfel's surface

We restrict our attention to the *Picard modular surfaces* with compactification $\overline{X}$ and boundary divisor $D \subseteq \overline{X}$ which were discussed by Picard [28], Hirzebruch [14] and Holzapfel [15, 16]. These surfaces are compactifications of ball quotients $X = \mathbb{B}/\Gamma$ where Γ is a subgroup of $SU(2,1;\mathcal{O})$ with $\mathcal{O} = \mathbb{Z} \oplus \mathbb{Z}\omega$, $\omega = \exp(2\pi i/3)$, i.e., $\mathcal{O}$ is the ring of Eisenstein numbers. In the case $\Gamma = SU(2,1;\mathcal{O})$, studied already by Picard, the quotient $\mathbb{B}/\Gamma$ is

$\mathbb{P}^2 -$ 4 points, an open set of which is $U = \mathbb{P}^2 - \Delta$ and Δ is a configuration of 6 lines (not a normal crossing divisor). U is a natural parameter space for a family of *Picard curves*

$$y^3 = x(x-1)(x-s)(x-t)$$

of genus 3 branched over 5 (ordered) points $0, 1, s, t, \infty$ in $\mathbb{P}^1$. The parameters s, t are coordinates in the affine set U. If one looks at the subgroup

$$\Gamma' = \Gamma \cap SL(3, \mathbb{C}),$$

then $X = \mathbb{B}/\Gamma'$ has a natural compactification $\overline{X}$ with a smooth boundary divisor D consisting of 4 disjoint elliptic curves $E_0 + E_1 + E_2 + E_3$, see [15, 16]. This surface $\overline{X}$ is birational to a covering of $\mathbb{P}^2 - \Delta$ and hence carries a family of curves over it. If we pass to yet another subgroup $\Gamma'' \subset \Gamma$ of finite index, then we obtain a Picard modular surface

$$\overline{X} = \widetilde{\widetilde{E \times E}}$$

with boundary D a union of 6 elliptic curves which are the strict transforms of the following 6 curves

$$T_1, T_\omega, T_{\omega^2}, E \times \{Q_0\}, E \times \{Q_1\}, E \times \{Q_2\}$$

on $E \times E$ in the notation of [15, page 257]. This is the surface we will study in this paper. The properties of the modular group Γ'' are described in [15, remark V.5]. In particular it acts freely on the ball. $\overline{X}$ is the blowup of $E \times E$ in the three points (Q_0, Q_0) (the origin), (Q_1, Q_1) and (Q_2, Q_2) of triple intersection. Note that E has the equation $y^2 z = x^3 - z^3$. On E we have an action of ω via $(x : y : z) \mapsto (\omega x : y : z)$. E maps to $\mathbb{P}^1$ using the projection

$$p : E \to \mathbb{P}^1, \quad (x : y : z) \mapsto (y : z).$$

This action has 3 fixpoints $Q_0 = (0 : 1 : 0)$ (the origin), $Q_1 = (0 : i : 1)$ and $Q_2 = (0 : -i : 1)$ which are triple ramification points of p. Therefore one has $3Q_0 = 3Q_1 = 3Q_2$ in $CH^1(E)$.

In order to proceed, we need to know something about the Picard group of $\overline{X}$.

Lemma 10.5.1. *In* $\mathrm{NS}(E \times E)$ *one has the relation*

$$T_1 + T_\omega + T_{\omega^2} = 3(0 \times E) + 3(E \times 0).$$

Proof Since E has complex multiplication by $\mathbb{Z}[\omega]$, the Néron–Severi group has rank 4 and divisors T_1, T_ω, $0 \times E$ and $E \times 0$ form a basis of $\mathrm{NS}(E \times E)$. Using the intersection matrix of this basis, the claim follows. □

The following statement is needed later:

Lemma 10.5.2. *The log–canonical divisor is divisible by three:*

$$K_{\overline{X}} + D = 3L$$

for some line bundle L.

Proof If we denote by $\sigma : \overline{X} \to E \times E$ the blowup in the three points (Q_0, Q_0), (Q_1, Q_1) and (Q_2, Q_2), then we denote by $Z = Z_1 + Z_2 + Z_3$ the union of all exceptional divisors. We get:

$$\sigma^* T = D_1 + Z,\ \sigma^* T_\omega = D_2 + Z,\ \sigma^* T_{\omega^2} = D_3 + Z,$$

and

$$\sigma^* E \times Q_0 = D_4 + Z_1,\ \sigma^* E \times Q_1 = D_5 + Z_2,\ \sigma^* E \times Q_2 = D_6 + Z_3.$$

Now look at the line bundle $K_{\overline{X}} + D$. Since

$$K_{\overline{X}} + D = \sigma^* K_{E \times E} + Z + D = Z + D,$$

we compute

$$K_{\overline{X}} + D = \sum_{i=1}^{6} D_i + \sum_{j=1}^{3} Z_j.$$

The first sum,

$$D_1 + D_2 + D_3 = -3Z + \sigma^*(T_1 + T_\omega + T_{\omega^2}) = -3Z + 3\sigma^*(0 \times E + E \times 0).$$

is divisible by 3. Using $3Q_0 = 3Q_1 = 3Q_2$, the rest can be computed in $\mathrm{NS}(\overline{X})$ as

$$D_4 + D_5 + D_6 + Z = \sigma^*(E \times 0 + E \times Q_1 + E \times Q_2) = 3\sigma^*(E \times 0).$$

Therefore the class of $K_{\overline{X}} + D$ in $\mathrm{NS}(\overline{X})$ is given by

$$K_{\overline{X}} + D = -3Z + 3\sigma^*(0 \times E) + 6\sigma^*(E \times 0)$$

and divisible by 3. Since $\mathrm{Pic}^0(\overline{X})$ is a divisible group, $K_{\overline{X}} + D$ is divisible by 3 in $\mathrm{Pic}(\overline{X})$ and we get a line bundle L with $K_{\overline{X}} + D = 3L$ whose class in $\mathrm{NS}(\overline{X})$ is given by

$$L = \sigma^*(0 \times E) - Z + 2\sigma^*(E \times 0).$$

If we write

$$\sigma^*(0 \times E) = D_0 + Z_1,$$

we obtain the equation

$$L = D_0 + D_5 + D_6$$

in $\mathrm{NS}(\overline{X})$. Note that D_0 intersects both D_5 and D_6 in one point. All D_i, $i = 1, \ldots, 6$ have negative selfintersection and are disjoint. □

It is not difficult to see that L is a nef and big line bundle since $\overline{X}$ has logarithmic Kodaira dimension 2 [15]. L is trivial on all components of D by the adjunction formula, since they are smooth elliptic curves.

The rest of this section is about the rank 6 local system $\mathbb{V} = R^1 p_* \mathbb{Z}$ on X. The following Lemma was known to Picard [28], he wrote down 3×3 monodromy matrices with values in the Eisenstein numbers:

Lemma 10.5.3. *$\mathbb{V}$ is a direct sum of two local systems $\mathbb{V} = \mathbb{V}_1 \oplus \mathbb{V}_2$ of rank 3. The decomposition is defined over the Eisenstein numbers.*

Proof The cohomology $H^1(C)$ of any Picard curve C has a natural $\mathbb{Z}/3\mathbb{Z}$ Galois action. Since the projective line has $H^1(\mathbb{P}^1, \mathbb{Z}) = 0$, the local system $\mathbb{V} \otimes \mathbb{C}$ decomposes into two 3–dimensional local systems

$$\mathbb{V} = \mathbb{V}_1 \oplus \mathbb{V}_2$$

which are conjugate to each other and defined over the Eisenstein numbers.

Both local systems $\mathbb{V}_1, \mathbb{V}_2$ are irreducible and non–constant. □

10.5.2 Higgs bundles on Holzapfel's surface

Now we will study the categorical correspondence between local systems and *Higgs bundles*. It turns out that it is sometimes easier to deal with one resp. the other.

Definition 10.5.4. A Higgs bundle on a smooth variety Y is a holomorphic vector bundle E together with a holomorphic map

$$\vartheta : E \to E \otimes \Omega^1_Y$$

which satisfies $\vartheta \wedge \vartheta = 0$, i.e., an $\mathrm{End}(E)$ valued holomorphic 1–form on Y.

Each Higgs bundle induces a complex of vector bundles:

$$E \to E \otimes \Omega^1_Y \to E \otimes \Omega^2_Y \to \ldots \to E \otimes \Omega^d_Y.$$

Higgs cohomology is the cohomology of this complex. The *Simpson* correspondence on a projective variety Y gives an equivalence of categories between polystable Higgs bundles with vanishing Chern classes and semisimple local systems $\mathbb{V}$ on Y [29, Sect. 8]. This correspondence is very difficult to describe in general and uses a deep existence theorem for harmonic metrics. For quasi–projective Y this may be generalized provided that the appropriate harmonic metrics exist, which is still not known until today. There is however the case of VHS (Variations of Hodge structures) where the harmonic metric is the Hodge metric and is canonically given. For example if we have a smooth, projective family $f : A \to X$ as in our example and $\mathbb{V} = R^m f_* \mathbb{C}$ is a direct image sheaf, then the corresponding Higgs bundle is

$$E = \bigoplus_{p+q=m} E^{p,q}$$

where $E^{p,q}$ is the p-th graded piece of the Hodge filtration $F^\bullet$ on $\mathcal{H} = \mathbb{V} \otimes \mathcal{O}_X$. The Higgs operator ϑ is then given by the graded part of the Gauss–Manin connection, i.e., the cup product with the Kodaira–Spencer class. In the non–compact case there is also a corresponding log–version for Higgs bundles, where Ω^1_Y is replaced by $\Omega^1_Y(\log D)$ for some normal crossing divisor $D \subset Y$ and E by the Deligne extension. Therefore we have to assume that the monodromies around the divisors at infinity are unipotent and not only quasi–unipotent as in [17, Sect. 2]. This is the case in Holzapfel's example, in fact above we have already checked that the log–canonical divisor $K_{\overline{X}} + D$ is divisible by three. We refer to [29] and [17] for the general theory. In our case let $E = E^{1,0} \oplus E^{0,1}$ be the Higgs bundle corresponding to $\mathbb{V}_1$ with Higgs field

$$\vartheta : E \to E \otimes \Omega^1_{\overline{X}}(\log D).$$

This bundle is called the *uniformizing bundle* in [29, Sect. 9].

Let us return to Holzapfel's example. We may assume that $E^{1,0}$ is 2–dimensional and $E^{0,1}$ is 1–dimensional, otherwise we permute $\mathbb{V}_1$ and $\mathbb{V}_2$.

Lemma 10.5.5. *$\vartheta : E^{1,0} \to E^{0,1} \otimes \Omega^1_{\overline{X}}(\log D)$ is an isomorphism.*

Proof For the generic fiber this is true for rank reasons. At the boundary D this is a local computation using the definition of the Deligne extension. This has been shown in greater generality in [17, Sect. 2-4] (cf. also [20, Sect. 4]), therefore we do not give any more details. □

Let us summarize what we have shown for Holzapfel's surface $\overline{X}$:

Corollary 10.5.6. *$K_{\overline{X}}(D)$ is nef and big and there is a nef and big line bundle L with*

$$L^{\otimes 3} \cong K_{\overline{X}}(D).$$

The uniformizing bundle E has components

$$E^{1,0} = \Omega^1_{\overline{X}}(\log D) \otimes L^{-1}, \quad E^{0,1} = L^{-1}.$$

The Higgs operator ϑ is the identity as a map $E^{1,0} \to E^{0,1} \otimes \Omega^1_{\overline{X}}(\log D)$ and it is trivial on $E^{0,1}$.

10.6 Vanishing of intersection cohomology

Let X be Holzapfel's surface from the previous section. We now want to discuss the vanishing of intersection cohomology

$$H^1(X, \mathbb{W})$$

for irreducible, non–constant local systems $\mathbb{W} \subseteq R^i p_* \mathbb{Q}$. Let $\mathbb{V}_1$ be as in the previous section. Denote by (E, ϑ) the corresponding Higgs bundle with

$$E = \left(\Omega^1_{\overline{X}}(\log D) \otimes L^{-1}\right) \oplus L^{-1}$$

and Higgs field

$$\vartheta : E \to E \otimes \Omega^1_{\overline{X}}(\log D).$$

Our goal is to compute the intersection cohomology of $\mathbb{V}_1$. We use the isomorphism between L^2– and intersection cohomology for $\mathbb{C}$–VHS, a theorem of Cattani, Kaplan and Schmid together with the isomorphism between L^2–cohomology and L^2–Higgs cohomology from [17, Thm. A/B]. Therefore for computations of intersection cohomology we may use L^2–Higgs cohomology. We refer to [17] for a general introduction to all cohomology theories involved.

Theorem 10.6.1. *The intersection cohomology $H^q(X, \mathbb{V}_1)$ vanishes for $q \neq 2$. By conjugation the same holds for $\mathbb{V}_2$.*

Proof We need only show this for $q = 1$, since $\mathbb{V}_1$ has no invariant sections, hence $H^0(X, \mathbb{V}_1) = 0$ and the other vanishings follow via duality

$$H^q(X, \mathbb{V}_1) \cong H^{2\dim(X)-q}(X, \mathbb{V}_2)$$

from the analogous statement for $\mathbb{V}_2$. The following theorem provides the necessary technical tool. □

Theorem 10.6.2 ([17, Thm. B]). *The intersection cohomology $H^q(X, \mathbb{V}_1)$ can be computed as the q–th hypercohomology of the complex*

$$0 \to \Omega^0(E)_{(2)} \to \Omega^1(E)_{(2)} \to \Omega^2(E)_{(2)} \to 0$$

on $\overline{X}$, where E is as above. This is a subcomplex of

$$E \xrightarrow{\vartheta} E \otimes \Omega^1_{\overline{X}}(\log D) \xrightarrow{\vartheta} E \otimes \Omega^2_{\overline{X}}(\log D).$$

In the case where D is smooth, this is a proper subcomplex with the property

$$\Omega^1(E)_{(2)} \subseteq \Omega^1_{\overline{X}} \otimes E.$$

Proof This is a special case of the results in [17]. The subcomplex is explicitly described in section 10.8 of our paper. □

Lemma 10.6.3. *Let E be as above with L nef and big. Then the vanishing*

$$H^0(\Omega^1_{\overline{X}}(\log D) \otimes \Omega^1_{\overline{X}} \otimes L^{-1}) = 0$$

implies the statement of theorem 10.6.1.

Proof We first compute the cohomology groups for the complex of vector bundles and discuss the L^2–conditions later. Any logarithmic Higgs bundle $E = \oplus E^{p,q}$ coming from a VHS has differential

$$\vartheta : E^{p,q} \to E^{p-1,q+1} \otimes \Omega^1_{\overline{X}}(\log D).$$

In our case $E = E^{1,0} \oplus E^{0,1}$ and the restriction of ϑ to $E^{0,1}$ is zero. The differential

$$\vartheta : E^{1,0} \to E^{0,1} \otimes \Omega^1_{\overline{X}}(\log D)$$

is the identity. Therefore the complex

$$(E^\bullet, \vartheta) : E \xrightarrow{\vartheta} E \otimes \Omega^1_{\overline{X}}(\log D) \xrightarrow{\vartheta} E \otimes \Omega^2_{\overline{X}}(\log D)$$

looks like:

$$\begin{array}{ccc}
 & \left(\Omega^1_{\overline{X}}(\log D) \otimes L^{-1}\right) & \oplus\, L^{-1} \\
 & \downarrow\cong & \downarrow \\
 & \left(\Omega^1_{\overline{X}}(\log D)^{\otimes 2} \otimes L^{-1}\right) \oplus \left(L^{-1} \otimes \Omega^1_{\overline{X}}(\log D)\right) & 0 \\
 & \downarrow & \\
\left(\Omega^1_{\overline{X}}(\log D) \otimes L^{-1} \otimes \Omega^2_{\overline{X}}(\log D)\right) \oplus \left(L^{-1} \otimes \Omega^2_{\overline{X}}(\log D)\right). & &
\end{array}$$

Therefore it is quasi–isomorphic to a complex

$$L^{-1} \xrightarrow{0} S^2\Omega^1_{\overline{X}}(\log D) \otimes L^{-1} \xrightarrow{0} \Omega^1_{\overline{X}}(\log D) \otimes \Omega^2_{\overline{X}}(\log D) \otimes L^{-1}$$

with trivial differentials. As L is nef and big, we have

$$H^0(L^{-1}) = H^1(L^{-1}) = 0.$$

Hence we get

$$\mathbb{H}^1(\overline{X}, (E^\bullet, \vartheta)) \cong H^0(\overline{X}, S^2\Omega^1_{\overline{X}}(\log D) \otimes L^{-1})$$

and $\mathbb{H}^2(\overline{X}, (E^\bullet, \vartheta))$ is equal to

$$H^0(\overline{X}, K_{\overline{X}} \otimes L)^\vee \oplus H^0(\overline{X}, \Omega^1_{\overline{X}}(\log D) \otimes \Omega^2_{\overline{X}}(\log D) \otimes L^{-1}) \oplus H^1(\overline{X}, S^2\Omega^1_{\overline{X}}(\log D) \otimes L^{-1}).$$

If we now impose the L^2–conditions and use the complex $\Omega^*_{(2)}(E)$ instead of $(E^\bullet, \vartheta)$, the resulting cohomology groups are subquotients of the groups described above. Since

$$\Omega^1(E)_{(2)} \subseteq \Omega^1_{\overline{X}} \otimes E$$

we conclude that the vanishing

$$H^0(\overline{X}, \Omega^1_{\overline{X}}(\log D) \otimes \Omega^1_{\overline{X}} \otimes L^{-1}) = 0$$

is sufficient to deduce the vanishing of intersection cohomology. $\square$

Now we verify the vanishing statement.

Lemma 10.6.4. *In the example above we have*

$$H^0(\Omega^1_{\overline{X}}(\log D) \otimes \Omega^1_{\overline{X}} \otimes L^{-1}) = 0.$$

Proof Let $\sigma : \overline{X} \to E \times E$ be the blow up of the 3 points of intersection. Then one has an exact sequence

$$0 \to \sigma^*\Omega^1_{E\times E} \to \Omega^1_{\overline{X}} \to i_*\Omega^1_Z \to 0,$$

where Z is the union of all (disjoint) exceptional divisors. Now $\Omega^1_{E\times E}$ is the trivial sheaf of rank 2. Therefore $\Omega^1_{\overline{X}}(\log D) \otimes \Omega^1_{\overline{X}} \otimes L^{-1}$ has as a subsheaf 2 copies of $\Omega^1_{\overline{X}}(\log D) \otimes L^{-1}$. The group

$$H^0(\overline{X}, \Omega^1_{\overline{X}}(\log D) \otimes L^{-1})$$

is zero by the Bogomolov–Sommese vanishing theorem (see [7, Cor. 6.9]), since L is nef and big. In order to prove the assertion it is hence sufficient to show that

$$H^0(Z, \Omega^1_{\overline{X}}(\log D) \otimes \Omega^1_Z \otimes L^{-1}) = 0.$$

But Z is a disjoint union of $\mathbb{P}^1$'s. In our example we have $K_{\overline{X}}(D) \otimes \mathcal{O}_Z \cong \mathcal{O}_Z(3)$ since $(L.Z) = 1$ and therefore $\Omega^1_Z \otimes L^{-1} \cong \mathcal{O}_Z(-3)$. Now we use in addition the conormal sequence

$$0 \to N_Z^* \to \Omega^1_{\overline{X}}(\log D)|_Z \to \Omega^1_Z(\log(D \cap Z)) \to 0.$$

Note that $N_Z^* = \mathcal{O}_Z(1)$. Twisting this with $\Omega^1_Z \otimes L^{-1} \cong \mathcal{O}_Z(-3)$ gives an exact sequence

$$0 \to \mathcal{O}_Z(-2) \to \Omega^1_{\overline{X}}(\log D) \otimes \Omega^1_Z \otimes L^{-1} \to \mathcal{O}_Z(-1) \to 0.$$

On global sections this proves the assertion. □

So far we have only shown the vanishing of $H^q(X, \mathbb{V}_1)$ and hence of $H^q(X, \mathbb{V})$ for $q \neq 2$. In order to apply the method of Gordon, Hanamura and Murre, we also have to deal with the case $\Lambda^i \mathbb{V}$.

Theorem 10.6.5. *Let ρ be an irreducible, non–constant representation of $\pi_1(X)$, which is a direct factor in $\Lambda^k(\mathbb{V}_1 \oplus \mathbb{V}_2)$ for $k \leq 2$. Then the intersection cohomology group*

$$H^1(X, \mathbb{V}_\rho)$$

is zero.

Proof Let us first compute all such representations: if $k = 1$ we have only $\mathbb{V}_1$ and its dual. If $k = 2$, we have the decomposition

$$\Lambda^2(\mathbb{V}_1 \oplus \mathbb{V}_2) = \Lambda^2 \mathbb{V}_1 \oplus \Lambda^2 \mathbb{V}_2 \oplus \mathrm{End}(\mathbb{V}_1).$$

Since $\mathbb{V}_1$ is 3–dimensional, $\Lambda^2 \mathbb{V}_1 \cong \mathbb{V}_2$ and therefore the only irreducible, non–constant representation that is new here is $\mathrm{End}^0(\mathbb{V}_1)$, the trace–free endomorphisms of $\mathbb{V}_1$. Since we have already shown the vanishing $H^1(X, \mathbb{V}_{1,2})$, it remains to treat $H^1(X, \mathrm{End}^0(\mathbb{V}_1))$. The vanishing of $H^1(X, \mathrm{End}^0(\mathbb{V}_1))$ is a general and well–known statement: The representation $\mathrm{End}^0(\mathbb{V}_1)$ has regular highest weight and therefore contributes only to the middle dimension H^2. A reference for this is [19, Main Thm.], cf. [4, ch. VII] and [30]. □

The vanishing of $H^1(X, \mathrm{End}^0(\mathbb{V}_1))$ has the following amazing consequence, which does not seem easy to prove directly using purely algebraic methods. In the compact case this has been shown by Miyaoka, cf. [24].

Lemma 10.6.6. *In our situation we have*

$$H^0_{L^2}(\overline{X}, S^3 \Omega^1_{\overline{X}}(\log D)(-D) \otimes L^{-3}) = 0.$$

Proof Write down the Higgs complex for $\mathrm{End}^0(E)$. In degree one, a term which contains

$$S^3\Omega^1_{\overline{X}}(\log D)(-D) \otimes L^{-3}$$

occurs. Since H^1 vanishes, this cohomology group must vanish too. □

Finally we want to discuss the case $k = 3$. Unfortunately here the vanishing techniques do not work in general. But we are able to at least give a bound for the dimension of the remaining cohomology group. Namely for $k = 3$, one has

$$\Lambda^3(\mathbb{V}_1 \oplus \mathbb{V}_2) = \Lambda^3\mathbb{V}_1 \oplus \Lambda^3\mathbb{V}_2 \oplus (\Lambda^2\mathbb{V}_1 \otimes \mathbb{V}_2) \oplus (\Lambda^2\mathbb{V}_2 \otimes \mathbb{V}_1).$$

Here the only new irreducible and non–constant representation is

$$S^2\mathbb{V}_1 \subseteq \mathbb{V}_1 \otimes \mathbb{V}_1$$

and its dual. We would like to compute $H^1(X, S^2\mathbb{V}_1)$ using a variant of the symmetric product of the L^2–complexes $\Omega^*(S)_{(2)}$ as described in the appendix. The Higgs complex without L^2–conditions looks as follows:

$$\begin{array}{c}
\left(S^2\Omega^1_{\overline{X}}(\log D) \otimes L^{-2}\right) \oplus \left(\Omega^1_{\overline{X}}(\log D) \otimes L^{-2}\right) \oplus L^{-2} \\
\downarrow \\
\left(S^2\Omega^1_{\overline{X}}(\log D) \otimes L^{-2} \otimes \Omega^1_{\overline{X}}(\log D)\right) \oplus \left(\Omega^1_{\overline{X}}(\log D) \otimes L^{-2} \otimes \Omega^1_{\overline{X}}(\log D)\right) \oplus \left(L^{-2} \otimes \Omega^1_{\overline{X}}(\log D)\right) \\
\downarrow \\
\left(S^2\Omega^1_{\overline{X}}(\log D) \otimes L^{-2} \otimes \Omega^2_{\overline{X}}(\log D)\right) \oplus \left(\Omega^1_{\overline{X}}(\log D) \otimes L^{-2} \otimes \Omega^2_{\overline{X}}(\log D)\right) \oplus \left(L^{-2} \otimes \Omega^2_{\overline{X}}(\log D)\right)
\end{array}$$

Again many pieces of differentials in this complex are isomorphisms or zero. For example the differential

$$S^2\Omega^1_{\overline{X}}(\log D) \otimes L^{-2} \otimes \Omega^1_{\overline{X}}(\log D) \to \Omega^1_{\overline{X}}(\log D) \otimes L^{-2} \otimes \Omega^2_{\overline{X}}(\log D)$$

is a projection map onto a direct summand, since for every vector space W we have the identity

$$S^2W \otimes W = S^3W \oplus (W \otimes \Lambda^2 W).$$

Therefore the Higgs complex for $S^2(E)$ is quasi–isomorphic to

$$L^{-2} \xrightarrow{0} S^3\Omega^1_{\overline{X}}(\log D) \otimes L^{-2} \xrightarrow{0} S^2\Omega^1_{\overline{X}}(\log D) \otimes L^{-2} \otimes \Omega^2_{\overline{X}}(\log D).$$

We conclude that the first cohomology is given by

$$H^0(\overline{X}, S^3\Omega^1_{\overline{X}}(\log D) \otimes L^{-2}).$$

If we additionally impose the L^2–conditions (see appendix), then we see that the first Higgs cohomology of $S^2(E, \vartheta)$ vanishes, provided that we have

$$H^0(\overline{X}, S^2\Omega^1_{\overline{X}}(\log D) \otimes \Omega^1_{\overline{X}} \otimes L^{-2}) = 0.$$

Using

$$0 \to \sigma^* \Omega^1_{E\times E} \to \Omega^1_{\overline{X}} \to i_* \Omega^1_Z \to 0$$

we obtain an exact sequence

$$0 \to H^0(\overline{X}, S^2\Omega^1_{\overline{X}}(\log D) \otimes L^{-2}) \to H^0(\overline{X}, S^2\Omega^1_{\overline{X}}(\log D) \otimes \Omega^1_{\overline{X}} \otimes L^{-2}) \to$$
$$\to H^0(Z, S^2\Omega^1_{\overline{X}}(\log D) \otimes \Omega^1_Z \otimes L^{-2}).$$

A generalization of [24, example 3] leads to the vanishing

$$H^0(\overline{X}, S^2\Omega^1_{\overline{X}}(\log D) \otimes L^{-2}) = 0.$$

Since $\Omega^1_{\overline{X}}(\log D)|_Z = \mathcal{O}_Z(1) \oplus \mathcal{O}_Z(2)$, we get

$$H^0(Z, S^2\Omega^1_{\overline{X}}(\log D) \otimes \Omega^1_Z \otimes L^{-2}) = \mathbb{C}^3,$$

because $\Omega^1_{\overline{X}}(\log D) \otimes \Omega^1_Z \otimes L^{-2} = \mathcal{O}_Z \oplus 2\mathcal{O}_Z(-1) \oplus \mathcal{O}_Z(-2)$. However we are not able to decide whether these 3 sections lift to $\overline{X}$. When we restrict to forms with fewer poles, then the vanishing will hold for a kind of cuspidal cohomology.

10.7 Proof of the Main Theorem

In paragraph 10.7.1 we prove our main theorem, in paragraph 10.7.2 we give some indication on the proof of the motivic decomposition conjecture in our case, however the details will be published in a forthcoming paper. We thus will drop the assumption on the motivic decomposition conjecture in Theorem 10.7.2.

10.7.1 From Relative to Absolute

We now state and prove our main theorem. Let $p : \overline{\mathcal{A}} \longrightarrow \overline{X}$ be the compactified family over Holzapfel's surface. Assume the motivic decomposition conjecture 10.1.1 ([5], [10, Conj. 1.4]) for $\overline{\mathcal{A}}/\overline{X}$. In the proof we will need an auxiliary statement which was implicitely proven in section 6:

Lemma 10.7.1. *Let $x \in X$ be a base point. Then $\pi_1^{\mathrm{top}}(X, x)$ acts on the Betti cohomology group $H^{2j}(\mathcal{A}_x(\mathbb{C}), \mathbb{Q})$. Then, for $0 \le j \le d = 3$, the cycle class map $CH^j(\mathcal{A}) \to H^{2j}(\mathcal{A}_x(\mathbb{C}), \mathbb{Q})^{\pi_1^{top}(X,x)}$ is surjective.*

Proof By Lemma 5.3 the sheaf $R^1p_*\mathbb{C}$ is a sum of two irreducible representation of $\pi_1^{top}(X,x)$. By the proof of Theorem 6.5., $R^2p_*\mathbb{C}$ decomposes into a one–dimensional constant representation and three irreducible ones. The constant part corresponds to the identity in $\mathrm{End}(\mathbb{V}_1)$ and therefore to the polarization class on the fibers, which is a Hodge class. Therefore the invariant classes in $H^2(\mathcal{A}_x(\mathbb{C}),\mathbb{Q})$, and by duality also in $H^4(\mathcal{A}_x(\mathbb{C}),\mathbb{Q})$, consist of Hodge classes and are hence in the image of the cycle class map by the Hodge conjecture for divisors (and curves). □

Now we can prove our main theorem:

Theorem 10.7.2. *Assuming the motivic decomposition conjecture 10.1.1, the total space of the family $p:\overline{\mathcal{A}}\longrightarrow\overline{X}$ supports a partial set of Chow–Künneth projectors π_i for $i\neq 4,5,6$.*

Proof The motivic decomposition conjecture 10.1.1 states that we have a relative Chow–Künneth decomposition with projectors Π^i_α on strata X_α which is compatible with the topological decomposition theorem [3]

$$\sum_{j,\alpha}\Psi^j_\alpha:\mathbb{R}\overline{p}_*\mathbb{Q}_{\overline{a}}\xrightarrow{\cong}\bigoplus_{j,\alpha}IC_{X_\alpha}(\mathcal{V}^j_\alpha)[-j-\dim(X_\alpha)].$$

Now we want to pass from relative Chow–Künneth decompositions to absolute ones. We use the notation of [11] and for the reader's convenience we recall everything. Let P^i/X and P^i_α/X be the mutually orthogonal projectors adding up to the identity $\Delta(a/X)\in CH_{\dim(A)}(a\times_X a)$ such that

$$(P^i/X)_*\mathbb{R}p_*\mathbb{Q}_a=IC_X(R^ip_*\mathbb{Q}_a)[-i],\ (P^i_\alpha/X)_*\mathbb{R}p_*\mathbb{Q}_a=IC_{X_\alpha}(\mathcal{V}^i_\alpha)[-i-\dim(X_\alpha)],$$

where the sheaves $\mathcal{V}^i_\alpha$ are local systems supported over the cusps. The projectors P^i_α/X on the boundary strata decompose further into Chow–Künneth components, since the boundary strata consist of smooth elliptic curves and the stratification has the product type fibers described in theorem 10.4.1. Let us now summarize what we know about the local systems $R^ip_*\mathbb{C}$ on the open stratum X_0 from section 6: $R^1p_*\mathbb{C}$ is irreducible and has no cohomology except in degree 2 by Theorem 10.6.5. $R^2p_*\mathbb{C}$ contains a trivial subsystem and the remaining complement has no cohomology except in degree 2 again by Theorem 10.6.5. $R^3p_*\mathbb{C}$ also contains a trivial subsystem and its complement has cohomology possibly in degrees 1, 2, 3, see section 6. By duality similar properties hold for $R^ip_*\mathbb{C}$ with $i=4,5,6$.

Using these properties together with Lemma 10.7.1 we can follow closely the proof of Thm. 1.3 in [11]: First construct projectors $(P^{2r}/X)_{\text{alg}}$ which are constituents of (P^{2r}/X) for $0 \le r \le 3$. This follows directly from Lemma 10.7.1 as in Step II of [11, section 1.7.]. Step III from [11, section 1.7.] is valid by the vanishing observations above. As in Step IV of loc. cit. this implies that we have a decomposition into motives in $\mathrm{CH}\mathcal{M}(k)$

$$M^{2r-1} = (a, P^{2r-1}, 0) \; (1 \le r \le d), \; M^{2r}_{\text{trans}} = (a, P^{2r}_{\text{trans}}, 0) \; (0 \le r \le d),$$

$$M^{2r}_{\text{alg}} = (a, P^{2r}_{\text{alg}}, 0) \; (0 \le r \le d),$$

plus additional boundary motives M^j_α for each stratum X_α. As in Step V of [11] we can split M^{2r}_{alg} further. The projectors constructed in this way define a set of Chow–Künneth projectors π_i for $i \ne 4, 5, 6$, since the relative projectors which contribute to more than one cohomology only affect cohomological degrees $4, 5$ and 6. □

Remark 10.7.3. If $H^1(X, S^2\mathbb{V}_1)$ vanishes or consists of algebraic $(2,2)$ Hodge classes only, then we even obtain a complete Chow–Künneth decomposition in the same way, since algebraic Hodge (p,p)–classes define Lefschetz motives $\mathbb{Z}(-p)$ which can be split off by projectors in a canonical way. Therefore the Hodge conjecture on a would imply a complete Chow–Künneth decomposition. However the Hodge conjecture is not very far from proving the total decomposition directly.

10.7.2 Motivic decomposition conjecture

The goal of this paragraph is to sketch the proof of the motivic decomposition conjecture 10.1.1 in the case we treat in this paper. The complete details for the following argument will be published in a future publication. First note that since $\mathcal{A}$ is an abelian variety we can use the work of Deninger and Murre ([6]) on Chow-Künneth decompositions of Abelian schemes to obtain relative Chow-Künneth projectors for $\mathcal{A}/X$. To actually get relative Chow-Künneth projectors for $\overline{\mathcal{A}}/\overline{X}$, we observe the following.

Recall our results in section 10.4. We showed that the special fibres over the smooth elliptic cusp curves D_i are of the form $Y_s = E \times \mathbb{P}^1 \times \mathbb{P}^1$. We do not need the cycle class map

$$CH_*(Y_s \times Y_s) \to H_*(Y_s \times Y_s)$$

to be an isomorphism as in [10, Thm. I]. Since the boundary strata on $\overline{X}$ are smooth elliptic curves it is sufficient to know the Hodge conjecture for the special fibres. But the special fibres are composed of elliptic curves and rational varieties by our results in section 10.4. Therefore the methods in [10] can be refined to work also in this case and we can drop the assumption in theorem 10.7.2.

Remarks 10.7.4. We hope to come back to this problem later and prove the motivic decomposition conjecture for all Picard families. The existence of absolute Chow–Künneth decompositions however seems to be out of reach for other examples since vanishing results will hold only for large arithmetic subgroups, i.e., small level.

10.8 Appendix: Algebraic L^2-sub complexes of symmetric powers of the uniformizing bundle of a two–dimensional complex ball quotient

$\overline{X}$ a 2-dim projective variety with a normal crossing divisor D, $X = \overline{X} - D$; assume that the coordinates near the divisor are z_1, z_2.

Consider the uniformizing bundle of a 2-ball quotient

$$E = \left(\Omega^1_{\overline{X}}(\log D) \otimes \mathcal{K}^{-1/3}_{\overline{X}}(\log D)\right) \oplus \mathcal{K}^{-1/3}_{\overline{X}}(\log D)$$

We consider two cases: 1) D is a smooth divisor (the case we need) and 2) D is a normal crossing divisor.

Case 1 Assume that D is defined by $z_1 = 0$. Taking v as the generating section of $\mathcal{K}^{-1/3}_{\overline{X}}(\log D)$, $\frac{dz_1}{z_1} \otimes v, dz_2 \otimes v$ as the generating sections of $\Omega^1_{\overline{X}}(\log D) \otimes \mathcal{K}^{-1/3}_{\overline{X}}(\log D)$, then the Higgs field

$$\vartheta : E \to E \otimes \Omega^1_{\overline{X}}(\log D)$$

is defined by setting $\vartheta(\frac{dz_1}{z_1} \otimes v) = v \otimes \frac{dz_1}{z_1}$, $\vartheta(dz_2 \otimes v) = v \otimes dz_2$, and $\vartheta(v) = 0$.

Clearly, if ϑ is written as $N_1 \frac{dz_1}{z_1} + N_2 dz_2$, then $N_1(\frac{dz_1}{z_1} \otimes v) = v$, $N_1(dz_2 \otimes v) = 0$, $N_1(v) = 0$, $N_2(\frac{dz_1}{z_1} \otimes v) = 0$, $N_2(dz_2 \otimes v) = v$, $N_2(v) = 0$; the kernel of N_1 is the subsheaf generated by $dz_2 \otimes v$

and v. Using the usual notation, we then have

$$\begin{aligned}
\mathrm{Gr}_1 W(N_1) &= \text{generated by } \frac{dz_1}{z_1} \otimes v \\
\mathrm{Gr}_0 W(N_1) &= \text{generated by } dz_2 \otimes v \\
\mathrm{Gr}_{-1} W(N_1) &= \text{generated by } v.
\end{aligned}$$

So with $\{x\}$ representing the line bundle generated by an element x, one has

$$\begin{aligned}
\Omega^0(E)_{(2)} &= z_1\{\frac{dz_1}{z_1} \otimes v\} + \{dz_2 \otimes v\} + \{v\} \\
&= \mathrm{Ker} N_1 + z_1 E; \\
\Omega^1(E)_{(2)} &= \frac{dz_1}{z_1} \otimes (z_1\{\frac{dz_1}{z_1} \otimes v\} + z_1\{dz_2 \otimes v\} + z_1\{v\}) \\
&\quad + dz_2 \otimes (z_1\{\frac{dz_1}{z_1} \otimes v\} + \{dz_2 \otimes v\} + \{v\}) \\
&= \frac{dz_1}{z_1} \otimes z_1 E + dz_2 \otimes (\mathrm{Ker} N_1 + z_1 E); \\
\Omega^2(E)_{(2)} &= \frac{dz_1}{z_1} \wedge dz_2 \otimes (z_1\{\frac{dz_1}{z_1} \otimes v\} + z_1\{dz_2 \otimes v\} + z_1\{v\}) \\
&= \frac{dz_1}{z_1} \wedge dz_2 \otimes z_1 E.
\end{aligned}$$

Case 2 : As before, taking v as the generating section of $\mathcal{K}_{\overline{X}}^{-1/3}(\log D)$, $\frac{dz_1}{z_1} \otimes v$, $\frac{dz_2}{z_2} \otimes v$ as the generating sections of $\Omega^1_{\overline{X}}(\log D) \otimes \mathcal{K}_{\overline{X}}^{-1/3}(\log D)$, then the Higgs field

$$\vartheta : E \to E \otimes \Omega^1_{\overline{X}}(\log D)$$

is defined by setting $\vartheta(\frac{dz_1}{z_1} \otimes v) = v \otimes \frac{dz_1}{z_1}$, $\vartheta(\frac{dz_2}{z_2} \otimes v) = v \otimes \frac{dz_2}{z_2}$, and $\vartheta(v) = 0$.

Clearly, if ϑ is written as $N_1 \frac{dz_1}{z_1} + N_2 \frac{dz_2}{z_2}$, then $N_1(\frac{dz_1}{z_1} \otimes v) = v$, $N_1(\frac{dz_2}{z_2} \otimes v) = 0$, $N_1(v) = 0$, $N_2(\frac{dz_1}{z_1} \otimes v) = 0$, $N_2(\frac{dz_2}{z_2} \otimes v) = v$, $N_2(v) = 0$; the kernel of N_1 (resp. N_2) is the subsheaf generated by

$\frac{dz_2}{z_2} \otimes v$ (resp. $\frac{dz_1}{z_1} \otimes v$) and v. We then have

$$\begin{aligned}
\mathrm{Gr}_1 W(N_1) &= \text{generated by } \frac{dz_1}{z_1} \otimes v \\
\mathrm{Gr}_0 W(N_1) &= \text{generated by } \frac{dz_2}{z_2} \otimes v \\
\mathrm{Gr}_{-1} W(N_1) &= \text{generated by } v \\
\mathrm{Gr}_1 W(N_2) &= \text{generated by } \frac{dz_2}{z_2} \otimes v \\
\mathrm{Gr}_0 W(N_2) &= \text{generated by } \frac{dz_1}{z_1} \otimes v \\
\mathrm{Gr}_{-1} W(N_2) &= \text{generated by } v \\
\mathrm{Gr}_1 W(N_1 + N_2) &= \text{generated by } (\frac{dz_1}{z_1} + \frac{dz_2}{z_2}) \otimes v \\
\mathrm{Gr}_0 W(N_1 + N_2) &= \text{generated by } (\frac{dz_1}{z_1} - \frac{dz_2}{z_2}) \otimes v \\
\mathrm{Gr}_{-1} W(N_1 + N_2) &= \text{generated by } v
\end{aligned}$$

So, one has

$$\begin{aligned}
\Omega^0(E)_{(2)} &= z_1\{\frac{dz_1}{z_1} \otimes v\} + z_2\{\frac{dz_2}{z_2} \otimes v\} + \{v\} \\
&= \mathrm{Ker} N_1 \cap \mathrm{Ker} N_2 + z_2 \mathrm{Ker} N_1 + z_1 \mathrm{Ker} N_2; \\
\Omega^1(E)_{(2)} &= \frac{dz_1}{z_1} \otimes (z_1\{\frac{dz_1}{z_1} \otimes v\} + z_1 z_2\{\frac{dz_2}{z_2} \otimes v\} + z_1\{v\}) \\
&\quad + \frac{dz_2}{z_2} \otimes (z_2\{\frac{dz_2}{z_2} \otimes v\} + z_1 z_2\{\frac{dz_1}{z_1} \otimes v\} + z_2\{v\}) \\
&= \frac{dz_1}{z_1} \otimes (z_1 \mathrm{Ker} N_2 + z_1 z_2 \mathrm{Ker} N_1) \\
&\quad + \frac{dz_2}{z_2} \otimes (z_2 \mathrm{Ker} N_1 + z_1 z_2 \mathrm{Ker} N_2); \\
\Omega^2(E)_{(2)} &= \frac{dz_1}{z_1} \wedge \frac{dz_2}{z_2} \otimes z_1 z_2 E,
\end{aligned}$$

For the above two cases, it is to easy to check that $\vartheta(\Omega^0(E)_{(2)}) \subset \Omega^1(E)_{(2)}$ and $\vartheta(\Omega^1(E)_{(2)}) \subset \Omega^2(E)_{(2)}$. Thus, together $\vartheta \wedge \vartheta = 0$, we have the complex $(\{\Omega^i(E)_{(2)}\}_{i=0}^2, \vartheta)$

$$0 \to \Omega^0(E)_{(2)} \to \Omega^1(E)_{(2)} \to \Omega^2(E)_{(2)} \to 0$$

with ϑ as the boundary operator.

Now we take the 2nd-order symmetric power of (E,ϑ), we obtain a new Higgs bundle $\mathcal{S}^2(E,\vartheta)$ (briefly, the Higgs field is still denoted by ϑ) as follows,

$$\begin{aligned}\mathcal{S}^2(E,\vartheta) = \mathcal{S}^2\left(\Omega^1_{\overline{X}}(\log D)\right) \otimes \mathcal{K}^{-2/3}_{\overline{X}}(\log D) \oplus \\ \oplus\, \Omega^1_{\overline{X}}(\log D) \otimes \mathcal{K}^{-2/3}_{\overline{X}}(\log D) \oplus \mathcal{K}^{-2/3}_{\overline{X}}(\log D).\end{aligned}$$

The Higgs field ϑ maps $\mathcal{S}^2(E,\vartheta)$ into $\mathcal{S}^2(E,\vartheta) \otimes \Omega^1_{\overline{X}}(\log D)$ and $\mathcal{S}^2(E,\vartheta) \otimes \Omega^1_{\overline{X}}(\log D)$ into $\mathcal{S}^2(E,\vartheta) \otimes \Omega^2_{\overline{X}}(\log D)$ so that one has a complex with the differentiation ϑ as follows

$$(*) \qquad \begin{aligned} 0 \to \mathcal{S}^2(E,\vartheta) \to \mathcal{S}^2(E,\vartheta) \otimes \Omega^1_{\overline{X}}(\log D) \\ \to \mathcal{S}^2(E,\vartheta) \otimes \Omega^2_{\overline{X}}(\log D) \to 0;\end{aligned}$$

more precisely, one has

$$\begin{aligned}
&\vartheta\left(\mathcal{S}^2\left(\Omega^1_{\overline{X}}(\log D)\right) \otimes \mathcal{K}^{-2/3}_{\overline{X}}(\log D)\right) \subset \left(\Omega^1_{\overline{X}}(\log D) \otimes \mathcal{K}^{-2/3}_{\overline{X}}(\log D)\right) \\
&\qquad\qquad\qquad\qquad \otimes \Omega^1_{\overline{X}}(\log D) \\
&\vartheta\left(\Omega^1_{\overline{X}}(\log D) \otimes \mathcal{K}^{-2/3}_{\overline{X}}(\log D)\right) \subset \mathcal{K}^{-2/3}_{\overline{X}}(\log D) \otimes \Omega^1_{\overline{X}}(\log D) \\
&\vartheta\left(\mathcal{K}^{-2/3}_{\overline{X}}(\log D)\right) = 0
\end{aligned}$$

and

$$\begin{aligned}
&\vartheta\left(\left(\mathcal{S}^2\left(\Omega^1_{\overline{X}}(\log D)\right) \otimes \mathcal{K}^{-2/3}_{\overline{X}}(\log D)\right) \otimes \Omega^1_{\overline{X}}(\log D)\right) \\
&\subset \left(\Omega^1_{\overline{X}}(\log D) \otimes \mathcal{K}^{-2/3}_{\overline{X}}(\log D)\right) \otimes \Omega^2_{\overline{X}}(\log D) \\
&\vartheta\left(\left(\Omega^1_{\overline{X}}(\log D) \otimes \mathcal{K}^{-2/3}_{\overline{X}}(\log D)\right) \otimes \Omega^1_{\overline{X}}(\log D)\right) \subset \mathcal{K}^{-2/3}_{\overline{X}}(\log D) \otimes \Omega^2_{\overline{X}}(\log D) \\
&\vartheta\left(\mathcal{K}^{-2/3}_{\overline{X}}(\log D) \otimes \Omega^1_{\overline{X}}(\log D)\right) = 0
\end{aligned}$$

Note: Let V be a $SL(2)$-module, then

$$\mathcal{S}^2 V \otimes V \simeq \mathcal{S}^3 V \oplus V \otimes \wedge^2 V.$$

In general, one needs to consider the representations of $GL(2)$; in such a case, we can take the determinant of the representation in question, and then go back to a representation of $SL(2)$.

$$\begin{aligned}
\mathcal{S}^2(E,\vartheta) &= \mathcal{S}^2\big(\Omega^1_{\overline{X}}(\log D)\big)\otimes\mathcal{K}^{-2/3}_{\overline{X}}(\log D)\\
&\oplus\Omega^1_{\overline{X}}(\log D)\otimes\mathcal{K}^{-2/3}_{\overline{X}}(\log D)\\
&\oplus\mathcal{K}^{-2/3}_{\overline{X}}(\log D)\\
\mathcal{S}^2(E,\vartheta)\otimes\Omega^1_{\overline{X}}(\log D) &= \big(\mathcal{S}^2\big(\Omega^1_{\overline{X}}(\log D)\big)\otimes\mathcal{K}^{-2/3}_{\overline{X}}(\log D)\big)\otimes\Omega^1_{\overline{X}}(\log D)\\
&\oplus\big(\Omega^1_{\overline{X}}(\log D)\otimes\mathcal{K}^{-2/3}_{\overline{X}}(\log D)\big)\otimes\Omega^1_{\overline{X}}(\log D)\\
&\oplus\mathcal{K}^{-2/3}_{\overline{X}}(\log D)\otimes\Omega^1_{\overline{X}}(\log D)\\
\mathcal{S}^2(E,\vartheta)\otimes\Omega^1_{\overline{X}}(\log D) &= \big(\mathcal{S}^2\big(\Omega^1_{\overline{X}}(\log D)\big)\otimes\mathcal{K}^{-2/3}_{\overline{X}}(\log D)\big)\otimes\Omega^2_{\overline{X}}(\log D)\\
&\oplus\big(\Omega^1_{\overline{X}}(\log D)\otimes\mathcal{K}^{-2/3}_{\overline{X}}(\log D)\big)\otimes\Omega^2_{\overline{X}}(\log D)\\
&\oplus\mathcal{K}^{-2/3}_{\overline{X}}(\log D)\otimes\Omega^2_{\overline{X}}(\log D)
\end{aligned}$$

Assuming that the divisor D is smooth, we next want to consider the L^2-holomorphic Dolbeault sub-complex of the above complex (*):

$$\begin{aligned}
0\to(\mathcal{S}^2(E,\vartheta))_{(2)}&\to(\mathcal{S}^2(E,\vartheta)\otimes\Omega^1_{\overline{X}}(\log D))_{(2)}\\
&\to(\mathcal{S}^2(E,\vartheta)\otimes\Omega^2_{\overline{X}}(\log D))_{(2)}\to 0,
\end{aligned}$$

and explicitly write down $(\mathcal{S}^2(E,\vartheta)\otimes\Omega^i_{\overline{X}}(\log D))_{(2)}$.

Note that taking symmetric power for L^2-complex does not have obvious functorial properties in general.

We will continue to use the previous notations. For simplicity, we will further set $v_1=\frac{dz_1}{z_1}\otimes v$ and $v_2=dz_2\otimes v$; we also denote $e_1\otimes e_2+e_2\otimes e_1$ by $e_1\odot e_2$, the symmetric product of the vectors e_1 and e_2.

Thus, $\mathcal{S}^2\big(\Omega^1_{\overline{X}}(\log D)\big)\otimes\mathcal{K}^{-2/3}_{\overline{X}}(\log D)$, as a sheaf, is generated by $v_1\odot v_1$, $v_1\odot v_2$, $v_2\odot v_2$; $\Omega^1_{\overline{X}}(\log D)\otimes\mathcal{K}^{-2/3}_{\overline{X}}(\log D)$ is generated by $v_1\odot v$, $v_2\odot v$; and $\mathcal{K}^{-2/3}_{\overline{X}}(\log D)$ is generated by $v\odot v$. Also, it is easy to check how N_1, N_2 act on these generators; as for N_1, we have (Note $N_1v_1=v, N_1v_2=0, N_1v=0$.)

$$\begin{aligned}
N_1(v_1\odot v_1)&=2v_1\odot v\\
N_1(v_1\odot v_2)&=v_2\odot v\\
N_1(v_2\odot v_2)&=0\\
N_1(v_1\odot v)&=v\odot v\\
N_1(v_2\odot v)&=0\\
N_1(v\odot v)&=0.
\end{aligned}$$

Clearly, N_1 maps $\mathcal{S}^2(\Omega^1_{\overline{X}}(\log D)) \otimes \mathcal{K}^{-2/3}_{\overline{X}}(\log D)$ into $\Omega^1_{\overline{X}}(\log D) \otimes \mathcal{K}^{-2/3}_{\overline{X}}(\log D)$, $\Omega^1_{\overline{X}}(\log D) \otimes \mathcal{K}^{-2/3}_{\overline{X}}(\log D)$ into $\mathcal{K}^{-2/3}_{\overline{X}}(\log D)$, and then $\mathcal{K}^{-2/3}_{\overline{X}}(\log D)$ to 0. So, N_1 is of index 3 on the 2nd-order symmetric power $\mathcal{S}^2E$(as is obvious from the abstract theory since N_1 is of index 2 on E); and we then have the following gradings

$$\begin{aligned}
\mathrm{Gr}_2 W(N_1) &= \text{generated by } \ v_1 \odot v_1 \\
\mathrm{Gr}_1 W(N_1) &= \text{generated by } \ v_1 \odot v_2 \\
\mathrm{Gr}_0 W(N_1) &= \text{generated by } \ v_1 \odot v, v_2 \odot v_2 \\
\mathrm{Gr}_{-1} W(N_1) &= \text{generated by } \ v_2 \odot v \\
\mathrm{Gr}_{-2} W(N_1) &= \text{generated by } \ v \odot v.
\end{aligned}$$

(Note that N_1, acting on E, has two invariant (irreducible) components, one being generated by v_1, v, the other by v_2, so that N_1 has three invariant components on $\mathcal{S}^2E$, as is explicitly showed in the above gradings.)

Now we can write down L^2-holomorphic sections of $\mathcal{S}^2E$, namely the sections generated by $v_1 \odot v, v_2 \odot v_2, v_2 \odot v, v \odot v$, and $z_1\mathcal{S}^2E$; in the invariant terms, they should be

$$(\mathcal{S}^2(E, \vartheta))_{(2)} = E \odot \mathrm{Im}N_1 + \mathcal{S}^2(\mathrm{Ker}N_1) + z_1\mathcal{S}^2E.$$

Now it is easy to also write down $(\mathcal{S}^2(E, \vartheta) \otimes \Omega^1_{\overline{X}}(\log D))_{(2)}$ and $(\mathcal{S}^2(E, \vartheta) \otimes \Omega^2_{\overline{X}}(\log D))_{(2)}$:

$$\begin{aligned}
(\mathcal{S}^2(E, \vartheta) \otimes \Omega^1_{\overline{X}}(\log D))_{(2)} &= \frac{dz_1}{z_1} \otimes (\mathcal{S}^2(\mathrm{Im}N_1) + z_1\mathcal{S}^2E) \\
&\quad + dz_2 \otimes (E \odot \mathrm{Im}N_1 + \mathcal{S}^2(\mathrm{Ker}N_1) + z_1\mathcal{S}^2E); \\
(\mathcal{S}^2(E, \vartheta) \otimes \Omega^2_{\overline{X}}(\log D))_{(2)} &= \frac{dz_1}{z_1} \wedge dz_2 \otimes (\mathcal{S}^2(\mathrm{Im}N_1) + z_1\mathcal{S}^2E).
\end{aligned}$$

Similary, one can determine the algebraic L^2-sub complex of $S^n(E, \vartheta)$ for any $n \in \mathbb{N}$.

10.9 Acknowledgement

It is a pleasure to dedicate this work to Jaap Murre who has been so tremendously important for the mathematical community. In particular we want to thank him for his constant support during so many years.

We are grateful to Bas Edixhoven, Jan Nagel and Chris Peters for organizing such a wonderful meeting in Leiden. Many thanks go to F. Grunewald, R.-P. Holzapfel, J.-S. Li and J. Schwermer for very helpful discussions.

References

[1] Angel, P. del and S. Müller-Stach: On Chow motives of 3–folds, Transactions of the AMS **352**, 1623–1633 (2000).

[2] Ash, A., D. Mumford, M. Rapoport and Y. Tai : *Smooth compactification of locally symmetric varieties*, Math. Sci. Press, Brookline, 1975.

[3] Beilinson, A., J. Bernstein and P. Deligne: *Faisceaux pervers, analyse et topologie sur les espaces singuliers*, Astérisque **100**, 7–171 (1982).

[4] Borel, A. and N. Wallach: Continuous cohomology, discrete subgroups, and representations of reductive groups, Annals of Math. Studies **94**, (1980).

[5] Corti, A. and M. Hanamura: Motivic decomposition and intersection Chow groups I, Duke Math. J. **103** (2000), no. 3, 459–522.

[6] Deninger, C. and J. Murre: Motivic decomposition of abelian schemes and the Fourier transform, J. Reine Angew. Math., **422** (1991), 201–219.

[7] Esnault, H. and E. Viehweg: *Lectures on Vanishing Theorems*, freely available under http://www.uni-essen.de/~mat903.

[8] Chai, C. and G. Faltings: *Degenerations of Abelian Varieties*, Ergebnisse der Mathematik, Band **22**, Springer Verlag, 1990.

[9] Gordon, B. B.: Canonical models of Picard modular surfaces, in *The zeta functions of Picard modular surfaces*, Univ. Montréal, Montreal, 1992, 1–29.

[10] Gordon, B. B., M. Hanamura and J. P. Murre: Relative Chow-Künneth projectors for modular varieties, J. Reine Angew. Math., **558** (2003), 1–14.

[11] Gordon, B. B., M. Hanamura and J. P. Murre: Chow-Künneth projectors for modular varieties, C. R. Acad. Sci. Paris, Ser. I, **335**, 745–750 (2002).

[12] Gordon, B. B., M. Hanamura and J. P. Murre: Absolute Chow-Künneth projectors for modular varieties, Crelle Journal **580**, 139–155 (2005).

[13] Hemperly, J.: The parabolic contribution to the number of linearly independent automorphic forms on a certain bounded domain, Am.Journ.Math **94**, 1972.

[14] Hirzebuch, F.: Chern numbers of algebraic surfaces- an example, Math. Annalen **266**, 351–356 (1984).

[15] Holzapfel, R.-P.: Chern numbers of algebraic surfaces – Hirzebruch's examples are Picard modular surfaces, Math. Nachrichten **126**, 255–273 (1986).

[16] Holzapfel, R.-P.: *Geometry and arithmetic around Euler partial differential equations*, Reidel Publ. Dordrecht (1987).

[17] Jost, J., Y.-H. Yang and K. Zuo: The cohomology of a variation of polarized Hodge structures over a quasi–compact Kähler manifold, preprint math.AG/0312145.

[18] Künnemann, K.: On the Chow motive of an abelian scheme, in *Motives (Seattle, WA, 1991)*, Providence, RI, 1994, Amer. Math. Soc., 189–205.

[19] Li, J.-S. and J. Schwermer: On the Eisenstein cohomology of arithmetic groups, Duke Math. J. **123** no. 1, 141–169 (2004).

[20] Looijenga, E.: Compactifications defined by arrangements I: The ball–quotient case, Duke Math. J. **118** (2003), no. 1, 151–187.

[21] Matsushima, Y. and G. Shimura: On the cohomology groups attached to certain vector valued differential forms on the product of upper half planes, Ann. of Math. **78**, 417–449 (1963).

[22] Miller. A.: The moduli of Abelian Threefolds with Complex Multiplication and Compactifications, Preprint.

[23] Miller. A.: Families of Abelian Threefolds with CM and their degenerations, Preprint.

[24] Miyaoka, Y.: Examples of stable Higgs bundles with flat connections, preprint.

[25] Mumford, D.: An analytic construction of degenerating abelian varieties over complete rings, Compositio Math. 24, 1972, 239-272. Appeared also as an appendix to [8].

[26] Murre, J. P.: Lectures on Motives, in *Proceedings Transcendental Aspects of Algebraic Cycles, Grenoble 2001*, LMS Lectures Note Series **313**, Cambridge Univ. Press (2004), 123–170.

[27] Murre, J. P.: On the motive of an algebraic surface, J. Reine Angew. Math. **409** (1990), 190–204.

[28] Picard, E.: Sur des fonctions de deux variables indépendentes analogues aux fonctions modulaires, Acta. Math. **2**, 114–135 (1883).

[29] Simpson, C.: Constructing variations of Hodge structure using Yang-Mills theory and applications to uniformization, J. Am. Math. Soc. **1**, No.4, 867–918 (1988).

[30] Vogan, D. and G. Zuckerman: Unitary representations with non–zero cohomology, Compositio Math. **53**, 51–90 (1984).

11

The Regulator Map for Complete Intersections

Jan Nagel

Université Lille 1, Mathématiques - Bât. M2,
F-59655 Villeneuve d'Ascq Cedex, France
nagel@math.univ-lille1.fr

To Professor Murre, with great respect.

11.1 Introduction

Since the introduction of the theory of infinitesimal variations of Hodge structure, infinitesimal methods have been successfully applied to a number of problems concerning the relationship between algebraic cycles and Hodge theory. One of the common techniques is to study infinitesimal invariants associated to families of algebraic cycles. This approach led to a proof of the infinitesimal Noether–Lefschetz theorem and was further developed by Green and Voisin in their study of the image of the Abel–Jacobi map for hypersurfaces in projective space. This work was reinterpreted and extended by Nori [18]. He proved a connectivity theorem for the universal family X_T of complete intersections of multidegree $(d_0, \ldots, d_r)$ on a polarised variety $(Y, \mathcal{O}_Y(1))$ inside the trivial family $Y_T = Y \times T$. Specifically, he proved that if the fibers of $X_T \to T$ are n–dimensional then $H^{n+k}(Y_T, X_T; \mathbb{Q}) = 0$ for all $k \leq n$ if $\min(d_0, \ldots, d_r)$ is sufficiently large. In [16] I proved an effective version of Nori's connectivity theorem; see [23] and [1] for related results in the case $Y = \mathbb{P}^N$.

For geometric applications of Nori's theorem one usually does not need the full strength of Nori's theorem; it often suffices to have $H^{n+k}(Y_T, X_T) = 0$ for all $k \leq c$, for some integer $c \leq n$. The theorems of Noether–Lefschetz and Green–Voisin can be deduced from Nori's theorem by taking $Y = \mathbb{P}^N$ and $(c, n) = (1, 2m)$ or $(c, n) = (2, 2m - 1)$. It turns out that in these cases, the degree bounds are sharp.

One can also use Nori's theorem to study the regulator maps on Bloch's higher Chow groups, as was noted in [21]. In [23] Voisin considered the extreme case $c = n$ of Nori's theorem for hypersurfaces in projective space and showed that also in this case the bound is sharp, by constructing interesting higher Chow cycles on hypersurfaces of low degree. One could therefore ask whether the degree bounds computed in [16, Thm. 3.13] are optimal for complete intersections in projective space. In this note we show that this is not the case, by studying the image of the regulator maps defined on the higher Chow groups $\mathsf{CH}^p(X,1)$ and $\mathsf{CH}^p(X,2)$. We improve the bounds computed in [16, Theorems. 4.4 and 4.6] using two methods: (1) a version of the Jacobi ring introduced in [1]; (2) a correspondence between the cohomology of complete intersections of quadrics and double coverings of projective space, combined with a version of Nori's theorem for cyclic coverings.

These results are treated in sections 2 and 3. We conclude with an improved result on the image of the regulator map (Theorem 11.4.2) which is optimal for $\mathsf{CH}^p(X,2)$.

Acknowledgment. A part of this paper was prepared during a visit to the Max–Planck Institut für Mathematik in Bonn in the spring of 2003. I would like to thank the institute for its hospitality and excellent working conditions.

11.2 Infinitesimal calculations

For the definition and basic properties of Bloch's higher Chow groups $\mathsf{CH}^p(X,q)$ we refer to [15]. There exist regulator maps

$$c_{p,q} : \mathsf{CH}^p(X,q) \to H^{2p-q}_{\mathcal{D}}(X, \mathbb{Z}(p))$$

that generalise the classical Deligne cycle class map; see [10] for an explicit description of these maps using integration currents. The starting point is the following result; cf. [16, Lemma 4.1] and the references cited there.

Proposition 11.2.1. *Let $U \subset \prod_{i=0}^{r} \mathbb{P}H^0(\mathbb{P}^N, \mathcal{O}_{\mathbb{P}}(d_i))$ be the open subset parametrising smooth complete intersections of dimension n and multidegree $(d_0, \ldots, d_r)$ in $\mathbb{P}^N$, and let $X_U \subset \mathbb{P}^N_U = \mathbb{P}^N \times U$ be the universal family. If*

$$H^k_{\mathcal{D}}(\mathbb{P}^N \times T, X_T) = 0$$

for all $k \leq 2p - q + 1$ and for every smooth morphism $T \to U$, then the image of the regulator map

$$c_{p,q} : \mathsf{CH}^p(X,q) \otimes \mathbb{Q} \to H^{2p-k}_{\mathcal{D}}(X, \mathbb{Q}(p))$$

is contained in the image of the restriction map $H^{2p-q}_{\mathcal{D}}(\mathbb{P}^N, \mathbb{Q}(p)) \to H^{2p-q}_{\mathcal{D}}(X, \mathbb{Q}(p))$ if X is very general.

Using an effective version of Nori's connectivity theorem, we computed degree bounds for the cases $q = 1$ and $q = 2$.

Theorem 11.2.2. *Put*

$$\delta_{\min} = \min(d_0, \ldots, d_r), \ \delta_{\max} = \max(d_0, \ldots, d_r).$$

If $(n, q) \in \{(2m, 1), (2m-1, 2)\}$, *the conclusion of Proposition 11.2.1 holds if the following conditions are satisfied.*

(C_0) $\sum_{i=0}^{r} d_i + (m-1)\delta_{\min} \geq n + r + 3$;
(C_1) $\sum_{i=0}^{r} d_i + m\delta_{\min} \geq n + r + 2 + \delta_{\max}$.

Proof See [16, Theorems 4.4 and 4.6]. □

Corollary 11.2.3. *If* $(n, q) = (2m, 1)$, *the conclusion of Proposition 11.2.1 holds, with the possible exception of the following cases.*

i) $X = V(2) \subset \mathbb{P}^{2m+1}$, $X = V(3) \subset \mathbb{P}^3$, $X = V(4) \subset \mathbb{P}^3$, $X = V(3) \subset \mathbb{P}^5$;
ii) $X = V(d, 2) \subset \mathbb{P}^{2m+2}$, $d \geq 2$;
iii) $X = V(2, 2, 2) \subset \mathbb{P}^{2m+3}$.

Corollary 11.2.4. *If* $(n, q) = (2m-1, 2)$, *the conclusion of Proposition 11.2.1 holds, with the possible exception of the following cases.*

i) $X = V(2) \subset \mathbb{P}^{2m}$, $X = V(3) \subset \mathbb{P}^2$;
ii) $X = V(2, 2) \subset \mathbb{P}^{2m+1}$, $m \geq 1$.

Remark 11.2.5. Similar degree bounds can be worked out for $q \geq 3$. They coincide with the bounds of Corollary 11.2.3 (q odd) or Corollary 11.2.4 (q even). We have refrained from studying these cases as there is no description of $\mathsf{CH}^p(X, q)$ using Gersten–Quillen resolutions if $q \geq 3$.

To see whether the bounds of Corollaries 11.2.3 and 11.2.4 can be improved, we recall the idea of the proof of Theorem 11.2.2. Using mixed Hodge theory, one checks that it suffices to show

$$F^{m+1}H^{n+2}(\mathbb{P}^N_T, X_T) = 0 \tag{11.1}$$

in both cases. Let $f : X_T \to T$ be the structure morphism, and put $\mathcal{H}^{p,q}(X_T) = R^q f_* \Omega^p_{X_T/T}$. Let $\mathcal{H}^{p,q}_{\mathrm{pr}}(X_T)$ be the subbundle corresponding to primitive cohomology. By spectral sequence arguments one shows that the condition (11.1) is satisfied if the complex

$$0 \to \mathcal{H}^{p,n-p}_{\mathrm{pr}}(X_T) \to \Omega^1_T \otimes \mathcal{H}^{p-1,n-p+1}_{\mathrm{pr}}(X_T) \to \mathcal{H}^{p-2,n-p+2}_{\mathrm{pr}}(X_T) \tag{11.2}$$

is exact for all $p \geq m+1$.

Put $E = \bigoplus_{i=0}^{r} \mathcal{O}_{\mathbb{P}}(d_i)$, $P = \mathbb{P}(E^{\vee})$ and set $\xi_E = \mathcal{O}_P(1)$. Let Σ be the sheaf of differential operators of order ≤ 1 on sections of ξ_E. Let X be a complete intersection defined by a section $s = (f_0, \ldots, f_r)$, and let σ be the corresponding section of $H^0(P, \xi_E)$. Contraction with the 1–jet $j^1(\sigma)$ defines maps $K_P \otimes \Sigma \otimes \xi_E^{a-1} \to K_P \otimes \xi_E^a$ for all $a \geq 1$. Put

$$
\begin{aligned}
J(K_P \otimes \xi_E^a) &= \mathrm{Im}(H^0(P, K_P \otimes \Sigma \otimes \xi_E^{a-1}) \to H^0(P, K_P \otimes \xi_E^a)) \\
R(K_P \otimes \xi_E^a) &= H^0(P, K_P \otimes \xi_E^a)/J(K_P \otimes \xi_E^a).
\end{aligned}
$$

Proposition 11.2.6. *We have an isomorphism* $H^{n-p,p}_{\mathrm{pr}}(X) \cong R(K_P \otimes \xi_E^{p+r+1})$.

Proof Cf. [5, Section 10.4] and the references cited there. □

The proof of Theorem 11.2.2 proceeds as follows. By semicontinuity it suffices to check the exactness of (11.2) pointwise. One can reduce to the case $T = H^0(\mathbb{P}^N, E) - \Delta$, where Δ is the discriminant locus. Hence the tangent space to T is $V = H^0(\mathbb{P}^N, E)$. Applying these reductions to the dual of the complex (11.2), we see that it suffices to check the exactness of

$$\Lambda^2 V \otimes H^{n-p+2,p-2}_{\mathrm{pr}}(X_t) \to V \otimes H^{n-p+1,p-1}_{\mathrm{pr}}(X_t) \to H^{n-p,p}_{\mathrm{pr}}(X_t) \to 0. \quad (11.3)$$

By Proposition 11.2.6, this complex is isomorphic to

$$R_\bullet = (\Lambda^2 V \otimes R(K_P \otimes \xi_E^{p+r-1}) \to V \otimes R(K_P \otimes \xi_E^{p+r}) \to R(K_P \otimes \xi_E^{p+r+1}) \to 0).$$

Let $S_\bullet$ (resp. $J_\bullet$) be the complexes obtained by replacing the terms $R(K_P \otimes \xi_E^q)$ in $R_\bullet$ by $H^0(P, K_P \otimes \xi_E^q)$ (resp. $J(K_P \otimes \xi_E^q)$). The exact sequence of complexes $0 \to J_\bullet \to S_\bullet \to R_\bullet \to 0$ shows that $H_1(R_\bullet) = H_0(R_\bullet) = 0$ if

$$H_1(S_\bullet) = H_0(S_\bullet) = 0 \ (1), \quad H_0(J_\bullet) = 0 \ (2).$$

Using Castelnuovo–Mumford regularity, one then shows that (C_0) implies (1) and (C_1) implies (2).

It is possible to improve condition (ii) of Corollary 11.2.3. To this end, let $U' \subset \prod_{i=1}^{r} \mathbb{P}H^0(\mathbb{P}^N, \mathcal{O}_{\mathbb{P}}(d_i))$ be the open subset parametrising smooth complete intersections of multidegree $(d_1, \ldots, d_r)$, and let Y_U be the pull-back of its universal family to $U \subset \prod_{i=0}^{r} \mathbb{P}H^0(\mathbb{P}^N, \mathcal{O}_{\mathbb{P}}(d_i))$. Given a smooth morphism $T \to U$, the inclusions of pairs

$$(Y_T, X_T) \subset (\mathbb{P}^N_T, X_T) \subset (\mathbb{P}^N_T, Y_T)$$

induces a long exact sequence

$$\to H^k(\mathbb{P}^N_T, Y_T) \to H^k(\mathbb{P}^N_T, X_T) \to H^k(Y_T, X_T) \to H^{k+1}(\mathbb{P}^N_T, X_T) \to$$

of cohomology groups. The vanishing of $H^k(Y_T, X_T)$ was investigated by Asakura and S. Saito [1]. For the vanishing of $H^{n+2}(Y_T, X_T)$ one introduces the bundles $\mathcal{H}^{p,q}(Y_T, X_T)$ of relative cohomology and studies exactness of the complex

$$0 \to \mathcal{H}^{p,n-p+1}(Y_T, X_T) \to \Omega^1_T \otimes \mathcal{H}^{p-1,n-p+2}(Y_T, X_T) \\ \to \Omega^2_T \otimes \mathcal{H}^{p-2,n-p+3}(Y_T, X_T).$$

Exactness of this complex is reduced to exactness of

$$\Lambda^2 V \otimes R'(K_P \otimes \xi_E^{p+r-1}) \to V \otimes R'(K_P \otimes \xi_E^{p+r}) \to R'(K_P \otimes \xi_E^{p+r+1}) \to 0 \tag{11.4}$$

where R' is the Jacobi ring defined in [1, §1].

Theorem 11.2.7 (Asakura–Saito). *Put* $d_{\max} = \max(d_1, \ldots, d_r)$. *The complex* (11.4) *is exact if*

(C_0) $\sum_{i=0}^r d_i + (m-1)\delta_{\min} \geq n + r + 3$;
(C_1') $\sum_{i=0}^r d_i + m\delta_{\min} \geq n + r + 2 + d_{\max}$.

Proof See [1, Thm. 9-3 (ii)]. □

Corollary 11.2.8. *The conclusion of Proposition 11.2.1 holds if* $(n, q) = (2m, 1)$ *and* $(d_0, d_1) = (2, d)$ *if* $d \geq 4$.

Proof If $(d_0, d_1) = (2, d)$ then Y_T is a family of odd–dimensional quadrics. As these quadrics have no primitive cohomology, a Leray spectral sequence argument shows that $H^k(\mathbb{P}^N_T, Y_T) = 0$ for all k. Hence we obtain isomorphisms $H^k(\mathbb{P}^N_T, X_T) \cong H^k(Y_T, X_T)$ for all k. Using Theorem 11.2.7 we obtain $H^k(\mathbb{P}^N_T, X_T) = 0$ for all $k \leq 2m + 2$. Hence $H^k_{\mathcal{D}}(\mathbb{P}^N_T, X_T) = 0$ for all $k \leq 2m + 2$. □

11.3 Complete intersections of quadrics

In this section we show how to exclude the exceptional cases of Corollary 11.2.3 (iii) and Corollary 11.2.4 (ii) using a correspondence between the cohomology of a double covering of projective space and the cohomology of a complete intersection of quadrics, and a version of Nori's theorem for double coverings (and more generally cyclic coverings) of projective space.

We start with Corollary 11.2.3 (iii). Let $X = V(Q_0, Q_1, Q_2) \subset \mathbb{P}^{2m+3}$ be a smooth complete intersection of three quadrics. Given $(\lambda_0 : \lambda_1 : \lambda_2) \in \mathbb{P}^2$, write $Q_\lambda = \lambda_0 Q_0 + \lambda_1 Q_1 + \lambda_2 Q_2$. (By abuse of notation, we use the same notation for a quadric Q, its defining equation and its associated symmetric matrix.) Let

$$\mathcal{X} = \{(x,\lambda) \in \mathbb{P}^{2m+3} \times \mathbb{P}^2 | x \in Q_\lambda\}$$

be the associated quadric bundle over $\mathbb{P}^2$, and let

$$C = \{\lambda \in \mathbb{P}^2 | \mathrm{corank}(Q_\lambda) \geq 1\}$$

be the discriminant curve.

The passage from the complete intersection X to the hypersurface $\mathcal{X} \subset \mathbb{P}^{2m+3} \times \mathbb{P}^2$ induces an isomorphism on middle dimensional primitive cohomology. This result is sometimes referred to as the Cayley trick; cf. [7, §6].

Proposition 11.3.1 (Cayley trick). *We have an isomorphism of Hodge structures* $H^{2m}_{\mathrm{pr}}(X)(-2) \cong H^{2m+4}_{\mathrm{pr}}(\mathcal{X})$.

The isomorphism of the previous Proposition is induced by the correspondence

$$\begin{array}{ccc} \Gamma_1 = f^{-1}(X) = X \times \mathbb{P}^2 & \rightarrow & \mathcal{X} \\ \downarrow f & & \\ X. & & \end{array}$$

As a smooth quadric in $\mathbb{P}^{2m+3}$ contains two families of $(m+1)$–planes, there exists a family Γ_2 of $(m+1)$–planes contained in the fibers of $f : \mathcal{X} \to \mathbb{P}^2$. The base of this family is a double covering $\pi : S \to \mathbb{P}^2$ that is ramified over the discriminant curve C.

Theorem 11.3.2 (O'Grady). *The correspondence* $\Gamma = \Gamma_2 \circ {}^t\Gamma_1$ *induces an isomorphism of Hodge structures* $H^2_{\mathrm{pr}}(S,\mathbb{Q}) \cong H^{2m}_{\mathrm{pr}}(X,\mathbb{Q})$.

Proof See [19]; cf. [11] for a more general result, valid for arbitrary even–dimensional quadric bundles over $\mathbb{P}^2$. □

The discriminant curve C is defined by the homogeneous polynomial

$$F(\lambda_0,\lambda_1,\lambda_2) = \det(\lambda_0 Q_0 + \lambda_1 Q_1 + \lambda_2 Q_2)$$

of degree $2m+4$; if the quadrics Q_0, Q_1 and Q_2 are general, C is smooth. Consider the vector bundle $E = \oplus^3 \mathcal{O}_{\mathbb{P}}(2)$ on $\mathbb{P}^{2m+3}$, and the map

$$H : H^0(\mathbb{P}^{2m+3}, E) \to H^0(\mathbb{P}^2, \mathcal{O}_{\mathbb{P}}(2m+4))$$

that sends a net of quadrics to the equation of its discriminant curve. The map H induces a rational map

$$h : \mathbb{P}H^0(\mathbb{P}^{2m+3}, E) - - > \mathbb{P}H^0(\mathbb{P}^2, \mathcal{O}_{\mathbb{P}}(2m+4)).$$

Lemma 11.3.3. *Let $U \subset \mathbb{P}H^0(\mathbb{P}^2, \mathcal{O}_{\mathbb{P}}(2m+4))$ be the open subset parametrising smooth curves of degree $2m+4$. There exists a Zariski open subset $U' \subset \mathbb{P}H^0(\mathbb{P}^{2m+3}, E) - \Delta$ such that $h : U' \to U$ is a smooth morphism.*

Proof By a classical theorem of Dixon [8], a general smooth plane curve of even degree can be realised as the discriminant curve of a net of quadrics; see [2, Prop. 4.2 and Remark 4.4] for a modern proof. Hence the map h is dominant, and the assertion follows from [9, III, Lemma 10.5]. □

In the sequel we shall need a relative version of Theorem 11.3.2. Over U we have the universal family $S_U \to U$ whose fiber over $[F] \in U$ is the double covering of $\mathbb{P}^2$ ramified over the curve $V(F)$. Let $S_T = S_U \times_U T$ be the pullback of this family to T along the map $h : T \to U$. Over T we have the family of quadric bundles $f_T : \mathcal{X}_T \to T$ associated to the family of complete intersections $X_T \to T$. We have relative correspondences

$$\begin{array}{ccccc} \Gamma_{1,T} & \to & \mathcal{X}_T & \quad \Gamma_{2,T} & \to \mathcal{X}_T \\ \downarrow & & & \downarrow & \\ X_T & & & S_T. & \end{array}$$

To state the relative version of Theorem 11.3.2 we need some notation. Given a map $f : Y \to X$ of topological spaces, let $M(f)$ be the mapping cylinder of f. The map f factors as $Y \hookrightarrow M(f) \xrightarrow{\sim} X$ where the second map is a homotopy equivalence. Define

$$H^k(X, Y; \mathbb{Z}) = H^k(M(f), Y; \mathbb{Z}).$$

The groups $H^k(X, Y)$ fit into a long exact sequence

$$\to H^{k-1}(Y) \to H^k(X, Y) \to H^k(X) \xrightarrow{f^*} H^k(Y) \to \tag{11.5}$$

of cohomology groups. If f is the inclusion of a subspace, they coincide with the usual relative cohomology of the pair (X, Y).

Since we have isomorphisms

$$H^2_{\text{pr}}(S) \cong H^3(\mathbb{P}^2, S), \quad H^{2m}_{\text{pr}}(X) \cong H^{2m+1}(\mathbb{P}^{2m+3}, X),$$

Theorem 11.3.2 can be restated as an isomorphism $H^3(\mathbb{P}^2, S) \cong H^{2m+1}(\mathbb{P}^{2m+3}, X)$.

Theorem 11.3.4. *The relative correspondence* $\Gamma_T = \Gamma_{2,T} \circ {}^t\Gamma_{1,T}$ *induces an isomorphism* $H^{k+2}(\mathbb{P}^2_T, S_T) \cong H^{2m+k}(\mathbb{P}^{2m+3}_T, X_T)$ *for all* $k \geq 0$.

Proof Let

$$f_T : X_T \to T,\ g_T : S_T \to T,\ \varphi_T : \mathbb{P}^{2m+3}_T \to T,\ \psi_T : \mathbb{P}^2_T \to T$$

be the projections onto the base T. By the Lefschetz hyperplane theorem and the Barth–Lefschetz theorem for cyclic coverings of projective space [12, Thm. 2.1] we have

$$R^q(f_T)_*\mathbb{Q} \cong R^q(\varphi_T)_*\mathbb{Q}, \quad q \neq 2m$$
$$R^q(g_T)_*\mathbb{Q} \cong R^q(\psi_T)_*\mathbb{Q}, \quad q \neq 2.$$

Set $\mathbb{V} = \operatorname{coker}(R^{2m}(\varphi_T)_*\mathbb{Q} \to R^{2m}(f_T)_*\mathbb{Q})$, $\mathbb{W} = \operatorname{coker}(R^2(\psi_T)_*\mathbb{Q} \to R^2(g_T)_*\mathbb{Q})$. The correspondence Γ_T induces a homomorphism of local systems $\Gamma_{T,*} : \mathbb{W} \to \mathbb{V}$. By the proper base change theorem and Theorem 11.3.2, $\Gamma(t)_*$ is an isomorphism for all $t \in T$. Hence $\Gamma_{T,*} : \mathbb{W} \to \mathbb{V}$ is an isomorphism. Combining the long exact sequence (11.5) with the Lefschetz/Barth–Lefschetz isomorphisms we obtain

$$H^{2m+k}(\mathbb{P}^{2m+3}_T, X_T) \cong H^{k-1}(T, \mathbb{V}), \quad H^{k+2}(\mathbb{P}^2_T, S_T) \cong H^{k-1}(T, \mathbb{W}),$$

and the result follows. □

The vanishing of $H^*(\mathbb{P}^2_T, S_T)$ follows from an effective version of Nori's connectivity theorem for cyclic coverings of projective space, which can be seen as a generalisation of the result of Müller–Stach on the Abel–Jacobi map [13]. An outline of the proof can be found in [17].

Theorem 11.3.5. *Let* $U \subset H^0(\mathbb{P}^n, \mathcal{O}_{\mathbb{P}}(d.e))$ *be the open subset parametrising cyclic coverings* $Y \to \mathbb{P}^n$ *of degree* e *that ramify over a divisor* $D \subset \mathbb{P}^n$ *of degree* $d.e$*, and let* $Y_U \to U$ *be the universal family. Let* $T \to U$ *be a smooth morphism, and* $c \leq n$ *an integer. We have*

$$F^\mu H^{n+k}(\mathbb{P}^n_T, Y_T) = 0$$

for all $k \leq c$ *if* $(\mu - c)e + 1)d \geq n + c$.

Corollary 11.3.6. *The conclusion of Proposition 11.2.1 holds if* $(n, q) = (2m, 1)$, $r = 2$ *and* $(d_0, d_1, d_2) = (2, 2, 2)$ *if* $m \geq 2$.

Proof Let U' be the open subset of $\mathbb{P}(\oplus^3 H^0(\mathbb{P}^{2m+3}, \mathcal{O}_{\mathbb{P}}(2))$ introduced in Lemma 11.3.3, and let $T \to U'$ be a smooth morphism. By Theorem 11.3.5

and Theorem 11.3.4 we obtain $H^{2m+1}(\mathbb{P}_T^{2m+3}, X_T) = H^{2m+2}(\mathbb{P}_T^{2m+3}, X_T) = 0$. □

In a similar way one can remove the exceptions in Corollary 11.2.4 (ii) for $m \geq 2$, using the following theorem of M. Reid [20].

Theorem 11.3.7 (Reid). *Let $X = V(2,2) \subset \mathbb{P}^{2m+1}$ be a general smooth complete intersection of two quadrics. There is a family of m–planes in the fibers of the associated quadric bundle $\mathcal{X} \to \mathbb{P}^1$ that defines a correspondence Γ between $\mathcal{X}$ and a hyperelliptic curve C, branched over a divisor $D \subset \mathbb{P}^1$ of degree $2m+2$. This correspondence induces an isomorphism*

$$\Gamma_* : H^1(C) \to H^{2m-1}(X).$$

11.4 Exceptional cases

We end with a discussion of the remaining exceptional cases. We start with the case $q = 1$. There are a number of trivial exceptions coming from the Noether–Lefschetz theorem. Consider the commutative diagram

$$\begin{array}{ccc} \mathsf{CH}^m(X) \otimes \mathbb{C}^* & \xrightarrow{\mu} & \mathsf{CH}^{m+1}(X,1) \\ \downarrow{\scriptstyle c_{m+1,1}} & & \downarrow{\scriptstyle c_{m+1,1}} \\ \mathrm{Hdg}^m(X) \otimes \mathbb{C}^* & \xrightarrow{\mu_{\mathcal{D}}} & H_{\mathcal{D}}^{2m+1}(X, \mathbb{Z}(m+1)). \end{array}$$

The composition of $\mu_{\mathcal{D}}$ with the projection

$$H_{\mathcal{D}}^{2m+1}(X, \mathbb{Z}(m+1)) = \frac{H^{2m+1}(X, \mathbb{C})}{F^{m+1}H^{2m+1}(X, \mathbb{C}) + H^{2m+1}(X, \mathbb{Z})} \to \frac{H^{m,m}(X)}{\mathrm{Hdg}^m(X)}$$

is an injective map

$$\mathrm{Hdg}^m(X) \otimes \mathbb{C}^* = \frac{\mathrm{Hdg}^m(X) \otimes \mathbb{C}}{\mathrm{Hdg}^m(X)} \hookrightarrow \frac{H^{m,m}(X)}{\mathrm{Hdg}^m(X)}.$$

Hence $\mu_{\mathcal{D}}$ is injective, and we obtain an injective map from $\mathrm{Hdg}^m_{\mathrm{pr}}(X)$ to the cokernel of $i^* : H_{\mathcal{D}}^{2m+1}(\mathbb{P}^N, \mathbb{Z}(m+1)) \to H_{\mathcal{D}}^{2m+1}(X, \mathbb{Z}(m+1))$. This remark covers the cases

$$X = V(2) \subset \mathbb{P}^{2m+1}, \; X = V(3) \subset \mathbb{P}^3, \; X = V(2,2) \subset \mathbb{P}^{2m+2}.$$

The cycles that we considered above are *decomposable*, i.e., belong to the image of the map μ. The other counterexamples in low degree come from indecomposable higher Chow cycles. On K3 surfaces one can produce indecompable higher Chow cycles using rational nodal curves [4]; see [22],

[14] and [6] for earlier results in this direction. Collino [6] gave examples of indecomposable higher Chow cycles on cubic fourfolds.

Remark 11.4.1. Note that both in the case of K3 surfaces and of cubic fourfolds we are dealing with a Hodge structure V of weight 2 with $\dim V^{2,0} = 1$. (In the case of a cubic fourfold X, take $V = H^4(X, \mathbb{C})(1)$.) Hence one might ask whether the existence of indecomposable higher Chow cycles on these varieties is related to the Kuga–Satake construction; cf. [22, 4.4-4.5].

The remaining exceptional case for $q = 1$ is $X = V(3,2) \subset \mathbb{P}^{2m+2}$, $m \geq 2$ (if $m = 1$, X is a K3 surface). I do not know what happens in this case. With the notation of section 2, we can show that $H_0(S_\bullet) = 0$ and $H_1(S_\bullet) \neq 0$. (The latter result can be seen by decomposing the terms of the complex $S_\bullet$ into irreducible $\mathrm{SL}(V)$–modules.) Hence $H_1(R_\bullet) = 0$ if and only if the map $H_1(J_\bullet) \to H_1(S_\bullet)$ is surjective; it seems hard to verify this condition.

For $q = 2$ the situation is much simpler. The only cases to consider are

$$X = V(2) \subset \mathbb{P}^2,\ X = V(3) \subset \mathbb{P}^2, X = V(2,2) \subset \mathbb{P}^3.$$

In the first case the conclusion of Proposition 11.2.1 trivially holds, since the target of the regulator map is zero. The remaining two cases are elliptic curves. Bloch [3] showed that the image of

$$c_{2,2} : \mathsf{CH}^2(X, 2) \to H^2_{\mathcal{D}}(X, \mathbb{Z}(2))$$

is nonzero for elliptic curves. Since $H^2_{\mathcal{D}}(\mathbb{P}^N, \mathbb{Z}(2)) = 0$, the result follows.

The results in this note can be summarised as follows.

Theorem 11.4.2. *Let X be a smooth complete intersection in $\mathbb{P}^N$ with inclusion map $i : X \to \mathbb{P}^N$.*

1) If $\dim X = 2m$ and X is very general, the image of the regulator map

$$c_{m+1,1} : \mathsf{CH}^{m+1}(X, 1) \to H^{2m+1}_{\mathcal{D}}(X, \mathbb{Z}(m+1))/i^* H^{2m+1}_{\mathcal{D}}(\mathbb{P}^N, \mathbb{Z}(m+1))$$

is a torsion group, with the possible exception of the cases

i) $X = V(2) \subset \mathbb{P}^{2m+1}$, $X = V(d) \subset \mathbb{P}^3$ $(d \leq 4)$, $X = V(3) \subset \mathbb{P}^5$;
ii) $X = V(2,2) \subset \mathbb{P}^{2m+2}$, $X = V(3,2) \subset \mathbb{P}^{2m+2}$, $m \geq 1$;
iii) $X = V(2,2,2) \subset \mathbb{P}^5$.

2) If $\dim X = 2m - 1$ and X is very general, the image of the map

$$c_{m+1,2} : \mathsf{CH}^{m+1}(X, 2) \to H^{2m+1}_{\mathcal{D}}(X, \mathbb{Z}(m+1))/i^* H^{2m+1}_{\mathcal{D}}(\mathbb{P}^N, \mathbb{Z}(m+1))$$

is a torsion group, unless X is an elliptic curve.

References

[1] Asakura, M. and S. Saito: Generalized Jacobian rings for complete intersections preprint math.AG/0203147

[2] Beauville, A.: Determinantal hypersurfaces. Dedicated to William Fulton on the occasion of his 60th birthday Michigan Math. J. **48** (2000), 39–64.

[3] Bloch, S.: *Higher regulators, algebraic K-theory, and zeta functions of elliptic curves* CRM Monograph Series. **11** Providence, RI: American Mathematical Society (2000)

[4] Chen, X. and J. Lewis: The Hodge-$\mathfrak{D}$-conjecture for $K3$ and abelian surfaces J. Algebraic Geom. **14** (2005) 213–240.

[5] Carlson, J., S. Müller-Stach and C. Peters: *Period mappings and period domains* Cambridge Studies in Advanced Mathematics **85** Cambridge University Press, Cambridge (2003)

[6] Collino, A.: Indecomposable motivic cohomology classes on quartic surfaces and on cubic fourfolds, in *Algebraic K-theory and its applications (Trieste, 1997)* 370–402, World Sci. Publishing, River Edge, NJ, (1999)

[7] Cox, D.: Recent developments in toric geometry, Proc. Sympos. Pure Math., **62**, Part 2 Amer. Math. Soc., Providence, RI (1997) 389–436

[8] Dixon, A. C.: Note on the reduction of a ternary quartic to a symmetric determinant Proc. Camb. Phil. Soc. **11** (1902) 350–351

[9] Hartshorne, R.: *Algebraic geometry.* Graduate Texts in Mathematics **52** Springer-Verlag, New York-Heidelberg (1977)

[10] Kerr, M., J. Lewis and S. Müller-Stach: The Abel-Jacobi map for higher Chow groups preprint math.AG/0409116.

[11] Laszlo, Y.: Théorème de Torelli générique pour les intersections complètes de trois quadriques de dimension paire Invent. Math. **98** (1989) 247-264

[12] Lazarsfeld, R.: A Barth-type theorem for branched coverings of projective space Math. Ann. **249** (1980) 153–162

[13] Müller–Stach, S.: Syzygies and the Abel-Jacobi map for cyclic coverings Manuscripta Math **82** (1994) 433–443

[14] Müller–Stach, S.: Constructing indecomposable motivic cohomology classes on algebraic surfaces J. Algebraic Geom. **6** (1997) 513–543

[15] Müller-Stach, S.: Algebraic cycle complexes: Basic properties, in *The arithmetic and geometry of algebraic cycles Vol. 1* Gordon, B. Brent (ed.) et al., Kluwer Academic Publishers (2000) 285-305

[16] Nagel, J.: Effective bounds for Hodge–theoretic connectivity J. Alg. Geom. **11** (2002) 1–32.

[17] Nagel, J.: The image of the regulator map for complete intersections of three quadrics preprint MPI 03-46, 2003

[18] Nori, M.V.: Algebraic cycles and Hodge-theoretic connectivity Invent. Math. **111** (1993) 349–373

[19] O'Grady, K.: The Hodge structure of the intersection of three quadrics in an odd-dimensional projective space Math. Ann. **273** (1986) 277–285.

[20] Reid, M.: *The intersection of two quadrics*, Ph.D. Thesis, Cambridge 1972 (unpublished). Available at www.maths.warwick.ac.uk/~miles/3folds/qu.ps.

[21] Voisin, C.: Variations of Hodge structure and algebraic cycles, in *Proceedings ICM '94, Vol. I*, Chatterji, S. D. (ed.) Basel: Birkhäuser (1995) 706-715

[22] Voisin, C.: Remarks on zero-cycles of self-products of varieties, in *Moduli of vector bundles* Maruyama, Masaki (ed.), New York, NY, Marcel Dekker. Lect. Notes Pure Appl. Math. **179** (1996) 265-285

[23] Voisin, C.: Nori's connectivity theorem and higher Chow groups. J. Inst. Math. Jussieu **1** (2002) 307–329.

12

Hodge Number Polynomials for Nearby and Vanishing Cohomology

C.A.M. Peters

Department of Mathematics, University of Grenoble
UMR 5582 CNRS-UJF, 38402-Saint-Martin d'Hères, France
chris.peters@ujf-grenoble.fr

J.H.M. Steenbrink

Institute for Mathematics, Astrophysics and Particle Physics, Radboud University Nijmegen Toernooiveld, NL-6525 ED Nijmegen, The Netherlands
J.Steenbrink@math.ru.nl

To our friend Jaap Murre

Introduction

The behaviour of the cohomology of a degenerating family of complex projective manifolds has been intensively studied in the nineteen-seventies by Clemens, Griffiths, Schmid and others. See [Gr] for a nice overview. Recently, the theory of motivic integration, initiated by Kontsevich and developed by Denef and Loeser, has given a new impetus to this topic. In particular, in the case of a one-parameter degeneration it has produced an object ψ_f in the Grothendieck group of complex algebraic varieties, called the *motivic nearby fibre* [B05], which reflects the limit mixed Hodge structure of the family in a certain sense. The purpose of this paper is twofold. First, we prove that the motivic nearby fibre is well-defined without using the theory of motivic integration. Instead we use the Weak Factorization Theorem [AKMW]. Second, we give a survey of formulas containing numerical invariants of the limit mixed Hodge structure, and in particular of the vanishing cohomology of an isolated hypersurface singularity, without using the theory of mixed Hodge structures or of variations of Hodge structure.

We hope that in this way this interesting topic becomes accessible to a wider audience.

12.1 Real Hodge structures

A real Hodge structure on a finite dimensional real vector space V consists of a direct sum decomposition

$$V_{\mathbb{C}} = \bigoplus_{p,q \in \mathbb{Z}} V^{p,q}, \text{ with } V^{p,q} = \overline{V^{q,p}}$$

on its complexification $V_{\mathbb{C}} = V \otimes \mathbb{C}$. The corresponding *Hodge filtration* is given by

$$F^p(V) = \bigoplus_{r \geq p} V^{r,s}.$$

The numbers

$$h^{p,q}(V) := \dim V^{p,q}$$

are the Hodge numbers of the Hodge structure. If for some integer k we have $h^{p,q} = 0$ for all (p,q) with $p+q \neq k$ the Hodge structure is pure of weight k. Any real Hodge structure is the direct sum of pure Hodge structures. The polynomial

$$\begin{aligned} P_{\mathrm{hn}}(V) &= \sum_{p,q \in \mathbb{Z}} h^{p,q}(V) u^p v^q \\ &= \sum h^{p,k-p}(V) u^p v^{k-p} \in \mathbb{Z}[u, v, u^{-1}, v^{-1}] \end{aligned} \tag{12.1}$$

is its associated *Hodge number polynomial.*† A classical example of a weight k Hodge real structure is furnished by the rank k (singular) cohomology group $H^k(X)$ (with $\mathbb{R}$- coefficients) of a compact Kähler manifold X.

Various multilinear algebra operations can be applied to Hodge structures as we now explain. Suppose that V and W are two real vector spaces with a Hodge structure of weight k and ℓ respectively. Then:

(i) $V \otimes W$ has a Hodge structure of weight $k + \ell$ given by

$$F^p(V \otimes W)_{\mathbb{C}} = \sum_m F^m(V_{\mathbb{C}}) \otimes F^{p-m}(W_{\mathbb{C}}) \subset V_{\mathbb{C}} \otimes_{\mathbb{C}} W_{\mathbb{C}}$$

and with Hodge number polynomial given by

$$P_{\mathrm{hn}}(V \otimes W) = P_{\mathrm{hn}}(V) P_{\mathrm{hn}}(W). \tag{12.2}$$

† There are other conventions in the literature; for instance, some authors put a sign $(-1)^{p+q}$ in front of the coefficient $h^{p,q}(V)$ of $u^p v^q$.

(ii) On $\operatorname{Hom}(V, W)$ we have a Hodge structure of weight $\ell - k$:

$$F^p \operatorname{Hom}(V, W)_{\mathbb{C}} = \{f : V_{\mathbb{C}} \to W_{\mathbb{C}} \mid fF^n(V_{\mathbb{C}}) \subset F^{n+p}(W_{\mathbb{C}}) \quad \forall n\}$$

with Hodge number polynomial

$$P_{\mathrm{hn}}(\operatorname{Hom}(V, W)(u, v) = P_{\mathrm{hn}}(V)(u^{-1}, v^{-1})P_{\mathrm{hn}}(W)(u, v). \qquad (12.3)$$

In particular, taking $W = \mathbb{R}$ with $W_{\mathbb{C}} = W^{0,0}$ we get a Hodge structure of weight $-k$ on the dual $V^\vee$ of V with Hodge number polynomial

$$P_{\mathrm{hn}}(V^\vee)(u, v) = P_{\mathrm{hn}}(V)(u^{-1}, v^{-1}). \qquad (12.4)$$

The category $\mathfrak{hs}$ of real Hodge structures leads to a ring, the *Grothendieck ring* $K_0(\mathfrak{hs})$ which is is the free group on the isomorphism classes $[V]$ of real Hodge structures V modulo the subgroup generated by $[V] - [V'] - [V'']$ where

$$0 \to V' \to V \to V'' \to 0$$

is an exact sequence of pure Hodge structures and where the complexified maps preserve the Hodge decompositions. Because the Hodge number polynomial (12.1) is clearly additive and by (12.2) behaves well on products the Hodge number polynomial defines a ring homomorphism

$$P_{\mathrm{hn}} : \mathrm{K}_0(\mathfrak{hs}) \to \mathbb{Z}[u, v, u^{-1}, v^{-1}].$$

As remarked before, pure Hodge structures of weight k in algebraic geometry arise as the (real) cohomology groups $H^k(X)$ of smooth complex projective varieties. We combine these as follows:

$$\chi_{\mathrm{Hdg}}(X) := \sum (-1)^k [H^k(X)] \in \mathrm{K}_0(\mathfrak{hs}); \qquad (12.5)$$

$$e_{\mathrm{Hdg}}(X) := \sum (-1)^k P_{\mathrm{hn}}(H^k(X)) \in \mathbb{Z}[u, v, u^{-1}, v^{-1}] \qquad (12.6)$$

which we call the Hodge-Grothendieck character and the Hodge-Euler polynomial of X respectively.

Let us now recall the definition of the naive Grothendieck group $\mathrm{K}_0(\mathrm{Var})$ of (complex) algebraic varieties. It is the quotient of the free abelian group on isomorphism classes $[X]$ of algebraic varieties over $\mathbb{C}$ with the so-called *scissor relations* $[X] = [X - Y] + [Y]$ for $Y \subset X$ a closed subvariety. The cartesian product is compatible with the scissor relations and induces a product structure on $\mathrm{K}_0(\mathrm{Var})$, making it into a ring. There is a nice set of generators and relations for $\mathrm{K}_0(\mathrm{Var})$. To explain this we first recall:

Lemma 12.1.1. *Suppose that X is a smooth projective variety and $Y \subset X$ is a smooth closed subvariety. Let $\pi : Z \to X$ be the blowing-up with centre Y and let $E = \pi^{-1}(Y)$ be the exceptional divisor. Then*

$$\chi_{\mathrm{Hdg}}(X) - \chi_{\mathrm{Hdg}}(Y) = \chi_{\mathrm{Hdg}}(Z) - \chi_{\mathrm{Hdg}}(E);$$
$$e_{\mathrm{Hdg}}(X) - e_{\mathrm{Hdg}}(Y) = e_{\mathrm{Hdg}}(Z) - e_{\mathrm{Hdg}}(E).$$

Proof By [GH, p. 605]

$$0 \to H^k(X) \to H^k(Z) \oplus H^k(Y) \to H^k(E) \to 0$$

is exact. □

Theorem 12.1.2 ([B04, Theorem 3.1]). *The group* $\mathrm{K}_0(\mathrm{Var})$ *is isomorphic to the free abelian group generated by the isomorphism classes of smooth complex projective varieties subject to the relations* $[\varnothing] = 0$ *and* $[Z] - [E] = [X] - [Y]$ *where X, Y, Z, E are as in Lemma 12.1.1.*

It follows that for every complex algebraic variety X there exist projective smooth varieties $X_1, \ldots, X_r, Y_1, \ldots, Y_s$ such that

$$[X] = \sum_i [X_i] - \sum_j [Y_j] \text{ in } \mathrm{K}_0(\mathrm{Var})$$

and so, using Lemma 12.1.1 we have:

Corollary 12.1.3. *The Hodge Euler character extends to a ring homomorphism*

$$\chi_{\mathrm{Hdg}} : K_0(\mathrm{Var}) \to K_0(\mathfrak{hs})$$

and the Hodge number polynomial extends to a ring homomorphism

$$e_{\mathrm{Hdg}} : \mathrm{K}_0(\mathrm{Var}) \to \mathbb{Z}[u, v, u^{-1}, v^{-1}]$$

Remark 12.1.4. By Deligne's theory [Del71], [Del74] there is a mixed Hodge structure on the real vector spaces $H^k(X)$. For our purposes, since we are working with real coefficients, a mixed Hodge structure is just a real Hodge structure, i.e. a direct sum of real Hodge structures of various weights, and so the Hodge character and Hodge number polynomial are defined for any real mixed Hodge structure. However, ordinary cohomology does not behave well with respect to the scissor relation; we need compactly supported cohomology $H^k_c(X; \mathbb{R})$. But these also carry a Hodge structure and we have the following explicit expression for the above characters.

$$\chi_{\mathrm{Hdg}}(X) = \sum (-1)^k [H^k_c(X))];$$
$$e_{\mathrm{Hdg}}(X) = \sum (-1)^k P_{\mathrm{hn}}(H^k_c(X)).$$

Example 12.1.5. **1)** Let U be a smooth, but not necessarily compact complex algebraic manifold. Such a manifold has a *good compactification* X, i.e. X is a compact complex algebraic manifold and $D = X - U$ is a normal crossing divisor, say $D = D_1 \cup \cdots D_N$ with D_j smooth and irreducible. We introduce

$$\begin{aligned} D_I &= D_{i_1} \cap D_{i_2} \cap \cdots \cap D_{i_m}, \quad I = \{i_1, \ldots, i_m\}; \\ a_I &: D_I \hookrightarrow X \end{aligned}$$

and we set

$$\begin{aligned} D(0) &= X; \\ D(m) &= \textstyle\coprod_{|I|=m} D_I, \quad m = 1, \ldots, N; \\ a_m &= \textstyle\coprod_{|I|=m} a_I : D(m) \to X. \end{aligned}$$

Then each connected component of $D(m)$ is a complex submanifold of X of codimension m. Note that

$$[U] = \sum_m (-1)^m [D(m)] \in \mathrm{K}_0(\mathrm{Var}).$$

Hence

$$\begin{aligned} \chi_{\mathrm{Hdg}}(U) &= \sum_m (-1)^m \chi_{\mathrm{Hdg}}(D(m)); \\ e_{\mathrm{Hdg}}(U) &= \sum_m (-1)^m e_{\mathrm{Hdg}}(D(m)). \end{aligned}$$

2) If X is compact the construction of cubical hyperresolutions $(X_I)_{\varnothing \neq I \subset A}$ of X from [GNPP] leads to the expression

$$[X] = \sum_{\varnothing \neq I \subset A} (-1)^{|I|-1} [X_I].$$

and we find:

$$\begin{aligned} \chi_{\mathrm{Hdg}}(X) &= \sum_{\varnothing \neq I \subset A} (-1)^{|I|-1} \chi_{\mathrm{Hdg}}(X_I); \\ e_{\mathrm{Hdg}}(X) &= \sum_{\varnothing \neq I \subset A} (-1)^{|I|-1} e_{\mathrm{Hdg}}(X_I). \end{aligned}$$

The scissor-relations imply that the inclusion-exclusion principle can be applied to a disjoint union X of locally closed subvarieties $X_1, \ldots, X_m$:

$$\chi_{\mathrm{Hdg}}(X) = \sum_{i=1}^{m} \chi_{\mathrm{Hdg}}(X_i)$$

and a similar expression holds for the Hodge Euler polynomials.

Example 12.1.6. Let $T^n = (\mathbb{C}^*)^n$ be an n- dimensional algebraic torus. Then $e_{\mathrm{Hdg}}(T^1) = uv-1$ so $e_{\mathrm{Hdg}}(T^n) = (uv-1)^n$. Consider an n- dimensional toric variety X. It is a disjoint union of T^n- orbits. Suppose that X has s_k orbits of dimension k. Then

$$e_{\mathrm{Hdg}}(X) = \sum_{k=0}^{n} s_k e_{\mathrm{Hdg}}(T^k) = \sum_{k=0}^{n} s_k (uv-1)^k.$$

If X has a pure Hodge structure (e.g. if X is compact and has only quotient singularities) then this formula determines the Hodge numbers of X.

12.2 Nearby and vanishing cohomology

In this section we consider a relative situation. We let X be a complex manifold, $\Delta \subset \mathbb{C}$ the unit disk and $f : X \to \Delta$ a holomorphic map which is smooth over the punctured disk Δ^*. We say that f is a *one-parameter degeneration.* Let us assume that $E = \bigcup_{i \in I} E_i = f^{-1}(0)$ is a divisor with strict normal crossings on X. We have the *specialization diagram*

$$\begin{array}{ccccc} X_\infty & \xrightarrow{k} & X & \xleftarrow{i} & E \\ \downarrow{\scriptstyle \tilde{f}} & & \downarrow{\scriptstyle f} & & \downarrow \\ \mathfrak{h} & \xrightarrow{\mathrm{e}} & \Delta & \leftarrow & \{0\} \end{array}$$

where $\mathfrak{h}$ is the complex upper half plane, $\mathrm{e}(z) := \exp(2\pi \mathrm{i} z)$ and where

$$X_\infty := X \times_{\Delta^*} \mathfrak{h}.$$

We let e_i denote the multiplicity of f along E_i and choose a positive integer multiple e of all e_i. We let $\tilde{f} : \tilde{X} \to \Delta$ denote the normalization of the pull-back of X under the map $\mu_e : \Delta \to \Delta$ given by $\tau \mapsto \tau^e = t$. It fits into a commutative diagram describing the *e-th root* of f:

$$\begin{array}{ccc} \tilde{X} & \xrightarrow{\rho} & X \\ {\scriptstyle \tilde{f}}\downarrow & & \downarrow{\scriptstyle f} \\ \Delta & \xrightarrow{\mu_e} & \Delta. \end{array}$$

We put

$$D_i = \rho^{-1}(E_i)\,,\; D_J = \bigcap_{i \in J} D_j,\; D(m) = \coprod_{|J|=m} D_J.$$

Then $D_J \to E_J$ is a cyclic cover of degree $\gcd(e_j \mid j \in J)$. The maps $D_J \to E_J$ do not depend on the choice of the integer e and so in particular this is true for the varieties D_J. See e.g. [Ste77] or [B05] for a detailed study of the geometry of this situation. The special fibre $\tilde{X}_0 = \tilde{f}^{-1}(0)$ is now a complex variety equipped with the action of the cyclic group of order e. Let us introduce the associated Grothendieck-group:

Definition 12.2.1. We let $\mathrm{K}_0^{\hat{\mu}}(\mathrm{Var})$ denote the Grothendieck group of complex algebraic varieties with an action of a finite order automorphism modulo the subgroup generated by expressions $[\mathbb{P}(V)] - [\mathbb{P}^n \times X]$ where V is a vector bundle of rank $n+1$ over X with action which is linear over the action on X. See [B05, Sect. 2.2] for details.

As a motivation, we should remark that in the ordinary Grothendieck group the relation $[\mathbb{P}(V)] = [\mathbb{P}^n \times X]$ follows from the fact that any algebraic vector bundle is trivial over a Zariski-open subset and the fact that the scissor-relations hold. The above relations extend to the case where one has a group action.

Definition 12.2.2. Suppose that the fibres of f are projective varieties. Following [B05, Ch. 2] we define the *motivic nearby fibre* of f by

$$\psi_f := \sum_{m\geq 1} (-1)^{m-1}[D(m) \times \mathbb{P}^{m-1}] \in \mathrm{K}_0^{\hat{\mu}}(\mathrm{Var})$$

and the *motivic vanishing fibre* by

$$\phi_f := \psi_f - [E] \in \mathrm{K}_0^{\hat{\mu}}(\mathrm{Var}). \tag{12.7}$$

Remark 12.2.3. If we let E_I^0 be the open subset of E_I consisting of points which are exactly on the E_j with $j \in I$, D_I^0 the corresponding subset of D_I and $D^0(i) = \coprod_{|I|=i} D_I^0$. We have $D_I = \coprod_{J\supset I} D_J^0$ and the scissor relations imply that

$$\begin{aligned}
\sum_{i\geq 1}(-1)^{i-1}[D(i) \times \mathbb{P}^{i-1}] &= \sum_{i\geq 1}(-1)^{i-1}\sum_{j\geq i}\binom{j}{i}[D^0(j) \times \mathbb{P}^{i-1}] \\
&= \sum_{j\geq 1}[D^0(j)] \times \sum_{i=1}^{j}(-1)^{i-1}\binom{j}{i}[\mathbb{P}^{i-1}] \\
&= \sum_{j\geq 1}(-1)^{j-1}[D^0(j) \times (\mathbb{C}^*)^{j-1}].
\end{aligned}$$

This expression for the motivic nearby fibre has been used in [L].

The motivic nearby fibre turns out to be a relative bimeromorphic invariant:

Lemma 12.2.4. *Suppose that $g : Y \to X$ is a bimeromorphic proper map which is an isomorphism over $X - E$. Assume that $(f \circ g)^{-1}(0)$ is a divisor with strict normal crossings. Then*

$$\psi_f = \psi_{fg}.$$

Proof In [B05], the proof relies on the theory of motivic integration [DL]. We give a different proof, based on the *weak factorization theorem* [AKMW]. This theorem reduces the problem to the following situation: g is the blowing-up of X in a connected submanifold $Z \subset E$, with the following property. Let $A \subset I$ be those indices i for which $Z \subset E_i$. Then Z intersects the divisor $\bigcup_{i \notin A} E_i$ transversely, hence $Z \cap \bigcup_{i \notin A} E_i$ is a divisor with normal crossings in Z.

We fix the following notation. We let $L := [\mathbb{A}^1]$ and $P_m := [\mathbb{P}^m] = \sum_{i=0}^m L^i$. For $J \subset I$ we let $j = |J|$. So, using the product structure in $K_0(\text{Var})$, we have

$$\psi_f = \sum_{\varnothing \neq J \subset I} (-1)^{j-1} [D_J] P_{j-1}.$$

Let $E' = \bigcup_{i \in I'} E'_i$ be the zero fibre of fg. We have $I = I' \cup \{*\}$ where E'_* is the exceptional divisor of g and E'_i is the proper transform in Y of E_i. Form $\rho' : D' \to E'$, the associated ramified cyclic covering. For $J \subset I$ we let $J' = J \cup \{*\}$. Note that E'_* has multiplicity equal to $\sum_{i \in A} e_i$. Without loss of generality we may assume that e is also a multiple of this integer. We have two kinds of j'- uple intersections $D_{J'}$: those which only contain D'_j, $j \neq *$ and those which contain D'_*. So,

$$\psi_{fg} = [D'_*] + \sum_{\varnothing \neq J \subset I} (-1)^{j-1} ([D'_J] P_{j-1} - [D'_{J'}] P_j).$$

We are going to calculate the difference between ψ_{fg} and ψ_f. Let $B \subset I - A$. Let $c = \text{codim}(Z, X)$. Then for all $K \subset A$ we have that $\text{codim}(Z \cap E_{K \cup B}, E_{K \cup B}) = c - k$, and $D'_{K \cup B}$ is the blowing up of $D_{K \cup B}$ with centre $Z \cap E_{K \cup B} = Z \cap E_B =: Z_B$ and exceptional divisor $D'_{K \cup B'}$. Hence, if we let $W_B = \rho'^{-1}(Z_B)$, we have

$$[D'_{K \cup B}] = [D_{K \cup B}] + [D'_{K \cup B'}] - [W_B] = [D_{K \cup B}] + [W_B](P_{c-k-1} - 1).$$

Hence

$$\psi_{fg} - \psi_f = \sum_B c_B[W_B]$$

for suitable coefficients c_B. Note that $Z_\varnothing = Z$ and that $D'_* = W_\varnothing \times P_{c-1}$ and so if $B = \varnothing$ we get

$$\begin{aligned} c_\varnothing &= P_{c-1} + \sum_{\varnothing \neq K \subset A} (-1)^{k-1} \left((P_{c-k-1} - 1)P_{k-1} - P_{c-k-1}P_k\right) \\ &= P_{c-1} + \sum_{\varnothing \neq K \subset A} (-1)^k (L^k P_{c-k-1} + P_{k-1}) = P_{c-1} \sum_{K \subset A} (-1)^k = 0. \end{aligned}$$

In case $B \neq \varnothing$ we get

$$\begin{aligned} c_B &= \sum_{K \subset A} (-1)^{k+b-1} \left((P_{c-k-1} - 1)P_{k+b-1} - P_{c-k-1}P_{k+b}\right) \\ &= \sum_{K \subset A} (-1)^{k+b} (L^{k+b} P_{c-k-1} + P_{k+b-1}) = P_{c+b-1} \sum_{K \subset A} (-1)^{b+k} = 0. \end{aligned}$$

□

Let us now pass to the nearby fibre in the Hodge theoretic sense. In [Schm] and [Ste76] a mixed Hodge structure on $H^k(X_\infty)$ was constructed; its weight filtration is the *monodromy weight filtration* which we now explain. The loop winding once counterclockwise around the origin gives a generator of $\pi(\Delta^*, *)$, $* \in \Delta^*$. Its action on the fibre X_* over $*$ is well defined up to homotopy and on $H^k(X_*) \simeq H^k(X_\infty)$ it defines the *monodromy automorphism* T. Let $N = \log T_u$ be the logarithm of the unipotent part in the Jordan decomposition of T. Then W is the unique increasing filtration on $H^k(X_\infty)$ such that $N(W_j) \subset W_{j-2}$ and $N^j : \mathrm{Gr}^W_{k+j} \to \mathrm{Gr}^W_{k-j}$ is an isomorphism for all $j \geq 0$. In fact, from [Schm, Lemma 6.4] we deduce:

Lemma 12.2.5. *There is a Lefschetz-type decomposition*

$$\mathrm{Gr}^W H^k(X_\infty) = \bigoplus_{\ell=0}^{k} \bigoplus_{r=0}^{\ell} N^r P_{k+\ell},$$

where $P_{k+\ell}$ is pure of weight $k + \ell$. The endomorphism N has $\dim P_{k+m-1}$ *Jordan blocs of size m.*

The Hodge filtration is constructed in [Schm] as a limit of the Hodge filtrations on nearby smooth fibres in a certain sense, and in [Ste76] using the relative logarithmic de Rham complex. With X_t a smooth fibre of f it follows that

$$\dim F^m H^k(X_t) = \dim F^m H^k(X_\infty)$$

and hence

$$e_{\mathrm{Hdg}}(X_\infty)_{|v=1} = e_{\mathrm{Hdg}}(X_t)_{|v=1}. \tag{12.8}$$

We next remark that the nearby cycle sheaf $i^* Rk_* k^* \mathbb{R}_X$ can be used to put a Hodge structure on the cohomology groups $H^k(X_\infty)$, while the hypercohomology of the vanishing cycle sheaf $\phi_f = \mathrm{Cone}[i^*\mathbb{R}_X \to \psi_f \mathbb{R}_X]$ (it is a *complex* of sheaves of real vector spaces on E) likewise admits a Hodge structure. In fact, the spectral sequence of [Ste76, Cor. 4.20] shows that

$$\chi_{\mathrm{Hdg}}(X_\infty) = \chi_{\mathrm{Hdg}}(\psi_f)$$

and then the definition shows that

$$\chi_{\mathrm{Hdg}}(X_\infty) - \chi_{\mathrm{Hdg}}(E) = \chi_{\mathrm{Hdg}}(\phi_f).$$

In fact this formula motivates the nomenclature "motivic nearby fibre" and "motivic vanishing cycle".

As a concluding remark, the semisimple part T_s of the monodromy is an automorphism of the mixed Hodge structure on $H^k(X_\infty)$.

We can use these remarks to deduce information about the Hodge numbers on $H^k(X_\infty)$ from information about the geometry of the central fibre as we shall illustrate now.

Example 12.2.6. 1) Let $F, L_1, \ldots, L_d \in \mathbb{C}[X_0, X_1, X_2]$ be homogeneous forms with $\deg F = d$ and $\deg L_i = 1$ for $i = 1, \ldots, d$, such that $F \cdot L_1 \cdots L_d = 0$ defines a reduced divisor with normal crossings on $\mathbb{P}^2(\mathbb{C})$. We consider the space

$$X = \{([x_0, x_1, x_2], t) \in \mathbb{P}^2 \times \Delta \mid \prod_{i=1}^{d} L_i(x_0, x_1, x_2) + tF(x_0, x_1, x_2) = 0\}$$

where Δ is a small disk around $0 \in \mathbb{C}$. Then X is smooth and the map $f : X \to \Delta$ given by the projection to the second factor has as its zero fibre the union $E_1 \cup \cdots \cup E_d$ of the lines E_i with equation $L_i = 0$. These lines are in general position and have multiplicity one. We obtain

$$\psi_f = (d - \binom{d}{2})[\mathbb{P}^1]$$

so

$$e_{\mathrm{Hdg}}(\psi_f) = (1 - \binom{d-1}{2})(1 + uv)$$

and substituting $v = 1$ in this formula we get the formula $g = \binom{d-1}{2}$ for the genus of a smooth plane curve of degree d. The monodromy on $H^1(X_\infty)$ has g Jordan blocks of size 2, so is "maximally unipotent".

2) If we consider a similar example, but replace $\mathbb{P}^2$ by $\mathbb{P}^3$ and curves by surfaces, lines by planes, then the space X will not be smooth but has ordinary double points at the points of the zero fibre where two of the planes meet the surface $F = 0$. There are $d\binom{d}{2}$ of such points, d on each line of intersection. If we blow these up, we obtain a family $f : \tilde{X} \to \Delta$ whose zero fibre $D = E \cup F$ is the union of components E_i, $i = 1, \ldots, d$ which are copies of $\mathbb{P}^2$ blown up in $d(d-1)$ points, and components $F_j, j = 1, \ldots, d\binom{d}{2}$ which are copies of $\mathbb{P}^1 \times \mathbb{P}^1$. Thus

$$e_{\mathrm{Hdg}}(D(1)) = d(1 + (d^2 - d + 1)uv + u^2v^2) + d\binom{d}{2}(1 + uv)^2.$$

The double point locus $D(2)$ consists of the $\binom{d}{2}$ lines of intersections of the E_i together with the $d^2(d-1)$ exceptional lines in the E_i. So

$$e_{\mathrm{Hdg}}(D(2)) = d(d-1)(d + \frac{1}{2})(1 + uv).$$

Finally $D(3)$ consists of the $\binom{d}{3}$ intersection points of the E_i together with one point on each component F_j, so

$$e_{\mathrm{Hdg}}(D(3)) = \binom{d}{3} + d\binom{d}{2} = \frac{1}{3}d(d-1)(2d-1)$$

We get

$$e_{\mathrm{Hdg}}(\psi_f) = \left(\binom{d-1}{3} + 1\right)(1 + u^2v^2) + \frac{1}{3}d(2d^2 - 6d + 7)uv$$

in accordance with the Hodge numbers for a smooth degree d surface:

$$h^{2,0} = h^{0,2} = \binom{d-1}{3}, \quad h^{1,1} = \frac{1}{3}d(2d^2 - 6d + 7).$$

The monodromy on $H^2(X_\infty)$ has $\binom{d-1}{3}$ Jordan blocs of size 3 and $\frac{1}{2}d^3 - d^2 + \frac{1}{2}d + 1$ blocks of size 1.

3) Consider a similar smoothing of the union of two transverse quadrics in $\mathbb{P}^3$. The generic fibre is a smooth K3-surface and after blowing up the 16 double points of the total space we obtain the following special fibre:

   (i) $E(1)$ has two components which are blowings up of $\mathbb{P}^1 \times \mathbb{P}^1$ in 16 points, and 16 components isomorphic to $\mathbb{P}^1 \times \mathbb{P}^1$; hence $e_{\mathrm{Hdg}}(E(1)) = 18(1 + uv)^2 + 32uv$.

(ii) $E(2)$ consists of the 32 exceptional lines together with the strict transform of the intersection of the two quadrics, which is an elliptic curve; hence $e_{\mathrm{Hdg}}(E(2)) = 33(1+uv) - u - v$;

(iii) $E(3)$ consists of 16 points: one point on each exceptional $\mathbb{P}^1 \times \mathbb{P}^1$, so $e_{\mathrm{Hdg}}(E(3)) = 16$.

We get

$$e_{\mathrm{Hdg}}(X_\infty) = 1 + u + v + 18uv + u^2v + uv^2 + u^2v^2.$$

Putting $v = 1$ we get $2 + 20u + 2u^2$, in agreement with the Hodge numbers $(1, 20, 1)$ on the H^2 of a K3-surface. The monodromy has two Jordan blocs of size 2 and 18 blocs of size 1.

12.3 Equivariant Hodge number polynomials

We have seen that the mixed Hodge structure on the cohomology of the nearby fibre of a one-parameter degeneration comes with an automorphism of finite order. This leads us to consider the category $\mathfrak{hs}_{\mathbb{R}}^{\hat{\mu}}$ of pairs (H, γ) consisting of a real Hodge structure (i.e. direct sum of pure real Hodge structures of possibly different weights) H and an automorphism γ of finite order of this Hodge structure.

We are going to consider a kind of tensor product of two such objects, which we call *convolution* (see [SchS], where this operation was defined for mixed Hodge structures and called *join*). We will explain this by settling an equivalence of categories between $\mathfrak{hs}_{\mathbb{R}}^{\hat{\mu}}$ and a category $\mathfrak{fhs}$ of so-called *fractional Hodge structures.* (called *Hodge structures with fractional weights* in [L]; it is however not the weights which are fractional, but the indices of the Hodge filtration!

Definition 12.3.1. (See [L]). A *fractional Hodge structure* of weight k is a real vector space H of finite dimension, equipped with a decomposition

$$H_{\mathbb{C}} = \bigoplus_{a+b=k} H^{a,b}$$

where $a, b \in \mathbb{Q}$, such that $H^{b,a} = \overline{H^{a,b}}$. A *fractional Hodge structure* is defined as a direct sum of pure fractional Hodge structures of possibly different weights.

Lemma 12.3.2. *We have an equivalence of categories* $G : \mathfrak{hs}^{\hat{\mu}} \to \mathfrak{fhs}$.

Proof Let (H,γ) be an object of $\mathfrak{hs}^{\hat{\mu}}$ pure of weight k. We define $H_a = \mathrm{Ker}(\gamma - \exp(2\pi i a); H_{\mathbb{C}})$ for $0 \le a < 1$ and for $0 < a < 1$ put

$$\tilde{H}^{p+a,k+1-a-p} = H_a^{p,k-p}, \quad \tilde{H}^{p,k-p} = H_0^{p,k-p}$$

This transforms (H,γ) into a direct sum $\tilde{H} =: G(H,\gamma)$ of fractional Hodge structures of weights $k+1$ and k respectively. Conversely, for a fractional Hodge structure $\tilde{H}$ of weight k one has a unique automorphism γ of finite order which is multiplication by $\exp(2\pi i b)$ on $\tilde{H}^{b,k-b}$.

Note that this equivalence of categories does not preserve tensor products! Hence it makes sense to make the following

Definition 12.3.3. The *convolution* $(H',\gamma') * (H'',\gamma'')$ of two objects in $\mathfrak{hs}^{\hat{\mu}}$ is the object corresponding to the tensor product of their images in $\mathfrak{fhs}$:

$$G\left((H',\gamma') * (H'',\gamma'')\right) = G(H',\gamma') \otimes G(H'',\gamma'').$$

Note that the Hodge number polynomial map P_{hn} extends to a ring homomorphism

$$P_{\mathrm{hn}}^{\hat{\mu}} : K_0(\mathfrak{fhs}) \to R := \lim_{\leftarrow} \mathbb{Z}[u^{\frac{1}{n}}, v^{\frac{1}{n}}, u^{-1}, v^{-1}]$$

We denote its composition with the functor G by the same symbol. Hence

$$P_{\mathrm{hn}}^{\hat{\mu}} : K_0^{\hat{\mu}}(\mathfrak{hs}) \to R$$

transforms convolutions into products. We equally have an equivariant Hodge-Grothendieck character

$$\chi_{\mathrm{Hdg}}^{\hat{\mu}} : K_0^{\hat{\mu}}(\mathrm{Var}) \to K_0^{\hat{\mu}}(\mathfrak{hs})$$

and an equivariant Hodge-Euler characteristic

$$e_{\mathrm{Hdg}}^{\hat{\mu}} : K_0^{\hat{\mu}}(\mathrm{Var}) \to R.$$

Let $f : X \to \mathbb{C}$ be a projective morphism where X is smooth, of relative dimension n, with a single isolated critical point x such that $f(x) = 0$. Construct ψ_f by replacing the zero fibre X_0 by a divisor with normal crossings as above. Then the Milnor fibre of F of f at x has the homotopy type of a wedge of spheres of dimension n. Its cohomology is also equipped with a mixed Hodge structure. Recalling definition 12.7 of the motivic vanishing fibre, it can be shown that

$$P_{\mathrm{hn}}^{\hat{\mu}}(\tilde{H}^n(F)) = (-1)^n e_{\mathrm{Hdg}}^{\hat{\mu}}(\phi_f).$$

Write $P_{\mathrm{hn}}^{\hat{\mu}}(\tilde{H}^n(F)) = \sum_{\alpha \in \mathbb{Q},\, w \in \mathbb{Z}} m(\alpha, w) u^{\alpha} v^{w-\alpha}$. In the literature several numerical invariants have been attached to the singularity $f : (X,x) \to (\mathbb{C},0)$. These are all related to the numbers $m(\alpha,w)$ as follows:

i) The *characteristic pairs* [Ste77, Sect. 5].

$$\mathrm{Chp}(f,x) = \sum_{\alpha,w} m(\alpha,w)\cdot(n-\alpha,w).$$

ii) The *spectral pairs* [N-S]:

$$\mathrm{Spp}(f,x) = \sum_{\alpha\notin\mathbb{Z},w} m(\alpha,w)\cdot(\alpha,w) + \sum_{\alpha\in\mathbb{Z},w} m(\alpha,w)\cdot(\alpha,w+1).$$

iii) The singularity spectrum in Saito's sense [Sa]:

$$\mathrm{Sp}_{\mathrm{Sa}}(f,x) = P^{\hat{\mu}}_{\mathrm{hn}}(\tilde{H}^n(F))(t,1).$$

iv) The singularity spectrum in Varchenko's sense [Var]:

$$\mathrm{Sp}_{\mathrm{V}}(f,x) = t^{-1}P^{\hat{\mu}}_{\mathrm{hn}}(\tilde{H}^n(F))(t,1).$$

Note that the object ϕ_f depends only on the germ of f at the critical point x. Let us as a final remark rephrase the original Thom-Sebastiani theorem (i.e. for the case of isolated singularities):

Theorem 12.3.4. *Consider holomorphic germs* $f:(\mathbb{C}^{n+1},0)\to(\mathbb{C},0)$ *and* $g:(\mathbb{C}^{m+1},0)\to(\mathbb{C},0)$ *with isolated singularity. Then the germ* $f\oplus g:(\mathbb{C}^{n+1}\times\mathbb{C}^{m+1},(0,0))\to\mathbb{C}$ *with* $(f\oplus g)(x,y):=f(x)+g(y)$ *has also an isolated singularity, and*

$$\chi^{\hat{\mu}}_{\mathrm{Hdg}}(\phi_{f\oplus g}) = -\chi^{\hat{\mu}}_{\mathrm{Hdg}}(\phi_f) * \chi^{\hat{\mu}}_{\mathrm{Hdg}}(\phi_g)$$

so

$$e^{\hat{\mu}}_{\mathrm{Hdg}}(\phi_{f\oplus g}) = -e^{\hat{\mu}}_{\mathrm{Hdg}}(\phi_f)\cdot e^{\hat{\mu}}_{\mathrm{Hdg}}(\phi_g)\in R.$$

Remark 12.3.5. This theorem has been largely generalized, for functions with arbitrary singularities, and even on the level of motives. Denef and Loeser [DL] defined a convolution product for Chow motives, and Looijenga [L] defined one on $\mathcal{M}^{\hat{\mu}} = K_0^{\hat{\mu}}(\mathrm{Var})[L^{-1}]$, both with the property that the Hodge Euler polynomial commutes with convolution and that the Thom-Sebastiani property holds already on the level of varieties/motives.

References

[AKMW] Abramovich, D., K. Karu, K. Matsuki and J. Wlodarczyk: Torification and factorization of birational maps, J. Amer. Math. Soc. **15**, 531–572 (2002).

[B04] Bittner, F.: The universal Euler characteristic for varieties of characteristic zero, Comp. Math. **140**, 1011–1032 (2004)

[B05] Bittner, F.: On motivic zeta functions and the motivic nearby fibre, Math. Z. **249**, 63–83 (2005)

[Del71] Deligne, P.: Théorie de Hodge II, Publ. Math. I.H.E.S, **40**, 5–58 (1971)

[Del74] Deligne, P.: Théorie de Hodge III, Publ. Math., I. H. E. S, **44**, 5–77 (1974)

[DK] Danilov, V., and A. Khovanski: Newton polyhedra and an algorithm for computing Hodge-Deligne numbers Math., U. S. S. R. Izvestia, **29**, 274–298 (1987)

[DL] Denef, J. and F. Lœser: Motivic exponential integrals and a motivic Thom-Sebastiani theorem, Duke Math. J. **99**, 285–309 (1999)

[Gr] Griffiths, Ph. ed.: *Topics in Transcendental Geometry*, Annals of Math. Studies **106**, Princeton Univ. Press 1984.

[GH] Griffiths, Ph. and J. Harris: *Principles of Algebraic Geometry*, Wiley 1978.

[GN] Guillén, F. and V. Navarro Aznar: Sur le théorème local des cycles invariants, Duke Math. J., **61**, 133–155 (1990)

[GNPP] Guillén, F., V. Navarro Aznar, P. Pascual-Gainza and F. Puerta: *Hyperrésolutions cubiques et descente cohomologique*, Springer Lecture Notes in Math., **1335**, (1988)

[L] Looijenga, E.: Motivic measures. Séminaire Bourbaki, 52ème année, 1999–2000, no. 874

[N-S] Nemethi, A. and J. H. M. Steenbrink: Spectral Pairs, Mixed Hodge Modules, and Series of Plane Curve Singularities, New York J. Math. **1**, 149–177 (1995)

[Sa] Saito, M.: On the exponents and the geometric genus of an isolated hypersurface singularity. AMS Proc. Symp. Pure Math. **40** Part 2, 465–472 (1983)

[Schm] Schmid, W.: Variation of Hodge structure: the singularities of the period mapping, Invent. Math., **22**, 211–319 (1973)

[SchS] Scherk, J. and J. H. M. Steenbrink: On the Mixed Hodge Structure on the Cohomology of the Milnor Fibre, Mathematische Annalen **271**, 641–665 (1985)

[Ste76] Steenbrink, J. H. M.: Limits of Hodge structures, Inv.Math., **31**, 229–257 (1976)

[Ste77] Steenbrink, J. H. M.: Mixed Hodge structures on the vanishing cohomology, in *Real and Complex Singularities, Oslo, 1976*, Sijthoff-Noordhoff, Alphen a/d Rijn, 525–563 (1977)

[Var] Varchenko, A. N.: Asymptotic mixed Hodge structure in the vanishing cohomology. Izv. Akad. Nauk SSSR, Ser. Mat. **45**, 540–591 (1981) (in Russian). [English transl.: Math. USSR Izvestija, **18:3**, 469–512 (1982)]

13

Direct Image of Logarithmic Complexes and Infinitesimal Invariants of Cycles

Morihiko Saito

RIMS Kyoto University, Kyoto 606-8502 Japan

msaito@kurims.kyoto-u.ac.jp

To Jacob Murre

Abstract

We show that the direct image of the filtered logarithmic de Rham complex is a direct sum of filtered logarithmic complexes with coefficients in variations of Hodge structures, using a generalization of the decomposition theorem of Beilinson, Bernstein and Deligne to the case of filtered D-modules. The advantage of using the logarithmic complexes is that we have the strictness of the Hodge filtration by Deligne after taking the cohomology group in the projective case. As a corollary, we get the total infinitesimal invariant of a (higher) cycle in a direct sum of the cohomology of filtered logarithmic complexes with coefficients, and this is essentially equivalent to the cohomology class of the cycle.

Introduction

Let X, S be complex manifolds or smooth algebraic varieties over a field of characteristic zero. Let $f : X \to S$ be a projective morphism, and D be a divisor on S such that f is smooth over $S - D$. We have a filtered locally free $\mathcal{O}$-module (V^i, F) on $S - D$ underlying a variation of Hodge structure whose fiber V_s^i at $s \in S - D$ is the cohomology of the fiber $H^i(X_s, \mathbb{C})$. If D is a divisor with normal crossings on S, let $\widetilde{V}^i$ denote the Deligne extension [7] of V^i such that the the eigenvalues of the residue of the connection are contained in $[0, 1)$. The Hodge filtration F is naturally extended to $\widetilde{V}^i$ by [25]. We have the logarithmic de Rham complex

$$\mathrm{DR}_{\log}(\widetilde{V}^i) = \Omega_S^\bullet(\log D) \otimes_{\mathcal{O}} \widetilde{V}^i,$$

which has the Hodge filtration F^p defined by $\Omega_S^j(\log D) \otimes_{\mathcal{O}} F^{p-j}\widetilde{V}^i$. In general, V^i can be extended to a regular holonomic $\mathcal{D}_S$-module M^i on which a local defining equation of D acts bijectively. By [23], M^i and hence the de Rham complex $\mathsf{DR}(M^i)$ have the Hodge filtration F. If $Y := f^*D$ is a divisor with normal crossings on X, then $\Omega_X^\bullet(\log Y)$ has the Hodge filtration F defined by the truncation σ (see [8]) as usual, i.e. $F^p\Omega_X^\bullet(\log Y) = \Omega_X^{\bullet \geq p}(\log Y)$.

Theorem 1. *Assume $Y = f^*D$ is a divisor with normal crossings. There is an increasing split filtration L on the filtered complex $\mathbf{R}f_*(\Omega_X^\bullet(\log Y), F)$ such that we have noncanonical and canonical isomorphisms in the filtered derived category:*

$$\mathbf{R}f_*(\Omega_X^\bullet(\log Y), F) \simeq \bigoplus_{i\in\mathbb{Z}}(\mathsf{DR}(M^i), F)[-i],$$
$$\mathrm{Gr}_i^L \mathbf{R}f_*(\Omega_X^\bullet(\log Y), F) = (\mathsf{DR}(M^i), F)[-i].$$

If D is a divisor with normal crossings, we have also

$$\mathbf{R}f_*(\Omega_X^\bullet(\log Y), F) \simeq \bigoplus_{i\in\mathbb{Z}}(\mathsf{DR}_{\log}(\widetilde{V}^i), F)[-i],$$
$$\mathrm{Gr}_i^L \mathbf{R}f_*(\Omega_X^\bullet(\log Y), F) = (\mathsf{DR}_{\log}(\widetilde{V}^i), F)[-i].$$

This follows from the decomposition theorem (see [2]) extended to the case of the direct image of $(\mathcal{O}_X, F)$ as a filtered $\mathcal{D}$-module, see [22]. Note that Hodge modules do not appear in the last statement if D is a divisor with normal crossings. The assertion becomes more complicated in the non logarithmic case, see Remark (i) in 13.2.5. A splitting of the filtration L is given by choosing the first noncanonical isomorphism in the filtered decomposition theorem, see (13.12). A canonical choice of the splitting is given by choosing a relatively ample class, see [9].

Let $\mathsf{CH}^p(X - Y, n)$ be Bloch's higher Chow group, see [3]. In the analytic case, we assume for simplicity that $f : (X, Y) \to (S, D)$ is the base change of a projective morphism of smooth complex algebraic varieties $f' : (X', Y') \to (S', D')$ by an open embedding of complex manifolds $S \to S'_{\mathrm{an}}$, and an element of $\mathsf{CH}^p(X - Y, n)$ is the restriction of an element of $\mathsf{CH}^p(X' - Y', n)$ to $X - Y$. If $n = 0$, we may assume that it is the restriction of an analytic cycle of codimension p on X. From Theorem 1, we can deduce

Corollary 1. *With the above notation and assumption, let $\xi \in \mathsf{CH}^p(X - Y, n)$. Then, choosing a splitting of the filtration L in Theorem 1 (or more precisely, choosing the first noncanonical isomorphism in the filtered decomposition theorem* (13.12)*, we have the total infinitesimal invariant*

$$\delta_{S,D}(\xi) = (\delta^i_{S,D}(\xi)) \in \bigoplus_{i\geq 0}\mathbf{H}^i(S, F^p\mathsf{DR}(M^{2p-n-i})),$$
$$(\textit{resp. } \overline{\delta}_{S,D}(\xi) = (\overline{\delta}^i_{S,D}(\xi)) \in \bigoplus_{i\geq 0}\mathbf{H}^i(S, \mathrm{Gr}^p_F \mathsf{DR}(M^{2p-n-i})),)$$

where $\delta^i_{S,D}(\xi)$ *(respectively.* $\overline{\delta}^i_{S,D}(\xi)$*) is independent of the choice of a splitting if the* $\delta^j_{S,D}(\xi)$ *(respectively.* $\overline{\delta}^j_{S,D}(\xi)$*) vanish for* $j < i$*. In the case* D *is a divisor with normal crossings, the assertion holds with* $\mathsf{DR}(M^{2p-n-i})$ *replaced by* $\mathsf{DR}_{\log}(\widetilde{V}^{2p-n-i})$*.*

This shows that the infinitesimal invariants in [14, 13, 27, 5, 1, 24] can be defined naturally in the cohomology of filtered logarithmic complexes with coefficients in variations of Hodge structures if D is a divisor with normal crossings, see 13.2.4 for the compatibility with [1]. Note that if S is Stein or affine, then $\mathbf{H}^i(S, F^p\mathsf{DR}_{\log}(\widetilde{V}^q))$ is the i-th cohomology group of the complex whose j-th component is $\Gamma(S, \Omega^j_S(\log D)\otimes_{\mathcal{O}} F^{p-j}\widetilde{V}^q)$. If D is empty, then an inductive definition of $\delta^i_{S,D}(\xi)$, $\overline{\delta}^i_{S,D}(\xi)$ was given by Shuji Saito [24] using the filtered Leray spectral sequence together with the E_2-degeneration argument in [6]. He also showed that the infinitesimal invariants depend only on the cohomology class of the cycle. If S is projective, then it follows from [8] that the total infinitesimal invariant $(\delta^i_{S,D}(\xi))$ is equivalent to the cycle class of ξ in $H^{2p-n}_{\mathsf{DR}}(X-Y)$ by the strictness of the Hodge filtration, and the filtration L comes from the Leray filtration on the cohomology of $X-Y$, see Remark (iii) in 13.2.5.

Corollary 1 is useful to study the behavior of the infinitesimal invariants near the boundary of the variety. If D is empty, let $\delta^i_S(\xi)$ denote $\delta^i_{S,D}(\xi)$. We can define $\delta^i_{\mathsf{DR},S}(\xi)$ as in [19] by omitting F^p before DR in Corollary 1 where $D = \varnothing$.

Corollary 2. *Assume* S *is projective. Let* $U = S - D$*. Then for each* $i \geq 0$*,* $\delta^i_{S,D}(\xi)$*,* $\overline{\delta}^i_{S,D}(\xi)$*,* $\delta^i_U(\xi)$ *and* $\delta^i_{\mathsf{DR},U}(\xi)$ *are equivalent to each other, i.e. one of them vanishes if and only if the others do.*

Indeed, $(\delta^i_{\mathsf{DR},U}(\xi))$ is determined by $(\delta^i_U(\xi))$, and $(\delta^i_U(\xi))$ by $(\delta^i_{S,D}(\xi))$. Moreover, $(\delta^i_{S,D}(\xi))$ is equivalent to $(\delta^i_{\mathsf{DR},U}(\xi))$ by the strictness of the Hodge filtration [8] applied to (X, Y) together with Theorem 1, see §13.2.3. For the relation with $\overline{\delta}^i_{S,D}(\xi)$, see §13.2.1. Note that the equivalence between $\delta^i_U(\xi)$ and $\delta^i_{\mathsf{DR},U}(\xi)$ in the case of algebraic cycles (i.e. $n = 0$) was first found by J.D. Lewis and Shuji Saito in [19] (assuming a conjecture of Brylinski and Zucker and the Hodge conjecture and using an L^2-argument). The above arguments seem to be closely related to their question, see also Remark (i) in §13.2.5 below.

As another corollary of Theorem 1 we have

Corollary 3. *Assume that* f *induces an isomorphism over* $S - D$*, and* $Y = f^*D$ *is a divisor with normal crossings on* X*. Then*

$$R^i f_*\Omega^p_X(\log Y) = 0 \quad \text{if } i + p > \dim X.$$

This follows immediately from Theorem 1 since $M^i = 0$ for $i \neq 0$. Corollary 3 is an analogue of the vanishing theorem of Kodaira-Nakano. However, this does not hold for a non logarithmic complex (e.g. if f is a blow-up with a point center). This corollary was inspired by a question of A. Dimca.

I would like to thank Dimca, Lewis and Shuji Saito for good questions and useful suggestions.

In Section 13.1, we prove Theorem 1 after reviewing some basic facts on filtered differential complexes. In Section 13.2 we explain the application of Theorem 1 to the infinitesimal invariants of (higher) cycles. In Section 13.3 we give some examples using Lefschetz pencils.

13.1 Direct image of logarithmic complexes

13.1.1 Filtered differential complexes

Let X be a complex manifold or a smooth algebraic variety over a field of characteristic zero. Let $D^bF(\mathcal{D}_X)$ (respectively. $D^bF(\mathcal{D}_X)^r$) be the bounded derived category of filtered left (respectively. right) $\mathcal{D}_X$-modules. Let $D^bF(\mathcal{O}_X, \mathsf{Diff})$ be the bounded derived category of filtered differential complexes (L, F) where F is exhaustive and locally bounded below (i.e. $F_p = 0$ for $p \ll 0$ locally on X), see [22, 2.2], We have an equivalence of categories

$$\mathsf{DR}^{-1} : D^bF(\mathcal{O}_X, \mathsf{Diff}) \to D^bF(\mathcal{D}_X)^r, \tag{13.1}$$

whose quasi-inverse is given by the de Rham functor DR^r for right $\mathcal{D}$-modules, see 13.1.2 below. Recall that, for a filtered $\mathcal{O}_X$-module (L, F), the associated filtered right $\mathcal{D}$-module $\mathsf{DR}^{-1}(L, F)$ is defined by

$$\mathsf{DR}^{-1}(L, F) = (L, F) \otimes_{\mathcal{O}} (\mathcal{D}, F), \tag{13.2}$$

and the morphisms $(L, F) \to (L', F)$ in $MF(\mathcal{O}_X, \mathsf{Diff})$ correspond bijectively to the morphisms of filtered $\mathcal{D}$-modules $\mathsf{DR}^{-1}(L, F) \to \mathsf{DR}^{-1}(L', F)$. More precisely, the condition on $(L, F) \to (L', F)$ is that the composition

$$F_pL \to L \to L' \to L'/F_qL'$$

is a differential operator of order $\leq p - q - 1$. The proof of (13.1) can be reduced to the canonical filtered quasi-isomorphism for a filtered right $\mathcal{D}$-module (M, F)

$$\mathsf{DR}^{-1} \circ \mathsf{DR}^r(M, F) \to (M, F),$$

which follows from a calculation of a Koszul complex.

Note that the direct image f_* of filtered differential complexes is defined by the sheaf-theoretic direct image $\mathbf{R}f_*$, and this direct image is compatible with the direct image f_* of filtered $\mathcal{D}$-modules via (13.1), see [22, 2.3.], So we get

$$\mathbf{R}f_* = \mathsf{DR}^r \circ f_* \circ \mathsf{DR}^{-1} : D^bF(\mathcal{O}_X, \mathsf{Diff}) \to D^bF(\mathcal{O}_S, \mathsf{Diff}), \tag{13.3}$$

where we use DR^r for right $\mathcal{D}$-modules (otherwise there is a shift of complex).

13.1.2 De Rham complex

The de Rham complex $\mathsf{DR}^r(M,F)$ of a filtered right $\mathcal{D}$-module (M,F) is defined by

$$(\mathsf{DR}^r(M,F))^i = \bigwedge^{-i}\Theta_X \otimes_{\mathcal{O}} (M, F[-i]) \quad \text{for } i \le 0. \tag{13.4}$$

Here $(F[-i])_p = F_{p+i}$ in a compatible way with $(F[-i])^p = F^{p-i}$ and $F_p = F^{-p}$. Recall that the filtered right $\mathcal{D}$-module associated with a filtered left $\mathcal{D}$-module (M,F) is defined by

$$(M,F)^r := (\Omega_X^{\dim X}, F) \otimes_{\mathcal{O}} (M,F), \tag{13.5}$$

where $\mathrm{Gr}_p^F \Omega_X^{\dim X} = 0$ for $p \ne -\dim X$. This induces an equivalence of categories between the left and right $\mathcal{D}$-modules. The usual de Rham complex $\mathsf{DR}(M,F)$ for a left $\mathcal{D}$-module is defined by

$$(\mathsf{DR}(M,F))^i = \Omega_X^i \otimes_{\mathcal{O}} (M, F[-i]) \quad \text{for } i \ge 0, \tag{13.6}$$

and this is compatible with (13.4) via (13.5) up to a shift of complex, i.e.

$$\mathsf{DR}(M,F) = \mathsf{DR}^r(M,F)^r[-\dim X]. \tag{13.7}$$

13.1.3 Logarithmic complex

Let X be as in §13.1.1, and Y be a divisor with normal crossings on X. Let (V,F) be a filtered locally free $\mathcal{O}$-module underlying a polarizable variation of Hodge structure on $X-Y$. Let $(\widetilde{V},F)$ be the Deligne extension of (V,F) to X such that the eigenvalues of the residue of the connection are contained in $[0,1)$. Then we have the filtered logarithmic de Rham complex $\mathsf{DR}_{\log}(\widetilde{V},F)$ such that F^p of its i-th component is

$$\Omega_X^i(\log Y) \otimes F^{p-i}\widetilde{V}.$$

If $(M,F) = (\mathcal{O}_X, F)$ with $\mathrm{Gr}_p^F \mathcal{O}_X = 0$ for $p \ne 0$, then

$$\mathsf{DR}_{\log}(\mathcal{O}_X, F) = (\Omega_X^\bullet(\log X), F).$$

Let $\widetilde{V}(*Y)$ be the localization of $\widetilde{V}$ by a local defining equation of Y. This is a regular holonomic left $\mathcal{D}_X$-module underlying a mixed Hodge module, and has the Hodge filtration F which is generated by the Hodge filtration F on $\widetilde{V}$, i.e.

$$F_p\widetilde{V}(*Y) = \textstyle\sum_\nu \partial^\nu F^{-p+|\nu|}\widetilde{V},$$

where $F_p = F^{-p}$ and $\partial^\nu = \prod_i \partial_i^{\nu_i}$ with $\partial_i = \partial/\partial x_i$. Here $(x_1, \dots, x_n)$ is a local coordinate system such that Y is contained in $\{x_1 \cdots x_n = 0\}$. By [23, 3.11] we have a filtered quasi-isomorphism

$$\mathsf{DR}_{\log}(\widetilde{V}, F) \xrightarrow{\sim} \mathsf{DR}(\widetilde{V}(*Y), F). \tag{13.8}$$

This generalizes the filtered quasi-isomorphism in [7]

$$(\Omega_X^\bullet(\log Y), F) \xrightarrow{\sim} \mathsf{DR}(\mathcal{O}_X(*Y), F). \tag{13.9}$$

Note that the direct image of the filtered $\mathcal{D}_X$-module $(\widetilde{V}(*Y), F)$ by $X \to pt$ in the case X projective (or proper algebraic) is given by the cohomology group of the de Rham complex $\mathsf{DR}(\widetilde{V}(*Y), F)$ (up to a shift of complex) by definition, and the Hodge filtration F on the direct image is strict by the theory of Hodge modules. So we get

$$\begin{aligned} F^p\mathbf{H}^i(X - Y, \mathsf{DR}(V)) &:= F^p\mathbf{H}^i(X, \mathsf{DR}(\widetilde{V}(*Y))) \\ = \mathbf{H}^i(X, F^p\mathsf{DR}(\widetilde{V}(*Y))) &= \mathbf{H}^i(X, F^p\mathsf{DR}_{\log}(\widetilde{V})). \end{aligned} \tag{13.10}$$

13.1.4 Decomposition theorem

Let $f : X \to S$ be a projective morphism of complex manifolds or smooth algebraic varieties over a field of characteristic zero. Then the decomposition theorem of Beilinson, Bernstein and Deligne [2] is extended to the case of Hodge modules ([22, 23]), and we have noncanonical and canonical isomorphisms

$$\left.\begin{aligned} f_*(\mathcal{O}_X, F) &\simeq \textstyle\bigoplus_{j\in\mathbb{Z}} \mathcal{H}^j f_*(\mathcal{O}_X, F)[-j] \quad \text{in } D^bF(\mathcal{D}_S), \\ \mathcal{H}^j f_*(\mathcal{O}_X, F) &= \textstyle\bigoplus_{Z\subset S} (M_Z^j, F) \quad \text{in } MF(\mathcal{D}_S), \end{aligned}\right\} \tag{13.11}$$

where Z are irreducible closed analytic or algebraic subsets of S, and (M_Z^j, F) are filtered $\mathcal{D}_S$-modules underlying a pure Hodge module of weight $j+\dim X$ and with strict support Z, i.e. M_Z^j has no nontrivial sub nor quotient module whose support is strictly smaller than Z. (Here $MF(\mathcal{D}_S)$ denotes the category of filtered left $\mathcal{D}_S$-modules.) Indeed, the second canonical isomorphism follows from the strict support decomposition which is part of the definition of pure Hodge modules, see [22, 5.1.6.]. The first noncanonical isomorphism follows from the strictness of the Hodge filtration and the relative hard Lefschetz theorem for the direct image (see [22, 5.3.1]) using the E_2-degeneration argument in [6] together with the equivalence of categories $D^bF(\mathcal{D}_S) \simeq D^bG(\mathcal{B}_S)$. Here $\mathcal{B}_S = \bigoplus_{i\in\mathbb{N}} F_i\mathcal{D}_S$ and $D^bG(\mathcal{B}_S)$ is the derived category of bounded complexes of graded left $\mathcal{B}_S$-modules $M_\bullet^\bullet$ such that $M_i^j = 0$ for $i \ll 0$ or $|j| \gg 0$, see [22, 2.1.12]. We need a derived category associated to some abelian category in order to apply the argument in [6] (see also [9]). In the algebraic case, we can also apply [6] to the derived category of mixed Hodge modules on S and it is also possible to use [23, 4.5.4] to show the first noncanonical isomorphism.

If f is smooth over the complement of a divisor $D \subset S$ and $Y := f^*D$ is a divisor with normal crossings, then the filtered direct image $f_*(\mathcal{O}_X(*Y), F)$ is strict (see [23, 2.15]), and we have noncanonical and canonical isomorphisms

$$\left.\begin{aligned} f_*(\mathcal{O}_X(*Y), F) &\simeq \textstyle\bigoplus_{j\in\mathbb{Z}} \mathcal{H}^j f_*(\mathcal{O}_X(*Y), F)[-j] \quad \text{in } D^bF(\mathcal{D}_S), \\ \mathcal{H}^j f_*(\mathcal{O}_X(*Y), F) &= (M_S^j(*D), F) \quad \text{in } MF(\mathcal{D}_S). \end{aligned}\right\} \tag{13.12}$$

Here $(M_S^j(*D), F)$ is the 'localization' of (M_S^j, F) along D which is the direct image of $(M_S^j, F)|_U$ by the open embedding $U := S - D \to S$ in the category of filtered $\mathcal{D}$-modules underlying mixed Hodge modules. (By the Riemann-Hilbert correspondence, this gives the direct image in the category of complexes with constructible cohomology because D is a divisor.) The Hodge filtration F on the direct image is determined by using the V-filtration of Kashiwara and Malgrange, and $(M_S^j(*D), F)$ is the unique extension of $(M_S^j, F)|_U$ which underlies a mixed Hodge module on S and whose underlying $\mathcal{D}_S$-module is the direct image in the category of regular holonomic $\mathcal{D}_S$-modules, see [23, 2.11]. So the second canonical isomorphism follows because the left-hand side satisfies these conditions. (Note that (M_Z^j, F) for $Z \neq S$ vanishes by the localization, because $Z \subset D$ if $(M_Z^j, F) \neq 0$.) The first noncanonical isomorphism follows from the strictness of the Hodge filtration and the relative hard Lefschetz theorem by the same argument as above.

13.1.5 Proof of Theorem 1

Let $r = \dim X - \dim S$. By (13.3)(13.9) and (13.12), we have isomorphisms

$$\left.\begin{aligned} \mathbf{R}f_*(\Omega_X^\bullet(\log Y), F) &= \mathrm{DR}^r \circ f_* \circ \mathrm{DR}^{-1}(\Omega_X^\bullet(\log Y), F) \\ &= \mathrm{DR}^r \circ f_*(\mathcal{O}_X(*Y), F)[-\dim X] \\ &\simeq \textstyle\bigoplus_{i\in\mathbb{Z}} \mathrm{DR}(M_S^i(*D), F)[-r-i], \end{aligned}\right\} \tag{13.13}$$

where the shift of complex by r follows from the difference of the de Rham complex for left and right $\mathcal{D}$-modules. Furthermore, letting L be the filtration induced by τ on the complex of filtered $\mathcal{D}_S$-modules $f_*(\mathcal{O}_X(*Y), F)[-r]$, we have a canonical isomorphism

$$\mathrm{Gr}_i^L f_*(\mathcal{O}_X(*Y), F)[-r] = (M_S^{i-r}(*D), F)[-i], \tag{13.14}$$

and the first assertion follows by setting $M^i = M_S^{i-r}(*D)$. The second assertion follows from the first by (13.8). This completes the proof of Theorem 1.

13.2 Infinitesimal invariants of cycles

13.2.1 Cycle classes

Let X be a complex manifold, and let $\mathcal{C}^{\bullet,\bullet}$ denote the double complex of vector spaces of currents on X. The associated single complex is denoted by $\mathcal{C}^{\bullet}$. Let F be the Hodge filtration by the first index of $\mathcal{C}^{\bullet,\bullet}$ (using the truncation σ in [8]). Let ξ be an analytic cycle of codimension p on X. Then it is well known that ξ defines a closed current in $F^p\mathcal{C}^{2p}$ by integrating the restrictions of C^∞ forms with compact supports on X to the smooth part of the support of ξ (and using a triangulation or a resolution of singularities of the cycle). So we have a cycle class of ξ in $H^{2p}(X, F^p\Omega_X^\bullet)$.

Assume X is a smooth algebraic variety over a field k of characteristic zero. Then the last assertion still holds (where $\Omega_X^\bullet$ means $\Omega_{X/k}^\bullet$), see [11]. Moreover, for the higher Chow groups, we have the cycle map (see [4, 10, 12, 15, 16])

$$cl : \mathsf{CH}^p(X, n) \to F^p H_{\mathsf{DR}}^{2p-n}(X),$$

where the Hodge filtration F is defined by using a smooth compactification of X whose complement is a divisor with normal crossings, see [8]. This cycle map is essentially equivalent to the cycle map to $\mathrm{Gr}_F^p H_{\mathsf{DR}}^{2p-n}(X)$ because we can reduce to the case $k = \mathbb{C}$ where we have the cycle map

$$\mathrm{cl} : \mathsf{CH}^p(X, n) \to \mathrm{Hom}_{\mathsf{MHS}}(\mathbb{Q}, H^{2p-n}(X, \mathbb{Q})(p)),$$

and morphisms of mixed Hodge structures are strictly compatible with the Hodge filtration F.

13.2.2 Proof of Corollary 2

By §13.2.1 the cycle class of ξ belongs to

$$H^{2p-n}(X, F^p\Omega_X^\bullet(\log Y)).$$

By Theorem 1, this gives the total infinitesimal invariant

$$\delta_{S,D}(\xi) = (\delta_{S,D}^{2p-n-i}(\xi)) \in \bigoplus_{i\in\mathbb{Z}} \mathbf{H}^{2p-n-i}(S, F^p\mathsf{DR}(M^i)),$$

and similarly for $\overline{\delta}_{S,D}(\xi)$. So the assertion follows.

13.2.3 Proof of Corollary 2

Choosing the first noncanonical isomorphism in the filtered decomposition theorem (13.12), we get canonical morphisms compatible with the direct sum decompositions

$$\begin{aligned}\bigoplus_{i\geq 0}\mathbf{H}^i(S, F^p\mathsf{DR}(M^{q-i})) &\to \bigoplus_{i\geq 0}\mathbf{H}^i(S-D, F^p\mathsf{DR}(M^{q-i}))\\ &\to \bigoplus_{i\geq 0}\mathbf{H}^i(S-D, \mathsf{DR}(M^{q-i})),\end{aligned}$$

and these are identified with the canonical morphisms

$$\begin{aligned}\mathbf{H}^q(X, F^p\Omega_X^\bullet(\log Y)) &\to \mathbf{H}^q(X-Y, F^p\Omega_{X-Y}^\bullet)\\ &\to \mathbf{H}^q(X-Y, \Omega_{X-Y}^\bullet).\end{aligned}$$

By Deligne [8], the composition of the last two morphisms is injective because of the strictness of the Hodge filtration, see also § 13.1.4. So we get the equivalence of $\delta_{S,D}^i(\xi)$, $\delta_U^i(\xi)$, $\delta_{\mathsf{DR},U}^i(\xi)$. The equivalence with $\overline{\delta}_{S,D}^i(\xi)$ follows from § 13.2.1.

13.2.4 Compatibility with the definition in [1]

When D is empty, the infinitesimal invariants are defined in [1] by using the extension groups of filtered $\mathcal{D}$-modules together with the forgetful functor from the category of mixed Hodge modules to that of filtered $\mathcal{D}$-modules. Its compatibility with the definition in this paper follows from the equivalence of categories (13.1) and the compatibility of the direct image functors (13.3).

Note that for $(L, F) \in D^bF(\mathcal{O}_X, \mathsf{Diff})$ in the notation of (1.1), we have a canonical isomorphism

$$\mathrm{Ext}^i((\Omega_X^\bullet, F), (L, F)) = \mathbf{H}^i(X, F_0L), \tag{13.15}$$

where the extension group is taken in $D^bF(\mathcal{O}_X, \mathsf{Diff})$. Indeed, the left-hand side is canonically isomorphic to

$$\begin{aligned}&\mathrm{Ext}^i(\mathsf{DR}^{-1}(\Omega_X^\bullet, F), \mathsf{DR}^{-1}(L, F))\\ &= \mathbf{H}^i(X, F_0\mathcal{H}om_{\mathcal{D}}(\mathsf{DR}(\mathcal{D}_X, F), \mathsf{DR}^{-1}(L, F))),\\ &= \mathbf{H}^i(X, F_0\mathsf{DR}^r\mathsf{DR}^{-1}(L)),\end{aligned}$$

and the last group is isomorphic to the right-hand side of (13.15) which is independent of a representative of (L, F). If X is projective, then this assertion follows also from the adjoint relation for filtered $\mathcal{D}$-modules.

If X is smooth projective and Y is a divisor with normal crossings, then the cycle class can be defined in

$$\begin{aligned}\mathrm{Ext}^{2p}((\Omega_X^\bullet, F), \Omega_X^\bullet(\log Y), F[p])) &= \mathbf{H}^{2p}(X, F^p\Omega_X^\bullet(\log Y)) \\ &= F^p\mathbf{H}^{2p}(X, \Omega_X^\bullet(\log Y)).\end{aligned}$$

13.2.5 Remarks

i) If we use (13.11) instead of (13.12) we get an analogue of Theorem 1 for non logarithmic complexes. However, the assertion becomes more complicated, and we get noncanonical and canonical isomorphisms

$$\left.\begin{aligned}\mathbf{R}f_*(\Omega_X^\bullet, F) &\simeq \textstyle\bigoplus_{i\in\mathbb{Z}, Z\subset S}(\mathsf{DR}(M_Z^{i-r}), F)[-i]. \\ \mathrm{Gr}_i^L \mathbf{R}f_*(\Omega_X^\bullet, F) &= \textstyle\bigoplus_{Z\subset S}(\mathsf{DR}(M_Z^{i-r}), F)[-i].\end{aligned}\right\} \tag{13.16}$$

This implies an analogue of Corollary 1. If D is a divisor with normal crossings, we have a filtered quasi-isomorphism for $Z = S$

$$(\mathsf{DR}_{\log}(\widetilde{M}_S^{i-r}), F) \xrightarrow{\sim} (\mathsf{DR}(M_S^{i-r}), F), \tag{13.17}$$

where $\mathsf{DR}_{\log}(\widetilde{M}_S^{i-r})$ is the intersection of $\mathsf{DR}(M_S^{i-r})$ with $\mathsf{DR}_{\log}(\widetilde{V}_S^i)$. This seems to be related with a question of Lewis and Shuji Saito, see also [19].

ii) If $\dim S = 1$, we can inductively define the infinitesimal invariants in Corollary 1 by an argument similar to [24] using [26].

iii) Assume S is projective and D is a divisor with normal crossings. Then the Leray filtration for $X \to S \to pt$ is given by the truncation τ on the complex of filtered $\mathcal{D}_S$-modules $f_*(\mathcal{O}_X(*Y), F)$, and gives the Leray filtration on the cohomology of $X - Y$ (induced by the truncation τ as in [8]). Indeed, the graded pieces $\mathcal{H}^j f_*(\mathcal{O}_X(*Y), F)$ of the filtration τ on S coincide with $(\widetilde{V}^{j+r}(*D), F)$, and give the open direct images by $U \to S$ of the graded pieces (V^{j+r}, F) of the filtration τ on U as filtered $\mathcal{D}$-modules underlying mixed Hodge modules. Note that the morphism $U \to S$ is open affine so that the direct image preserves regular holonomic $\mathcal{D}$-modules.

13.3 Examples

13.3.1 Lefschetz pencils

Let Y be a smooth irreducible projective variety of dimension n embedded in a projective space $\mathcal{P}$ over $\mathbb{C}$. We assume that $Y \neq \mathcal{P}$ and Y is not contained in a hyperplane of $\mathcal{P}$ so that the hyperplane sections of Y are parametrized by the dual projective spaces $\mathcal{P}^\vee$. Let $D \subset \mathcal{P}^\vee$ denote the discriminant. This is the image of a projective bundle over Y (consisting of hyperplanes tangent to Y), and hence D is irreducible. At a smooth point of D, the corresponding hyperplane section of Y has only one ordinary double point. We assume that the associated vanishing cycle is not zero in the cohomology of general hyperplane section X. This is equivalent to the non surjectivity of $H^{n-1}(Y) \to H^{n-1}(X)$.

A Lefschetz pencil of Y is a line $\mathbb{P}^1$ in $\mathcal{P}$ intersecting the discriminant D at smooth points of D (corresponding to hyperplane sections having only one ordinary double point). We have a projective morphism $\pi : \tilde{Y} \to \mathbb{P}^1$ such that $\tilde{Y}_t := \pi^{-1}(t)$ is the hyperplane section corresponding $t \in \mathbb{P}^1 \subset \mathcal{P}$ and $\tilde{Y}$ is the blow-up of Y along a smooth closed subvariety Z of codimension 2 which is the intersection of $\tilde{Y}_t$ for any (or two of) $t \in \mathbb{P}^1$.

A Lefschetz pencil of hypersurface sections of degree d is defined by replacing the embedding of Y using $\mathcal{O}_Y(d)$ so that a hyperplane section corresponds to a hypersurface section of degree d. Here $\mathcal{O}_Y(d)$ for an integer d denotes the invertible sheaf induced by that on $\mathcal{P}$ as usual.

13.3.2 Hypersurfaces containing a subvariety

Let $Y, \mathcal{P}$ be as in § 13.3.1. Let $E \subset Y$ be a closed subvariety (which is not necessarily irreducible nor reduced). Let

$$E_{\{i\}} = \{x \in E : \dim T_x E = i\}.$$

Let $\mathcal{I}_E$ be the ideal sheaf of E in Y. Let δ be a positive integer such that $\mathcal{I}_E(\delta)$ is generated by global sections. By [18, 20] (or [21]) we have the following

$$\left.\begin{array}{c}\text{If } \dim Y > \max\{\dim E_{\{i\}} + i\} \text{ and } d \geq \delta, \text{then there is a} \\ \text{smooth hypersurface section of degree } d \text{ containing } E.\end{array}\right\} \qquad (13.18)$$

We have furthermore

$$\left.\begin{array}{c}\text{If } \dim Y > \max\{\dim E_{\{i\}} + i\} + 1 \text{ and } d \geq \delta + 1, \text{ then there is a} \\ \text{Lefschetz pencil of hypersurface sections of degree } d \text{ containing } E.\end{array}\right\} \qquad (13.19)$$

Indeed, we have a pencil such that $\tilde{Y}_t$ has at most isolated singularities, because $\tilde{Y}_t$ is smooth near the center Z which is the intersection of two generic hypersurface sections containing E, and hence is smooth, see [18, 20] (or [21]). Note that a local equation of $\tilde{Y}_t$ near Z is given by $f - tg$ if t is identified with an appropriate affine coordinate of $\mathbb{P}^1$ where f, g are global sections of $\mathcal{I}_E(d)$ corresponding to smooth hypersurface sections.

To get only ordinary double points, note first that the parameter space of the hypersurfaces containing E is a linear subspace of $\mathcal{P}^\vee$. So it is enough to show that this linear subspace contains a point of the discriminant D corresponding to an ordinary double point. Thus we have to show that an isolated singularity can be deformed to ordinary double points by replacing the corresponding section $h \in \Gamma(Y, \mathcal{I}_E(d))$ with $h + \sum_i t_i g_i$ where $g_i \in \Gamma(Y, \mathcal{I}_E(d))$ and the $t_i \in \mathbb{C}$ are general with sufficiently small absolute values. Since $d \geq \delta + 1$, we see that $\Gamma(Y, \mathcal{I}_E(d))$ generates the 1-jets at each point of the complement of E. So the assertion follows from the fact that for a function with an isolated singularity f, the singularities of $\{f + \sum_i t_i x_i = 0\}$ are ordinary double points if $t_1, \ldots, t_n$ are general, where $x_1, \ldots, x_n$ are local coordinates. (Note that f has an ordinary double point if and only if the morphism defined by $(\partial f/\partial x_1, \ldots, \partial f/\partial x_n)$ is locally biholomorphic at this point.)

13.3.3 Construction

For $Y, \mathcal{P}$ be as in § 13.3.1, let $i_{Y,\mathcal{P}} : Y \to \mathcal{P}$ denote the inclusion. Assume

$$i_{Y,\mathcal{P}}^* : H^j(\mathcal{P}) \to H^j(Y) \text{ is surjective for any } j \neq \dim Y, \tag{13.20}$$

where cohomology has coefficients in any field of characteristic zero. This condition is satisfied if Y is a complete intersection.

Let E_1, E_2 be m-dimensional irreducible closed subvarieties of Y such that

$$E_1 \cap E_2 = \varnothing, \quad \deg E_1 = \deg E_2.$$

Here $\dim Y = n = 2m + s + 1$ with $m \geq 0$, $s \geq 1$. Let $E = E_1 \cup E_2$. With the notation of (3.2), assume

$$d > \delta, \quad \dim Y > \max\{\dim E_{\{i\}} + i\} + s, \tag{13.21}$$

$$i_{X^{(j)},Y}^* : H^{n-j}(Y) \to H^{n-j}(X^{(j)}) \text{ is not surjective for } j \leq s, \tag{13.22}$$

where $X^{(j)}$ is a general complete intersection of multi degree $(d, \ldots, d)$ and of codimension j in Y. (This is equivalent to the condition that the vanishing cycles for a hypersurface $X^{(j)}$ of $X^{(j-1)}$ are nonzero.)

Let X be a general hypersurface of degree d in Y containing E, see (13.18). Let L denote the intersection of X with a general linear subspace of codimension $m + s$ in the projective space. Then $[E_a]$ $(a = 1, 2)$ and $c[L \cap X]$

have the same cohomology class in $H^{2m+2s}(X)$ for some $c \in \mathbb{Q}$, because $\dim H^{2m+2s}(X) = 1$ by the weak and hard Lefschetz theorems together with (13.20). Let

$$\xi_a = [E_a] - c[L \cap X] \in \mathsf{CH}^{m+s}(X)_{\mathbb{Q}} \ (a = 1, 2).$$

These are homologous to zero. It may be expected that one of them is non torsion, generalizing an assertion in [24]. More precisely, let S be a smooth affine rational variety defined over a finitely generated subfield k of $\mathbb{C}$ and parametrizing the smooth hypersurfaces of degree d containing E as above so that there is the universal family $\mathfrak{X} \to S$ defined over k (see [2, 28]). Assume X corresponds to a geometric generic point of S with respect to k, i.e. X is the geometric generic fiber for some embedding $k(S) \to \mathbb{C}$. Let

$$\xi_{a,\mathfrak{X}} = [E_a \times_k S] - c[L]_{\mathfrak{X}} \in \mathsf{CH}^{m+s}(\mathfrak{X})_{\mathbb{Q}},$$

where $[L]_{\mathfrak{X}}$ is the pull-back of $[L]$ by $\mathfrak{X} \to Y$. Since the local system $\{H^{2m+2s-j}(\mathfrak{X}_s)\}$ on S is constant for $j < s$ and S is smooth affine rational, we see that $\delta_S^j(\xi_{a,\mathfrak{X}}) = 0$ for $j < s$. Then it may be expected that $\delta_S^s(\xi_{a,\mathfrak{X}}) \neq 0$ for one of a, where S can be replaced by any non empty open subvariety. We can show this for $s = 1$ as follows. (For $s > 1$, it may be necessary to assume further conditions on d, etc.)

13.3.4 The case $s = 1$

Consider a Lefschetz pencil $\pi : \tilde{Y} \to \mathbb{P}^1$ such that $\tilde{Y}_t := \pi^{-1}(t)$ for $t \in \mathbb{P}^1$ is a hypersurface of degree d in Y containing E. Here $\tilde{Y}$ is the blow-up of Y along a smooth closed subvariety Z, and Z is the intersection of $\tilde{Y}_t$ for any $t \in \mathbb{P}^1$. Note that $\tilde{Y}_t$ has an ordinary double point for $t \in \Lambda \subset \mathbb{P}^1$, where Λ denotes the discriminant, see (13.19).

Since Z has codimension 2 in Y, we have the isomorphism

$$H^n(\tilde{Y}) = H^n(Y) \oplus H^{n-2}(Z), \tag{13.23}$$

so that the cycle class of $[E_a \times \mathbb{P}^1] - c[L]_{\tilde{Y}} \in \mathsf{CH}^{m+1}(\tilde{Y})_{\mathbb{Q}}$ in $H^n(\tilde{Y})$ is identified with the difference of the cycle class $cl_Z(E_a) \in H^{n-2}(Z)$ and the cycle class of L in $H^n(Y)$. Indeed, the injection $H^{n-2}(Z) \to H^n(\tilde{Y})$ in the above direct sum decomposition is defined by using the projection $Z \times \mathbb{P}^1 \to Z$ and the closed embedding $Z \times \mathbb{P}^1 \to \tilde{Y}$, and the injection $H^n(Y) \to H^n(\tilde{Y})$ is the pull-back by $\tilde{Y} \to Y$, see [17].

By assumption, one of the $cl_Z(E_a)$ is not contained in the non primitive part, i.e. not a multiple of the cohomology class of the intersection of general hyperplane sections. Indeed, if both are contained in the non primitive part, then $cl_Z(E_1) = cl_Z(E_2)$ and this implies the vanishing of the self intersection number $E_a \cdot E_a$ in Z.

We will show that the cycle class of $[E_a \times \mathbb{P}^1] - c[L]_{\tilde{Y}}$ does not vanish in the cohomology of $\pi^{-1}(U)$ for any non empty open subvariety of $\mathbb{P}^1$, in other words, it does not belong to the image of $\bigoplus_{t\in\Lambda'} H^n_{\tilde{Y}_t}(\tilde{Y})$ where Λ' is any finite subset of $\mathbb{P}^1$ containing Λ. (Note that the condition for the Lefschetz pencil is generic, and for any proper closed subvariety of the parameter space, there is a Lefschetz pencil whose corresponding line is not contained in this subvariety.)

Thus the assertion is reduced to the statement that $\dim H^n_{\tilde{Y}_t}(\tilde{Y})$ is independent of $t \in \mathbb{P}^1$ because this implies that the image of $H^n_{\tilde{Y}_t}(\tilde{Y}) \to H^n(\tilde{Y})$ is independent of t. (Note that the Gysin morphism $H^{n-2}(\tilde{Y}_t) \to H^n(\tilde{Y})$ for a general t can be identified with the direct sum of the Gysin morphism $H^{n-2}(\tilde{Y}_t) \to H^n(Y)$ and the restriction morphism $H^{n-2}(\tilde{Y}_t) \to H^{n-2}(Z)$ up to a sign, and the image of the last morphism is the non primitive part by the weak Lefschetz theorem.) By duality, this is equivalent to the statement that $R^n\pi_*\mathbb{Q}_{\tilde{Y}}$ is a local system on $\mathbb{P}^1$. Then it follows from the assumption that the vanishing cycles are nonzero, see (13.22).

References

[1] Asakura, M.: Motives and algebraic de Rham cohomology, in *The arithmetic and geometry of algebraic cycles (Banff, AB, 1998)*, CRM Proc. Lecture Notes, **24**, Amer. Math. Soc., Providence, RI, 2000, pp. 133–154.

[2] Beilinson, A., J. Bernstein and P. Deligne: Faisceaux pervers, Astérisque, vol. **100**, Soc. Math. France, Paris, 1982.

[3] Bloch, S.: Algebraic cycles and higher K-theory, Advances in Math., **61** (1986), 267–304.

[4] Bloch, S.: Algebraic cycles and the Beilinson conjectures, Contemporary Math. **58** (1) (1986), 65–79.

[5] Collino, A.: Griffiths' infinitesimal invariant and higher K-theory on hyperelliptic Jacobians, J. Alg. Geom. **6** (1997), 393–415.

[6] Deligne, P.: Théorème de Lefschetz et critères de dégénérescence de suites spectrales, Inst. Hautes Etudes Sci. Publ. Math. **35** (1968), 259–278.

[7] Deligne, P.: Equation différentielle à points singuliers réguliers, Lect. Notes in Math. vol. **163**, Springer, Berlin, 1970.

[8] Deligne, P.: Théorie de Hodge I, Actes Congrès Intern. Math., 1970, vol. 1, 425–430; II, Publ. Math. IHES, 40 (1971), 5–57; III, ibid., 44 (1974), 5–77.

[9] Deligne, P.: Décompositions dans la catégorie dérivée, in Motives (Seattle, WA, 1991), Proc. Sympos. Pure Math., **55**, Part 1, Amer. Math. Soc., Providence, RI, 1994, pp. 115–128.

[10] Deninger, C. and A. Scholl: The Beilinson conjectures, in Proceedings Cambridge Math. Soc. (eds. Coats and Taylor) **153** (1992), 173–209.

[11] El Zein, F.: Complexe dualisant et applications à la classe fondamentale d'un cycle, Bull. Soc. Math. France Mém. No. 58 (1978)

[12] Esnault, H. and E. Viehweg: Deligne-Beilinson cohomology, in *Beilinson's conjectures on Special Values of L-functions*, Academic Press, Boston, 1988, pp. 43–92.

[13] Green, M.L.: Griffiths' infinitesimal invariant and the Abel-Jacobi map, J. Differential Geom. **29** (1989), 545–555.

[14] Griffiths, P.A.: Infinitesimal variations of Hodge structure, III, Determinantal varieties and the infinitesimal invariant of normal functions, Compositio Math. **50** (1983), 267–324.

[15] Jannsen, U.: Deligne homology, Hodge-D-conjecture, and motives, in *Beilinson's conjectures on Special Values of L-functions*, Academic Press, Boston, 1988, pp. 305–372.

[16] Jannsen, U.: Mixed motives and algebraic K-theory, Lect. Notes in Math., vol. **1400**, Springer, Berlin, 1990.

[17] Katz, N.: Etude cohomologique des pinceaux de Lefschetz, in Lect. Notes in Math., vol. **340**, Springer Berlin, 1973, pp. 254–327.

[18] Kleiman, S. and A. Altman: Bertini theorems for hypersurface sections containing a subscheme, Comm. Algebra **7** (1979), 775–790.

[19] Lewis, J.D. and S. Saito, preprint.

[20] Otwinowska, A: Monodromie d'une famille d'hypersurfaces, preprint (math.AG/0403151).

[21] Otwinowska, A. and M. Saito: Monodromy of a family of hypersurfaces containing a given subvariety, preprint (math.AG/0404469).

[22] Saito, M.: Modules de Hodge polarisables, Publ. RIMS, Kyoto Univ. **24** (1988), 849–995.

[23] Saito, M.: Mixed Hodge Modules, Publ. RIMS Kyoto Univ. **26** (1990), 221–333.

[24] Saito, S.: Higher normal functions and Griffiths groups, J. Algebraic Geom. **11** (2002), 161–201.

[25] Schmid, W.: Variation of Hodge structure: the singularities of the period mapping, Inv. Math. **22** (1973), 211–319.

[26] Steenbrink, J.H.M.: Limits of Hodge structures, Inv. Math. **31** (1975/76), no. 3, 229–257.

[27] Voisin, C.: Variations de structure de Hodge et zéro-cycles sur les surfaces générales, Math. Ann. **299** (1994), 77–103.

[28] Voisin, C.: Transcendental methods in the study of algebraic cycles, in Lect. Notes in Math. vol. **1594**, pp. 153–222.

14

Correspondence of Elliptic Curves and Mordell-Weil Lattices of Certain Elliptic $K3$'s

Tetsuji Shioda

Department of Mathematics, Rikkyo University,
3-34-1 Nishi-Ikebukuro, Toshima-ku, Tokyo 171-8501, Japan.
shioda@rkmath.rikkyo.ac.jp

To Jacob Murre

Abstract

We study the Mordell-Weil lattice of certain elliptic K3 surfaces, related to the Kummer surface of a product abelian surface. Our aim is first to determine the precise structure of such a lattice, and second to give some explicit generators, in the case beyond rational elliptic surfaces.

14.1 Introduction

The main purpose of this paper is to study certain elliptic fibrations on the Kummer surface of a product abelian surface, both geometrically (equation-free) and with the use of equations, and to identify the elements of the Mordell-Weil lattice coming from algebraic cycles on the Kummer surface, especially from the correspondences of the factor elliptic curves.

We state here the main results in terms of Weierstrass equations, which should show some new feature of Mordell-Weil lattices for elliptic K3 surfaces, different from the well-studied case of rational elliptic surfaces. The background will be explained after the statements.

Theorem 14.1.1. *Let C_1, C_2 be two elliptic curves with the absolute invariant j_1, j_2, defined over an algebraically closed field k of characteristic $\neq 2, 3$. Let $F^{(1)}$ denote the elliptic curve over the rational function field $k(T)$*

$$y^2 = x^3 - 3\alpha x + (T + \frac{1}{T} - 2\beta), \tag{14.1}$$

where

$$\alpha = \sqrt[3]{j_1 j_2}, \ \beta = \sqrt{(1 - j_1)(1 - j_2)}. \tag{14.2}$$

Assume that $j_1 \neq j_2$ (i.e. C_1, C_2 are not isomorphic to each other). Then there is a natural isomorphism of $\mathrm{Hom}(C_1, C_2)$ *to the Mordell-Weil lattice $F^{(1)}(k(T))$, $\varphi \mapsto R_\varphi$, such that the height of R_φ $\langle R_\varphi, R_\varphi \rangle$ is equal to* $2\deg(\varphi)$. *In other words, there is a natural isomorphism of lattices:*

$$\mathrm{Hom}(C_1, C_2)[2] \simeq F^{(1)}(k(T)). \tag{14.3}$$

Theorem 14.1.2. *Let $F^{(2)}$ denote the elliptic curve over $k(t)$, obtained from $F^{(1)}$ by the base change $T = t^2$. Assume that $j_1 \neq j_2$. Then the Mordell-Weil lattice $F^{(2)}(k(t))$ contains a sublattice of finite index 2^h ($h =$* $\mathrm{rk}\,\mathrm{Hom}(C_1, C_2)$*) which is naturally isomorphic to the direct sum of latticse*

$$\mathrm{Hom}(C_1, C_2)[4] \oplus A_2^*[2]^{\oplus 2} \tag{14.4}$$

where A_2^* denotes the dual lattice of the root lattice A_2 (of rank 2).

N. B. **(1)** The absolute invariant is normalized so that $j = 1$ for $y^2 = x^3 - x$.
(2) Given a lattice L, we denote by $L[n]$ the lattice structure on L with the norm (or pairing) multiplied by n.
(3) For the root lattices, we refer to [4].

Here we briefly mention some background of the above results; more details will be given later in §14.2 and §14.3. Let $S^{(n)}$ be the elliptic surface associated to the elliptic curve $F^{(n)} (n = 1, 2)$. Then they are both K3 surfaces, and $S^{(2)}$ is isomorphic to the Kummer surface $S = \mathrm{Km}(C_1 \times C_2)$ of the product of two elliptic curves. This elliptic fibration on the Kummer surface and $S^{(1)}$ are discovered by Inose ([5]) in search for the notion of isogeny between singular K3 surfaces ([6]).

More recently, Kuwata ([8]) has made a nice observation on Inose's results; he introduces elliptic K3 surfaces corresponding to $F^{(n)} (n \leq 6)$ defined by the base change $T = t^n$, and shows that their Mordell-Weil rank can become as large as 18, the maximum in case of $\mathrm{char}(k) = 0$. Inspired by the work of Inose and Kuwata, we ([15]) have made a preliminary study on these elliptic K3 surfaces from the viewpoint of Mordell-Weil lattices.

Now, given any $\varphi \in \mathrm{Hom}(C_1, C_2)$, the image of its graph under the rational map from $C_1 \times C_2$ to S (and to $S^{(1)}$) gives a curve on S (or $S^{(1)}$), and this image curve determines a rational point $P_\varphi \in F^{(2)}(k(t))$ (or $R_\varphi \in F^{(1)}(k(T))$) by the formalism of Mordell-Weil lattices (see §14.2; compare [17]). The correspondence $\varphi \mapsto R_\varphi$ in Theorem 14.1.1 is "natural" in the sense that it is defined in this way. Theorems 14.1.1 and 14.1.2 are the refinement of some results announced in [15], covering also the case of arbitrary characteristic $\neq 2, 3$. As an application to the case of the higher rank, we mention the following:

Example 14.1.3. Assume $j_1 \neq 0, j_2 = 0$ $(C_2 : y^2 = x^3 - 1)$. Then the Mordell-Weil lattice $F^{(6)}(k(t))$ of the elliptic curve over $k(t)$

$$y^2 = x^3 + (t^{12} - 2\beta t^6 + 1), \quad \beta = 1 - j_1^2 \tag{14.5}$$

is of rank $r = h + 16$ and it contains a finite index sublattice isomorphic to the direct sum

$$L_0 \oplus E_8[2] \oplus D_4^*[4]^{\oplus 2}, \quad \text{rk } L_0 = h \tag{14.6}$$

where E_8, D_4, A_2 are root lattices and $*$ means the dual lattice. If $\text{char}(k) = 0$, then we have ($h = 2$ or 0)

$$L_0 = A_2[6d_0] \text{ or } 0 \tag{14.7}$$

$d_0 (\geq 2)$ being the degree of minimal isogeny $C_1 \to C_2$ (cf. §14.8). The generators of $k(t)$-rational points of this sublattice can be given explicitly in case $h = 0$ or if d_0 is small.

This paper is organized as follows. In §14.2, we review the formalism of Mordell-Weil lattices which is our main tool. In the next sections, we study the elliptic fibrations on the Kummer surface of a product abelian surface. After reviewing the so-called double Kummer pencils (§14.3), we study the Inose's pencil, first by a geometric method (§14.4) and second by introducing equations (§14.5). With these preparations, we prove our main results (and Theorems 14.1.1 and 14.1.2) in §14.6. Some comments for the case $j_1 = j_2$ (§14.7) and examples (§14.8) are given. We hope to come back to the higher rank case in some other occasion.

It is my pleasure to dedicate this paper to Professor Jacob Murre on the occasion of his 75th birthday. It was reported at the workshop on Algebraic Cycles held at Lorentz Center, Leiden, in his honor. The paper has been prepared partly during my stay at the Max-Planck-Institut für Mathematik, Bonn, in the summer of 2004. I would like to thank the MPI for the hospitality and excellent atmosphere, and my special thanks go to Professor Hirzebruch for everything he has done for me. Finally I thank the referee for his/her careful reading of the manuscripts and for useful suggestions.

14.2 Review of the MWL-formalism

Let us make a review of the basic formalism of Mordell-Weil lattices, fixing some notation (cf. [14]).

Let E/K be an elliptic curve over the function field $K = k(C)$ of a smooth projective curve C/k. The base field k is an algebraically closed field of arbitrary characteristic (later we assume $\mathrm{char}(k) \neq 2, 3$). Let $f : S \longrightarrow C$ be the associated elliptic surface (the Kodaira-Néron model of E/K); S is a smooth projective surface over k and E is the generic fibre of f. The set of K-rational points of E, $E(K)$, is in a natural bijective correspondence with the set of the sections of f. For $P \in E(K)$, we use the same symbol P to denote the corresponding section $P : C \to S$ and the symbol (P) to denote the image curve in S; thus for example (O) denotes the image of the zero-section in S. Let $\mathrm{Sing}(f)$ (resp. $\mathrm{Red}(f)$) denote the set of singular fibres (reducible singular fibres) of f. It is known that, if $\mathrm{Sing}(f) \neq \varnothing$, then $E(K)$ is finitely generated (Mordell-Weil theorem).

Let $\mathrm{NS}(S)$ be the Néron-Severi group of S which is defined as the group of divisors on the surface S modulo algebraic equivalence; the class of a divisor D is denoted by $[D]$ (or simply by D if no confusion will arise). Let $T = T(f)$ denote the subgroup generated by the classes of the zero-section (O), any fibre F and all the irreducible components of reducible fibres which are disjoint from (O). Then we have a natural isomorphism

$$E(K) \simeq \mathrm{NS}(S)/T. \tag{14.8}$$

The correspondence is given by $P \mapsto [(P)] \mod T$, and the inverse correspondence is induced by the following map of the divisor group of S to $E(K)$:

$$\mu(D) = \mu_f(D) = \mathrm{sum}(D|_E) \in E(K) \tag{14.9}$$

Namely, given a divisor D on S, restrict it to the generic fibre E and take the summation of its components ($\in E(\bar{K})$, $\bar{K}$ being the algebraic closure of K) with respect to the group law of E.

Now $\mathrm{NS}(S)$ forms an indefinite integral lattice with respect to the intersection pairing, and T forms a sublattice, called the *trivial sublattice*, which has an orthogonal decomposition $T = U \oplus \sum_{v \in \mathrm{Red}(f)} T_v$ where U is the unimodular rank 2 lattice generated by $(O), F$, and T_v is the lattice of rank $m_v - 1$ spanned by the irreducible components away from (O) of the reducible fibre $f^{-1}(v)$. Each T_v is a root lattice of type A, D, E, up to sign, by Kodaira [7]. The key idea of Mordell-Weil lattices is to define the lattice structure on the Mordell-Weil group via the intersection theory on the surface as follows. There is a unique homomorphism

$$\nu : E(K) \longrightarrow N_{\mathbb{Q}} = \mathrm{NS}(S) \otimes \mathbb{Q} \tag{14.10}$$

Table 14.1. *Values of local contribution*

T_v^-	A_1	E_7	A_2	E_6	A_{b-1}	D_{b+4}
type of F_v	III	III^*	IV	IV^*	$I_b (b \geq 2)$	$I_b^* (b \geq 0)$
$\mathrm{contr}_v(P)$	$1/2$	$3/2$	$2/3$	$4/3$	$i(b-i)/b$	$\begin{cases} 1 & (i=1) \\ 1+b/4 & (i>1) \end{cases}$
$\mathrm{contr}_v(P,Q)$ $(i<j)$	–	–	$1/3$	$2/3$	$i(b-j)/b$	$\begin{cases} 1/2 & (i=1) \\ (2+b)/4 & (i>1) \end{cases}$

satisfying the condition: for every $P \in E(K)$,

$$\nu(P) \perp T, \quad \nu(P) \equiv [(P)] \mod T_{\mathbb{Q}}. \tag{14.11}$$

Then $E(K)$ modulo torsion is embedded into the orthogonal complement of T in $N_{\mathbb{Q}}$, which is negative-definite by the Hodge index theorem. Therefore, by defining the *height pairing* on $E(K)$ by the formula:

$$\langle P, Q \rangle := -(\nu(P) \cdot \nu(Q)), \tag{14.12}$$

one obtains the structure of a positive-definite lattice on $E(K)/E(K)_{tor}$. It is called the *Mordell-Weil lattice* (abbreviated from now on as MWL) of the elliptic curve E/K or the elliptic surface $f : S \to C$.

The height formula takes the following explicit form:

$$\langle P, Q \rangle = \chi + (PO) + (QO) - (PQ) - \sum_{v \in \mathrm{Red}(f)} \mathrm{contr}_v(P, Q) \tag{14.13}$$

where χ is the arithmetic genus of S, (PQ) denotes the intersection number of the sections (P) and (Q), and $\mathrm{contr}_v(P,Q)$ is a local contribution at v. For later use, we copy the table from [14, (8.16)], Table 14.2, in which i, j are defined so that the section (P) (or (Q)) intersects the i-th (or j-th) irreducible component of the singular fibre $f^{-1}(v)$ under suitable numbering.

The determinant of Néron-Severi lattice and that of MWL are related by the formula:

$$\det \mathrm{NS}(S) = \det(E(K)/E(K)_{tor}) \cdot \det T / |E(K)_{tor}|^2. \tag{14.14}$$

Remarks. 1) Given the information of the trivial lattice T, it is easy to write down the explicit formula for $\nu(P) = (P) + \cdots$ satisfying the *Linear Algebra* condition (14.11). Indeed this is how the height formula (14.13) is derived in general. On the other hand, it is also possible to compute the height by applying the original definition

(14.12), especially when P is given as $P = \mu(D)$ for some divisor D. This method gives an algorithm suited for computer calculation which can be used for checking theoretical computation.

2) The structure of MWL is clarified in the case where S is a rational elliptic surface. In this case, the lattices in question form a hierarchy dominated by the root lattice E_8, the unique positive-definite even unimodular lattice of rank 8 (cf. [11]). Also it is easy in this case to give the generators of rational points, for example, and there are many interesting applications. Beyond this case, not much is known even in the next simplest case of elliptic K3 surfaces.

14.3 The Kummer pencils

In the subsequent sections, we study certain elliptic K3 surfaces related to the Kummer surface of a product abelian surface.

In general, let A be an abelian surface and let ι_A denote the inversion automorphism of A. We assume that $\mathrm{char}(k) \neq 2$. The Kummer surface $S = \mathrm{Km}(A)$ is a smooth K3 surface obtained from the quotient surface A/ι_A by resolving the 16 singular points corresponding to the points of order 2 on A. The Picard number is given by $\rho(S) = \rho(A) + 16$.

Now consider the case of a product abelian surface, i.e. $A = C_1 \times C_2$ where C_1, C_2 are elliptic curves. If we denote by h the rank of the free module $\mathrm{Hom}(C_1, C_2)$ of homomorphisms of C_1 to C_2, then

$$\rho(S) = h + 18, \quad h := \mathrm{rk}\ \mathrm{Hom}(C_1, C_2) \tag{14.15}$$

since we have $\rho(A) = 2 + h$. It is known that $H = \mathrm{Hom}(C_1, C_2)$ has the structure of a positive-definite lattice such that the norm of $\varphi \in H$ is $\deg(\varphi)$, the degree of the homomorphism (see the Remark below).

First we look at the Kummer pencil (cf. [6, §2]), i.e. the elliptic fibration

$$\pi_1 : S = \mathrm{Km}(C_1 \times C_2) \to \mathbb{P}^1 \tag{14.16}$$

induced from the projection of A to C_1. It has the 4 singular fibres of type I_0^*:

$$\pi_1^{-1}(\bar{v}_i) = 2F_i + \sum_{j \in I} A_{ij}. \tag{14.17}$$

Here we use the following notation. Let $I = \{0, 1, 2, 3\}$ and let $\{v_i | i \in I\} \subset C_1$ be the 2-torsion points (take v_0 = the origin); similarly for $\{v'_j | j \in I\} \subset C_2$. We denote by $\bar{v}_i$ the image point of v_i under $C_1 \to C_1/\iota_1 = \mathbb{P}^1$. The curves $F_i, G_j \subset S (i, j \in I)$ are the image of $v_i \times C_2, C_1 \times v'_j$ under the rational map of degree two $A \to S$. Further A_{ij} denotes the exceptional curve corresponding to $v_i \times v'_j$. All the 24 curves $\{F_i, G_j, A_{ij}\}$ on S are smooth rational curves with self-intersection number -2 (i.e. -2-curves). The intersection numbers among these curves are given as follows:

$$\begin{cases} (F_i \cdot F_j) = -2\delta_{ij}, & (G_i \cdot G_j) = -2\delta_{ij}, \quad (F_i \cdot G_j) = 0, \\ (A_{ij} \cdot A_{kl}) = -2\delta_{ik}\delta_{jl}, & (F_i \cdot A_{kl}) = \delta_{ik}, \quad (G_i \cdot A_{kl}) = \delta_{il}. \end{cases} \tag{14.18}$$

Note that each of the 4 curves G_j gives a section of π_1 since it intersects the fibre (14.17) with intersection multiplicity 1. Take $G_0 = (O)$ as the zero-section. Then the other sections G_j are of order 2. The generic fibre of π_1 is isomorphic to the constant elliptic curve C_2 over $k(C_1) \supset k(\mathbb{P}^1)$, but of course not over $k(\mathbb{P}^1)$.

Proposition 14.3.1. *Let $\mathcal{E}$ denote the generic fibre of π_1. Then we have*

$$\mathcal{E}(k(\mathbb{P}^1)) \simeq \mathrm{Hom}(C_1, C_2) \oplus (\mathbb{Z}/2\mathbb{Z})^2 \tag{14.19}$$

i.e. the Mordell-Weil lattice $\mathcal{E}(k(\mathbb{P}^1))/(\mathrm{tor})$ is isomorphic to the lattice $H := \mathrm{Hom}(C_1, C_2)$ with norm $\varphi \mapsto \deg(\varphi)$.

Proof This should be well known if we ignore the lattice structure, but for the sake of completeness, let us first check the isomorphism of both side as groups. Take any $P \in \mathcal{E}(k(\mathbb{P}^1))$, and regard it as a section $\sigma : \mathbb{P}^1 \to S$. Its pullback to C_1, $\tilde{\sigma} : C_1 \to A = C_1 \times C_2$, is of the form $\tilde{\sigma}(u) = (u, \alpha(u))$ ($u \in C_1$), where $\alpha : C_1 \to C_2$ is a morphism such that $\alpha(-u) = -\alpha(u)$. Hence we have $\alpha(u) = \varphi(u) + v'$ for some homomorphism $\varphi \in \mathrm{Hom}(C_1, C_2)$ and a 2-torsion point $v' \in C_2$. This establishes the bijection of both sides.
For any nonzero $\varphi \in \mathrm{Hom}(C_1, C_2)$, consider the image $\Gamma = \Gamma_\varphi$ of its graph under the rational map $A \to S$. Let $Q_\varphi = \mu(\Gamma_\varphi) \in \mathcal{E}(k(\mathbb{P}^1))$, with $\mu = \mu_f$ for $f = \pi_1$ defined by (2.2) and (3.2). It is easy to see that $\varphi \mapsto Q_\varphi$ is a homomorphism. We must prove that the height $\langle Q_\varphi, Q_\varphi \rangle$ is equal to $\deg(\varphi)$. To prove this, we use the height formula (14.13) for $P = Q_\varphi$; note that $\chi = 2$ (for a K3):

$$\langle P, P \rangle = 4 + 2(PO) - \sum \mathrm{contr}_v(P).$$

For a moment, admit Lemma 14.3.2 below. Then the term $(PO) = (\Gamma \cdot G_0)$ is given by (14.20). On the other hand, Table 14.2 shows that we have $\mathrm{contr}_v(P) = 1$ iff the section $(P) = \Gamma$ meets a non-identity component of I_0^*-fibre. Thus the sum of local contribution is 3 (or 2 or 0) according to the case (a) (or (b) or (c)) of the Lemma. This proves $\langle P, P \rangle = \deg(\varphi)$. □

Lemma 14.3.2. *Given $\varphi \in \mathrm{Hom}(C_1, C_2)$, $\varphi \neq 0$, let $d = \deg(\varphi)$ and set $\varphi(v_i) = v'_{j(i)} (i \in I)$; let $n = n(\varphi)$ denote the number of distinct $j(i)$. Then the curve $\Gamma = \Gamma_\varphi$ is a (-2)-curve on S, and it satisfies $\Gamma \cdot A_{i,j} = \delta_{i,j(i)}$ and $\Gamma \cdot F_i = 0$ for all $i \in I$. The intersection number of Γ with G_i is described in the table 14.20 below according to the three cases* a) $n = 4$, b $n = 2$ *or* c) $n = 1$, *which can be also characterized by the following properties:*
a) *d is odd,* c) *$\varphi = 2\varphi_1$ for some $\varphi_1 \in \mathrm{Hom}(C_1, C_2)$,* b) *otherwise. (In case* b), *we change ordering v'_j so that $\{j(i)|i \in I\} = \{0, 1\}$.)*

Then we have

	a)	b)	c)
$\Gamma \cdot G_0$	$(d-1)/2$	$(d-2)/2$	$(d-4)/2$
$\Gamma \cdot G_1$	$(d-1)/2$	$(d-2)/2$	$d/2$
$\Gamma \cdot G_2$	$(d-1)/2$	$d/2$	$d/2$
$\Gamma \cdot G_3$	$(d-1)/2$	$d/2$	$d/2$

(14.20)

Proof Let $\tilde{\Gamma}$ be the graph of φ in $A = C_1 \times C_2$. Then a 2-torsion point of A lies on $\tilde{\Gamma}$ if and only if it is of the form $v_i \times v'_{j(i)}$ for some $i \in I$. Under the rational map $A \to S$ of degree two, $\tilde{\Gamma}$ is mapped to Γ. Thus we have $\Gamma \cap A_{i,j} \neq \varnothing$ if and only if $j = j(i)$, in which case the intersection number is one. Hence the first assertion. As for the intersection of Γ with G_j, for example with G_0, we note that $\tilde{\Gamma} \cdot (C_1 \times v'_0)$ has degree $d = \deg(\varphi)$, of which $n' = 4/n$ simple intersection occur at the 2-torsion points. Hence we have $\Gamma \cdot G_0 = (d - n')/2$ on S, as asserted. We can argue similarly for other G_j. □

Theorem 14.3.3. *The Néron-Severi lattice* $\mathrm{NS}(S)$ *of the Kummer surface* $S = \mathrm{Km}(C_1 \times C_2)$ *is generated by the 24 curves F_i, G_j, A_{ij} together with $\Gamma_\varphi (\varphi \in \mathrm{Hom}(C_1, C_2))$. Its determinant is given by*

$$\det \mathrm{NS}(S) = 2^4 \cdot \det \mathrm{Hom}(C_1, C_2). \tag{14.21}$$

Proof The first assertion follows from (14.8) (applied to $E = \mathcal{E}$) and (14.19), since both trivial lattice T and the torsion part $(\mathbb{Z}/2\mathbb{Z})^2$ are generated by curves belonging to the 24 curves. For (14.21), apply the formula (14.14), where we have $\det T = 4^4$ (as $T = U \oplus D_4^{\oplus 4}$) and $|E(K)_{\mathrm{tor}}| = 2^2$. □

Remark. The prototype of the above arguments is the well known fact:

$$\mathrm{NS}(A) = T_0 \oplus T_0^{\perp}, \quad (T_0^{\perp})[-1] \simeq \mathrm{Hom}(C_1, C_2)[2] \tag{14.22}$$

where $A = C_1 \times C_2$ and T_0 is the sublattice of $\mathrm{NS}(A)$ generated by $C_1 \times v'_0$ and $v_0 \times C_2$. It relates the correspondence theory of curves to geometry of surfaces, and its most remarkable application is Weil's proof of the Riemann hypothesis for curves over a finite field ([19]). At any rate, it shows that $\mathrm{Hom}(C_1, C_2)[2]$ (with $2\deg(\varphi)$ as the norm of φ) is a (positive-definite) integral lattice and $\det \mathrm{NS}(A) = 2^h \det \mathrm{Hom}(C_1, C_2)$. Note that $\mathrm{Hom}(C_1, C_2)$ itself is not necessarily an integral lattice.

Corollary 14.3.4. *Let N_0 denote the sublattice of $\mathrm{NS}(S)$ generated by the 24 curves $\{F_i, G_j, A_{ij}\}$. Then (i) N_0 is an indefinite lattice of rank 18 and $\det 2^4$. (ii) Assume that C_1 and C_2 are not isogeneous to each other. Then we have $\mathrm{NS}(S) = N_0$, namely $\mathrm{NS}(S)$ is generated by the 24 curves F_i, G_j, A_{ij}.*

This is obvious from Theorem 14.3.3. It is classically well known (e.g.[12]).

Proposition 14.3.5. *The map $\varphi \mapsto \Gamma_\varphi$ induces a surjective homomorphism from $\mathrm{Hom}(C_1, C_2)$ to $NS(S)/N_0$.*

Proof It suffices to show that

$$\Gamma_{\varphi+\psi} \equiv \Gamma_\varphi + \Gamma_\psi \mod N_0, \tag{14.23}$$

since the surjectivity follows from Theorem 14.3.3. Consider the divisor $D = \Gamma_{\varphi+\psi} - (\Gamma_\varphi + \Gamma_\psi) + \Gamma_0$ on S. The restriction of D to the generic fibre $\mathcal{E}$ of the Kummer pencil π_1 gives $\mu(D) = Q_{\varphi+\psi} - (Q_\varphi + Q_\psi) = O$, μ being the map (14.9). Hence D is algebraically equivalent to a sum of irreducible components of fibres of π_1, which proves (14.23). □

14.4 Inose's pencil

Next we define Inose's pencil on $S = \mathrm{Km}(C_1 \times C_2)$. Take the following divisors on S:

$$\begin{cases} \Phi_1 = G_1 + G_2 + G_3 + 2(A_{01} + A_{02} + A_{03}) + 3F_0, \\ \Phi_2 = F_1 + F_2 + F_3 + 2(A_{10} + A_{20} + A_{30}) + 3G_0, \end{cases} \tag{14.24}$$

where F_i, G_j, A_{ij} are the -2-curves used in §3. They are disjoint and they have the same type as a singular fibre of type IV^*. Recall that a divisor on a K3 surface X is a fibre of some elliptic fibration on X if it has the same type as one of the Kodaira's list of singular fibres (cf.[7], [12]). Therefore there is an elliptic fibration, say $f : S \to \mathbb{P}^1$, such that $f^{-1}(0) = \Phi_1$ and $f^{-1}(\infty) = \Phi_2$. We call it *Inose's pencil* on the Kummer surface $S = \mathrm{Km}(C_1 \times C_2)$. Note that the divisors Φ_1, Φ_2 are interchanged when the order of the factors C_1, C_2 is changed. Let E be the generic fibre of f; it will be identified with the elliptic curve $F^{(2)}/k(t)$ of Theorem 14.1.2 later.

Each of the 9 curves $A_{ij}(i,j \neq 0)$ defines a section of f; for instance, A_{33} intersects the fibre Φ_1 transversally at the simple component G_3. Let us choose $A_{33} = (O)$ as the zero-section. To avoid confusion, we let $Q_{ij} \in E(k(t))$ denote the section such that $(Q_{ij}) = A_{ij}$.

Throughout § 14.4, we make the assumption:

(#) f *has no other reducible fibres than* Φ_1, Φ_2.

Lemma 14.4.1. *Under (#), the Mordell-Weil lattice $L = E(k(t))$ of the Inose's pencil has rank $4+h$, and the 9 sections Q_{ij} form a sublattice L_1 of rank 4 isomorphic to $A_2^*[2]^{\oplus 2}$.*

Proof Under the assumption (#), the trivial lattice T is isomorphic to $U \oplus E_6^{\oplus 2}$, of rank 14, and hence the Mordell-Weil rank is equal to $\rho(S) - 14 = 4 + h$ by (14.8), (14.15). Also $E(k(t))$ is torsion-free by the height formula. Let us compute the height of $Q = Q_{ij}(i,j \neq 0)$ by (14.13). The curve $(Q) = A_{ij}$ hits the singular fibre Φ_1 (of type IV^*) at a non-identity component iff $i = 1, 2$. By Table 14.2, the local contribution is equal to

$$\mathrm{contr}_v(Q) = 4/3 \text{ for } i = 1,2, \text{ and } = 0 \text{ for } i = 3.$$

By replacing i by j, we get the corresponding value at the fibre Φ_2. Hence we have $\langle Q, Q \rangle = 4 - 4/3 - 4/3 = 4/3$. Similarly we can compute the height pairing $\langle Q, Q' \rangle$ for $Q \neq Q'$. Thus we see that both $\{Q_{11}, Q_{22}\}$ and $\{Q_{12}, Q_{21}\}$ span a mutually orthogonal sublattice (isomorphic to $A_2^*[2]$) with the Gram matrix $\begin{pmatrix} 4/3 & 2/3 \\ 2/3 & 4/3 \end{pmatrix}$. □

Now we turn our attention to the curves $\Gamma_\varphi (\varphi \in H)$ to study the remaining rank h part of the Mordell-Weil lattice of f. Let

$$P_\varphi = \mu_f(\Gamma_\varphi) \in E(k(t)), \tag{14.25}$$

with μ_f as in (14.9). To compute the height $\langle P_\varphi, P_\varphi \rangle$, we cannot directly apply the height formula (14.13) as before, because we do not know the local contribution terms. Thus we need to go back to the original definition (14.12).

Lemma 14.4.2. *Given $\varphi \in \mathrm{Hom}(C_1, C_2)$ with $d = \deg(\varphi)$, the height $\langle P_\varphi, P_\varphi \rangle$ has the following value:*

$$\langle P_\varphi, P_\varphi \rangle = \begin{cases} 2d^2 - d + 1 & (a) \\ 2d^2 - 2d + 4/3 & (b) \\ 2d^2 + d & (c) \end{cases} \tag{14.26}$$

in the respective case a), b), c) *for φ stated in Lemma 14.3.2.*

Proof Set $P = P_\varphi$. Consider the case (c) where φ is divisible by 2 in H; in particular d is divisible by 4. Solving the linear algebra condition (14.11) for $(P) = \Gamma_\varphi$, we find by a direct computation that

$$\begin{aligned}\nu(P) = \Gamma_\varphi - 3d\Phi - \frac{3d}{2}(O) + d(F_1 + F_2 + G_1 + G_2) + 2dF_0 + (3d-1)G_0 \\ + \frac{3d}{2}(A_{01} + A_{02}) + dA_{03} + 2d(A_{10} + A_{20}) + \frac{3d}{2}A_{30} \qquad (14.27)\end{aligned}$$

where Φ denotes a fibre class of f. Then by (14.13), we compute $\langle P, P\rangle = -(\nu(P)^2)$ using Lemma 14.3.2 and verify that it is equal to $2d^2 + d$, as asserted. The other cases (a), (b) can be verified in the same way. $\square$

It follows from above that $P_\varphi \neq 0$ for $\varphi \neq 0$, but we cannot say that the map $\varphi \to P_\varphi$ is an injective map from $H = \mathrm{Hom}(C_1, C_2)$ to the Mordell-Weil lattice, since we only know (Prop. 14.3.5 that the map $\varphi \to \Gamma_\varphi$ induces a group homomorphism $H \to \mathrm{NS}(S)/N_0$.

To remedy this situation, let us proceed as follows. Consider the orthogonal complement of L_1 in L (with the notation of Lemma 14.4.1):

$$L_1' = L_1^\perp. \qquad (14.28)$$

Obviously the $\mathbb{Q}$-vector space $V = L \otimes \mathbb{Q}$ is an orthogonal direct sum of $V_1 = L_1 \otimes \mathbb{Q}$ and $V_1' = L_1' \otimes \mathbb{Q}$, although the lattice L itself is not in general equal to the direct sum $L_1 \oplus L_1'$. Decompose $P_\varphi \in V$ as a sum of the V_1-component and V_1'-component:

$$P_\varphi = [P_\varphi]^+ + [P_\varphi]^-, \quad [P_\varphi]^+ \in V_1, [P_\varphi]^- \in V_1'. \qquad (14.29)$$

Lemma 14.4.3. *The V_1-component of $P_\varphi \in L$ is represented by the following element:*

$$[P_\varphi]^+ = \begin{cases} \frac{d}{2}(Q_{12} + Q_{21}) + \frac{d-1}{2}(Q_{11} + Q_{22}) & (a) \\ \frac{d-1}{2}(Q_{12} + Q_{21} + Q_{22}) + \frac{d}{2}Q_{11} & (b) \\ \frac{d}{2}(Q_{11} + Q_{12} + Q_{21} + Q_{22}) & (c) \end{cases} \qquad (14.30)$$

In particular, it is an element of L_1 if and only if φ is in the case (c), i.e. $\varphi = 2\varphi_1$ for some $\varphi_1 \in \mathrm{Hom}(C_1, C_2)$.

Proof The first part is verified by a linear algebra computation. For the second part, note that in case (c), $d = \deg(\varphi) = 4\deg(\varphi_1)$ is divisible by 4. Thus $d/2$ is an integer, and $[P_\varphi]^+ \in L_1$. In case (a), d is odd, and we can easily see that $(Q_{12} + Q_{21})$ is not divisible by 2 in L. The case (b) can be treated in a similar way. $\square$

Lemma 14.4.4. *Depending on the case of $\varphi \in H$, we have*

$$\langle [P_\varphi]^+, [P_\varphi]^+ \rangle = \begin{cases} 2d^2 - 2d + 1 & (a) \\ 2d^2 - 3d + 4/3 & (b) \\ 2d^2 & (c) \end{cases} \tag{14.31}$$

and, for any $\varphi \in H$,

$$\langle [P_\varphi]^-, [P_\varphi]^- \rangle = d. \tag{14.32}$$

Proof Using Lemma 14.4.1, check first that both $(Q_{12}+Q_{21})$ and $(Q_{11}+Q_{22})$ have height 4 and they are orthogonal. By Lemma 14.4.3, we see for instance in case (a) that the "height" of $[P_\varphi]^+$ is equal to

$$(\frac{d}{2})^2 \cdot 4 + (\frac{d-1}{2})^2 \cdot 4 = 2d^2 - 2d + 1.$$

Other cases are similar, and this proves (14.31). Now it follows from (14.29) that

$$\langle [P_\varphi]^-, [P_\varphi]^- \rangle = \langle P_\varphi, P_\varphi \rangle - \langle [P_\varphi]^+, [P_\varphi]^+ \rangle$$

Comparing (14.26) and (14.31), we conclude (14.32) that the "height" of $[P_\varphi]^-$ is equal to d in all cases. □

Proposition 14.4.5. *Let*

$$R_\varphi := 2[P_\varphi]^- \in L_1'. \tag{14.33}$$

Then the map $\varphi \mapsto R_\varphi$ gives an imbedding of the lattice $H[4]$ into L_1'.

Proof Let $N = \mathrm{NS}(S)$. By (14.8), we have $L \cong N/T$. Under this isomorphism, we have $L_1 \cong N_0/T$, since N_0 is generated by the 24 curves (cf. Cor.14.3.4) of which 15 (resp. 9) give generators of T (resp. L_1). It follows that $L/L_1 \cong N/N_0$, and the map $\varphi \mapsto P_\varphi$ induces a group homomorphism $H \to L/L_1$ by Proposition 14.3.5. On the other hand, the orthogonal projection $V \to V_1'$ induces the homomorphism $L/L_1 \to \frac{1}{2}L_1'$ sending $P_\varphi \mod L_1$ to $[P_\varphi]^-$. By composing the two maps, we obtain a homomorphism

$$H \to \frac{1}{2}L_1', \quad \varphi \mapsto [P_\varphi]^- \tag{14.34}$$

preserving the norm (or height) by (14.32). In other words, the map $\varphi \mapsto R_\varphi$ gives an injective homomorphism $H \to L_1'$ such that

$$\langle R_\varphi, R_\varphi \rangle = 4\deg(\varphi). \tag{14.35}$$

This proves the assertion. □

14.5 Defining equation of Inose's pencil

Now we introduce the equations to make more detailed analysis. Suppose that the elliptic curve $C_l (l = 1, 2)$ is given by the Weierstrass equation:

$$C_l : y_l^2 = f_l(x_l) = x_l^3 + \cdots \quad = \prod_{k=1}^{3} (x_l - a_{l,k}). \tag{14.36}$$

The 2-torsion points of C_1, C_2 are given by $v_i = (a_{1,i}, 0), v'_j = (a_{2,j}, 0)$.

The function

$$t = y_2/y_1 \tag{14.37}$$

on $A = C_1 \times C_2$ is invariant under ι_A, and it defines an elliptic fibration on the Kummer surface S whose generic fibre is isomorphic to the plane cubic curve over $k(t)$ defined by

$$f_1(x_1)t^2 = f_2(x_2). \tag{14.38}$$

The following result is essentially due to Inose [5], for which we give a simplified proof bellow (cf. [15]):

Proposition 14.5.1. *The elliptic fibration on the Kummer surface S induced by $t = y_2/y_1$ is isomorphic to the Inose's pencil. The Weierstrass form of the cubic curve* (14.38) *is given by*

$$E^{(2)} : y^2 = x^3 - 3\alpha t^4 x + t^4(t^4 - 2\beta t^2 + 1) \tag{14.39}$$

where α, β are defined by (1.2), i.e. $\alpha = \sqrt[3]{j_1 j_2}$, $\beta = \sqrt{(1 - j_1)(1 - j_2)}$. There are two singular fibres of type IV^* at $t = 0$ and ∞, and the other singular fibres are given, in an abridged form, as follows:

i) $I_1 \times 8$ *if* $j_1 \neq j_2, j_1 j_2 \neq 0$,
ii) $II \times 4$ *if* $j_1 \neq j_2, j_1 j_2 = 0$,
iii) $I_2 \times 2 + I_1 \times 4$ *if* $j_1 = j_2 \neq 0, 1$,
iv) $I_2 \times 4$ *if* $j_1 = j_2 = 1$,
v) $IV \times 2$ *if* $j_1 = j_2 = 0$.

Proof To prove the first assertion, we claim that the divisor of the function t on S is equal to

$$(t) = \Phi_1 - \Phi_2 \tag{14.40}$$

where Φ_1, Φ_2 are the divisors defined by (14.24).

Indeed, by (14.36) and (14.37), we have

$$(t^2) = (f_2(x_2)/f_1(x_1)) = \sum_{k=1}^{3}(x_2 - a_{2,k}) - \sum_{i=1}^{3}(x_1 - a_{1,i}). \tag{14.41}$$

Since the function x_1 defines the first Kummer pencil π_1 (3.2), we have

$$(x_1 - a_{1,i}) = \pi_1^{-1}(v_i) - \pi_1^{-1}(v_0) = 2F_i + \sum_{j\in I} A_{i,j} - (2F_0 + \sum_{j\in I} A_{0,j}) \tag{14.42}$$

by (3.3). Writing down the corresponding fact for the the second Kummer pencil π_2, we have

$$(x_2 - a_{2,j}) = \pi_2^{-1}(v'_j) - \pi_2^{-1}(v'_0) = 2G_j + \sum_{i\in I} A_{i,j} - (2G_0 + \sum_{i\in I} A_{i,0}). \tag{14.43}$$

Then, by (14.24), (14.41), (14.42) and (14.43), we can easily check that

$$(t^2) = 2(\Phi_1 - \Phi_2). \tag{14.44}$$

This implies our claim (14.40), proving that the function t defines Inose's pencil.

Next, setting $T = t^2$, we consider the linear pencil of plane cubic curves

$$f_1(x_1)T = f_2(x_2). \tag{14.45}$$

The base points $(x_1, x_2) = (v_k, v'_l)$ define nine $k(T)$-rational points of the generic member, which can be transformed to a Weierstrass form over $k(T)$ such that one of the points, say (v_3, v'_3), is mapped to the point at infinity (cf. [2]). By carrying out the computation, one obtains an equation is of the form:

$$E^{(1)} : y^2 = x^3 + AT^2x + T^2B(T) \tag{14.46}$$

where A is a constant and $B(T)$ is a quadratic polynomial which depend on the coefficients of f_1, f_2. By replacing x, y, T by suitable constant multiples, they can be normalized so that

$$A = -3\alpha, B(T) = T^2 - 2\beta T + 1, \tag{14.47}$$

with $\alpha = \sqrt[3]{j_1 j_2}, \beta = \sqrt{(1-j_1)(1-j_2)}$ as in (14.2). (Note that the choice of the cube root or square root is irrelevant, since they give rise to isomorphic Weierstrass equations.)

Going back to (14.38), we see that the Weierstrass form of this plane cubic is given by $E^{(2)} = E^{(1)}|_{T=t^2}$ defined by (14.39). The singular fibres are easily determined by using [7] or [18]. (Also it is a simple consequence of the following lemma, since the map $t \mapsto T = t^2$ is a double cover ramified only at $t = 0$ and ∞.) □

Lemma 14.5.2. *The elliptic surface corresponding to $E^{(1)}$ is a rational surface. It has two singular fibres of type IV at $T = 0$ and ∞, and other singular fibres are given as follows:*

i) $I_1 \times 4$ *if* $j_1 \neq j_2, j_1 j_2 \neq 0$,
ii) $II \times 2$ *if* $j_1 \neq j_2, j_1 j_2 = 0$,
iii) $I_2 + I_1 \times 2$ *if* $j_1 = j_2 \neq 0, 1$,
iv) $I_2 \times 2$ *if* $j_1 = j_2 = 1$, *(v)* IV *if* $j_1 = j_2 = 0$.

Proof The discriminant $\Delta(E^{(1)})$ is given by $T^4(B(T)^2 - 4\alpha^3 T^2)$ up to constant. Then the verification is a simple exercise using [7] or [18]. □

Proposition 14.5.3. *The Mordell-Weil lattice $E^{(1)}(k(T))$ is isomorphic to $(A_2^*)^{\oplus 2}$ if $j_1 \neq j_2$, and to $A_2^* \oplus \langle 1/6 \rangle$, or $\langle 1/6 \rangle^{\oplus 2}$ or $A_2^* \oplus \mathbb{Z}/3\mathbb{Z}$ if $j_1 = j_2$, in the respective case (iii) or (iv) or (v). $E^{(1)}(k(T))$ is generated by the rational points of the form $x = aT, y = T(cT + d)$. If $j_1 \neq j_2$, there are 12 such points which are the 12 minimal vectors of height (or norm) $2/3$ in $(A_2^*)^{\oplus 2}$.*

Proof Assume $j_1 \neq j_2$. By the height formula (14.13), a point $P = (x, y) \in E^{(1)}(k(T))$ has the minimal norm $2/3$ if and only if $(PO) = 0$ and (P) passes through the non-identity component of each of the two reducible fibres of type IV^*. The first condition $(PO) = 0$ implies that the coordinates x, y of P are polynomial of degree ≤ 2 or 3 (cf.[14, §10]) and the second condition implies that their constant terms as well as the highest terms should vanish. Hence the result follows. The same method can be applied for the case $j_1 = j_2$. □

We note some consequence of Proposition 14.5.1:

Corollary 14.5.4. *The Mordell-Weil rank $r^{(2)} = \mathrm{rk}\ E^{(2)}(k(t))$ is equal to $4 + h$ if $j_1 \neq j_2$, and to $2 + h$ (resp. $h = 2$) in case (iii) (resp. (iv) or (v)).*

Proof The rank $r^{(2)}$ is equal to the Picard number of S minus the rank of the trivial lattice (cf. (14.8)), so the verification is immediate. □

Corollary 14.5.5. *The torsion subgroup of $E^{(2)}(k(t))$ is trivial in case (i)-(iv) and $\mathbb{Z}/3\mathbb{Z}$ in case (v).*

Proof This follows from the classification results of Shimada [13] (cf. also [10], [9]). □

Corollary 14.5.6. *The condition (#) in § 14.4 holds if and only if $j_1 \neq j_2$, i.e. C_1 and C_2 are not isomorphic to each other.*

14.6 MWL of Inose's pencil

We keep the notation from the previous sections. We assume the condition $j_1 \neq j_2$ in this section.

The Mordell-Weil lattice $L = E^{(2)}(k(t))$ obviously contains $E^{(1)}(k(T))$ with $T = t^2$, which can be identified with the sublattice L_1 of Lemma 14.4.1. Recall that the height of a point gets multiplied by the degree of the base change (see [14, Prop. 8.12]).

Let $\sigma : t \mapsto -t$ be the non-trivial automorphism of the quadratic extension $k(t)/k(T)$. It acts naturally on L and we have

$$L_1 = E^{(1)}(k(T)) = \{P \in L | P^\sigma = P\}. \tag{14.48}$$

On the other hand, letting $F^{(1)}$ denote the quadratic twist of $E^{(1)}/k(T)$ with respect to $k(t)/k(T)$ $(t^2 = T)$:

$$F^{(1)} : y^2 = x^3 - 3\alpha T^4 x + T^5(T^2 - 2\beta T + 1), \tag{14.49}$$

we have

$$F^{(1)}(k(T)) \xrightarrow{\sim} \{P \in E^{(2)}(k(t)) | P^\sigma = -P\} =: L_1'' \subset L \tag{14.50}$$

where $Q = (x(T), y(T)) \in F^{(1)}(k(T))$ corresponds to $P = (x(t^2)/t^2, y(t^2)/t^3)$. Note that L_1'' is orthogonal to L_1 (and $L_1'' \cap L_1 = 0$) since there is no 2-torsion (in fact, L is torsion free under the assumption). Moreover, as is standard in this situation, we have for any $P \in L$

$$2P = (P + P^\sigma) + (P - P^\sigma), \quad (P + P^\sigma) \in L_1, \quad (P - P^\sigma) \in L_1''. \tag{14.51}$$

It follows that $L_1 + L_1'' \supset 2L$ and $L_1'' = L_1'$ is the orthogonal complement of L_1, (4.5). Also V_1 (or V_1') in §4 is the eigenspace with eigenvalue 1 (or -1) for the action of σ on $V = L \otimes \mathbb{Q}$.

Theorem 14.6.1. *Assume $j_1 \neq j_2$. Then the Mordell-Weil lattice $L = E^{(2)}(k(t))$ contains the sublattice $L_1 \oplus L_1'$ with finite index $I = 2^h$ where*

$$L_1 = E^{(1)}(k(T))[2] \cong A_2^*[2]^{\oplus 2}, \quad \text{rk } L_1 = 4, \det L_1 = \frac{2^4}{3^2}, \tag{14.52}$$

$$L_1' \cong F^{(1)}(k(T))[2] \cong H[4], \quad \text{rk } L_1' = h, \det L_1' = 2^{2h} \cdot \delta \tag{14.53}$$

*in which h (or δ) denotes as before the rank (or det) of $H = $* Hom$(C_1, C_2)$.

Proof The only facts yet to be proven in the the above statements are the following:

$$\text{i)} \quad I = 2^h, \qquad \text{ii)} \quad L_1' \cong H[4]. \tag{14.54}$$

Letting ν be the index of $H[4]$ in L_1' in Proposition 14.4.5, we have

$$\det L_1' = 4^h \cdot \delta/\nu^2. \tag{14.55}$$

Hence we have

$$\det L = \det(L_1 \oplus L_1')/I^2 = 2^4/3^2 \cdot 4^h\delta/(\nu^2 I^2). \tag{14.56}$$

On the other hand, by applying (14.14) to $E = E^{(2)}, S = \mathrm{Km}(C_1 \times C_2)$ and $T = U \oplus E_6^2$, we have $\det NS(S) = \det L \cdot 3^2$, which gives by Theorem 14.3.3

$$\det L = 2^4 \cdot \delta/3^2. \tag{14.57}$$

By comparing (14.56) and (14.57), we have

$$I \cdot \nu = 2^h. \tag{14.58}$$

The next lemma shows $I = 2^h$, and hence $\nu = 1$ by (14.58), which is equivalent to the claim (ii) in (14.54). This proves both claims of (14.54). □

Lemma 14.6.2. *The map $\varphi \mapsto P_\varphi$ induces an isomorphism*

$$H/2H \cong L/(L_1 + L_1'). \tag{14.59}$$

In particular, the index $I = [L : L_1 + L_1']$ is equal to 2^h.

Proof The map in question induces a surjective homomorphism of H to L/L_1 (as shown in the proof of Proposition 14.4.5), and hence $H \to L/(L_1 + L_1')$ is also a surjection. By (6.4), the latter map induces a surjection $H/2H \to L/(L_1 + L_1')$, which is also an injection by Lemma 14.4.3. □

Theorem 14.6.3. *Let $S^{(1)}$ denote the elliptic surface associated with $F^{(1)}/k(T)$. Then it is a K3 surface, and it has two singular fibres of type II^* at $T = 0, \infty$, and the other singular fibres are given as follows:*

i) $I_1 \times 4$ if $j_1 \neq j_2, j_1 j_2 \neq 0$,
ii) $II \times 2$ if $j_1 \neq j_2, j_1 j_2 = 0$, (iii) $I_2 + I_1 \times 2$ if $j_1 = j_2 \neq 0, 1$,
iii) $I_2 \times 2$ if $j_1 = j_2 = 1$, (v) IV if $j_1 = j_2 = 0$.

The Mordell-Weil rank $r^{(1)} = \mathrm{rk}\ F^{(1)}(k(T))$ is equal to h if $j_1 \neq j_2$, and to $h-1$ (resp. $h-2=0$) in case iii) *(resp.* iv) *or* v)*).*

Assume $j_1 \neq j_2$. Then the Mordell-Weil lattice $F^{(1)}(k(T))$ is isomorphic to $H[2] = \mathrm{Hom}(C_1, C_2)[2]$.

Proof The singular fibres are checked in the same way as in Lemma 14.5.2 for $E^{(1)}/k(T)$ which is the twist of $F^{(1)}/k(T)$, and it shows that $S^{(1)}$ is a K3 surface since the Euler number (or the order of the discriminant) is 24.

As for the rank formula, it follows from Corollary 14.5.4 and

$$r^{(1)} = \text{rk } E^{(2)}(k(t)) - \text{rk } E^{(1)}(k(T)). \tag{14.60}$$

The final assertion is just a restatement of the fact $L'_1 \cong H[4]$ proven in Theorem 14.6.1, (14.53), in view of the height behavior under the base change (here, of degree two). □

Now Theorem 14.1.1 or 14.1.2 in the Introduction (§14.1) follow from the above Theorem 14.6.3 or 14.6.1. (Note that $F^{(1)}/k(T)$ in §(14.1) and §14.6 are the same up to simple coordinate change. Also $F^{(2)}/k(t)$ and $E^{(2)}/k(t)$ are isomorphic.) They are formulated in terms of elliptic curves only, without reference to a K3 or Kummer surface, but the latter is essential for the proof as seen above.

14.7 Comments on the case $j_1 = j_2$

We have excluded the case $j_1 = j_2$ for the sake of simplicity in some of the above discussion. In this case, we have "extra" reducible fibres in Inose's pencil (see Proposition 14.5.1) and the Mordell-Weil rank drops. We can clarify this situation by the use of the curve Γ_φ (§14.3) for the " isomorphism correspondence".

Proposition 14.7.1. *Assume $j_1 = j_2$, i.e. C_1 and C_2 are isomorphic elliptic curves. Let $\varphi : C_1 \to C_2$ be any isomorphism. Then the curve Γ_φ (the image of the graph of φ in $A = C_1 \times C_2$ under the rational map $A \to S$) is an irreducible component of an extra reducible fibre.*

Proof We can assume that $C_1 = C_2$ and it is defined by (5.1). Recall that the elliptic fibration $f : S \to \mathbb{P}^1$ is given by the function (14.37):

$$t = y_2/y_1. \tag{14.61}$$

Suppose $\varphi : (x_1, y_1) \mapsto (x_2, y_2)$ is an automorphism of C_1.

i) If $\varphi = \text{id}$ is the identity, then we have $t = y_2/y_1 = 1$ on its graph. Hence the curve Γ_{id} is contained in the fibre over $t = 1$, $f^{-1}(1)$. Similarly, if $\varphi = -\text{id}$ is the inversion, we have $y_2 = -y_1$ so that $t = -1$. Hence $\Gamma_{-\text{id}} \subset f^{-1}(-1)$. In this way, we get two reducible fibres of type I_2 at $t = 1, -1$ for a general value of j_1, namely for $j_1 \neq 0, 1$.
ii) Let $C_1 : y_1^2 = x_1^3 - x_1 (j_1 = 1)$ and suppose $\varphi : (x_1, y_1) \mapsto (-x_1, \pm i y_1)$. Then we have $t = \pm i$ $(i = \sqrt{-1})$. In this case, we get four reducible fibres of type I_2 at $t = 1, -1, i, -i$.

iii) Let $C_1 : y_1^2 = x_1^3 - 1$ ($j_1 = 0$) and suppose $\varphi : (x_1, y_1) \mapsto (\omega x_1, \pm y_1)$ ($\omega^3 = 1$). Then we have $t = \pm 1$. In this case, we get two singular fibres of type IV at $t = 1, -1$. The three curves Γ_φ for three values of ω give the three irreducible components for a type IV-fibre.

This completes the proof. □

14.8 Examples

First, the general case of Theorem 14.6.1 and 14.6.3 is very simple.

Example 14.8.1. Assume that C_1, C_2 are non-isogeneous elliptic curves. Then

$$F^{(2)}(k(t)) \cong A_2^*[2], \quad F^{(1)}(k(T)) = \{0\} \tag{14.62}$$

The generators of $k(t)$-rational points are given by Proposition 14.5.3 by setting $T = t^2$.

Next a special case of Theorem 14.6.3 implies:

Example 14.8.2. Suppose that $C_2 : y^2 = x^3 - 1$ ($j_2 = 0$) and $j_1 = j$ is arbitrary. We have $\alpha = 0, \beta = \sqrt{1-j}$, and $F^{(1)}$ has the equation:

$$F_j = F^{(1)} : y^2 = x^3 + T^5(T^2 - 2\beta T + 1), \quad j = 1 - \beta^2. \tag{14.63}$$

Then

1) F_j has a $k(T)$-rational point ($\neq O$) if and only if $j = j(C_1)$ for some elliptic curve C_1 isogeneous (but not isomorphic) to C_2, and thus
2) $F_j(k(T)) \neq \{O\}$ holds only for countably many values of $j \in k$.

   Let us write down further properties of the elliptic curve $F_j/k(T)$. For simplicity, assume $\mathrm{char}(k) = 0$ and let $F_j(k(T)) \neq \{O\}$.
3) The Mordell-Weil lattice $F_j(k(T))$ is isomorphic to $A_2[d_0]$ where $d_0 = \deg(\varphi_0)$ is the minimal isogeny $\varphi_0 : C_1 \to C_2$. (A_2 is the root lattice.)
4) The minimal height is $2d_0 \geq 4$.
5) There is a rational point P of height $\langle P, P \rangle = 4$ if and only if C_1 has degree 2 isogeny to C_2. In this case, $P = (\xi, \eta)$ is an "integral point" with both $\xi, \eta \in k[T]$ with $\deg(\xi) = 4, \deg(\eta) = 6$. (This follows from the height formula.)
6) Such C_1 is unique up to isomorphism and one has $j = j(C_1) = 125/4$. Then F_j is given by

$$F_j : y^2 = x^3 + T^5(T^2 - 11\sqrt{-1}T + 1). \tag{14.64}$$

which is equivalent (up to coordinate change) to the following equation:

$$y^2 = x^3 + T^5(T^2 - 11T - 1). \tag{14.65}$$

7) There are three different methods to find an integral point $P = (\xi, \eta)$ of height 4 of the form in (5). [N.B. The resulting integral point is essentially the same by the uniqueness in (5) or (6).]
   (i) straightforward computer search ([3]),
   (ii) explicit computation of $P_\varphi = \mu(\Gamma_\varphi)$ for the degree two isogeny φ, explained in the present paper (the result was announced in [15]),
   (iii) use the idea of "Shafarevich partner" ([16]). We outline the third method below, because this simple device can be useful in more general situations.

8) Take a rational elliptic surface with 4 singular fibres $I_1 \times 2, I_5 \times 2$ (see e.g. [1]). Assume that the fibres of type I_5 are at $T = 0, \infty$. Write down the minimal Weierstrass equation of the generic fibre as follows:

$$Y^2 = X^3 - 3\xi X - 2\eta, \quad \xi, \eta \in k[T] \tag{14.66}$$

Then the discriminant is equal to $\xi^3 - \eta^2$ up to constant, but at the same time it should take the form $cT^5(T - v_1)(T - v_2)$ by the data of singular fibres. This shows that $P = (\xi, \eta)$ gives rise to an integral point of required form.

Note that this argument can be reversed to prove the existence and uniqueness of the rational elliptic surface with singular fibres $I_1 \times 2, I_5 \times 2$, from the knowledge of such an integral point of height 4.

References

[1] Beauville, A.: Les familles stables de courbes elliptiques sur $\mathbb{P}^1$, C. R. Acad. Sci. Paris, **294** (I), 657–660 (1982).
[2] Cassels, J.W.S.: *Lectures on Elliptic Curves* Cambridge Univ. Press (1991).
[3] Chahal, J., M. Meijer and J. Top: Sections on certain $j = 0$ elliptic surfaces, Comment. Math. Univ. St. Pauli **49** (2000).
[4] Conway, J. and N. Sloane: Sphere Packings, Lattices and Groups, Springer-Verlag (1988); 2nd ed.(1993); 3rd ed.(1999).
[5] Inose, H. : Defining equations of singular K3 surfaces and a notion of isogeny in *Intl. Symp. on Algebraic Geometry/ Kyoto 1977* Kinokuniya, 495–502 (1978).
[6] Inose, H. and T. Shioda: On singular K3 surfaces, in: *Complex Analysis and Algebraic Geometry*, Iwanami Shoten and Cambridge Univ. Press, 119–136 (1977).

[7] Kodaira, K.: On compact analytic surfaces II-III, Ann. of Math. **77**, 563–626 (1963); 78, 1–40 (1963); *Collected Works*, III, 1269–1372, Iwanami and Princeton Univ. Press (1975).
[8] Kuwata, M.: Elliptic K3 surfaces with given Mordell-Weil rank, Comment. Math. Univ. St. Pauli **49** (2000).
[9] Nishiyama, K.: On Jacobian fibrations on some K3 surfaces and their Mordell-Weil groups, Japan J. Math. **22**, 293–347 (1996).
[10] Oguiso, K.: On Jacobian fibrations on the Kummer surfaces of product of non-isogeneous elliptic curves, J. Math. Soc. Japan **41**, 651–680 (1989).
[11] Oguiso, K. and T. Shioda: The Mordell–Weil lattice of a rational elliptic surface, Comment. Math. Univ. St. Pauli **40**, 83–99 (1991).
[12] Piateckii-Shapiro, I. and I. R. Shafarevich: A Torelli theorem for algebraic surfaces of type K3, Math. USSR Izv. **5**, 547–587 (1971).
[13] Shimada, I.: On elliptic K3 surfaces, Michigan Math. J. **47** (2000), 423–446.
[14] Shioda, T.: On the Mordell-Weil lattices, Comment. Math. Univ. St. Pauli **39**, 211–240 (1990).
[15] — : A note on K3 surfaces and sphere packings, Proc.Japan Acad. 76A, 68–72 (2000).
[16] — : Elliptic surfaces and Davenport-Stothers triples, Comment. Math. Univ. St. Pauli (to appear).
[17] — : Classical Kummer surfaces and Mordell-Weil Lattices, Proc. KIAS Conf. Algebraic Geometry 2004 (to appear).
[18] Tate, J: Algorithm for determining the type of a singular fiber in an elliptic pencil, SLN 476, 33–52 (1975).
[19] Weil, A.: Variétés abéliennes et courbes algébriques, Hermann, Paris (1948/1973).

15

Motives from Diffraction

Jan Stienstra
Mathematisch Instituut, Universiteit Utrecht, the Netherlands
stien@math.uu.nl

Dedicated to Jaap Murre and Spencer Bloch

Abstract

We look at geometrical and arithmetical patterns created from a finite subset of $\mathbb{Z}^n$ by diffracting waves and bipartite graphs. We hope that this can make a link between Motives and the Melting Crystals/Dimer models in String Theory.

15.1 Introduction

Why is it that, occasionally, mathematicians studying Motives *and physicists searching for a* Theory of Everything *seem to be looking at the same examples, just from different angles? Should the Theory of Everything include properties of Numbers? Does Physics yield realizations of Motives which have not been considered before in the cohomological set-up of motivic theory?* Calabi-Yau varieties of dimensions 1 and 2, being elliptic curves and K3-surfaces, have a long and rich history in number theory and geometry. Calabi-Yau varieties of dimension 3 have played an important role in many developments in String Theory. The discovery of *Mirror Symmetry* attracted the attention of physicists and mathematicians to Calabi-Yau's near the *large complex structure limit* [13, 20]. Some analogies between String Theory and Arithmetic Algebraic Geometry near this limit were discussed in [16, 17, 18]. Recently new models appeared, called *Melting Crystals* and *Dimers* [14, 11], which led to interesting new insights in String Theory, without going near the large complex structure limit. The present paper is an attempt to find motivic aspects of these new models. We look at geometrical and enumerative patterns associated with a finite subset $\mathfrak{A}$ of $\mathbb{Z}^n$. The geometry comes from waves diffracting on $\mathfrak{A}$ and from a periodic weighted bipartite graph generated by $\mathfrak{A}$. The latter is related to the dimers (although here we can not say more about this relation). Since the tori involved in these models are naturally dual to each other there seems to be some sort of mirror symmetry between the diffraction and the graph pictures. The enumerative patterns count lattice points on the diffraction pattern, points on varieties over finite fields and paths on the graph. They are expressed through a sequence of polynomials $\mathsf{B}_N(z)$ with coefficients in $\mathbb{Z}$ and via a limit for $z \in \mathbb{C}$:

$$\mathsf{Q}(z) = \lim_{N\to\infty} |\mathsf{B}_N(z)|^{-N^{-n}} . \tag{15.1}$$

Limit formulas like (15.1) appear frequently and in very diverse contexts in the literature, e.g. for *entropy in algebraic dynamical systems* in [7, Theorem 4.9], for *partition functions per fundamental domain in dimer models* in [11, Theorem 3.5], for *integrated density of states* in [9, p. 206]. Moreover, $\mathsf{Q}(z)$ appears as *Mahler measure* in [4], as the exponential of a *period in Deligne cohomology* in [6, 15], and in *instanton counts* in [18]; see the remark at the end of Section 15.5. With some additional restrictions $\mathfrak{A}$ provides the toric data for a family of Calabi-Yau varieties and various well-known results about Calabi-Yau varieties near the large complex structure limit can be derived from the Taylor series expansion of $\log \mathsf{Q}(z)$ near $z = \infty$; see the Remark at the end of Section 15.6. In the present paper we are not so much interested in the large complex structure limit. Instead we focus on the polynomials $\mathsf{B}_N(z)$ and the limit formula (15.1). *This does not require conditions of 'Calabi-Yau type'.*

When waves are diffracted at some finite set $\mathfrak{A}$ of points in a plane, the diffraction pattern observed in a plane at large distance is, according to the Frauenhofer model, the absolute value squared of the Fourier transform of $\mathfrak{A}$. There is no mathematical reason to restrict this model to dimension 2. Also the points may have weights ≥ 1. So, we take a finite subset $\mathfrak{A}$ of $\mathbb{Z}^n$ and positive integers c_a ($\mathsf{a} \in \mathfrak{A}$). These data can be summarized as a distribution $\mathsf{D} = \sum_{\mathsf{a}\in\mathfrak{A}} c_\mathsf{a}\delta_\mathsf{a}$, where δ_a denotes the Dirac delta distribution, evaluating test functions at the point a. The Fourier transform of D is the function $\widehat{\mathsf{D}}(\mathsf{t}) = \sum_{\mathsf{a}\in\mathfrak{A}} c_\mathsf{a} e^{-2\pi i\langle \mathsf{t},\mathsf{a}\rangle}$ on $\mathbb{R}^n$; here $\langle , \rangle$ is the standard inner product on $\mathbb{R}^n$. The diffraction pattern consists of the level sets of the function

$$|\widehat{\mathsf{D}}(\mathsf{t})|^2 = \sum_{\mathsf{a},\mathsf{b}\in\mathfrak{A}} c_\mathsf{a} c_\mathsf{b} e^{2\pi i\langle \mathsf{t},\mathsf{a}-\mathsf{b}\rangle} = \sum_{\mathsf{a},\mathsf{b}\in\mathfrak{A}} c_\mathsf{a} c_\mathsf{b} \cos(2\pi\langle \mathsf{t},\mathsf{a}-\mathsf{b}\rangle) .$$

This function is periodic with period lattice $\Lambda^\vee$ dual to the lattice Λ spanned over $\mathbb{Z}$ by the differences $\mathsf{a}-\mathsf{b}$ with $\mathsf{a},\mathsf{b} \in \mathfrak{A}$. *Throughout this note we assume that Λ has rank n.* Looking at the intersections of the diffraction pattern

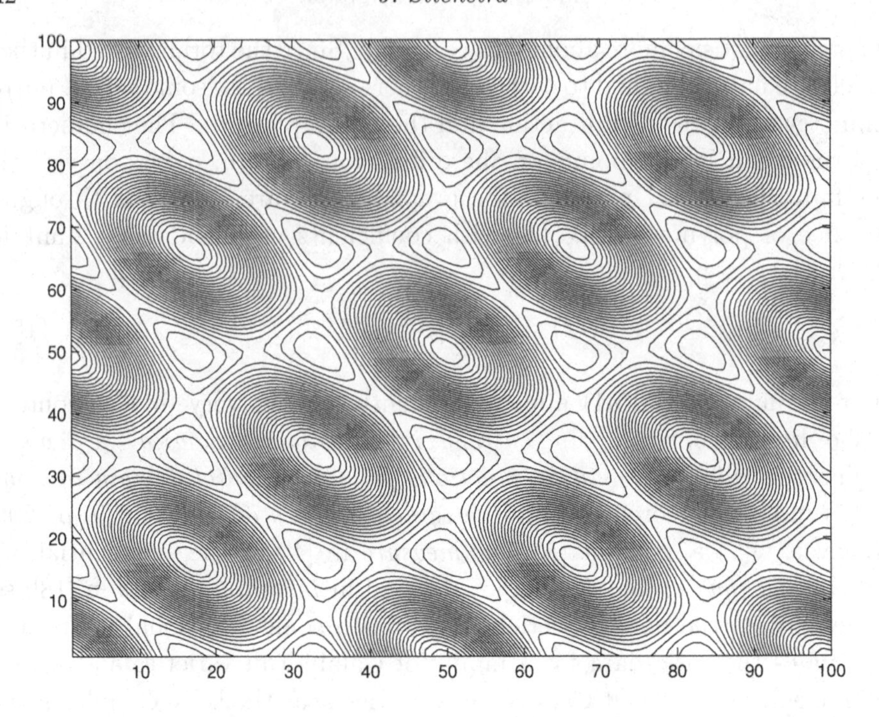

Fig. 15.1. *Diffraction pattern for* $\mathfrak{A} = \{(1,0), (0,1), (-1,-1)\} \subset \mathbb{Z}^2$, *all* $c_{\mathsf{a}} = 1$.

with the lattices $\frac{1}{N}\Lambda^\vee$ we introduce the enumerative data

$$\mathrm{mult}_N(r) := \sharp\{\mathsf{t} \in \tfrac{1}{N}\Lambda^\vee/\,\Lambda^\vee \mid |\widehat{\mathsf{D}}(\mathsf{t})|^2 = r\,\} \quad \text{for} \quad N \in \mathbb{N},\, r \in \mathbb{R}\,, \tag{15.2}$$

and use these to define polynomials $\mathsf{B}_N(z)$ as follows:

Definition 15.1.1.

$$\mathsf{B}_N(z) := \prod_{r\in\mathbb{R}} (z-r)^{\mathrm{mult}_N(r)}\,. \tag{15.3}$$

One could also introduce the generating function $\mathsf{F}(z,T) := \sum_{N\in\mathbb{N}} \mathsf{B}_N(z)\, T^{N^n}$ but except for the classical number theory of the case $n=1$ (see Section 15.7.1), and the observation that Formula (15.1) gives $\mathsf{Q}(z)$ as the radius of convergence of $\mathsf{F}(z,T)$ as a complex power series in T, we do not yet have appealing results about $\mathsf{F}(z,T)$.

For the graph model we start from the same data: the finite set $\mathfrak{A} \subset \mathbb{Z}^n$, the weights c_{a} ($\mathsf{a} \in \mathfrak{A}$) and the lattice Λ spanned by the differences $\mathsf{a}-\mathsf{b}$ with a, $\mathsf{b} \in \mathfrak{A}$. We must now assume that

$$\mathfrak{A} \cap \Lambda = \varnothing.$$

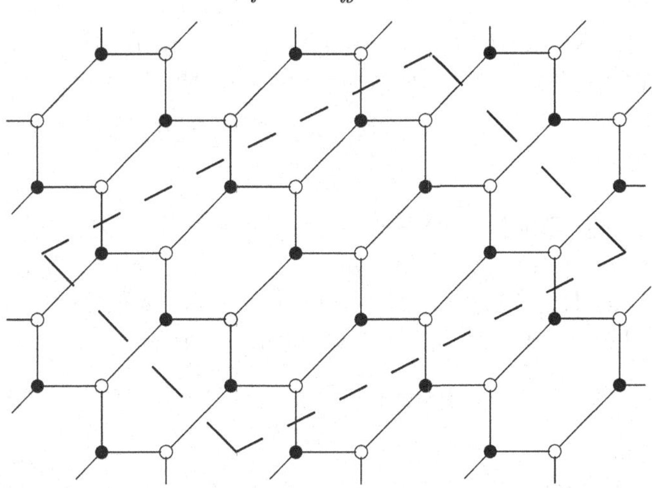

Fig. 15.2. *A fundamental parallelogram of the lattice* 3Λ *and a piece of the bipartite graph* Γ *for* $\mathfrak{A} = \{(1,0), (0,1), (-1,-1)\} \subset \mathbb{Z}^2$, *and all* $c_{\mathsf{a}} = 1$. *By identifying opposite sides of the parallelogram one obtains the graph* Γ_3.

One can then construct a weighted bipartite graph Γ as follows. Bipartite graphs have two kinds of vertices, often called black and white. The set of *black vertices* of Γ is Λ. The set of *white vertices* of Γ is $\mathfrak{A}+\Lambda$. Note that $\mathfrak{A}+\Lambda$ is just one single coset of Λ in $\mathbb{Z}^n$. In Γ there is an (oriented) edge from vertex v_1 to vertex v_2 if and only if v_1 is black, v_2 is white and $\mathsf{v}_2-\mathsf{v}_1 \in \mathfrak{A}$. If $\mathsf{v}_2-\mathsf{v}_1 = \mathsf{a} \in \mathfrak{A}$ the edge is said to be of *type* a and gets *weight* c_{a}. The graph Γ is Λ-periodic and for every $N \in \mathbb{N}$ one has the finite graph $\Gamma_N := \Gamma/N\Lambda$, which is naturally embedded in the torus $\mathbb{R}^n/N\Lambda$. By a *closed path* of length $2k$ on Γ or Γ_N we mean a sequence of edges $(e_1, e_2, \ldots, e_{2k-1}, e_{2k})$ such that for $i = 1, \ldots, k$ the intersection $e_{2i-1} \cap e_{2i}$ contains a white vertex and $e_{2i} \cap e_{2i+1}$ contains a black vertex; here $e_{2k+1} = e_1$. By the weight of such a path we mean the product of the weights of the edges $e_1, e_2, \ldots, e_{2k-1}, e_{2k}$. We denote the set of closed paths of length $2k$ on Γ_N by $\Gamma_N(2k)$. Enumerating the closed paths on Γ_N according to length and weight we prove in Section 15.4 that this leads to a new interpretation of the polynomials $B_N(z)$:

Theorem 15.1.2.

$$B_N(z) = z^{N^n} \exp\left(-\sum_{k\geq 1} \sum_{\gamma\in\Gamma_N(2k)} \text{weight}(\gamma) z^{-k} \right). \tag{15.4}$$

□

Formulas (15.3) and (15.4) transfer the enumerative data between the two models. We pass to algebraic geometry with the Laurent polynomial

$$W(x_1,\ldots,x_n) = \sum_{\mathsf{a},\mathsf{b}\in\mathfrak{A}} c_{\mathsf{a}} c_{\mathsf{b}} \mathsf{x}^{\mathsf{a}-\mathsf{b}} \in \mathbb{Z}[x_1^{\pm 1},\ldots,x_n^{\pm 1}], \qquad (15.5)$$

which satisfies $|\widehat{\mathsf{D}}(\mathsf{t})|^2 = W(e^{2\pi i t_1},\ldots,e^{2\pi i t_n})$; here $\mathsf{x}^\lambda := \prod_{j=1}^n x_j^{\lambda_j}$ if $\lambda = (\lambda_1,\ldots,\lambda_n) \in \Lambda$. For $N \in \mathbb{N}$ let μ_N denote the group of N-th roots of unity and let $\mu_N^\Lambda := \mathrm{Hom}(\Lambda, \mu_N)$ be the group of homomorphisms from the lattice Λ to μ_N. Thus the defining formula (15.3) can be rewritten as:

$$\mathsf{B}_N(z) = \prod_{\mathsf{x}\in\mu_N^\Lambda} (z - W(\mathsf{x}))\,. \qquad (15.6)$$

Written in the form (15.6) the polynomials $\mathsf{B}_N(z)$ appear as direct generalizations of quantities introduced by Lehmer [12] for a 1-variable (i.e. $n = 1$) polynomial $W(x)$. Using (15.6) one can easily show (Proposition 15.5.1) that the polynomials $\mathsf{B}_N(z)$ have integer coefficients and that $\mathsf{B}_{N'}(z)$ divides $\mathsf{B}_N(z)$ in $\mathbb{Z}[z]$ if N' divides N in $\mathbb{Z}$. Thus for $h \in \mathbb{Z}$ also $\mathsf{B}_N(h)$ is an integer. Lehmer was particularly interested in the prime factorization of these integers in case $n = 1$ [12, 7, 15]. Also for general $n \geq 1$ these prime factorizations must be interesting, for instance because they relate to counting points on varieties over finite fields; see Section 15.5 for details. *Thus prime factorization gives a third occurrence of* $\mathsf{B}_N(z)$ *in enumerative problems, related to counting points on varieties over finite fields.* $\mathsf{Q}(z)$ appears in [4, 6, 15] as *Mahler measure* with ties to special values of L-functions. It would be nice if the limit formula (15.1) together with the prime factorization of the numbers $\mathsf{B}_N(z)$ (with $z \in \mathbb{Z}$) could shed new light on these very intriguing ties.

In *Section 15.2* we study the density distribution of the level sets in the diffraction pattern. Passing from measures to complex functions with the Hilbert transform we find one interpretation of $\mathsf{Q}(z)$, $\mathsf{B}_N(z)$ and (15.1). In *Section 15.3* we briefly discuss another interpretation in connection with the spectrum of a discretized Laplace operator. In *Section 15.4* we prove Theorem 15.1.2. In *Section 15.5* we pass to toric geometry, where the diffraction pattern reappears as the intersection of a real torus with a family of hypersurfaces in a complex torus and where $\log \mathsf{Q}(z)$ becomes a period integral, while the prime factorization of $\mathsf{B}_N(z)$ for $z \in \mathbb{Z}$ somehow relates to counting points on those hypersurfaces over finite fields. In *Section 15.6* we discuss sequences of integers which appear as moments of measures, path counts on graphs and coefficients in Taylor expansions of solutions of Picard-Fuchs differential equations. Finally, in *Section 15.7* we present some concrete examples.

15.2 The diffraction pattern

The function $|\widehat{\mathsf{D}}(\mathsf{t})|^2$ is periodic with period lattice $\Lambda^\vee$ dual to the lattice Λ:

$$\begin{aligned}\Lambda^\vee &:= \{\mathsf{t} \in \mathbb{R}^n \mid \langle \mathsf{t}, \mathsf{a} - \mathsf{b}\rangle \in \mathbb{Z}\,,\ \forall \mathsf{a}, \mathsf{b} \in \mathfrak{A}\}\,,\\ \Lambda &:= \mathbb{Z}\text{–Span}\{\mathsf{a} - \mathsf{b} \mid \mathsf{a}, \mathsf{b} \in \mathfrak{A}\}.\end{aligned}$$

Throughout this note we assume that the lattices Λ and $\Lambda^\vee$ have rank n.

Because of this periodicity $|\widehat{\mathsf{D}}(\mathsf{t})|^2$ descends to a function on $\mathbb{R}^n/\Lambda^\vee$. Defining for $\mathsf{t} \in \mathbb{R}^n$ the function $e_\mathsf{t} : \mathbb{R}^n \to \mathbb{C}$ by $e_\mathsf{t}(\mathsf{v}) = e^{2\pi i \langle \mathsf{t}, \mathsf{v}\rangle}$ we obtain an isomorphism of real tori

$$\mathbb{R}^n/\Lambda^\vee \simeq \mathbb{U}^\Lambda\,, \qquad \mathsf{t} \mapsto e_\mathsf{t}\,, \tag{15.7}$$

where $\mathbb{U}^\Lambda := \mathrm{Hom}(\Lambda, \mathbb{U})$ is the torus of group homomorphisms from the lattice Λ to the unit circle $\mathbb{U} := \{x \in \mathbb{C} \mid |x| = 1\}$. Recall that a group homomorphism $\psi : \Lambda \to \mathbb{U}$ induces an algebra homomorphism ψ_* from the group algebra $\mathbb{C}[\Lambda]$ to $\mathbb{C}$. Thus $\mathbb{C}[\Lambda]$ is the natural algebra of functions on $\mathbb{U}^\Lambda$ and ψ_* evaluates functions at the point ψ of $\mathbb{U}^\Lambda$. The inclusion $\Lambda \subset \mathbb{Z}^n$ identifies $\mathbb{C}[\Lambda]$ with the subalgebra of the algebra of Laurent polynomials $\mathbb{C}[x_1^{\pm 1}, \ldots, x_n^{\pm 1}]$, which consists of $\mathbb{C}$-linear combinations of the monomials $\mathsf{x}^\lambda := \prod_{j=1}^n x_j^{\lambda_j}$ with $\lambda = (\lambda_1, \ldots, \lambda_n) \in \Lambda$. Thus, via (15.7), the function $|\widehat{\mathsf{D}}(\mathsf{t})|^2$ coincides with the Laurent polynomial $W(x_1, \ldots, x_n)$ defined in (15.5). Positivity of the coefficients c_a implies that the function $|\widehat{\mathsf{D}}(\mathsf{t})|^2$ attains its maximum exactly at the points $\mathsf{t} \in \Lambda^\vee$. In terms of the torus $\mathbb{U}^\Lambda$ and the function $W : \mathbb{U}^\Lambda \to \mathbb{R}$ this means that W attains its maximum exactly at the origin 1 of the torus group $\mathbb{U}^\Lambda$:

$$\forall \mathsf{x} \in \mathbb{U}^\Lambda - \{1\} : \quad W(\mathsf{x}) < W(1) = C^2 \qquad \text{with} \quad C := \sum_{\mathsf{a} \in \mathfrak{A}} c_\mathsf{a}\,.$$

Some important aspects of the density distribution in the diffraction pattern are captured by the function

$$V : \mathbb{R} \to \mathbb{R}\,, \qquad V(r) := \frac{\mathrm{volume}\{\mathsf{t} \in \mathbb{R}^n/\Lambda^\vee \mid |\widehat{\mathsf{D}}(\mathsf{t})|^2 \le r\}}{\mathrm{volume}(\mathbb{R}^n/\Lambda^\vee)}\,. \tag{15.8}$$

We view the derivative $dV(r)$ of V as a measure on $\mathbb{R}$. In our analysis it will be important that the measure $dV(r)$ is also the push forward of the standard measure $dt_1\, dt_2 \ldots dt_n$ on $\mathbb{R}^n$ by the function $|\widehat{\mathsf{D}}(\mathsf{t})|^2$.

Another insight into the diffraction pattern comes from its intersection with the torsion subgroup of $\mathbb{U}^\Lambda$. For $N \in \mathbb{N}$ let $\mu_N \subset \mathbb{U}$ denote the group of N-th roots of unity. Then the group of N-torsion points in $\mathbb{U}^\Lambda$ is $\mu_N^\Lambda := \mathrm{Hom}(\Lambda, \mu_N)$ and (15.2) can be rewritten as

$$\mathrm{mult}_N(r) = \sharp(\mu_N^\Lambda \cap W^{-1}(r)) \qquad \text{for} \quad N \in \mathbb{N},\, r \in \mathbb{R}.$$

Moreover we set, in analogy with (15.8),

$$V_N(r) := \frac{1}{N^n}\sharp\{\mathsf{x} \in \mu_N^\Lambda \mid W(\mathsf{x}) \leq r\} \qquad \text{for} \quad N \in \mathbb{N},\, r \in \mathbb{R}.$$

The derivative of the step function $V_N : \mathbb{R} \to \mathbb{R}$ is the distribution

$$dV_N(r) \;=\; N^{-n}\sum_r \mathrm{mult}_N(r)\,\delta_r, \tag{15.9}$$

which assigns to a continuous function f on $\mathbb{R}$ the value

$$\int_{\mathbb{R}} f(r)dV_N(r) := N^{-n}\sum_r \mathrm{mult}_N(r)\,f(r) = N^{-n}\sum_{\mathsf{x}\in\mu_N^\Lambda} f(W(\mathsf{x})). \tag{15.10}$$

One thus finds a limit of distributions

$$\lim_{N\to\infty} dV_N(r) \;=\; dV(r)\,; \tag{15.11}$$

by definition, this means that for every continuous function f on $\mathbb{R}$

$$\lim_{N\to\infty}\int_{\mathbb{R}} f(r)dV_N(r) \;=\; \int_{\mathbb{R}} f(r)dV(r)\,. \tag{15.12}$$

Measure theory is connected with complex function theory by the Hilbert transform. The Hilbert transform of the measure $dV(r)$ is the function $-\frac{1}{\pi}\mathsf{H}(z)$ defined by

$$\mathsf{H}(z) \;:=\; \int_{\mathbb{R}} \frac{1}{z-r}dV(r) \qquad \text{for} \quad z \in \mathbb{C} - \mathfrak{I}\,; \tag{15.13}$$

here $\mathfrak{I} := \overline{\{r \in \mathbb{R} \,|\, 0 < V(r) < 1\}} \;\subset\; [0, C^2]$ is the support of the measure $dV(r)$. The measure can be recovered from its Hilbert transform because for $r_0 \in \mathbb{R}$

$$\frac{dV}{dr}(r_0) = \frac{1}{2\pi i}\lim_{\epsilon\in\mathbb{R},\epsilon\downarrow 0}\left(\mathsf{H}(r_0 + i\epsilon) \;-\; \mathsf{H}(r_0 - i\epsilon)\right).$$

Another way of writing the connection between $dV(r)$ and $\mathsf{H}(z)$ is

$$\frac{1}{2\pi i}\oint_\gamma f(z)\mathsf{H}(z)dz \;=\; \int_{\mathbb{R}}\left(\frac{1}{2\pi i}\oint_\gamma \frac{f(z)}{z-r}dz\right)dV(r) \;=\; \int_{\mathbb{R}} f(r)dV(r) \tag{15.14}$$

for holomorphic functions f defined on some open neighborhood $U$of the interval $\mathfrak{I}$ in $\mathbb{C}$ and closed paths γ in $U - \mathfrak{I}$ encircling $\mathfrak{I}$ once counter clockwise.

Next we consider the function

$$\mathsf{Q} : \mathbb{C} - \mathfrak{J} \longrightarrow \mathbb{R}_{>0}, \qquad \mathsf{Q}(z) := \exp\left(- \int_{\mathbb{R}} \log|z - r| dV(r) \right). \tag{15.15}$$

This function satisfies $\frac{d}{dz} \log \mathsf{Q}(z) = -\mathsf{H}(z)$ and thus (15.14) can be rewritten as

$$\frac{-1}{2\pi i} \oint_{\gamma} f(z) d \log \mathsf{Q}(z) = \int_{\mathbb{R}} f(r) dV(r).$$

This means that, at least intuitively, the functions $\mathsf{Q}(z)$ *and* $e^{-V(r)}$ *correspond to each other via some kind of comparison isomorphism.* In order to find the analogue of (15.11) in terms of functions on $\mathbb{C} - \mathfrak{J}$ we apply (15.10) to the function $f(r) = \log|z - r|$ on $\mathbb{R}$ with fixed $z \in \mathbb{C} - \mathfrak{J}$:

$$\int_{\mathbb{R}} \log|z - r| dV_N(r) = N^{-n} \log|\mathsf{B}_N(z)| \tag{15.16}$$

where $\mathsf{B}_N(z) = \prod_{r \in \mathfrak{J}} (z - r)^{\mathrm{mult}_N(r)} = \prod_{\mathsf{x} \in \mu_N^{\Lambda}} (z - W(\mathsf{x}))$ as in (15.3) and (15.6). Combining (15.12), (15.15) and (15.16) we find the limit announced in (15.1):

Proposition 15.2.1. $\mathsf{Q}(z) = \lim_{N \to \infty} |\mathsf{B}_N(z)|^{-N^{-n}}$ *for every* $z \in \mathbb{C} - \mathfrak{J}$. □

15.3 The Laplacian perspective

Convolution with the distribution D gives the operator $\mathsf{D}f(\mathsf{v}) := \sum_{\mathsf{a} \in \mathfrak{A}} c_{\mathsf{a}} f(\mathsf{v} - \mathsf{a})$ on the space of $\mathbb{C}$-valued functions on $\mathbb{R}^n$. Let $\overline{\mathsf{D}}f(\mathsf{v}) := \sum_{\mathsf{a} \in \mathfrak{A}} c_{\mathsf{a}} f(\mathsf{v} + \mathsf{a})$ and

$$\Delta := \mathsf{D}\overline{\mathsf{D}}, \qquad \Delta f(\mathsf{v}) = \sum_{\mathsf{a}, \mathsf{b} \in \mathfrak{A}} c_{\mathsf{a}} c_{\mathsf{b}} f(\mathsf{v} + \mathsf{a} - \mathsf{b}).$$

For a sufficiently differentiable function f on $\mathbb{R}^n$ the Taylor expansion

$$\Delta f(\mathsf{v}) = C^2 f(\mathsf{v}) + \tfrac{1}{2} \sum_{i,j=1}^{n} \sum_{\mathsf{a}, \mathsf{b} \in \mathfrak{A}} c_{\mathsf{a}} c_{\mathsf{b}} (a_i - b_i)(a_j - b_j) \frac{\partial^2 f}{\partial v_i \partial v_j}(\mathsf{v}) + \dots$$

shows that the difference operator $\Delta - C^2$ is a *discrete approximation of the Laplace operator* corresponding to the Hessian of the function $|\widehat{\mathsf{D}}(\mathsf{t})|^2$ at its maximum.

Remark. In [9] Gieseker, Knörrer and Trubowitz investigate Schrödinger equations in solid state physics via a discrete approximation of the Laplacian. In their situation the Schrödinger operator is the discretized Laplacian *plus* *a periodic potential function.* So from the perspective of [9] the present note deals only with the (simple) case of zero potential. On the other hand we consider more general discretization schemes and possibly higher dimensions.

We now turn to the spectrum of Δ. For $\mathsf{t} \in \mathbb{R}^n$ the function $e_{\mathsf{t}} : \mathbb{R}^n \to \mathbb{C}$ given by $e_{\mathsf{t}}(\mathsf{v}) = e^{2\pi i \langle \mathsf{t}, \mathsf{v} \rangle}$ is an eigenfunction for Δ with eigenvalue $|\widehat{\mathsf{D}}(\mathsf{t})|^2$:

$$\Delta e_{\mathsf{t}}(\mathsf{v}) = \sum_{\mathsf{a},\mathsf{b} \in \mathfrak{A}} c_{\mathsf{a}} c_{\mathsf{b}} e^{2\pi i \langle \mathsf{t}, \mathsf{v}+\mathsf{a}-\mathsf{b} \rangle} = |\widehat{\mathsf{D}}(\mathsf{t})|^2 \, e_{\mathsf{t}}(\mathsf{v}) \,.$$

Take a positive integer N. The space of $\mathbb{C}$-valued C^∞-functions on $\mathbb{R}^n$ which are periodic for the sublattice $N\Lambda$ of Λ is spanned by the functions e_{t} with t in the dual lattice $\frac{1}{N}\Lambda^\vee$. The characteristic polynomial of the restriction of Δ to this space is therefore (see (15.3) and (15.6))

$$\prod_{\mathsf{t} \in \frac{1}{N}\Lambda^\vee / \Lambda^\vee} (z - |\widehat{\mathsf{D}}(\mathsf{t})|^2) = \prod_{\mathsf{x} \in \mu_N^\Lambda} (z - W(\mathsf{x})) = \prod_{r \in \mathbb{R}} (z - r)^{\mathrm{mult}_N(r)} = \mathsf{B}_N(z).$$

With (15.9) *and* (15.11) *the measure $dV(r)$ can now be interpreted as the density of the eigenvalues of Δ on the space of $\mathbb{C}$-valued C^∞-functions on $\mathbb{R}^n$ which are periodic for some sublattice $N\Lambda$ of Λ.*

15.4 Enumeration of paths on a periodic weighted bipartite graph

In this section we prove Theorem 15.1.2. Recall from the Introduction just before Theorem 15.1.2 the various ingredients: the finite set $\mathfrak{A} \subset \mathbb{Z}^n$, the weights c_{a}, the lattice Λ and the graphs Γ and Γ_N. Recall also the closed paths on Γ_N, their lengths and weights, and the set $\Gamma_N(2k)$ of closed paths of length $2k$ on Γ_N. Consider a path $(e_1, e_2, \ldots, e_{2k-1}, e_{2k})$ on Γ with edge e_i going from black to white if i is odd, respectively from white to black if i is even. Let s denote the starting point of the path (i.e. the black vertex of edge e_1). Let for $j = 1, \ldots, k$ edge e_{2j-1} be of type a_j and edge e_{2j} of type b_j. Then the end point of the path (i.e. the black vertex of e_{2k}) is $\mathsf{s} + \sum_{j=1}^k (\mathsf{a}_j - \mathsf{b}_j)$. The weight of the path is $\prod_{j=1}^k c_{\mathsf{a}_j} c_{\mathsf{b}_j}$. The path closes on Γ_N if and only if $\sum_{j=1}^k (\mathsf{a}_j - \mathsf{b}_j) \in N\Lambda$. Next recall from (15.5) that $W(\mathsf{x}) = \sum_{\mathsf{a},\mathsf{b} \in \mathfrak{A}} c_{\mathsf{a}} c_{\mathsf{b}} \mathsf{x}^{\mathsf{a}-\mathsf{b}}$ and set

$$\mathsf{m}_k^{(N)} := N^{-n} \sum_{\mathsf{x} \in \mu_N^\Lambda} W(\mathsf{x})^k \,. \tag{15.17}$$

So $\mathsf{m}_k^{(N)}$ is the sum of the coefficients of those monomials in $W(\mathsf{x})^k$ with exponent in $N\Lambda$. In view of the above considerations $\mathsf{m}_k^{(N)}$ is therefore equal to the sum of the weights of the paths on Γ which start at s, have length $2k$ and close in Γ_N. Since on Γ_N there are N^n black vertices and on a path of length $2k$ there are k black vertices we conclude

$$\frac{N^n}{k}\mathsf{m}_k^{(N)} = \sum_{\gamma\in\Gamma_N(2k)} \text{weight}(\gamma)\,. \tag{15.18}$$

From (15.6) one sees for $|z| > C^2$

$$\begin{aligned} N^{-n}\log B_N(z) &= \log z + N^{-n}\sum_{\mathsf{x}\in\mu_N^\Lambda}\log(1-W(\mathsf{x})z^{-1}) \\ &= \log z - \sum_{k\geq 1}\frac{\mathsf{m}_k^{(N)}}{k}z^{-k}\,. \end{aligned} \tag{15.19}$$

Combining (15.18) and (15.19) we find

$$B_N(z) = z^{N^n}\exp\left(-\sum_{k\geq 1}\sum_{\gamma\in\Gamma_N(2k)}\text{weight}(\gamma)\,z^{-k}\right).$$

This finishes the proof of Theorem 15.1.2. $\square$

Remark. In the Laplacian perspective $N^n\mathsf{m}_k^{(N)}$ is the trace of the operator Δ^k on the space of $\mathbb{C}$-valued C^∞-functions on $\mathbb{R}^n$ which are periodic for the sublattice $N\Lambda$ of Λ. The polynomial $B_N(z)$ is the characteristic polynomial of Δ on this space. Formula (15.19) gives the well-known relation between the characteristic polynomial of an operator and the traces of its powers.

Remark. One may refine the above enumerations by keeping track of the homology class to which the closed path belongs. That means that instead of (15.17) one extracts from the polynomial $W(\mathsf{x})^k$ the sub polynomial consisting of terms with exponent in $N\Lambda$. Such a refinement of the enumerations with homology data appears also in the theory of dimer models (cf. [11]), but its meaning for the diffraction pattern is not clear.

15.5 Algebraic geometry.

The polynomial $\mathsf{B}_N(z) = \prod_{\mathsf{x}\in\mu_N^\Lambda}(z - W(\mathsf{x}))$ has coefficients in the ring of integers of the cyclotomic field $\mathbb{Q}(\mu_N)$ and is clearly invariant under the Galois group of $\mathbb{Q}(\mu_N)$ over $\mathbb{Q}$. Consequently, the coefficients of $\mathsf{B}_N(z)$

lie in $\mathbb{Z}$. The same argument applies to the polynomial $\mathsf{B}_N(z)\mathsf{B}_{N'}(z)^{-1} = \prod_{\mathsf{x}\in\mu_N^\Lambda-\mu_{N'}^\Lambda}(z - W(\mathsf{x}))$ if N' divides N. Thus we have proved

Proposition 15.5.1. *For every $N \in \mathbb{N}$ the coefficients of $\mathsf{B}_N(z)$ lie in $\mathbb{Z}$. If N' divides N in $\mathbb{Z}$, then $\mathsf{B}_{N'}(z)$ divides $\mathsf{B}_N(z)$ in $\mathbb{Z}[z]$.* □

Fix a prime number p and a positive integer $\nu \in \mathbb{Z}_{>0}$. Let $\mathbb{W}(\mathbb{F}_{p^\nu})$ denote the ring of Witt vectors of the finite field $\mathbb{F}_{p^\nu}$ (see e.g. [3]). So, $\mathbb{W}(\mathbb{F}_{p^\nu})$ is a complete discrete valuation ring with maximal ideal $p\mathbb{W}(\mathbb{F}_{p^\nu})$ and residue field $\mathbb{F}_{p^\nu}$. The Teichmüller lifting is a map $\tau : \mathbb{F}_{p^\nu} \longrightarrow \mathbb{W}(\mathbb{F}_{p^\nu})$ such that

$$x \equiv \tau(x) \bmod p\,, \qquad \tau(xy) = \tau(x)\tau(y) \qquad \forall x, y \in \mathbb{F}_{p^\nu}\,.$$

Every non-zero $x \in \mathbb{F}_{p^\nu}$ satisfies

$$x^{p^\nu - 1} = 1\,.$$

Thus there is an isomorphism $\mu_{p^\nu-1} \simeq \mathbb{F}_{p^\nu}^*$. Such an isomorphism composed with the Teichmüller lifting gives an embedding $j : \mu_{p^\nu-1} \hookrightarrow \mathbb{W}(\mathbb{F}_{p^\nu})$. Thus for $\mathsf{x} \in \mu_{p^\nu-1}^\Lambda$ we get

$$W(j(\mathsf{x})) \in \mathbb{W}(\mathbb{F}_{p^\nu})\,.$$

Recall the p-adic valuation on $\mathbb{Z}$: for $k \in \mathbb{Z}$, $k \neq 0$:

$$v_p(k) \;:=\; \max\{v \in \mathbb{Z} \mid p^v \text{ divides } k\}\,.$$

Proposition 15.5.2. *For p, ν as above and for $z \in \mathbb{Z}$ the p-adic valuation of the integer $\mathsf{B}_{p^\nu-1}(z)$ satisfies*

$$v_p(\mathsf{B}_{p^\nu-1}(z)) \geq \sharp\{\xi \in (\mathbb{F}_{p^\nu}^*)^n \mid W(\xi) = z \textit{ in } \mathbb{F}_{p^\nu}\,\}\,. \tag{15.20}$$

Proof From (15.6) we obtain the product decomposition, with factors in $\mathbb{W}(\mathbb{F}_{p^\nu})$,

$$\mathsf{B}_N(z) = \prod_{\xi\in(\mathbb{F}_{p^\nu}^*)^n} (z - W(\tau(\xi)))\,.$$

The result of the proposition now follows because

$$z - W(\tau(\xi)) \in p\mathbb{W}(\mathbb{F}_{p^\nu}) \qquad \Leftrightarrow \qquad W(\xi) = z \text{ in } \mathbb{F}_{p^\nu}$$

□

Remark. In (15.30) we give an example showing that in (15.20) we may have a strict inequality.

Remark about the relation with Mahler measure and L-functions.
The *logarithmic Mahler measure* $\mathsf{m}(F)$ and the *Mahler measure* $\mathsf{M}(F)$ of a Laurent polynomial $F(x_1, \ldots, x_n)$ with complex coefficients are:

$$\mathsf{m}(F) := \frac{1}{(2\pi i)^n} \oint\!\!\oint_{|x_1|=\ldots=|x_n|=1} \log|F(x_1, \ldots, x_n)| \frac{dx_1}{x_1} \cdot \ldots \cdot \frac{dx_n}{x_n},$$
$$\mathsf{M}(F) := \exp(\mathsf{m}(F)).$$

Boyd [4] gives a survey of many (two-variable) Laurent polynomials for which $\mathsf{m}(F)$ equals (numerically to many decimal places) a 'simple' non-zero rational number times the derivative at 0 of the L-function of the projective plane curve Z_F defined by the vanishing of F:

$$\mathsf{m}(F) \cdot \mathbb{Q}^* = L'(Z_F, 0) \cdot \mathbb{Q}^*. \tag{15.21}$$

Deninger [6] and Rodriguez Villegas [15] showed that the experimentally observed relations (15.21) agree with predictions from the Bloch-Beilinson conjectures. Rodriguez Villegas [15] provided actual proofs for a few special examples. Since the measure $dV(r)$ is the push forward of the measure $dt_1\, dt_2 \ldots dt_n$ on $\mathbb{R}^n$ by the function $|\widehat{\mathsf{D}}(\mathsf{t})|^2$, one can rewrite Formula (15.15) as:

$$-\log \mathsf{Q}(z) = \frac{1}{(2\pi i)^n} \int_{\mathbb{U}^n} \log|z - W(x_1, \ldots, x_n)| \frac{dx_1}{x_1} \frac{dx_2}{x_2} \cdots \frac{dx_n}{x_n}. \tag{15.22}$$

On the right hand side of (15.22) we now recognize the logarithmic Mahler measure of the Laurent polynomial $z - W(x_1, \ldots, x_n) \in \mathbb{C}[x_1^{\pm 1}, \ldots, x_n^{\pm 1}]$. For fixed $z \in \mathbb{Z}$ formulas (15.1) and (15.20) provide a link between $\mathsf{Q}(z)$ and counting points over finite fields on the variety with equation $W(x_1, \ldots, x_n) = z$. It may be an interesting challenge to further extend these ideas to a proof of a result like (15.21).

15.6 Moments.

Important invariants of the measure $dV(r)$ are its *moments* m_k ($k \in \mathbb{Z}_{\geq 0}$):

$$\begin{aligned} \mathsf{m}_k := \textstyle\int_{\mathbb{R}} r^k dV(r) &= \textstyle\int_0^1 \cdots \int_0^1 |\widehat{\mathsf{D}}(t_1, \ldots, t_n)|^{2k} dt_1 \ldots dt_n \\ &= \text{constant term of Fourier series } |\widehat{\mathsf{D}}(t_1, \ldots, t_n)|^{2k} \\ &= \text{constant term of Laurent polynomial } W(x_1, \ldots, x_n)^k. \end{aligned} \tag{15.23}$$

The relation between the moments and the functions $\mathsf{H}(z)$, $\mathsf{Q}(z)$ defined in (15.13) and (15.15) is: for $z \in \mathbb{R}$, $z > C^2$,

$$\mathsf{H}(z) = \sum_{k \geq 0} \mathsf{m}_k z^{-k-1}, \qquad \mathsf{Q}(z) = z^{-1} \exp\left(\sum_{k \geq 1} \frac{\mathsf{m}_k}{k} z^{-k} \right). \tag{15.24}$$

It is clear that the moments m_k of $dV(r)$ are non-negative integers. They satisfy all kinds of arithmetical relations. There are, for instance, recurrences like (15.28) and congruences like the following

Lemma 15.6.1. $\mathsf{m}_{kp^{\alpha+1}} \equiv \mathsf{m}_{kp^\alpha} \bmod p^{\alpha+1}$ *for every prime number* p *and* $k, \alpha \in \mathbb{Z}_{\geq 0}$.

Proof The Laurent polynomial $W(x_1, \ldots, x_n)$ has coefficients in $\mathbb{Z}$. Therefore

$$W(x_1, \ldots, x_n)^{kp^{\alpha+1}} \equiv W(x_1^p, \ldots, x_n^p)^{kp^\alpha} \bmod p^{\alpha+1}\mathbb{Z}[x_1^{\pm 1}, \ldots, x_n^{\pm 1}].$$

The lemma follows by taking constant terms. □

Theorems 1.1, 1.2, 1.3 in [1] together with the above lemma immediately yield the following integrality result for series and product expansions:

Corollary 15.6.2. *For* $z \in \mathbb{R}$, $z > C^2$

$$z^{-1} \exp\left(\sum_{k \geq 1} \frac{\mathsf{m}_k}{k} z^{-k} \right) = z^{-1} + \sum_{k \geq 1} A_k z^{-k-1} = z^{-1} \prod_{k \geq 1} (1 - z^{-k})^{-b_k} \tag{15.25}$$

with $A_k, b_k \in \mathbb{Z}$ *for all* $k \geq 1$. □

Remark. In [1, 8] the result of Corollary 15.6.2 is used to interpret $z\mathsf{Q}(z)$ as the Artin-Mazur zeta function of a dynamical system, provided the integers b_k are not negative. We have not yet found such a dynamical system within the present framework.

For $N \in \mathbb{N}$ the moments of the measure $dV_N(r)$ are, by definition,

$$\mathsf{m}_k^{(N)} := \int_{\mathbb{R}} r^k dV_N(r) = N^{-n} \sum_{\mathsf{x} \in \mu_N} W(\mathsf{x})^k .$$

These are the same numbers as in (15.17).

Proposition 15.6.3. *With the above notations we have*

$$\mathsf{m}_k^{(N)} \geq \mathsf{m}_k \geq 0 \qquad \textit{for all} \qquad N,\, k,$$

$$\mathsf{m}_k^{(N)} = \mathsf{m}_k \qquad \textit{if} \qquad N > k \max_{\mathsf{a},\mathsf{b} \in \mathfrak{A}} \max_{1 \leq j \leq n} |a_j - b_j| .$$

Proof $N^n\,(\mathsf{m}_k^{(N)} - \mathsf{m}_k)$ is the sum of the coefficients of all non-constant monomials in the Laurent polynomial $W(x_1,\ldots,x_n)^k$ with exponents divisible by N. Since all coefficients of $W(x_1,\ldots,x_n)$ are positive, this shows $\mathsf{m}_k^{(N)} \geq \mathsf{m}_k \geq 0$. Assume $N > k \max_{\mathsf{a},\mathsf{b}\in\mathfrak{A},\,1\leq j\leq n} |a_j - b_j|$. Then all exponents in the monomials of the Laurent polynomial $W(x_1,\ldots,x_n)^k$ are $> -N$ and $< N$. So only the exponent of the constant term is divisible by N. Therefore $\mathsf{m}_k^{(N)} = \mathsf{m}_k$. □

Note the natural interpretation (and proof) of this proposition in terms of closed paths on the graph Γ_N: closed paths on Γ_N which are too short are in fact projections of closed paths on Γ.

Corollary 15.6.4. *For* $N > \ell \max_{\mathsf{a},\mathsf{b}\in\mathfrak{A}} \max_{1\leq j\leq n} |a_j - b_j|$ *and* $|z| > C^2$*:*

$$\mathsf{Q}(z) \cdot |B_N(z)|^{N^{-n}} = \left|\exp\left(\sum_{k>\ell} \frac{\mathsf{m}_k - \mathsf{m}_k^{(N)}}{k} z^{-k}\right)\right| .$$

This not only gives an estimate for the rate of convergence of (15.1) *with respect to the usual absolute value on* $\mathbb{C}$*, but it also yields the following congruence of power series in* z^{-1}*:*

$$B_N(z)^{-N^{-n}} \equiv z^{-1} + \sum_{k\geq 1} A_k z^{-k-1} \mod z^{-\ell-1}$$

with A_k *as in* (15.25). □

Remark about the relation with the large complex structure limit.
Since the measure $dV(r)$ is the push forward of the measure $dt_1\, dt_2 \ldots dt_n$ on $\mathbb{R}^n$ by the function $|\widehat{\mathsf{D}}(\mathsf{t})|^2$, one can rewrite (15.13) as

$$\mathsf{H}(z) = \frac{1}{(2\pi i)^n} \int_{\mathbb{U}^n} \frac{1}{z - W(x_1,\ldots,x_n)} \frac{dx_1}{x_1} \frac{dx_2}{x_2} \cdots \frac{dx_n}{x_n}$$

for $z \in \mathbb{C} - \mathfrak{I}$. From this (and the residue theorem) one sees that $\mathsf{H}(z)$ is *a period of some differential form of degree* $n-1$ *along some* $(n-1)$*-cycle on the hypersurface in* $(\mathbb{C}^*)^n$ *given by the equation* $W(x_1,\ldots,x_n) = z$. As z varies we get a 1-parameter family of hypersurfaces. The function $\mathsf{H}(z)$ is a solution of the Picard-Fuchs differential equation associated with (that $(n-1)$-form on) this family of hypersurfaces. The Picard-Fuchs equation is equivalent with a recurrence relation for the coefficients m_k in the power series expansion (15.24) of $\mathsf{H}(z)$ near $z = \infty$. All this is standard knowledge about Calabi-Yau varieties near the large complex structure limit and there is an equally standard algorithm to derive from the Picard-Fuchs differential equation enumerative information about numbers of instantons (or rational curves); see for instance [20, 13, 18]. On the other hand, we have the enumerative data $\mathsf{m}_k^{(N)}$ of the present paper. In the limit for $N \to \infty$ these yield the moments m_k and, hence, the Picard-Fuchs differential equation and eventually the instanton numbers.

15.7 Examples

15.7.1 $n = 1$

Mahler measures of one variable polynomials have a long history with many interesting results; see the introductory sections of [4, 7, 15]. We limit our discussion to one example, without a claim of new results. This simple, yet non-trivial, example has $n = 1$, $\mathfrak{A} = \{-1, 1\} \subset \mathbb{Z}$, $c_{-1} = c_1 = 1$ and hence

$$|\widehat{\mathsf{D}}(t)|^2 = 2 + 2\cos(4\pi t)\,, \qquad W(x) = (x + x^{-1})^2\,.$$

The moments are

$$\mathsf{m}_k \;=\; \text{constant term of } (x + x^{-1})^{2k} \;=\; \binom{2k}{k}$$

and hence by (15.24): for $z \in \mathbb{R}$, $z > 4$,

$$\mathsf{H}(z) = \sum_{k \geq 0} \binom{2k}{k} \, z^{-k-1} \;=\; \frac{1}{\sqrt{z(z-4)}}\,,$$

$$\mathsf{Q}(z) = \exp\left(- \int \frac{dz}{\sqrt{z(z-4)}} \right) = \frac{1}{2}\left(z - 2 - \sqrt{z(z-4)} \right)\,.$$

Applying Formula (15.17) to the present example we find

$$\mathsf{m}_k^{(N)} \;=\; \sum_{j \equiv k \bmod N} \binom{2k}{j}\,,$$

which nicely illustrates Proposition 15.6.3. Setting $z = 2 + u + u^{-1}$ one finds for the polynomials $\mathsf{B}_N(z)$ defined in (15.3):

$$\begin{aligned}
\mathsf{B}_N(z) &= \prod_{x^2 \in \mu_N} (z - 2 - x^2 - x^{-2}) \;=\; u^{-N} \prod_{x^2 \in \mu_N} (u - x^2)(u - x^{-2}) \\
&= u^N + u^{-N} - 2 \\
&= \left(\frac{1}{2}\left(z - 2 - \sqrt{z(z-4)} \right) \right)^N + \left(\frac{1}{2}\left(z - 2 + \sqrt{z(z-4)} \right) \right)^N - 2 \\
&= -2 + 2^{1-N} \sum_j \binom{N}{2j} z^j (z-4)^j (z-2)^{N-2j}\,.
\end{aligned}$$

So, $\mathsf{B}_N(z)$ *is up to some shift and normalization the* N*-th Čebyšev polynomial.* The above computation also shows $\mathsf{B}_N(z) = \mathsf{Q}(z)^N + \mathsf{Q}(z)^{-N} - 2$ and thus, in agreement with (15.1),

$$\lim_{N\to\infty} \mathsf{B}_N(z)^{-N^{-1}} = \mathsf{Q}(z)\,.$$

For actual computation of $\mathsf{B}_N(z)$ in case $z \in \mathbb{Z}$ one can use the generating series identity:

$$\sum_{N\geq 1} \mathsf{B}_N(z)\,\frac{T^N}{N} = -\log\left(1-(z-4)\frac{T}{(1-T)^2}\right).$$

For $z = 6$ one finds (using PARI)

$$\begin{aligned}\sum B_N(6)T^N &= 2T+12T^2+50T^3+192T^4+722T^5+2700T^6+10082T^7\\ &\quad +37632T^8+140450T^9+524172T^{10}+1956242T^{11}\\ &\quad +7300800T^{12}+27246962T^{13}+101687052T^{14}\\ &\quad +379501250T^{15}+1416317952T^{16}+5285770562T^{17}+\ldots\end{aligned}$$

For primes p in the displayed range the number $B_{p-1}(6)$ is divisible by p^2 for $p \equiv \pm 1 \bmod 12$ and is not divisible by p for $p \equiv \pm 5 \bmod 12$ and is exactly divisible by p if $p = 2, 3$. We also checked $5^2 \mid B_{24}(6)$ and $7^2 \mid B_{48}(6)$. This agrees with the number of solutions of the equation $u + u^{-1} = 4$ in $\mathbb{F}_p$ and $\mathbb{F}_{p^2}$. If $z \in \mathbb{Z}, z > 4$, then $\mathsf{Q}(z) = \frac{1}{2}(z - 2 - \sqrt{z(z-4)})$ is a unit in the real quadratic field $\mathbb{Q}(\sqrt{z(z-4)})$. According to Dirichlet's class number formula it relates to the L-function of this real quadratic field:

$$\log(\mathsf{Q}(z)) = \frac{\sqrt{D}}{2h}L(1,\chi)$$

where D, h, χ are the discriminant, class number, character, respectively, of the real quadratic field $\mathbb{Q}(\sqrt{z(z-4)})$ (see e. g. [5]). The relations between Mahler measures and values of L-functions, which have been observed for some curves, are perfect analogues of the above class number formula (see [15]).

15.7.2 The honeycomb pattern

For a nice two-dimensional example we take $\mathfrak{A} = \{(1,0), (0,1), (-1,-1)\} \subset \mathbb{Z}^2$, $c_{(1,0)} = c_{(0,1)} = c_{(-1,-1)} = 1$ and hence

$$\begin{aligned}|\widehat{\mathsf{D}}(t_1,t_2)|^2 &= 3+2\cos(2\pi(t_1-t_2))+2\cos(2\pi(2t_1+t_2))\\ &\quad +2\cos(2\pi(t_1+2t_2)),\\ W(x_1,x_2) &= (x_1+x_2+x_1^{-1}x_2^{-1})(x_1^{-1}+x_2^{-1}+x_1x_2)\\ &= x_1x_2^{-1}+x_1^2x_2+x_1^{-1}x_2+x_1x_2^2+x_1^{-2}x_2^{-1}+x_1^{-1}x_2^{-2}+3\,.\end{aligned}$$

As a basis for the lattice Λ we take $(2,1)$ and $(-1,-2)$. This leads to coordinates $u_1=x_1^2x_2$ and $u_2=x_1^{-1}x_2^{-2}$ on the torus $\mathbb{U}^\Lambda$. In these coordinates the function W reads

$$\begin{aligned}W(u_1,u_2) &= u_1+u_1^{-1}+u_2+u_2^{-1}+u_1^{-1}u_2+u_1u_2^{-1}+3 \qquad (15.26)\\ &= (u_1+u_2+1)(u_1^{-1}+u_2^{-1}+1)\,.\end{aligned}$$

Figure 15.2 shows a piece of the graph Γ. Figure 15.1 shows some level sets of the function $|\widehat{\mathsf{D}}(t_1,t_2)|^2$. The dual lattice $\Lambda^\vee$ is spanned by $(1,0)$ and $(\frac{1}{3},\frac{1}{3})$. The maximum of the function $|\widehat{\mathsf{D}}(t_1,t_2)|^2$ equals 9 and is attained at the points of $\Lambda^\vee$. The minimum of the function $|\widehat{\mathsf{D}}(t_1,t_2)|^2$ equals 0 and is attained at the points of $(0,-\frac{1}{3})+\Lambda^\vee$ and $(\frac{1}{3},0)+\Lambda^\vee$. There are saddle points with critical value 1 at $(-\frac{1}{6},\frac{1}{3})+\Lambda^\vee$, $(\frac{1}{3},-\frac{1}{6})+\Lambda^\vee$ and $(\frac{1}{6},\frac{1}{6})+\Lambda^\vee$. In terms of the coordinates u_1, u_2 the maximum lies at $(u_1,\,u_2)=(1,1)$, the minima at $(e^{2\pi i/3},e^{4\pi i/3})$, $(e^{4\pi i/3},e^{2\pi i/3})$ and the saddle points at $(1,-1),(-1,1),(-1,-1)$. The algebraic geometry of this example concerns the 1-parameter family of elliptic curves with equation $z-W(u_1,u_2)=0$. In homogeneous coordinates $(U_0:U_1:U_2)$ on the projective plane $\mathbb{P}^2$, with $u_1=U_1U_0^{-1}$, $u_2=U_2U_0^{-1}$, this becomes a homogeneous equation of degree 3:

$$(U_0U_1+U_0U_2+U_1U_2)(U_0+U_1+U_2)-zU_0U_1U_2=0\,. \qquad (15.27)$$

Beauville [2] showed that there are exactly six semi-stable families of elliptic curves over $\mathbb{P}^1$ with four singular fibres. The pencil (15.27) is one of these six. It has singular fibres at $z=0,1,9,\infty$ with Kodaira types I_2,I_3,I_1,I_6, respectively. Note that the first three match the critical points and levels in the diffraction pattern. After blowing up the points $(1,0,0),(0,1,0),(0,0,1)$ of $\mathbb{P}^2$ one gets the Del Pezzo surface dP_3. The elliptic pencil (15.27) naturally lives on dP_3. It has six base points, corresponding to six sections of the pencil. Since the base points have a zero coordinate, these sections do not intersect the real torus $\mathbb{U}^\Lambda$. Equations (15.26) and (15.27) also appear in the literature in connection with the string theory of dP_3. Formula (15.23) and some manipulations of binomials give the moments:

$$\mathsf{m}_k=\sum_{j=0}^{k}\binom{k}{j}^2\binom{2j}{j}.$$

These numbers satisfy the recurrence relation (see [19, Table 7])

$$(k+1)^2 \mathsf{m}_{k+1} = (10k^2 + 10k + 3)\mathsf{m}_k - 9k^2 \mathsf{m}_{k-1}\,. \tag{15.28}$$

We refer to [19, Example $\mathcal{C}$] and to [18, Example ♯6] for relations of these numbers to modular forms and instanton counts. Golyshev [10] can derive the recurrence (15.28) from the quantum cohomology of dP_3. The numerical evidence for the relation (15.21) between Mahler measure and L-function in this example is given in [4, Table 2]. With Formulas (15.17) and (15.26) one easily calculates

$$\mathsf{m}_k^{(N)} = \sum_{i_1 \equiv i_2 \bmod N,\, j_1 \equiv j_2 \bmod N} \binom{k}{i_1}\binom{k-i_1}{j_1}\binom{k}{i_2}\binom{k-i_2}{j_2}\,.$$

Note that these formulas confirm $\mathsf{m}_k^{(N)} = \mathsf{m}_k$ for $k < N$. Equation (15.27) is clearly invariant under permutations of U_0, U_1, U_2. Therefore the diffraction pattern has this S_3-symmetry too. Since only the critical points have a non-trivial stabilizer in S_3 the multiplicities $\mathrm{mult}_N(r)$ in this example satisfy

$$\begin{array}{lll} \mathrm{mult}_N(9) = 1 & \forall N & \\ \mathrm{mult}_N(0) = 2 & \text{if} \;\; 3|N & \\ \mathrm{mult}_N(1) \equiv 3 \bmod 6 & \text{if} \;\; 2|N & \\ \mathrm{mult}_N(r) \equiv 0 \bmod 6 & \text{if} \;\; r \neq 0,\,1,\,9, & \forall N. \end{array} \tag{15.29}$$

We have computed the numbers $\mathrm{mult}_N(r)$ for some values of N. We found for instance

$$\mathsf{B}_6(z) = z^2\,(z-1)^{15}\,(z-3)^6\,(z-4)^6\,(z-7)^6\,(z-9)\,.$$

We computed $W(u_1, u_2) = (u_1 + u_2 + 1)(u_1^{-1} + u_2^{-1} + 1)$ for $u_1, u_2 \in \mathbb{F}_7^*$: the (i,j)-entry of the following 6×6-matrix is $W(i,j) \bmod 7$:

$$\begin{bmatrix} 2 & 3 & 0 & 3 & 0 & 1 \\ 3 & 3 & 4 & 0 & 1 & 1 \\ 0 & 4 & 0 & 1 & 4 & 1 \\ 3 & 0 & 1 & 3 & 4 & 1 \\ 0 & 1 & 4 & 4 & 0 & 1 \\ 1 & 1 & 1 & 1 & 1 & 1 \end{bmatrix}.$$

This yields the following count of points over $\mathbb{F}_7$:

$z \bmod 7$	:	0	1	2	3	4	5	6
$\sharp\{\xi \in (\mathbb{F}_7^*)^2 \mid W(\xi) = z \text{ in } \mathbb{F}_7\}$	:	8	15	1	6	6	0	0

Thus we see that the inequality in (15.20) can be strict:

$$v_7(\mathsf{B}_6(53)) = 12 > 6 = \sharp\{\xi \in (\mathbb{F}_7^*)^2 \mid W(\xi) = 53 \text{ in } \mathbb{F}_7\}. \qquad (15.30)$$

Acknowledgment. It is my pleasure to dedicate this paper to Jaap Murre and Spencer Bloch on the occasion of their 75th, respectively, 60th birthdays. Both have been very important for my formation as a mathematician, from PhD-student time till present.

References

[1] Arias de Reyna, J: Dynamical zeta functions and Kummer congruences, arXiv:math. NT/0309190.

[2] Beauville, A.: Les familles stables de courbes elliptiques sur $\mathbf{P}^1$ admettant quatre fibres singulières, C. R. Acad. Sc. Paris, t. **294** (1982), 657–660.

[3] Bourbaki, N.: *Éléments de mathématique: Algèbre commutative, chapitres 8 et 9*, Masson, Paris (1983).

[4] Boyd, D.: Mahler's measure and special values of L-functions, Experimental Math. vol. **7** (1998) 37–82.

[5] Borewicz, S. , Šafarevič, I.: *Zahlentheorie*, Birkhäuser Verlag Basel (1966).

[6] Deninger, C.: Deligne periods of mixed motives, K-theory and the entropy of certain $\mathbb{Z}^n$-actions, J. Amer. Math. Soc. **10** (1997) 259–281.

[7] Everest, G. , T. Ward: *Heights of polynomials and entropy in algebraic dynamics*, Springer-Verlag London (1999).

[8] Everest, G. , Y. Puri, T. Ward: Integer sequences counting periodic points, arXiv:math. NT/0204173.

[9] Gieseker, D. , H. Knörrer, E. Trubowitz: *The Geometry of Algebraic Fermi Curves*, Perspectives in Math. vol. **14**, Academic Press, San Diego (1993).

[10] Golyshev, V.: *private communication.*

[11] Kenyon, R., A. Okounkov, S. Sheffield: Dimers and Amoebae, arXiv:math-ph/0311005

[12] Lehmer, D.: Factorization of certain cyclotomic functions, Annals of Math. **34** (1933) 461–479.

[13] Morrison, D.: Mirror symmetry and rational curves on quintic threefolds: a guide for mathematicians, J. of the Amer. Math. Soc. **6** (1993), 223–247.

[14] Okounkov, A., N. Reshetikin, C. Vafa: Quantum Calabi-Yau and Classical Crystals, arXiv:hep-th/0309208.

[15] Rodriguez Villegas, F.: Modular Mahler measures I, in *Topics in number theory (University Park, PA, 1997)*, Ahlgren, S., G. Andrews, K. Ono (eds) 17–48, Math. Appl., **467**, Kluwer Acad. Publ., Dordrecht (1999).

[16] Stienstra, J.: Ordinary Calabi-Yau-3 crystals, in *Calabi-Yau Varieties and Mirror Symmetry*, N. Yui, J. D. Lewis (eds.), Fields Institute Communications vol. **38**, AMS (2003), 255–271.

[17] Stienstra, J.: The ordinary limit for varieties over $\mathbb{Z}[x_1, \ldots, x_r]$, in *Calabi-Yau Varieties and Mirror Symmetry*, N. Yui, J. D. Lewis (eds.), Fields Institute Communications vol. **38**, AMS (2003), 273–305.

[18] Stienstra, J.: Mahler Measure Variations, Eisenstein Series and Instanton Expansions, to appear in *Mirror Symmetry V*, AMS/International Press; also arXiv:math. NT/0502193.
[19] Stienstra, J., F. Beukers: On the Picard-Fuchs equation and the formal Brauer group of certain elliptic K3-surfaces, Math. Ann. **271** (1985) 269–304.
[20] S. -T. Yau (ed.): *Essays on Mirror Manifolds*, Hong Kong: International Press (1992).

Printed in the United States
By Bookmasters